Logic, Epistemology, and the Unity of Science

Founding Editor

John Symons

Volume 54

Logic, Epistemology, and the Unity of Science aims to reconsider the question of the unity of science in light of recent developments in logic. At present, no single logical, semantical or methodological framework dominates the philosophy of science. However, the editors of this series believe that formal frameworks, for example, constructive type theory, deontic logics, dialogical logics, epistemic logics, modal logics, and proof-theoretical semantics, have the potential to cast new light on basic issues in the discussion of the unity of science.

This series provides a venue where philosophers and logicians can apply specific systematic and historic insights to fundamental philosophical problems. While the series is open to a wide variety of perspectives, including the study and analysis of argumentation and the critical discussion of the relationship between logic and philosophy of science, the aim is to provide an integrated picture of the scientific enterprise in all its diversity.

This book series is indexed in SCOPUS.

For inquiries and submissions of proposals, authors can contact Christi Lue at christi.lue@springer.com

More information about this series at https://link.springer.com/bookseries/6936

Francesco Ademollo · Fabrizio Amerini ·
Vincenzo De Risi
Editors

Thinking and Calculating

Essays in Logic, Its History and Its Philosophical Applications in Honour of Massimo Mugnai

Springer

Editors
Francesco Ademollo
Dipartimento di Lettere e Filosofia
Università di Firenze
Florence, Italy

Fabrizio Amerini
Department of Humanities, Social Sciences, and Cultural Industries
Università di Parma
Parma, Italy

Vincenzo De Risi
Laboratoire SPHère
CNRS / Université Paris Cité
Paris, France

ISSN 2214-9775 ISSN 2214-9783 (electronic)
Logic, Epistemology, and the Unity of Science
ISBN 978-3-030-97302-5 ISBN 978-3-030-97303-2 (eBook)
https://doi.org/10.1007/978-3-030-97303-2

Mathematics Subject Classification: 03-03, 03-06, 03A05, 03B80

This Springer imprint is published by the registered company Springer Nature Switzerland AG
The registered company address is: Gewerbestrasse 11, 6330 Cham, Switzerland

Preface

This book has a twofold nature. It is *about* something; and it is *for* someone.

On the one hand, it is a collection of twenty-two essays in the history of logic—broadly conceived—written by outstanding specialists in the field. Our aim has been to display the vastness and depth of the developments of logic throughout the centuries, the large array of problems that have fallen within its purview, and the manifold relations that it has entertained with other disciplines. The topics do not cover every aspect of the history of logic, and a few key figures (for example Bolzano or Frege) happen not to be included. In this as in other respects, the volume does not aim to be a textbook providing a comprehensive survey. Rather, it highlights several less-frequented topics and provides a clear picture of some of the most promising developments in recent historiography. Its goal is more to advance cutting-edge research in the field, and display its breadth and diversity, than to encompass the field itself exhaustively.

On the other hand, the book is intended as a tribute to Massimo Mugnai, whose contribution to the history of medieval and modern logic, and to the understanding of the writings of Leibniz in particular, have shaped the field in the last four decades. So the volume also aims to illustrate the breadth of Massimo's work and influence. Some of the authors have been his students in Bari, in Florence, or at the Scuola Normale Superiore in Pisa (where he moved in 2002 and stayed until his retirement in 2017); some have been his colleagues; all have been among his many interlocutors, collaborators, and friends.

Here we shall not try to list Massimo's many scholarly and academic achievements; a comprehensive account up to 2013 was provided by Richard Arthur in that year's issue of the *Leibniz Review*, and we are delighted to refer to it as well as to the complete bibliography which is included at the end of this volume. We will, however, say at least this: Massimo's special combination of intellectual curiosity, huge erudition, strong philosophical interests, an argumentative and outspoken nature, and caustic wit has awed and inspired generations of students and is admired and held dear by his peers and friends. This book is a token of gratitude for many years of philosophical conversation with him.

The book is arranged chronologically. It consists of a section on antiquity (with contributions by Crivelli, Ademollo, De Risi, D'Agostino and Piazza); one on medieval and late scholasticism (with contributions by Martin, Ciola, Binini, Panaccio, Amerini, Mondadori, Knuuttila and Strobino); one on Leibniz (Ugaglia, Antognazza, Di Bella, Arthur, Pasini, Normore); and a final section on contemporary topics (Centrone and Minari, Cantini, Lando, Bagnoli, Belardinelli).

The large variety of the topics treated would, however, also allow for a systematic rather than chronological ordering. It is possible to follow different threads across the chapters, and readers interested in specific themes will easily identify different paths. Scholars and students who work on the logic of deduction and its history will find the chapter by D'Agostino and Piazza very stimulating, while those who are rather interested in the philosophy of language may prefer to start with the chapters on medieval theories of meaning (Martin, Panaccio, Knuuttila and Strobino). Someone whose main interest lies in metaphysics and its history might want to look at the chapters on modality (Ciola, Binini, Mondadori, Lando, Belardinelli), or those on essentialism (Crivelli, Ademollo, Antognazza, Di Bella), or on the problem of universals (Ademollo, Normore). Others, whose focus is the history and philosophy of science, may prefer to turn to the chapters on the interactions between logic and mathematics (De Risi, Centrone and Minari, Cantini) or those on infinity and time (Ugaglia, Arthur). Those primarily interested in epistemology might be attracted to the historical enquiries by De Risi, Amerini, Knuuttila and Strobino, Pasini; and someone who does moral philosophy will find much food for thought in the paper by Bagnoli. We hope therefore that not only professional historians of logic, but also specialists of different fields will be interested in the book and profit from reading it.

An important feature of the book is indeed the relevance given to the philosophical applications of logic and its interaction with other disciplines. All contributors have made a substantial effort to stress the importance of logic for the general development of science and philosophy, from antiquity to the modern age, and to explore some aspects of such fruitful interaction in today's research. In fact several of us were not originally trained as historians of logic, but rather as classicists, historians of mathematics and physics, metaphysicians, epistemologists, philosophers of language, or moral philosophers, and for reasons internal to their respective fields came to feel the necessity to work on logic and its history. We hope that the result of such a collaboration among very different scientific personalities may help to explain the importance that logic has held throughout the centuries with respect to many areas of human life and culture.

Earlier drafts of four chapters—by Maria Rosa Antognazza, Calvin Normore, Fabrizio Mondadori, and Carla Bagnoli—were presented, and the project of this *Festschrift* was announced, during a conference celebrating Massimo's retirement which was organized on 21 May 2019 at the Scuola Normale with the title *Logic, Ethics and Modalities*. We are very grateful to the Scuola, and in particular to Mario Piazza, for their generous hospitality on that occasion. We also wish that we were able to thank all four speakers—but here our gratitude is mixed with sadness; for Fabrizio Mondadori, one of Massimo's oldest and dearest friends died in February 2021. He had already sent us a provisional draft of his paper, but did not have the time to revise it. We decided to publish it in that form, correcting only a few typographical trifles and adding a brief prefatory note. (We are also very sad to add that Simo Knuuttila, another of Massimo's friends and of the book's contributors, passed away in June 2022, when he had already been able to correct the second proofs of the chapter which he co-authored with Riccardo Strobino.)

Other former students, colleagues, or friends of Massimo who gave us help and advice in various ways, or attended the Pisa conference, or wished to participate in the volume but were eventually unable to do so, include Jennifer Ashworth, Sergio Bernini, Andrea Borghini, Luigi Cataldi Madonna, Paolo Fait, Paolo Freguglia, Hykel Hosni, Mariano Giaquinta, Marko Malink, Paolo Mancosu, Enrico Moriconi, Daniele Mundici, Roberto Palaia, Donald Rutherford, Tommaso Tempesti, and Achille Varzi. Our warmest thanks go to all of them; to several distinguished scholars who kindly accepted to serve as anonymous readers, according to their areas of specialization, and improved every chapter with their insightful comments; and above all to our fellow contributors, who eagerly joined us in this undertaking.

Florence, Italy — Francesco Ademollo
Parma, Italy — Fabrizio Amerini
Paris, France — Vincenzo De Risi

Contents

Classical Antiquity

The Method of Models in Plato's *Statesman* 3
Paolo Crivelli
References 28

Anti-Platonism in Aristotle's *Categories* 31
Francesco Ademollo
1 Introduction: Aristotle's 'Meta-Ontology' 31
2 Fundamental and Non-fundamental Entities 34
3 Primary Substances as Ultimate Subjects 35
4 Self-Contradictory Universals? 37
5 Identity through Time 39
6 Substances and Accidents 40
7 Dependence 41
8 Explanatory Priority in *Categories* 12–13 43
9 Anti-Platonism 45
10 Universals in the *Posterior Analytics* 46
11 Essentialism and Anti-essentialism 49
References 51

Aristotle on Common Axioms 53
Vincenzo De Risi
1 Introduction: Common Axioms and Universal Science 53
2 Logical and Mathematical Axioms 58
3 Generic Unity and Unity by Analogy 62
4 Specialized Common Axioms: The Schematic Interpretation 66
5 The Inferential Interpretation of Common Axioms 70
6 Axioms, Logic and the Categories 77
References 81

Chrysippus' Logic in a Natural Deduction Setting 83
Marcello D'Agostino and Mario Piazza
1 Introduction 83
2 Chrysippus' Connectives 84
3 Chrysippus' Logic as a Natural Deduction System 87
4 Shallow and Deep Arguments 94
5 Bridging the Gap with Classical Logic 96
6 Conclusion 98
References 98

The Middle Ages and the Scholastic Tradition

"*Generaliter De Nullo Enuntiabili Aliquid Scio*": Meaning and Propositional Content in the *Ars Meliduna* 103
Christopher J. Martin
1 Introduction 103
2 The Ars Meliduna on Semantics of Terms 105
3 The Ars Meliduna on Assertables 111
4 Conclusion 124
References 124

Complete Forms, Individuals and Alternate World Histories: Gilbert of Poitiers 127
Graziana Ciola
1 Introduction 127
2 Porretan Ontology—An Outline 129
3 "Individuum" 137
4 On Creation, Necessity and Possible Worlds 139
5 Identicals and Counterparts 143
6 Some Closing Remarks 145
References 147

Turning Potentialities into Possibilities: Early Medieval Approaches to the Metaphysics of Modality 151
Irene Binini
1 The "Potency-Based" Account of Possibility 151
2 Anselm on the Predication of Antecedent Possibilities 155
3 Early Twelfth-Century Logicians on the Signification of Modal Terms 158
4 A New Understanding of Possibility 164
5 Conclusion 166
References 168

Ockham on Abstract Pseudo-Names 171
Claude Panaccio
1 Concrete and Abstract Terms 172
2 Pseudo-Names 173
3 Contextual Analysis 175
4 Avicenna on Horseness 178
References 182

Ockham and Chatton on the Origin of Logical Concepts 185
Fabrizio Amerini
1 Origin and Nature of the Logical Concepts in Ockham 186
2 Walter Chatton on the Origin of Logical Concepts 199
3 Conclusion 201
References 202

William of Heytesbury and Peter of Mantua on Demonstrative Pronouns in Epistemic Contexts 205
Riccardo Strobino and Simo Knuuttila
1 Epistemic Principles in Heytesbury's First Argument (13vaẓ–14vp, Trans. 446–455) 206
2 Heytesbury's Solutions of Further Counter-Arguments 209
3 The Signification of Demonstrative Pronouns in Epistemic Contexts in Heytesbury 212
4 Peter of Mantua on Knowing and Doubting 214
5 Epistemic Principles in Peter of Mantua's *De scire et dubitare* 215
6 Analysis of Peter of Mantua's *first dubium* 216
7 Conclusion 227
References 229

Poncius contra (Dicta Mastrii contra (Dicta Poncii)) 231
Fabrizio Mondadori
1 Interpreting an Interpretation 231
2 "Possibilis ex Se Formaliter" and "Possibilis Principiative per Intellectum Divinum" 232
3 Counterpossibles and Adversative Conjunctions 240
4 "Esse diminutum" 243
5 "A se" 245
6 Independence 250

Leibniz

Possibility vs Iterativity: Leibniz and Aristotle on the Infinite 255
Monica Ugaglia
1 Introduction 255
2 Potentiality as Iterativity: Aristotle's Infinite 256
3 Potentiality *as* Possibility: Leibniz's Mathematical Infinite 260

4 Potentiality *Without* Possibility: Leibniz's Physical Infinite 263
References 268

Pure Positivity in Leibniz 271
Maria Rosa Antognazza
1 Introduction 271
2 The Role of Pure Positivity in the Ontological Argument 271
3 Pure Positivity, Perfection, and Pure Act 273
4 Pure Positivity, Being, and Reality 275
5 Pure Positivity and the Absolute 277
6 Pure Positivity, Plurality, and the Infinite 278
7 Conclusion 281
References 282

Essentialism, Super-Essentialism and/or Anti-Essentialism in Leibniz 285
Stefano Di Bella
1 Is There Anything Essential to an Individual? 285
2 Super-Essentialism or Anti-Essentialism? A Possible Ambivalence in the "Complete Concept" Doctrine 287
3 Individual Essences? 288
4 Criticizing "Old-Fashioned Essentialism": "Second Substance" Terms and Natural Kinds 289
5 Criticizing the New Essentialism: Varieties of Essentialism and Two Types of Leibnizian Essence 291
6 The Essential and the Intrinsic: An (Attempted) Way Out of the Modal Problem? 293
7 Making Sense of Intrinsicality 294
8 Change, Time and the "Essence"/"Nature" Pair 295
References 297

Leibniz's Metaphysics of Change: Vague States and Physical Continuity 299
Richard T. W. Arthur
References 321

Is Leibniz's 'Lex Iustitiae' a Logical Law? 323
Enrico Pasini
1 The Analysts' Justice 323
2 Logical Laws and Laws of Justice 327
3 Algebraic Laws, Homogeneity and Homeoptosis 331
4 Conclusion 336
References 338

Leibniz among the Nominalists ... 341
Calvin G. Normore
1 The Medieval Nominalist Tradition ... 341
2 Nominalism ... 342
3 Abstract and Concrete Terms ... 343
4 Connotative Terms and the Language of Thought ... 345
5 Ontology ... 345
6 Appellation and Predication ... 347
7 *Abstracta Logica* ... 348
8 Relations ... 349
9 Conclusion ... 351
References ... 351

Modern Logic and Its Applications

Oskar Becker and the Modal Translation of Intuitionistic Logic ... 355
Stefania Centrone and Pierluigi Minari
1 Introduction ... 355
2 Gödel's Result ... 356
3 Oskar Becker and the Search for a "System of Closed Modalities" ... 359
4 Becker's Idea of a Modal Interpretation of Intuitionistic Logic ... 365
5 Conclusions ... 366
References ... 367

Reflecting and Unfolding ... 369
Andrea Cantini
1 On Reflective Closure ... 369
2 Feferman-Strahm's Unfolding ... 373
3 On the Implicit Commitment Thesis ICT ... 378
4 Conclusion ... 381
References ... 381

Metaphysical Modality, without Possible Worlds ... 385
Giorgio Lando
1 Introduction ... 385
2 Metaphysical Modality as Absolute Modality ... 388
3 McFetridge's Thesis and the Sources of Modality ... 391
4 Metaphysical Modality as Essential Modality ... 394
5 Possible Worlds and Metaphysical Modality ... 399
6 Conclusion: Terminological Issues ... 405
References ... 407

Counterpart Semantics at Work: Independence and Incompleteness Results in Quantified Modal Logic ... 409
Francesco Belardinelli
1 Introduction ... 409
2 Kripke Semantics ... 411

3 QML systems 413
4 Counterpart Semantics 416
5 Independence and Incompleteness in Kripke Semantics 420
6 Concluding Remarks 430
References 431

The Form of Practical Reasoning 433
Carla Bagnoli
1 The Debate about Universality 436
2 A Constructivist Approach to Practical Knowledge 438
3 The Contested Role of Principles in Empirical Practical Reasoning 441
4 Gilbert Harman and the Epistemic Argument for Transductive Inferences 443
5 Problem: Does Transduction Convey Practical Knowledge? 444
6 The Requirement to Reconsider and the Form of Reasoning 445
7 The Constitutive Norm of Rational Agency 446
8 Rational Construction, and the Analogy with the Building Trade 447
9 Conclusion 448
References 449

Massimo Mugnai: Selected Publications 1973–2021 451

Classical Antiquity

The Method of Models in Plato's *Statesman*

Paolo Crivelli

Plato's *Statesman* is rich of methodological reflections. It is the dialogue where the method of division is most extensively applied, but also the one where this method most often fails to achieve its goal, namely finding a definition. These failures induce the dialogue's main speakers, namely a visitor from Elea (henceforth: 'the Visitor') and a young student named Socrates (henceforth: 'Young Socrates'), to consider alternative methodological avenues. One of these relies on models.

The central idea of the 'method of models' (as it is sometimes called, though the phrase is not Plato's) is that by considering a property in an easy context one will be in a good position to recognize its presence in difficult contexts, namely contexts that are 'obscure' and somehow 'hide' the presence of the property. The Visitor and Young Socrates first consider a model that will enable one better to grasp what a model is (a 'model of a model'): how teachers lead children to decipher occurrences of letters within syllables. Sometimes, children are able to decipher an occurrence of a letter in an easy syllable, but then, faced with an occurrence of the same letter in a difficult syllable, they are unable to decipher it. What the teacher does in such circumstances is to set the difficult syllable alongside the easy one, lead the children to realize that the occurrence they failed to decipher is of the same letter as the one they are able to decipher, and transfer the successful decipherment of one occurrence to the other. Something analogous can happen with the 'letters and syllables of reality'. The 'letters of reality' are probably simple kinds, while the 'syllables of reality' probably comprise both complex kinds (which consist of simple kinds) and perceptible particulars (or, to be more precise, perceptible-particulars-at-instants, which are bundles of simple kinds). The decipherment of the occurrences of the 'letters of reality' in a 'syllable of reality' that is a complex kind probably amounts to the mental act of defining the complex kind, while the (partial) decipherment of an

P. Crivelli (✉)
Département de Philosophie, Université de Genève, 5, rue de Candolle, CH-1211 Geneva, Switzerland
e-mail: paolo.crivelli@unige.ch

F. Ademollo et al. (eds.), *Thinking and Calculating*, Logic, Epistemology, and the Unity of Science 54, https://doi.org/10.1007/978-3-030-97303-2_1

occurrence of a 'letter of reality' in a 'syllable of reality' that is a perceptible particular probably amounts to the mental act of judging that the perceptible particular partakes of the kind that is the 'letter of reality'.[1]

A new method. In order to perform a correct division leading to the definition of statesmanship (which is the dialogue's official goal), the Visitor deems it appropriate to introduce a new method, one whose crucial feature is the appeal to a παράδειγμα (277d1–283a9). The Greek noun 'παράδειγμα' can be rendered both by 'model' and by 'example'.[2] Etymologically, 'παράδειγμα' suggests what is shown ('δείκνυμι') alongside ('παρά') something.[3] The translation 'model' alludes to the idea of imitation, 'example' to that of a particular case of something general. Neither of these ideas is essential to Plato's views about the items to which he applies 'παράδειγμα' in the *Statesman*'s pages presently under consideration. I nevertheless adopt 'model'.

The role of models in inquiry is prominent also in other dialogues. For instance, in the *Meno* (at 77a9–b1), Socrates offers some sample definitions (two of shape and one of colour) as models for Meno to imitate in his effort to define virtue. In the *Sophist*, the division leading to the definition of the angler is described as a model that will help one to carry out the arduous job of producing a division leading to a definition of the sophist (218d9, 221c6). Still in the *Sophist*, the art of producing plastic imitations (i.e. sculptures) is introduced as a model that will help one to understand the sophist's art as an imitative art (233d3).

The concept of a model also plays a prominent role in some discussions of forms in dialogues that antedate the *Statesman* and the *Sophist*. Forms are there described as models of which some perceptible particulars are imitations.[4] However, the employment of the concept of model with regard to forms appears to be quite different from the one presently under scrutiny.

Letters as models of models. In order to provide a general characterization of the method for acquiring knowledge by means of models (277d1–278e3), the Visitor resorts to a model: a model to demonstrate what a model is or how it is used. The model to which he resorts is that of letters (γράμματα, 277e3, or στοιχεῖα, 277e6).[5] He focuses on how children learn about letters as they occur in syllables. Faced with occurrences of letters in 'the shortest and easiest of syllables', children make true judgements and 'speak truths about them' (277e6–8); but, faced with other occurrences of the same letters in different syllables, they 'make mistakes both in

---

[1] Massimo Mugnai is not only one of the philosophers and historians of philosophy I admire most, but also a dear friend. It is an honour and a pleasure to be in a position to dedicate this study to him.

[2] Cf. LSJ *s.v.* 'παράδειγμα'.

[3] At one point (278b4), the Visitor seems to allude to this etymology (cf. below, n. 8 and text thereto).

[4] Cf. *R.* 6. 484c6–d10; *Prm.* 132c12–d4; 133c9–d2; *Ti.* 27d5–28b2; 29b1–c3; 48e2–49a4.

[5] On this section of the dialogue, cf. Miller (1980, 58–59); El Murr (2014, 51–52); Oberhammer (2016, 199–203).

judgement and in speech' (278a2–3). In order to lead the children to acquire knowledge, the teacher will bring them back to those syllables where they judged and spoke truly about the occurrences of letters, set these syllables[6] 'alongside those [*sc.* syllables] which are not yet known [παρὰ τὰ μήπω γιγνωσκόμενα]' (278a9–b1) to them (because they still make mistakes about the occurrences of letters in these syllables), and get them to realize that the same similarity, i.e. nature,[7] is present in the two syllables (or, as we would put it, that they contain tokens of the same types) (278a8–b3). In this way, the occurrences of letters 'that are truly judged about [δοξαζόμενα ἀληθῶς]' (278b3–4) by the children 'are shown by being set alongside [παρατιθεμένα δειχθῇ] all those [*sc.* occurrences of letters] of which there is ignorance [πᾶσι τοῖς ἀγνοουμένοις]' (278b4) (the expression 'παρατιθεμένα δειχθῇ' alludes to the etymology of 'παράδειγμα').[8] Once they have come to realize the relation of sameness obtaining between the occurrences of letters in the two syllables, the children no longer have different reactions (a true statement and a false one, or a true judgement and a false one) with respect to the occurrences of the same letters in the two different syllables. Rather, they have the same correct reaction (a true statement or a true judgement) with respect to all occurrences (here the teacher probably plays a role in order to avoid that the children have the same wrong reaction with respect to all occurrences). In particular, the occurrences of letters about which the children judge and speak truly, 'becoming thereby models' (278b4–5), enable them to address the occurrences of letters 'in all syllables [ἐν πάσαις ταῖς συλλαβαῖς]' (278b6) in a way that matches their relations of sameness and difference: the children address *this* occurrence of a letter in *this* syllable, which in fact is the same (or, as we would put it, is a token of the same type) as *that* occurrence of a letter (or those occurrences of a letter) in *that* syllable (either the same one or a different one), as the same; and they address *this* occurrence of a letter in *this* syllable, which in fact is different from (or, as we would put it, is a token of a different type with respect to) *that* occurrence of a letter (or those occurrences of letters) in *that* syllable (either the same one or a different one), as different. In other words, the children use the same name (e.g. 'Theta') to describe two occurrences of the same letter, and they use different names (e.g. 'Theta' and 'Tau') to describe two occurrences of different

[6] I take the object of 'τιθέναι' at 278a9 to be an understood 'ἐκεῖνα ἐν οἷς ταὐτὰ ταῦτα ὀρθῶς ἐδόξαζον' supplied from the immediately preceding sentence at 278a8–9 (cf. Stallbaum [1841, 218]; Campbell [1867, *Plt.* 82]; El Murr [2014, 46]): the teacher sets *the correctly deciphered syllables* alongside those that still induce mistakes. Diès (1950, 35) takes the object of 'τιθέναι' to be an understood 'αὐτούς' supplied from 278a6: the teacher sets *the children* in front of the syllables or occurrences of letters that still induce mistakes. Oberhammer (2016, 200) seems to take the object of 'τιθέναι' to be some phrase like 'ἃ ὀρθῶς ἐδόξαζον': the teacher sets *the correctly deciphered occurrences of letters* alongside those that still induce mistakes. Note that the only way in which one can set an occurrence of (say) 'Θ' in one syllable alongside its occurrence in another syllable is to set the two syllables alongside one another.

[7] I regard the occurrence of 'καί' at 278b2 as epexegetic: 'similarity' ('ὁμοιότης') here means 'nature' ('φύσις') and both expressions are probably synonyms of 'kind' ('γένος'): cf. 285b5.

[8] Cf. Campbell (1867, *Plt.* 82); Rowe (1995, 201); El Murr (2014, 52). The things 'that are truly judged about' and 'those of which there is ignorance' could also be syllables; but at 278a8–9 the children seem to judge correctly about occurrences of letters.

letters. The mental act of realizing that these two occurrences of letters are the same (i.e. of the same type) or different (i.e. of different types) is different from the mental act of recognizing the occurrences of letters in question (by using names like 'Theta' or 'Tau'): consider, by way of analogy, that one can discover that two sets have the same size (by finding a one-to-one function from one of them onto the other) without knowing what their size is.

His model of a model enables the Visitor to offer the following general description[9] of a model:

T1 ΞΕ. Οὐκοῦν τοῦτο μὲν ἱκανῶς συνειλήφαμεν, ὅτι 278c3
παραδείγματός γ' ἐστὶ τότε γένεσις, ὁπόταν ὂν ταὐτὸν ἐν
ἑτέρῳ διεσπασμένῳ δοξαζόμενον ὀρθῶς καὶ συναχθὲν περὶ 5
ἑκάτερον καὶ συνάμφω μίαν ἀληθῆ δόξαν ἀποτελῇ; 6

VIS. Have we then sufficiently grasped this, that there is the coming-to-be of a model when, being the same in a different thing that has been torn apart, by being judged about correctly and collected it brings about a single true judgement about each of the two and both together? (Pl. *Plt.* 278c3–6)

According to T1's description, x is a model of y just if (1) x is 'the same', i.e. of the same type, as y, (2) x is 'in a different thing', i.e. is contained in a complex different from the one that contains y, (3) what x is in has been 'torn apart', i.e. the complex that contains x has been analysed into its components, (4) x has been judged about correctly, (5) x has been 'collected', i.e. x has been subsumed under the same kind as y (so that x and y have been discovered to be of the same type), and (6) a single true judgement about each of x and y and both together arises in virtue of all this, i.e. the true judgement about x has given rise to a true judgement about y.[10]

Spelling, dictation, reading. What type of performance are children asked to carry out? It could be a vocal spelling of a syllable that has been uttered (this would be the performance requested by pronouncing the words, 'Tell me what the letters in the first syllable of the name "ΘΕΑΙΤΗΤΟΣ" are!'). Alternatively, it could be the writing down of a syllable that is being dictated ('Write the first syllable of the name "ΘΕΑΙΤΗΤΟΣ"!'). A third possibility is that the required performance could be that of reading out an inscription of a syllable ('Read this syllable!', uttered while pointing to the initial part of an inscription of the name 'ΘΕΑΙΤΗΤΟΣ')'.

Performances of all three types are mentioned in the last part of the *Theaetetus*. At 203a6–9, Socrates imagines a situation where someone asks the question, 'Tell me, Theaetetus, what is "ΣΩ"?', and Theaetetus answers, 'It's Sigma and Omega': this is a case of spelling of a syllable that has been uttered. At 207e7–208a2, Socrates imagines someone who tries to write down the name 'ΘΕΑΙΤΗΤΟΣ' and correctly begins by writing down its first syllable as 'ΘΕ', but then tries to write down the name 'ΘΕΟΔΩΡΟΣ' and wrongly begins by writing down its first syllable as 'TE':

[9] Oberhammer (2016, 202) speaks of a definition of model.

[10] On the difference between sentences containing 'each of the two' and sentences containing 'both together', cf. *Hp.Ma.* 299b8–303d10 with Crivelli (2012, 140).

this is a case of writing down a syllable that is being dictated. At 206a5–8, Socrates speaks of the time spent by Theaetetus 'in trying to distinguish, both by sight and by hearing, each individual letter in itself, in order that their position when they are said or written down should not bewilder you': this includes not only a performance based on an uttered syllable, but also one based on a written syllable, and therefore includes the activity of reading out an inscription of a syllable.[11] In Plato's time there was a single specialised teacher of reading and writing, the γραμματιστής, and the two activities were learnt by children at the same time.[12] The *Statesman*'s description (at 278b5–c1) of the ability acquired by the children as a capacity to 'address' the occurrences of letters in a way that matches their relations of sameness and difference points in the direction of spelling, but I suspect that Plato moves freely between the three types of performance (spelling, dictation, and reading).

The case mentioned in the second of the passages from the *Theaetetus* I just referred to, i.e. 207e7–208a2, is close to the one considered in the passage of the *Statesman* because it concerns two contexts that generate a correct and a wrong reaction. The *Theaetetus* passage is about two occurrences of the same letter, 'Θ', in two occurrences of the same syllable, 'ΘE', in different words, 'ΘEAITHTOΣ' and 'ΘEOΔΩPOΣ': a person writes down correctly the occurrence of 'Θ' in the occurrence of 'ΘE' in the word 'ΘEAITHTOΣ', but writes down incorrectly the occurrence of 'Θ' in the occurrence of 'ΘE' in the word 'ΘEOΔΩPOΣ'. Granted the plausible assumption that the situation described in the *Theaetetus* passage is an example of the one described in the *Statesman* passage, it follows that two occurrences of the same syllable in different words may be counted as different syllables. Note that in the *Theaetetus* Socrates goes on to claim, at 208a2–3, that the person who writes down correctly the first syllable of 'ΘEAITHTOΣ' but makes a mistake in writing down the first syllable of 'ΘEOΔΩPOΣ' does not know the syllable in question, 'ΘE'.

Recognition and identification. Spelling involves the mental act of recognition or identification, namely the mental act that one performs when one (correctly or incorrectly) recognizes or identifies someone or something.[13] Consider the mental act of recognizing or identifying a person. When one encounters a person, one sometimes has the impression of having met that person before, and one makes an effort to recognize or identify that person. Sometimes one's effort is successful, and one manages correctly to recognize or identify the person. On other occasions, one's effort is unsuccessful. One's failure can come about in two ways: either one is unable to come up with any verdict as to the identity of the person encountered, or one comes up with a wrong verdict and one incorrectly recognizes or identifies the person (this is a case of misidentification). Often, one expresses one's (correct or incorrect)

[11] Cf. El Murr (2014, 48).

[12] Cf. *Chrm.* 159c3–7; *Euthd.* 279e2–4; Joly (1986, 108–109).

[13] I do not treat 'to recognize' and 'to identify' as success verbs: one can, in principle, recognize incorrectly and identify incorrectly (a misidentification is still an identification).

recognition or identification of a person by means of a (true or false) statement of identity where one utters a demonstrative or a personal pronoun, followed by 'is', followed by what one takes to be the name of the person encountered (e.g. 'This is Socrates'). In spelling, acts of recognition or identification of occurrences of letters are typically formulated simply by uttering what one takes to be the names of the letters concerned (without demonstrative pronouns or forms of 'to be').[14]

Plato was very much interested in the mental act of recognition or identification. This is shown by the so-called 'Waxen Block' account of false judgement, which occupies a conspicuous section of the *Theaetetus* (190e5–196d2). Plato concentrates on judgements of recognition or identification of 'the things we have seen or heard or thought of ourselves' (191d6–7). He explains judgements of this sort as events where memory imprints stored in (something analogous to) a waxen block are assigned to 'perceptions and thoughts' (191d7). A judgement is true if the memory imprint assigned and the perception or thought to which it is assigned come from the same entity, it is false otherwise.[15] Plato explicitly mentions 'recognition [ἀναγνώρισις]' (193c5), and the same Greek verb, 'ἀναγιγνώσκω', can be translated both by 'to recognize' and 'to read'.[16] Something like recognition or identification seems to be mentioned also in a passage of the *Timaeus* (37a5–b1) that describes the cognitive processes of the world-soul as involving its rotating and discovering with what an object it has encountered is identical and from what that object is different. It is perhaps involved also in a passage of the *Sophist* (253d1–3) that describes the activity of dialectic as consisting in avoiding to regard an identical kind as different or a different kind as identical.[17]

In the *Statesman* passage that explains the method of models, the Visitor and Young Socrates concentrate on the mental act of recognizing or identifying occurrences of the same letters in different syllables: children correctly recognize or identify some occurrences of letters in certain syllables but do not correctly recognize or identify other occurrences of the same letters in other syllables. The correct recognition or identification gives rise to a correct act of spelling (or of writing down, or of reading); failure to perform a correct recognition or identification gives rise either to no act at all or to a mistaken act of spelling (or of writing down, or of reading). The recipe for correcting failures of this sort is to bring the children back to the cases where the recognition or identification was correct: the children correctly recognize or identify certain occurrences of a letter in certain syllables, and by doing this again and carrying out some pertinent comparisons they manage correctly to recognize or identify the occurrences of the same letter in the other syllables where they had previously been at a loss or made mistakes.

[14] For the names of letters, cf. *Cra.* 393d7–e10.

[15] Cf. 193c5–d2; 193d6–7; 193e1–2; 193e6–194a4; 194a6–8; 194b2–6.

[16] Cf. LSJ *s.v.* 'ἀναγιγνώσκω' i 2 and ii.

[17] Cf. Palumbo (1995, 178).

The letters and syllables of reality. After describing what happens to children with the letters and syllables of the alphabet, the Visitor goes on to consider analogous experiences we have with the 'letters of all things' and the 'syllables of objects', or, as I shall often call them, the 'letters and syllables of reality'[18]:

T2 ΞΕ. Θαυμάζοιμεν ἂν οὖν εἰ ταὐτὸν τοῦτο ἡμῶν ἡ ψυχὴ 278c8
φύσει περὶ τὰ τῶν πάντων στοιχεῖα πεπονθυῖα τοτὲ μὲν ὑπ' d
ἀληθείας περὶ ἓν ἕκαστον ἔν τισιν ἵσταται, τοτὲ δὲ περὶ
ἅπαντα ἐν ἑτέροις αὖ φέρεται, καὶ τὰ μὲν αὐτῶν ἁμῇ γέ πῃ
‹ἐκ› τῶν συγκράσεων ὀρθῶς δοξάζει, μετατιθέμενα δ' εἰς τὰς
τῶν πραγμάτων μακρὰς καὶ μὴ ῥᾳδίους συλλαβὰς ταὐτὰ
ταῦτα πάλιν ἀγνοεῖ;
ΝΕ. ΣΩ. Καὶ θαυμαστόν γε οὐδέν. 7

VIS. Would we then be surprised if our soul, suffering by nature this same experience about the letters of all things [*sc.* the 'letters of reality'], on some occasions were brought to a stand by truth about each one [*sc.* about each one of the 'letters of reality'] within some things [*sc.* within certain 'syllables of reality'] whereas on others were swept away about all of those [*sc.* about all of the 'letters of reality'] in other things [*sc.* in other 'syllables of reality'], and somehow or other were correctly to judge about some of them because of the mixtures [*sc.* because the 'syllables of reality' where they occur are easy], but were again to be ignorant about these same things if they are transferred in the large and non-easy syllables of objects [*sc.* in the difficult 'syllables of reality']?
Y.S. It would not be surprising at all. (Pl. *Plt.* 278c8–d7)

At 278d2 I read 'ἔν τισιν ἵσταται' with Hermann and David Robinson (the **β** family and W have 'ἕν τι συνίσταται' whereas the first hand of T appears to have read 'ἔν τισι ἵσταται').[19] At 278d4 I adopt Stallbaum's integration of 'ἐκ' to provide some anchorage for the otherwise floating genitive 'τῶν συγκράσεων'. Alternatively, one might adopt Campbell's suggestion of making 'τῶν συγκράσεων' dependent on 'ἁμῇ γέ πῃ' taken in a spatial sense, 'here and there among the mixtures', but such a usage is unparalleled. Rowe construes 'τῶν συγκράσεων' with 'αὐτῶν' and takes the resulting phrase to be governed by 'τὰ μέν', but it is difficult to see how all of this could mean 'some [*sc.* letters of reality] within mixtures': 'τὰ μὲν αὐτῶν τῶν συγκράσεων' can only mean 'some of the mixtures themselves', which however clashes with its twin 'μετατιθέμενα δ' εἰς τὰς ... συλλαβάς'.[20]

Passage T2 speaks of 'the letters', or 'elements', 'of all things' ('τὰ τῶν πάντων στοιχεῖα', 278d1, the noun 'στοιχεῖον' can mean both 'letter' and 'element') and of 'the syllables of objects' ('αἱ τῶν πραγμάτων συλλαβαί', 278d4–5). The Visitor

[18] The label 'of reality' is meant merely to indicate that what is in question are elements and composites at a metaphysical level (the noun 'reality' here does not translate any expression of Plato's Greek).

[19] Cf. Hermann (1851, xxx); Duke et al., (1995, 508).

[20] Cf. Stallbaum (1841, 219); Campbell (1867, *Plt.* 83); Rowe (1995, 202).

and Young Socrates agree that our souls can very well experience something analogous to what happens to children with ‘alphabetic’ letters and syllables: faced with an occurrence of a certain ‘letter’ (or ‘element’) in one ‘syllable’ (or ‘compound’) of reality, our souls recognize or identify it correctly, but faced with an occurrence of the same ‘letter’ (or ‘element’) in a different ‘syllable’ (or ‘compound’) of reality, they fail correctly to recognize or identify it. The recipe for overcoming this predicament is probably analogous to the (genuinely) alphabetic one adopted in the education of children. Thus, the occurrences of ‘letters’ (or ‘elements’) of reality which our souls fail correctly to recognize or identify must be set alongside those occurrences which they recognize or identify correctly. This brings our souls to realize what relations of identity or difference obtain between these occurrences of ‘letters’ (or ‘elements’) of reality (cf. 278b1–c1): our souls realize that *this* occurrence of a ‘letter’ (or ‘element’) of reality in *this* ‘syllable’ (or ‘compound’) of reality is the same as *that* occurrence of a ‘letter’ (or ‘element’) of reality in *that* ‘syllable’ (or ‘compound’) of reality, and they also realize that *this* occurrence of a ‘letter’ (or ‘element’) of reality in *this* ‘syllable’ (or ‘compound’) of reality is different from *those* occurrences of ‘letters’ (or ‘elements’) of reality in *that* ‘syllable’ (or ‘compound’) of reality. Once our souls have discovered these relations of identity and difference, they will correctly recognize or identify both occurrences of the same ‘letter’ (or ‘element’) in the two different ‘syllables’ (or ‘compounds’) of reality.[21]

Two possible types of letters and syllables of reality. It is not clear what the ‘letters’ (or ‘elements’) and the ‘syllables’ (or ‘compounds’) of reality are. At least two accounts of their nature are consistent with what the Visitor and Young Socrates say.

According to the first account, the ‘letters’ (or ‘elements’) and the ‘syllables’ (or ‘compounds’) of reality are (respectively) simple kinds and composite kinds (where, roughly speaking, the composition in question is ‘intensional composition’). Thus, for instance, the simple kinds animal, terrestrial, biped, and featherless are ‘letters’ (or ‘elements’) within the ‘syllable’ (or ‘compound’) that is the composite kind human. In general, all the kinds mentioned in another kind’s definition are ‘letters’ (or ‘elements’) within it as a ‘syllable’ (or ‘compound’).[22]

According to the second account, the ‘letters’ (or ‘elements’) and the ‘syllables’ (or ‘compounds’) of reality are (respectively) kinds and perceptible-particulars-at-instants. Some accidental attributes do not hold for ever of perceptible particulars as they are usually conceived of, namely as entities whose existence stretches over a period of time (which can be either long or short): Socrates was seated at some instants and not at others. The perceptible particulars that are the ‘syllables’ (or ‘compounds’) of reality are not perceptible particulars as they are usually conceived of, entities whose existence stretches over a period of time. Rather, they are ‘instantaneous perceptible particulars’, or ‘perceptible-particulars-at-instants’. The ‘letters’ (or ‘elements’) that are the constituents of any one of them are all and only the kinds

[21] Cf. Dixsaut (2001, 253).

[22] Cf. Skemp (1952, 162).

of which it partakes at its peculiar instant.[23] For instance, the instantaneous slice of Socrates at 3 p.m. of May 1st of 400 bc is a 'syllable' (or 'compound') of reality whose 'letters' (or 'elements') of reality are (perhaps) the kinds man, of-such-and-such-a-size, conversing, in-such-and-such-a-spot-of-the-market-place-in-Athens, standing, and all the other kinds of which Socrates was partaking at 3 p.m. of May 1st of 400 bc. Perceptible particulars as they are usually conceived of, entities whose existence stretches over a period of time, can be constructed as 'temporal worms' consisting of a series of 'instantaneous slices', i.e. the perceptible particulars at *t* for all instants *t* within certain periods (the periods during which the usual perceptible particulars exist).

The identity through time of perceptible particulars. No syllable survives the subtraction or the replacement of any of the occurrences of letters of which it is composed, nor does any syllable survive the addition of an occurrence of a letter: if in a syllable even a single occurrence of a letter is subtracted or replaced with an occurrence of a different letter, or even a single occurrence of a letter is added, then the 'original' syllable is destroyed and (in some cases) a 'new' one is born.[24] Accordingly, the second account presupposes that the 'career through time' of a perceptible particular as it is usually conceived of (i.e. as an entity whose existence stretches over a period of time, e.g. you or me) should be more properly described as a series of births and deaths of perceptible-particulars-at-instants,[25] and a perceptible particular as it is usually conceived of should be something like a construct, which as such does not, properly speaking, exist, but merely comes-to-be. Even what we would usually describe as a minimal qualitative change of a persistent substance is really the destruction of an earlier instantaneous perceptible particular and the birth of a later instantaneous perceptible particular. One may wonder what makes two perceptible-particulars-at-instants slices of the same perceptible particular as it is usually conceived of, i.e. as an entity whose existence stretches over a period of time. Two types of answer are possible: the first would appeal to a causal relation (one perceptible-particular-at-an-instant is the cause or the origin of the other), the second to a relation of similarity (one perceptible-particular-at-an-instant is appropriately similar to the other).

An account of this sort of the 'career through time' of a human being's body appears to be presented in a passage of the *Symposium* (207c9–e1). Here Diotima draws an analogy between the immortality of sorts which living beings strive for by means of reproduction, where in reality what happens is only that one living being that is destined to die begets another living being of the same species that will take its place, and what happens in a particular body in the course of an individual life,

---

[23] Several commentators credit Plato with the view that perceptible particulars are bundles of kinds or of kind-instances: cf. Burnet (1914, 165); Prauss (1968, 108–109); Burge (1971, 10); Mann (2000, 30, 124–125, 148, 179); Silverman (2002, 281). F.C. White (1977, 202–204) argues against attributing such a view to Plato, but his own preferred interpretation seems to involve unpalatable 'bare particulars'.

[24] If one subtracts the occurrence of 's' in the syllable 'strap', the result is 'trap', which is a new syllable. If one subtracts the occurrence of 'a' in 'strap', the result is 'strp', which is not a syllable.

[25] Cf. Mann (2000, 124–125, 157, 165).

where the replacement of the bodily components (e.g. hair, flesh, and blood) brings it about that the body 'never consists of the same things' (207d6–7) and therefore 'is always being renewed and in other respects passing away' (207d7–8).[26]

A similar picture emerges from two passages of the *Theaetetus* that pertain to the so-called 'Secret Doctrine of Protagoras'. In the first (158e5–160d4), Socrates portrays Protagoras' Secret Doctrine as maintaining that any episode which we would usually describe as an alteration of the same particular that continues to exist throughout the time taken by the alteration really involves different particulars at its beginning and end. For example, consider an episode which we would usually describe as a case where Socrates recovers his health and continues to exist throughout the time taken by his recovery: really, such an episode involves different particulars at its beginning and end, namely ill-Socrates and healthy-Socrates. Such an account presupposes that any episode that we would usually describe as an alteration of the same enduring particular really amounts to a destruction of one particular (the one at the beginning of the process) running in parallel with a generation of another particular (the one at the end of the process). In the second passage (166b6–c1), Socrates imagines Protagoras defending his own Secret Doctrine by saying that in situations where we would usually speak of a single man undergoing a change in his psychological state, we should, properly speaking, avoid expressions like 'the man' and use instead 'the men' to refer to the 'infinite men that come to be' (166b8–c1). Some confirmation that Protagoras' Secret Doctrine analyses change as a succession of co-occurring generations and destructions comes from the reference (at 152e5) to the comic poet Epicharmus as one of the 'wise' whose authority should bring support to Protagoras' Secret Doctrine. One might be surprised at finding Epicharmus mentioned alongside Parmenides, Protagoras, Heraclitus, Empedocles, and Homer (there is only one other reference to Epicharmus in a genuine dialogue of Plato).[27] However, one of Epicharmus' fragments suggests a plausible explanation of why he is mentioned. It describes a debtor who refuses to pay his debt by alleging that he is not the same individual as the one who contracted it: after all, there are some minor differences in mass between the present individual and the one who contracted the debt (he gained a bit of weight!), so, just as a number of pebbles is no longer the same if even just one pebble is added or subtracted, but a different number of pebbles comes about, so also in the present case two individuals are involved.[28] The view behind Epicharmus' scene was later taken up by the Academic skeptics to challenge the Stoics, who reacted to it.[29] It fits in well with the aspect of Protagoras'

[26] For detailed discussions of the metaphysical picture drawn in the *Symposium* passage, cf. Price (1989, 21–22); Ademollo (2018, 40–49).

[27] Cf. Pl. *Grg.* 505e1. There is also a reference in the spurious *Axiochus* (at 366c4).

[28] Cf. D.L. 3.11; Anon. *in Tht.* 71, 12–40; Plu. *Ser. Num. Vind.* 15. 559b.

[29] Cf. Plu. *Comm. Not.* 44. 1083a–c. The Stoic response is discussed by Sedley (1982, 259–271).

Secret Doctrine I have highlighted. Commentators disagree about the paternity of Protagoras' Secret Doctrine, but some[30] take it to be Plato's own position.[31]

Particulars as bundles of kinds. According to the metaphysical picture drawn by the second account of the 'letters' (or 'elements') and 'syllables' (or 'compounds') of reality, every perceptible-particular-at-an-instant is a bundle whose components are all and only the kinds of which it partakes at its peculiar instant. Its individuality, what makes it a particular, amounts to complete specification or determination, i.e. to the fact that for every range of reciprocally incompatible attributes one of whose members could be taken to contribute to single out that perceptible-particular-at-an-instant, exactly one member of that range is a component of it. Just as the letters of which a syllable is composed are not inherent in or 'hanging on' a putative substrate of the syllable, so also the kinds of which a perceptible-particular-at-an-instant partakes (at its peculiar instant) are not inherent in a putative substrate of it.

The second account of the nature of the 'letters' (or 'elements') and 'syllables' (or 'compounds') of reality cannot be easily dismissed. Five considerations speak in its favour.

(1) In the so-called 'Affinity Argument' for the immortality of the soul in the *Phaedo* (78b1–80c1), Socrates contrasts forms with perceptible particulars by mentioning several pairs of opposite traits. Two of these are the pair incomposite-composite (ἀσύνθετον-σύνθετον)[32] and the pair uniform-multiform (μονοειδές-πολυειδές)[33]: forms are incomposite and uniform whereas perceptible particulars are composite and multiform. Socrates is probably implying that perceptible particulars are composite because they are multiform, i.e. that they are composite because they are composed of the characteristics they enjoy, whereas forms are incomposite because they are uniform, i.e. that each of them is simple because it enjoys only one characteristic and therefore does not have many characteristics to be composed

---

[30] Cf. e.g. Burnet (1914, 241–242); Cornford (1935, 49). Not everyone agrees that Protagoras' Secret Doctrine is Plato's own position: cf. Irwin (1977, 2–3, 6).

[31] It might be objected that in the *Theaetetus* itself (at 179c1–184b2) Plato refutes the Heracliteanism of Protagoras' Secret Doctrine. I cannot offer here a detailed reconstruction of this refutation. I restrict myself to the following brief remarks: the refutation starts with a distinction between motion and alteration as two kinds of change; it then points out that the Heracliteanism under scrutiny assumes that things are changing in all ways and therefore undergo not only continuous motion but also continuous alteration; it then goes on to argue that given that perceptible qualities and the corresponding perceptions are motions, during the time during which things are undergoing these motions they must also undergo alterations that lead them to lose the characteristics (perceptible qualities and perceptions) which are those motions; the refutation then goes on to infer further absurdities that concern the nature of perceptible qualities and perceptions. These brief remarks show that the refutation relies on the premiss that perceptible qualities and the corresponding perceptions are motions. But the version of Heracliteanism that Plato endorsed might well not have been committed to this thesis. In particular, if perceptible particulars are bundles of kinds, these kinds themselves need not be motions.

[32] Cf. *Phd.* 78c1; c3; c7; c8.

[33] Cf. *Phd.* 78d5; 80b2; 80b4; 83e3.

of (the *Sophist*'s idea that kinds can partake of one another is still to come, in the *Phaedo* composition concerns only perceptible particulars).[34]

(2) In a passage of the *Theaetetus* often referred to as 'Socrates' Dream' (201c7–203b11), Socrates and Theaetetus introduce an atomistic ontology based on 'the primary elements [πρῶτα ... στοιχεῖα], as it were, of which we and everything else are composed' (201e1–2). Socrates explains this atomistic ontology by describing the behaviour of letters and syllables, which he characterizes as 'models' (202e4) of it. He says that 'the elements are [...] perceivable [αἰσθητά]' (202b6–7). Since earlier in the dialogue (at 156b7–c2) perceived items were identified with colours, sounds, and, in general, perceptible qualities, it is reasonable to infer that the Dream's atomistic ontology regards us and everything else as composed of perceptible qualities.[35] This is confirmed by 205d1–2, where Socrates refers back to the Dream's atomistic ontology and describes its primary elements as 'uniform and partless' (the adjective I am rendering by 'uniform' is 'μονοειδής', which Socrates used also in the passage of the *Phaedo* discussed in the last paragraph). Although he criticizes (at 203c1–206c2) an epistemological theory based on the Dream's atomistic ontology, Socrates does not reject the atomistic ontology itself.[36]

An atomistic ontology that regards us and everything else as composed of perceptible qualities surfaces also in two other passages of the *Theaetetus*. The earlier one belongs within the presentation of Protagoras' so-called Secret Doctrine:

T3 ΣΩ. δεῖ δὲ καὶ κατὰ μέρος οὕτω λέγειν καὶ περὶ 157b8
πολλῶν ἀθροισθέντων, ᾧ δὴ ἀθροίσματι ἄνθρωπόν τε
τίθενται καὶ λίθον καὶ ἕκαστον ζῷόν τε καὶ εἶδος. c

SO. One must speak in this way [*sc.* by avoiding formulations that would suggest that things are stable] with respect both to the parts [*sc.* single perceptible qualities][37] and to the many aggregated together – to which aggregate they give ‹the names of›[38] man, of stone, and of each animal and kind. (Pl. *Tht.* 157b8–c1)[39]

As I pointed out earlier,[37] some commentators take Protagoras' Secret Doctrine to be Plato's own position. The later passage, where Socrates

---

[34] The above interpretation of *Phd.* 78b1–80c1 draws on Prauss (1968, 100–108), to which I refer for a thorough defence of the exegesis (cf. also Heindorf [1809–1810, 100]; Bekker [1826, 5. 235]; Mann [2000, 107–120]).

[35] Here I am taking up one of the interpretative options proposed by McDowell (1973, 234, 238).

[36] An anonymous referee raises the question whether the Dream's claim that everything is composed of perceptible qualities suffices to provide support for the view that perceptible-particulars-at-instants are composed of all the attributes they partake of at their peculiar instants: for, among these attributes, there might be some that are not perceptible qualities. My impression is that the position put forward in the Dream belongs within the Secret Doctrine and that in this context the only attributes there are are perceptible qualities. The idea that perceptible-particulars-at-instants are composed of attributes they partake of is the hard one to accept: if it is accepted for those attributes that are perceptible qualities, it is likely to be accepted also for other attributes (provided one concedes that there are other attributes).

[37] Cf. above, n. 30 and text thereto.

is speaking in propria persona, is near the end of the dialogue:

T4 ΣΩ. Ἀλλ᾽ οὐ πρότερόν γε, οἶμαι, Θεαίτητος ἐν ἐμοὶ δοξασθήσεται, πρὶν ἂν ἡ σιμότης αὕτη τῶν ἄλλων σιμοτήτων ὧν ἐγὼ ἑώρακα διάφορόν τι μνημεῖον παρ᾽ ἐμοὶ ἐνσημηναμένη κατάθηται—καὶ τἆλλα οὕτω ἐξ ὧν εἶ σύ—ἥ με, καὶ ἐὰν αὔριον ἀπαντήσω, ἀναμνήσει καὶ ποιήσει ὀρθὰ δοξάζειν περὶ σοῦ.

SO. But, I think, there will not be in me a judgement about Theaetetus before this snub-nosedness has deposited in me, by being imprinted, a memory imprint different from the other snub-nosednesses which I have seen – and similarly with the other things of which you ‹consist›. This, even if I shall meet you tomorrow, will remind me and make me judge correctly about you. (Pl. *Tht.* 209c5–10)

The abstract noun-phrase 'this snub-nosedness' (209c6) is noteworthy because it can stand only for a quality. The remark that the memory imprint of 'this snub-nosedness' is different from 'the other snub-nosednesses which I have seen' (209c6–7) (or rather from their memory imprints) strongly suggests that it is a perceptible quality. Socrates indicates that Theaetetus consists 'of [ἐξ]' (209c8) 'this snub-nosedness' and 'other things' (209c8), i.e. further components, which are probably other perceptible qualities.[38] Note that in T4 memory imprints are of perceptible qualities (rather than of whole objects, as in the Waxen Block account of false judgement, at 190e5–196d2).

(3) In the first of the deductions that make up the *Parmenides*'s second part, Parmenides claims that 'if the one enjoyed any attribute apart from being one, it would enjoy the attribute of being more than one, which is impossile' (140a1–3). This claim probably relies on the assumption that the attributes which a thing enjoys are parts of it. Parmenides' reasoning is probably that if the one enjoyed some other attribute apart from that of being one, it would be a whole containing at least two parts (namely the two distinct attributes it enjoys), and it would for this reason be more than one.[39] In a similar vein, in the second deduction, Parmenides, after assuming that the one is, argues that the expressions 'is' and 'one' signify different things, namely being and the one, and that the one partakes of being (142b5–c7). He goes on to assert that since being and the one are reciprocally distinct but are both 'of' the one-that-is, it follows that the one-that-is is a whole of which being and the one are parts (142c7–d9).[40] These remarks suggest that an object is a whole whose parts are the kinds of which it partakes.

(4) In a passage of the *Philebus* Plato portrays Socrates as discussing both the method of division and the relationship between the kinds reached by division and the perceptible particulars that fall under them:

[38] Cf. McDowell (1973, 255).

[39] Cf. Owen (1970, 350).

[40] Cf. Owen (1970, 352–353); Schofield (1973, 29–36); Kenig Curd (1990, 20–2); *Sph.* 244d14–e1; 245b4–5; 245c1–2.

T5 ΣΩ. Θεῶν μὲν εἰς ἀνθρώπους δόσις, ὥς γε καταφαίνεται 16c5
ἐμοί, ποθὲν ἐκ θεῶν ἐρρίφη διά τινος Προμηθέως ἅμα
φανοτάτῳ τινὶ πυρί· καὶ οἱ μὲν παλαιοί, κρείττονες ἡμῶν
καὶ ἐγγυτέρω θεῶν οἰκοῦντες, ταύτην φήμην παρέδοσαν,
ὡς ἐξ ἑνὸς μὲν καὶ πολλῶν ὄντων τῶν ἀεὶ λεγομένων εἶναι,
πέρας δὲ καὶ ἀπειρίαν ἐν αὑτοῖς σύμφυτον ἐχόντων. δεῖν
οὖν ἡμᾶς τούτων οὕτω διακεκοσμημένων ἀεὶ μίαν ἰδέαν περὶ d
παντὸς ἑκάστοτε θεμένους ζητεῖν—εὑρήσειν γὰρ ἐνοῦσαν—
ἐὰν οὖν μεταλάβωμεν, μετὰ μίαν δύο, εἴ πως εἰσί, σκοπεῖν,
εἰ δὲ μή, τρεῖς ἤ τινα ἄλλον ἀριθμόν, καὶ τῶν ἓν ἐκείνων
ἕκαστον πάλιν ὡσαύτως, μέχριπερ ἂν τὸ κατ᾽ ἀρχὰς ἓν μὴ
ὅτι ἓν καὶ πολλὰ καὶ ἄπειρά ἐστι μόνον ἴδῃ τις, ἀλλὰ καὶ
ὁπόσα· τὴν δὲ τοῦ ἀπείρου ἰδέαν πρὸς τὸ πλῆθος μὴ προσ-
φέρειν πρὶν ἄν τις τὸν ἀριθμὸν αὐτοῦ πάντα κατίδῃ τὸν
μεταξὺ τοῦ ἀπείρου τε καὶ τοῦ ἑνός, τότε δ᾽ ἤδη τὸ ἓν e
ἕκαστον τῶν πάντων εἰς τὸ ἄπειρον μεθέντα χαίρειν ἐᾶν.
οἱ μὲν οὖν θεοί, ὅπερ εἶπον, οὕτως ἡμῖν παρέδοσαν σκοπεῖν
καὶ μανθάνειν καὶ διδάσκειν ἀλλήλους· οἱ δὲ νῦν τῶν ἀν-
θρώπων σοφοὶ ἓν μέν, ὅπως ἂν τύχωσι, καὶ πολλὰ θᾶττον 17a
καὶ βραδύτερον ποιοῦσι τοῦ δέοντος, μετὰ δὲ τὸ ἓν ἄπειρα
εὐθύς, τὰ δὲ μέσα αὐτοὺς ἐκφεύγει—οἷς διακεχώρισται τό τε
διαλεκτικῶς πάλιν καὶ τὸ ἐριστικῶς ἡμᾶς ποιεῖσθαι πρὸς
ἀλλήλους τοὺς λόγους.

SO. A gift of the gods to men, as it appears to me, was thrown by some god by means of a Prometheus together with some exceedingly bright fire. The ancients, superior to us and dwelling closer to the gods, passed down this prophetic saying, that the things ever said to be[44] are from one and many and have limit and unlimitedness naturally together within themselves. Since these things are organized in this way, we must look for a single form with regard to everything after positing it in each case, for we will find it because it is there. If we then share in it, after one we must see two, if they are there, otherwise three or some other number, and treat each of those ones in the same way, until we see that the initial one is not only one and many and unlimited, but also how many it is. One must not apply the form of the unlimited to the plurality before one has seen the total number of it, that which lies between the unlimited and the one, and only then give up on[45] each and every one[46] by letting it go into the unlimited. The gods, as I said, have entrusted us with inquiring and teaching and learning from one another in this way. The experts of the present day however produce one and many as chance has it and faster and slower than is due, and after the one they produce immediately the unlimited while intermediates elude them—whereas it is thanks to these that there is a distinction between our discussing with one another in a dialectical or eristic fashion. (Pl. *Phlb.* 16c5–17a5)

Passage T5 and its immediate context contain several cross-references to the pages of the *Phaedrus* that contain Plato's earliest presentation of division and collection[41]:

[41] Cf. Benitez (1989, 43–45); D. Frede (1997, 158); Benson (2010, 20–21).

Socrates' love for collections and divisions,[42] the divine nature of these procedures,[43] their association with dialectic,[44] and their relevance to art and knowledge.[45]

In T5 Socrates gives directions about how we ought to conduct our research. He begins with an injunction: at the start of our research 'we must look for a single form with regard to everything after positing it in each case' (16c10–d1). He justifies this injunction by stating that 'we will find it [*sc.* a single form] because it is there' (16d2). The form that Socrates prompts us to posit and find is probably the form under which fall all and only the entities comprised in our intended domain of research. Socrates then recommends that after positing and finding this single form, we 'see' (16d3) two or more further forms (as few as possible). He is probably instructing us to find two or more kinds immediately subordinate to the single kind that we have posited and found. He then urges us to apply the same procedure to each of the unities, i.e. kinds, which we have thus discovered. He demands that only after pursuing this procedure as far as possible we 'give up on each and every unity by letting it go into the unlimited' (16e1–2). He also invites us to keep track of the number of the forms we encountered.

The procedure recommended by Socrates comprises two phases: in the first, we must produce a complete classification of our intended domain of research on the basis of the method of division; in the second, we are asked to give up on each of the lowest kinds reached by the first phase's classification 'by letting it go into the unlimited' (16e2). Socrates does not explain what letting a kind go into the unlimited amounts to. I suspect that it involves applying something analogous to the method of division used in the first phase's classification in such a way as to generate perceptible-particulars-at-instants by combining the kinds they partake of (even if no finite mind could ever carry out the procedure that leads to perceptible-particulars-at-instants, the operation could in principle be performed by an ideal subject). My suspicion is caused by Socrates' remark that the procedure he is recommending will lead us to see that 'the initial unity is [...] one and many and unlimited' (16d5–6). Socrates seems to be indicating that just as the 'single form' (16d1) posited and found at the beginning is the two or more forms immediately subordinate to it, and also the four or more forms immediately subordinate to these, and so on, down to the finitely many lowest kinds at which the first phase's classification must stop, so also that initial single form is the indefinitely many[46] perceptible-particulars-at-instants that fall under its lowest kinds. Since it is the method of division that enables us to discover that the initial single form is the finitely many kinds reached in the course of the first phase's classification, it is reasonable to suspect that something analogous to the method of division should lead us to realize that the initial single form is the indefinitely

[42] Cf. *Phdr.* 266b3–4; *Phlb.* 16b5–6.

[43] Cf. *Phdr.* 266b5–7; *Phlb.* 16c5–8; 18b6–8.

[44] Cf. *Phdr.* 266b7–c1; *Phlb.* 17a3–5.

[45] Cf. *Phdr.* 277b5–c6; *Phlb.* 17b6–9, 17c7–d2.

[46] Shortly after the lines under consideration, Socrates raises the issue of when one can 'apply [προσφέρειν] the form of the unlimited to the plurality' (16d7–8): this requires that the unlimitedness in question should be unlimitedness in multiplicity (cf. Striker [1970, 20]).

many perceptible-particulars-at-instants that fall under its lowest kinds. The method would involve dividing progressively the lowest kinds by means of characteristics that are used like differentiae until perceptible-particulars-at-instants are completely specified (the procedure is merely analogous to division because it cannot be carried out completely, at least not by a finite mind). If passage T5 is indeed indicating that perceptible-particulars-at-instants are results that could at least in principle be reached by a procedure analogous to the division that is involved in classifying kinds, then perceptible-particulars-at-instants are made up of kinds much in the same way as any species reached by division is made up of its genus and its *differentia*.[47]

(5) According to the second account of the 'letters' (or 'elements') and 'syllables' (or 'compounds') of reality, Plato is explaining the relation of perceptible particulars to the kinds of which they partake as that of a compound to its elements, or of a mixture to its ingredients (consider the use of 'mixtures' at *Plt.* 278d4 < T2). Ancient sources credit Eudoxus and unidentified 'others' with the claim that perceptible particulars are mixtures whose ingredients are the forms of which they partake.[48] Since Eudoxus was close to the Academy around the time when the *Statesman* was written, it is tempting to speculate that Eudoxus developed his account of the relationship between perceptible particulars and forms on the basis of views presented by Plato in the *Statesman*.

Not every possible logically consistent bundle of kinds that is completely specified is a perceptible-particular-at-an-instant. For instance, there never was nor will ever be a perceptible-particular-at-an-instant that could be regarded as an instantaneous slice of Pegasus. Similarly, not all possible logically consistent bundles of kinds are kinds that could be reached in a classification of reality: for instance, the genus footed-animal is divided only by means of the *differentiae* biped and quadruped, not by means of the *differentiae* triped, pentaped, esaped, ..., and no classification of reality reaches species like triped-footed-animal, pentaped-footed-animal, esaped-footed-animal, ... This is either because there are no triped footed animals, pentaped footed animals, esaped footed animals, ... to classify, or because there is some biological impossibility involved in triped footed animals, pentaped footed animals, esaped footed animals, ... Similarly, there never was nor will ever be a completely specified bundle of kinds that could be regarded as a perceptible-particular-at-an-instant that is an instantaneous slice of Pegasus.

The last paragraph's results imply that bundles of kinds are not sets of kinds. For, apart from some exceptional cases connected with Russell's paradox, for every collection of entities there is a set whose elements are exactly the members of that collection: so, if bundles of kinds were sets of kinds, for every collection of kinds there would be the bundle of them, so there would also be a bundle of kinds that could be regarded as a perceptible-particular-at-an-instant that is an instantaneous slice of

[47] Cf. M. Frede (1983, 27), who suggests that particulars can be reached by the same procedure, namely division, whereby species are reached from their genera; Crivelli (2019, 46–52).

[48] Cf. Arist. *Metaph.* A9, 991^{a}12–19; M5, 1079^{b}15–23; Alex. Aphr. *in Metaph.* 97, 17–19, 27–8 = Arist. *de Ideis* fr. 119 Gigon 383^{b}5–6.

Pegasus.[49] What unifies the kinds that are parts of a bundle of kinds is probably something like a variably polyadic and possibly infinitary relation R (which may be regarded as the relation of compresence). R obtains between (two, three, …, or infinitely many) kinds and holds at instants. To exemplify, here is a sketch of a possible scenario: at a certain instant t_1, R obtains between the kinds $K_1, \ldots, K_n$, moreover R obtains also between the kinds $S_1, \ldots, S_m$ (and one or more from among $S_1, \ldots, S_m$ may well be among $K_1, \ldots, K_n$), but R does not obtain between the kinds $G_1, \ldots, G_i$ (and one or more from among $G_1, \ldots, G_i$ may well be among $S_1, \ldots, S_m$ or among $K_1, \ldots, K_n$); at a later instant t_2, R no longer obtains between $K_1, \ldots, K_n$, but R still obtains between $S_1, \ldots, S_m$, and R has come to obtain between $G_1, \ldots, G_i$.[50] For R to obtain between two or more kinds at a certain instant is the same as for the bundle of precisely those kinds to exist at that instant.

As pointed out earlier, in passage T5 Socrates distinguishes two phases in our inquiry: in the first, we are supposed to use the method of division in order to produce a complete classification of our intended domain of research; in the second, we are expected to apply something analogous to the method of division used in the first phase so as to reach perceptible-particulars-at-instants by combining the kinds they partake of. What determines the stopping point of the first phase, namely the lowest kinds subordinate to the single kind that covers exactly our intended domain of research? The most likely answer is that this stopping point, i.e. the lowest kinds subordinate to the single kind that covers exactly our intended domain of research, is fixed by the most specific, or most determinate, bundles of kinds that exist always, i.e. at all instants: some bundles of kinds exist now but not at any other instant, others exist for a certain stretch of time but not always, yet others exist always. The most specific, or most determinate, among the bundles of kinds that exist always may be plausibly identified with the lowest kinds reached by the classification that constitutes the first phase of the process described in T5.

The puzzle of multi-located kinds. An important feature of the second account of the 'letters' (or 'elements') and 'syllables' (or 'compounds') of reality is that it provides an elegant solution for a difficulty that troubled Plato in his late philosophy, a difficulty concerning how kinds are related to the perceptible particulars that fall under them. This difficulty is formulated with minor variations in the *Parmenides* (130e4–131e7 and 144c6–d4) and the *Philebus* (15b4–8).[51] In the version presented in the *Philebus*, it turns on a dilemma: any kind is in the many reciprocally separate perceptible particulars that fall under it either by having many different parts of itself wholly contained in them, in which case it is 'dispersed' (15b5–6) and is therefore 'many' (15b6), or by being wholly contained in each of them, in which case it is 'itself separately from itself' (15b6–7) and is therefore many because it is, so to speak, multiplied. So, in all cases, the kind is many. Since every kind is one, it is both one and many.

[49] Cf. Van Cleve (1985, 95–96).

[50] Cf. Van Cleve (1985, 98).

[51] Cf. S.E. *P.* 2.220–2.

Now, according to the second account of the 'letters' (or 'elements') and 'syllables' (or 'compounds') of reality, perceptible-particulars-at-instants are compounds whose components are the kinds of which they partake. Just as the same person can be a member of a chorus and of a chess-club (and there is nothing mysterious about this fact), so also the same kind can be a part of several perceptible-particulars-at-instants (and there is nothing mysterious about this fact either). Location attributes are among the kinds that contribute to make up perceptible-particulars-at-instants: location attributes like in-such-and-such-a-spot-of-the-market-place-in-Athens and in-such-and-such-a-spot-of-the-market-place-in-Megara can be present together with the kind human in different simultaneous perceptible-particulars-at-instants (e.g. in Socrates-at-3-p.m.-of-May-1st-of-400-bc and in Euclides-at-3-p.m.-of-May-1st-of-400-bc). The difficulty raised by the puzzle about the relation of a kind to the many reciprocally separate perceptible particulars that fall under it evaporates because a single kind's presence in many reciprocally separate perceptible particulars boils down to its belonging at the same time to different compounds together with different location attributes. It is worth remarking that in the *Philebus* the description of division offered in passage T5 seems to be introduced as part of a solution of the dilemma about the relation of kinds to the perceptible particulars that fall under them[52]: the interpretation of T5 I offered above enables one to see how division could indeed contribute to solving the dilemma.

Evaluation of the two accounts. The two accounts of the nature of the *Statesman*'s 'letters' (or 'elements') and 'syllables' (or 'compounds') of reality are not logically exclusive (one of them speaks of the composition of composite kinds, the other of the composition of perceptible particulars). Perhaps they should be jointly adopted: while the 'letters' (or 'elements') of reality are simple kinds, the 'syllables' (or 'compounds') of reality comprise both composite kinds, which consist of two or more simple kinds, and perceptible-particulars-at-instants, which consist of very many simple kinds, specifically of all the simple kinds of which they partake at their instants.

Note that in passage T2 the Visitor speaks of 'letters' (or 'elements') 'of all things' (278d1), with no restriction on the quantifier. This speaks in favour of the suggestion that the 'syllables' (or 'compounds') of reality should comprise both composite kinds and perceptible-particulars-at-instants.

The method of models applied. If we modify the account of the child's treatment of 'alphabetic' letters and syllables so as to adapt it to the letters and syllables 'of reality', we find ourselves in a position to credit Plato with the view that an application of the method of models consists in correctly recognizing or identifying certain occurrences of kinds within easy contexts, realizing their identity with other occurrences of them in difficult contexts (where one was unable correctly to recognize or identify them), and transferring the correct recognition or identification of them from the easy contexts to the difficult ones. The process of cognition turns out to be something like 'spelling' the syllables of reality, which amounts to recognizing

52 Cf. Crivelli (2019, 39–41).

and identifying the kinds within composite kinds or within perceptible-particulars-at-instants. If we apply the account of recognition and identification put forward in the Waxen Block account of false judgement in the *Theaetetus*,[53] the result is that the process of recognition or identification is actualized by associating one's grasp of one of the kinds ('letters' of reality) in some compound (a 'syllable' of reality, which can be either a kind or a perceptible particular) with some memory imprint that one has stored in the soul. The association can be done either correctly or incorrectly: if one associates one's grasp of one of the kinds with the right memory imprint, the result is a true judgement, whereas if one associates one's grasp of one of the kinds with the wrong memory imprint, the result is a false judgement. How the memory imprints have been acquired does not really matter to the account.

An important aspect of the analogy with the child's attempt to write down or spell out is that the starting point of the process of writing down or spelling out is an uttered syllable (rather than a written one). Uttered syllables present hearers with a special difficulty. The problem hearers face with uttered syllables, especially when they are beginning to learn a language (and this is of course the child's case), is often not that of recognizing or identifying a component of a complex sound, a component that they have already been able to separate from its immediate context, but that of discovering what the components are: if I am facing the flow of voice of a language that I do not master and is very different from those I am familiar with, my first and main difficulty is being able to single out the basic components, i.e. the uttered letters, within what strikes me as a continuous and unarticulated flow of sound. Something of this sort is likely to be at least part of the difficulty faced by inquirers who are trying to decipher the syllables of reality: their difficulty will often be not that of identifying or recognizing a letter of a syllable of reality, a letter that they have already isolated from its immediate context, but that of isolating the letters, of discovering how to analyse the syllable of reality. The method of models is probably supposed to help inquirers to overcome also this difficulty.

Deciphering the syllables of reality. Plato seems to be putting forward something like the view that to have knowledge is to be able to 'read the book of the world'.[54] But what is it to decipher a 'syllable of reality'?

As I pointed out earlier, there is some plausibility to the suggestion that in Plato's view there are two types of 'syllables of reality': composite kinds that consist of two or more simple kinds, and perceptible-particular-at-instants, each of which consists of all and only the kinds of which it partakes (at its peculiar instant). Accordingly, there are two different operations that fall under the rubric of deciphering a 'syllable of reality'. In what follows, I examine both of them in turn.

Deciphering composite kinds. A correct decipherment of a 'syllable of reality' that is a composite kind consisting of two or more simple kinds amounts to a correct recognition or identification of the simple kinds of which the composite kind consists and to a judgement to the effect that the composite kind is these kinds (without this additional judgement, the various correct recognitions or identifications could well

[53] Cf. above, text to n. 15.

[54] Cf. Casertano (1995, 146).

fail to be reciprocally linked in the way required by their forming a single 'syllable of reality'). These traits make it likely to assume that a correct decipherment of a 'syllable of reality' that is a composite kind consisting of two or more simple kinds is a definition of that composite kind.[55] The definition thus reached is the same as the one reached by a correct application of division: the method of models in not in competition with that of division; rather, it can help to implement the method of division by providing a route to discovering the kinds that function as differentiae.

The correct decipherment of a word must satisfy several conditions: all the letters of the word must be correctly recognized or identified (if someone uttered the word 'KATA' and asked me to spell it, I would make a mistake if I said 'It is Kappa and Tau'), only letters of the word must be correctly recognized or identified (in the imaginary situation just described, I would make a mistake if I said 'It is Kappa, Alpha, and Delta' or 'It is Kappa, Alpha, Tau, and Delta'), the letters must be correctly recognized or identified in the order in which they occur (in the imaginary situation considered, I would make a mistake if I said 'It is Kappa, Tau, and Alpha'), and different occurrences of the same letter must be mentioned separately (in the imaginary situation envisaged, I would make a mistake if I said 'It is Kappa, Alpha, and Tau'). The first two conditions for a correct alphabetic decipherment surely correspond to two conditions for a successful definition: all and only the simple kinds that are the ultimate components of a composite kind must be correctly recognized or identified. The third condition also arguably corresponds to a requirement for a successful definition: the simple kinds must be correctly recognized or identified in the right order (there is something awkward in saying 'A human is rational, mortal, and an animal'). The fourth condition also perhaps corresponds to a requirement for a successful definition: if there were a kind such as singing-slowly-while-walking-slowly, a successful definition should perhaps mention the kind slowness twice, both in connection with the kind singing and in connection with the kind walking. The characteristics corresponding to these four conditions (and perhaps others) are implicit in the use of 'is' involved in sentences expressing decipherments (like '"KATA" is Kappa, Alpha, Tau, and Alpha'): for instance, it is because of this use of 'is' that the sentence '"KATA" is Kappa and Alpha' is deemed false if it is taken to express a decipherment.

There is textual evidence for crediting Plato with the view that the correct decipherment of certain 'syllables of reality' is an act of defining. For, first, in a passage of the *Statesman* (285c8–d3 and d5–7) the Visitor compares the question 'of what letters any given name consists' (285c9–10) with 'the inquiry about the statesman' (285d5), an inquiry whose purpose is to define a kind. Secondly, in the passages of the *Theaetetus* where the decipherment of syllables is discussed, the question whereby such a decipherment is prompted is the definitional question, 'What is it?'.[56]

Definition as a mental act. Since the correct decipherment of a 'syllable of reality' that is a composite kind is a definition of it, and since such a decipherment is a mental act, the definitions in question are not (uttered or written) sentences, but

[55] Cf. Owen (1973, 360–361); El Murr (2014, 60–61).

[56] Cf. *Tht.* 203a8; 206e6–207a1; 207a5–7; 207b9–c1.

mental acts. This is remarkable because many philosophers often use 'definition' to denote not mental acts, but sentences of a special sort (such is for instance the usage attested in Aristotle).[57]

It might be assumed that for Plato the contrast between sentences and mental acts is not deep because in his view thought is inner voiceless speech.[58] But it is not clear what the language is which in Plato's view is employed in the soul's inner voiceless speech. Early Greek poets often portray characters speaking to themselves or having a debate with some part or faculty of themselves (in several passages from the *Iliad* Homer describes a hero as addressing his own θυμός and as being addressed by it).[59] Moreover, Isocrates compares the capacity to speak in public with that to have a conversation with oneself when deliberating.[60] It is hard to imagine that the poets or Isocrates could believe that such internal dialogues are going on in languages different from those normally used in vocal exchanges with others. If Plato is simply picking up the idea that thought is an inner voiceless conversation from the literary tradition, then the soul's inner voiceless dialogue can only be carried out in the language normally used by the thinkers in their vocal exchanges with others.

On the other hand, there are reasons for thinking that the soul's inner voiceless conversation is carried out (not in one of the natural languages normally used to communicate with others, but) in a language of mental items syntactically arranged in a way analogous to that in which words of ordinary natural languages are organized in spoken sentences.[61] Let me review some of these reasons.

1. One reason is provided by some passages of the *Theaetetus*. As I pointed out earlier,[62] in the Waxen Block account of false judgement (190e5–196d2) Plato explains judgements of recognition or identification as events where memory imprints are assigned to 'perceptions and thoughts' (191d7). Another passage in the same dialogue (189e4–190d7, less than one page before the section dedicated to the Waxen Block account) describes judgement as 'a statement silently addressed to oneself' (190a5–6). The two accounts of judgement, the one treating it as an assignment of memory imprints and the one treating it as an inner voiceless statement, are explicitly combined in a passage (195e9–196b3) at the end of the Waxen Block section. These passages jointly imply that the statements of the soul's inner voiceless conversation that are perceptual judgements of recognition or identification are assignments of memory imprints to perceptions, and this in turn suggests that judgement is an inner voiceless statement

[57] Cf. *Top.* 1.4, 102^{a}2–5; 6.11, 149^{a}1–2; *Metaph.* Z10, 1034^{b}20; 12, 1037^{b}25–6.

[58] Cf. *Tht.* 189e4–190a2; *Sph.* 264a9–b1.

[59] Cf. *Il.* 11. 403, 407; 17. 90, 97; 21. 552, 562; 22. 98, 122.

[60] Cf. *Ep.* 3.8; 15.256.

[61] Cf. Duncombe (2016, 109). Panaccio (1999, 34–36, 51–52) is instead inclined to deny that in Plato's view judgement and thought exhibit a syntactic structure analogous to that of spoken sentences of ordinary natural languages.

[62] Cf. above, text to n. 15.

whose 'words' are mental items that are arranged like the words of sentences of ordinary natural languages.[63]

2. The Waxen Block section's theme returns in a passage from the *Philebus* (38b12–13) where 'judgement and judging' are described as arising 'from memory and perception'. A shortly later passage speaks of mental phenomena that arise in connection with 'memory that comes to coincide with perception' (39a1). The idea that judgement is a self-addressed statement is formulated between these two passages (at 38d5–8).
3. A further reason for thinking that the soul's inner voiceless conversation is carried out in a language of mental items is provided by some Aristotelian texts. Both in his logical works, which are normally regarded as belonging to his early production, and in other treatises, Aristotle takes it for granted that thoughts are syntactically structured in a way analogous to uttered speech.[64] Such a position would be easily understandable if it had a Platonic pedigree. If this is correct, Plato's view could be that speech displays the articulation into names and verbs because it is an image of thought,[65] which already exhibits an analogous articulation of perceptions and memory imprints.

Although the evidence is not clear-cut, it seems to me reasonable to infer that Plato picked up from the earlier literary tradition the idea that thinkers have inner conversations with themselves but modified it by introducing the novelty that the language of these inner conversations is none of the languages normally spoken by people in their exchanges but a language of thought whose 'words' are mental items (e.g. perceptions and memory imprints). According to this plausible account, Plato is at the origin of a long philosophical tradition, among whose representatives rank William of Ockham and Jerry Fodor. In this case, the difference between Plato's account of definition as a mental act and that of the majority of philosophers as a sentence of a special sort remains deep.

An example from the *Republic*: defining justice. An example of an application of the method of models for the sake of defining a 'syllable of reality' that is a composite kind is perhaps to be found in some passages of the *Republic*.[66] These passages discuss the methodology to be employed to search for the definition of justice. Socrates describes an imaginary situation where we are trying to read some small and far-away letters and are unable to do it, but we realize that there are other letters that are large and near-by and therefore easily readable. In such a situation, we would proceed to decipher the easily readable letters and use them as a means to decipher those that we cannot easily read. Similarly, we want to discover the nature of justice in the individual human being (a difficult occurrence of a 'syllable'), but we are unable to do it. We therefore consider justice in the state (an easy occurrence of the same 'syllable'), and we examine what follows by transferring our findings

[63] Cf. Detel (1972, 43–44).

[64] Cf. *Int.* 1, 16^a9–16; 14, 23^a32–5; 24^b1–6; *APo.* 1.10, 76^b24–7; *de An.* 3.6, 430^a26–$^b6$; 430^b26–31; 8, 432^a11–12; *Metaph.* Γ4, 1006^b8–9; Panaccio (1999, 36–49); Duncombe (2016, 111–112).

[65] Cf. *Tht.* 206d1–6.

[66] Cf. *R.* 2. 368c7–369a4; 4. 434d6–435c3; Goldschmidt (1947, 54–56).

about the one to the other. In particular, it is difficult to decipher the occurrence of the 'syllable' justice in the 'word' justice-in-the-individual-human-being: it is difficult to recognize or identify the occurrences of the 'letters' harmony, three, and part in this occurrence of the 'syllable' justice. However, we have reason to suspect that there is another occurrence of the 'syllable' justice in the 'word' justice-in-the-state, and that this other occurrence of the 'syllable' justice is easier to decipher (the reason is that the adjective 'just' is applied to states as well as to individual human beings, cf. *R.* 4. 435b1–2). We therefore concentrate on the occurrence of the 'syllable' justice in the 'word' justice-in-the-state, and we are able correctly to decipher it by correctly recognizing or identifying the occurrences of the 'letters' harmony, three, and part in it. On the basis of this discovery, we go back to deciphering the occurrence of the 'syllable' justice in the 'word' justice-in-the-individual-human-being, and this leads us correctly to decipher it by correctly recognizing or identifying the occurrences of the 'letters' harmony, three, and part in it.

The situation can be clarified by going back to the example discussed in a passage of the *Theaetetus* (207e7–208a2). There, Socrates alerts Theaetetus to the case of a boy who writes down correctly the occurrence of the letter 'Θ' in the occurrence of the syllable 'ΘE' in the word 'ΘEAITHTOΣ', but writes down incorrectly the occurrence of the letter 'Θ' in the occurrence of the same syllable 'ΘE' in the word 'ΘEOΔΩPOΣ'. Once the teacher spots the boy's mistake, he will get the boy again correctly to recognize or identify the occurrence of the letter 'Θ' in the occurrence of the syllable 'ΘE' in the word 'ΘEAITHTOΣ', and this will eventually lead the boy correctly to recognize or identify also the occurrence of the letter 'Θ' in the occurrence of the syllable 'ΘE' in the word 'ΘEOΔΩPOΣ'. In the case discussed in the *Republic* passages, harmony, three, and parts correspond to the letter 'Θ', justice corresponds to the syllable 'ΘE', and justice-in-the-state and justice-in-the-individual-human-being correspond to the words 'ΘEAITHTOΣ' and 'ΘEOΔΩPOΣ'. Although they do not contain the expression 'model', the *Republic* passages appear to be describing and implementing a procedure very close to that of the method of models of the *Statesman* (the connection with which is also suggested by the analogy with the letters that are easy or hard to decipher).

Deciphering perceptible particulars. A correct decipherment of a 'syllable of reality' that is a perceptible-particular-at-an-instant, which consists of all and only the kinds of which it partakes (at its peculiar instant), is an impossible task (at least for a finite mind). For, the kinds of which a perceptible-particular-at-an-instant consists are too many for such a correct decipherment ever to be completed.[67]

Even if a complete correct decipherment of a 'syllable of reality' that is a perceptible-particular-at-an-instant is impossible, a partial correct decipherment is feasible. Such a partial correct decipherment simply amounts to a correct recognition or identification of one or more of the kinds of which it consists and to a judgement to the effect that these kinds are among those of which the perceptible-particular-at-an-instant consists, i.e. among the kinds of which it partakes (at its peculiar instant). For

[67] Cf. Woolf (2013, 204–206, 210).

instance, one of the many possible partial correct decipherments of the perceptible-particular-at-an-instant which is Theaetetus-at-*t*, where *t* is one of the instants when he is being questioned about true and false statement (at *Sph.* 262d13–263d5), is what leads to discovering that the kind seated is among the kinds of which Theaetetus-at-*t* consists, i.e. is one of the kinds of which he partakes at *t*.

No example of a (partial) decipherment of a 'syllable of reality' that is a perceptible-particular-at-an-instant appears explicitly to be discussed in Plato's dialogues. However, in a passage near the end of the *Theaetetus* (209c5–10 = T4), Socrates mentions the memory imprints of 'this snub-nosedness' (209c6) and of 'the other things of which you [*sc.* Theaetetus] ‹consist›' (209c8–9), i.e. memory imprints of perceptible qualities that are components of perceptible particulars.[68] Since, in the *Theaetetus*'s Waxen Block account of false judgement (190e5–196d2), the role of memory imprints is to enable the thinker to recognize or identify entities which he or she is perceiving or thinking of,[69] the passage near the end of the dialogue implies that one of the mental operations which a thinker may perform is that of recognizing or identifying perceptible qualities that are components of perceptible particulars. This is what a (partial) decipherment of a 'syllable of reality' that is a perceptible particular (or, to be more precise, a perceptible-particular-at-an-instant) amounts to.

One of the exegetical puzzles raised by the whole section of the *Theaetetus* that deals with the possibility of false judgement is that the main focus of these pages seems to be the possibility of false identity judgements, while at the beginning of the section (at 187c7–e8) Socrates seems to be concerned with the possibility of all false judgements (not only of false identity judgements). Commentators have wondered about this feature.[70] The suggestion that at least some predicative judgements are based on a recognition or identification of one or more of the traits of the entity to which the judgement refers might provide the beginning of a solution of this puzzle.

When is the method of models needed? Shortly before introducing the method of models, the Visitor expresses his dissatisfaction at the result reached through the divisions up to 277a2 by comparing it to a 'portrait [ζῷον]' (277b8) in which the artist has drawn the outlines but has not applied the colour (277b8–c3). This mention of a portrait suggests by association a reflection on the adequacy of painted images as vehicles for teaching. The Visitor distinguishes (277c3–6) two ways in which one can 'show any animal [δηλοῦν πᾶν ζῷον]' (277c4) (note the double use of 'ζῷον', meaning 'portrait' and 'animal', within five lines): one can do this either by means of painting or manual crafts, on the one hand, or 'by speech and discourse [λέξει καὶ λόγῳ]' (277c4) on the other. He adds that the second technique is more adequate 'for those who are able to follow' (277c5), while the first must be used 'for the others' (277c5). The first technique mentioned by the Visitor seems to amount to an explanation based on the use of painted or sculpted images as it is practiced with children in a classroom (e.g. an explanation of the meaning of 'lion', or of the nature of a lion, by showing a picture of a lion); the second seems to be an explanation based

[68] Cf. above, text to n. 41.

[69] Cf. above, text to n. 15.

[70] Cf. for instance McDowell (1973, 195); N.P. White (1976, 164); Burnyeat (1990, 71).

on verbal means (e.g. an explanation based on formulating a definition of a lion). Shortly afterwards, at the beginning of the presentation of the method of models, the Visitor remarks that 'it is a hard thing [...] to demonstrate any of the more important subjects without using models' (277d1–2). This remark is rather cryptic, but the Visitor expands on it later on (285d10–286b1). The later passage is the object of an exegetical controversy, but on the interpretation of it that I favour it points out that while certain kinds have perceptible similarities (αἰσθηταὶ ὁμοιότητες, 285e1) or images (εἴδωλα, 286a1), other kinds, and in particular the most important ones, have none: the kinds that have perceptible images can be explained by pointing to these images (he is probably alluding to the classroom explanations based on the use of painted or sculpted images), whereas the kinds that lack perceptible images can only be explained by purely verbal means, one of which is based on the method of models.[71]

The Visitor does not offer examples to explain what he has in mind. One such example is however probably provided by angling as opposed to the sophist's art and statesmanship. A picture of an angler can be used to offer something like an ostensive definition of angling: if I want to explain to a child what angling is and I have one or more pictures of fishermen who are practicing this activity (cf. *R.* 10. 598b8–c1), I can point to the pictures and say, 'This is what angling is'. On the other hand, there are no perceptible images of the kinds sophist and statesman that one may point to in order to provide anything like an ostensive definition of these arts: if I want to explain to someone what a sophist is and I have pictures (or even photographs) of sophists practicing their art, it will be of no avail to point to these pictures and say, 'This is what a sophist is'. The purely verbal means to be used to explain kinds that lack perceptible images probably include the method of models. For instance, if I want to explain to someone what a sophist is, I can do this only by purely verbal means, and in particular by a discussion that deploys the method of models (the model of an imitator comes in handy).

The discovery of the right models. There is a difficulty with the application of the method of models: how does an inquirer light on a suitable model? So long as the context is pedagogical, so that there is a teacher who leads a pupil to the acquisition of knowledge, the discovery of a suitable model will be the job of the teacher. In such a pedagogical context, no difficulty arises because the teacher already has knowledge and is therefore in a position to find a suitable model for the pupil. Now, it cannot be excluded that the method of models is to be applied only in a pedagogical context.[72] However, if the method is to be applied also in situations of genuine research, where no teacher has the competence needed to discover a suitable model, then matters are more complicated. For, if one is unable correctly to recognize or identify an occurrence of a kind *K* in a difficult 'syllable of reality' *S*, how will one select an alternative easy 'syllable of reality' *S'* where one is able correctly to recognize or identify an occurrence of *K*? One might even be unaware that *K* is what is at issue,

[71] Cf. Owen (1973, 349–358).

[72] Cf. R. Robinson (1953, 213); Miller (1980, 60).

in which case one would be deprived of a starting point for the selection of a suitable alternative easy 'syllable of reality'.

Perhaps, in cases of genuine inquiry, the framework within which the method of models is applied is different from the one that gives rise to the difficulty just described. In some cases of genuine inquiry, one is not simply trying to decipher 'syllables of reality' without having a specific kind *K* in mind, but it is clear from the start that one is aiming to recognize or identify occurrences of *K*. In such cases, one can train oneself by considering occurrences of *K* in 'syllables of reality' where one is able correctly to recognize or identify them, the hope being that this training will enable one correctly to recognize or identify occurrences of *K* in other 'syllables of reality' where one would have otherwise been unable correctly to recognize or identify them. In fact, when the method of models is applied at the beginning of the *Sophist*, its employment for the sake of practice is explicitly mentioned.[73] In other cases of genuine inquiry, one is trying to decipher a specific occurrence of a 'syllable of reality' and is unable to do it. One however has reason to suspect that a certain other occurrence of a 'syllable of reality' that is easier to decipher is an occurrence of the same 'syllable of reality' (the reason may have to do with linguistic usage, for instance with the fact that the same linguistic expression is commonly used for both occurrences). One therefore concentrates on the other occurrence of a 'syllable of reality' and thanks to one's findings there manages correctly to decipher the original occurrence.

References

Ademollo, F. (2018). On Plato's conception of change. *Oxford Studies in Ancient Philosophy, 55*, 35–83.

Bekker, I. (Ed. & Comm.). (1826). *Platonis Scripta Graece Omnia*. Priestley.

Benitez, E. E. (1989). *Forms in Plato's* Philebus. Van Gorcum.

Benson, H. H. (2010). Collection and division in the *Philebus*. In J. Dillon & L. Brisson (Eds.), *Plato's Philebus. Selected papers from the eighth symposium Platonicum* (pp. 19–24). Academia.

Brown, L. (1993). Understanding the *Theaetetus*. *Oxford Studies in Ancient Philosophy, 11*, 199–224.

Burge, E. L. (1971). The ideas as Aitiai in the Phaedo. *Phronesis, 16*, 1–13.

Burnet, J. (1914). *Greek philosophy: Thales to Plato*. Macmillan.

Burnyeat, M. (1976). Plato on the grammar of perceiving. *Classical Quarterly, 26*, 29–51.

Burnyeat, M. (Trans. & Comm.). (1990). *The Theaetetus of Plato*. Hackett.

Campbell, L. (Ed. & Comm.). (1867). *The Sophistes and Politicus of Plato*. Clarendon Press.

Campbell, L. (Ed. & Comm.). (1883). *The Theaetetus of Plato* (2nd ed.). Clarendon Press.

Casertano, G. (1995). Il problema del rapporto nome-cosa-discorso nel *Politico* (277–287). In C. Rowe (Ed.), *Reading the* Statesman. *Proceedings of the III Symposium Platonicum* (pp. 141–154). Academia.

Cornford, F. M. (1935). *Plato's theory of knowledge: The* Theaetetus *and the* Sophist *of Plato translated with a running commentary*. Kegan Paul, Trench, Trubner & Co., and Harcourt, Brace, and Co.

[73] Cf. *Sph.* 218d2, d5; *Men.* 75a8; *Plt.* 286a5; a8.

Crivelli, P. (2012). *Plato's account of falsehood: A study of the Sophist*. Cambridge University Press.

Crivelli, P. (2019). Division and classification: *Philebus* 14c–20a. In P. Dimas, R. E. Jones, & G. R. Lear (Eds.), *Plato's* Philebus*: A philosophical discussion* (pp. 34–54). Oxford University Press.

Dancy, R. M. (1987). Theaetetus' First baby: *Theaetetus* 151e–160e. *Philosophical Topics, 15*(2), 61–108.

Detel, W. (1972). *Platons Beschreibung des falschen Satzes im Theätet und Sophistes*. Vandenhoeck & Ruprecht.

Diès, A. (Ed. & Trans.). (1950). *Platon, Le politique* (2nd ed.). Les Belles Lettres.

Dixsaut, M. (2001). *Métamorphoses de la dialectique dans les dialogues de Platon*. Vrin.

Duke, E. A., Hicken, W. F., Nicoll, W. S. M., Robinson, D. B., & Strachan, J. C. G. (Eds.). (1995). *Platonis Opera* (i). Clarendon Press.

Duncombe, M. (2016). Thought as internal speech in Plato and Aristotle. *Logical Analysis and History of Philosophy—Philosophiegeschichte und logische Analyse, 19*, 105–125.

El Murr, D. (2014). *Savoir et gouverner: Essai sur la science politique platonicienne*. Vrin.

Frede, D. (1997). *Platon, Philebos*. Vandenhoeck & Ruprecht.

Frede, M. (1983). The title, unity, and authenticity of the Aristotelian *Categories*. In M. Frede (1987), *Essays in ancient philosophy* (pp. 11–28, 363–367). University of Minnesota Press.

Gigon, O. (Ed.). (1987). *Aristotelis Opera* (iii). de Gruyter.

Goldschmidt, V. (1947). *Le paradigme dans la dialectique platonicienne* (2nd ed. [repr.]). Vrin.

Hackforth, R. (Trans. & Comm.). (1945). *Plato's examination of pleasure: A translation of the Philebus, with introduction and commentary*. Cambridge University Press.

Heindorf, L. F. (Ed. & Comm.). (1809–1810). *Platonis Dialogi Selecti* (iv). Hitzig.

Hermann, C. F. (Ed.). (1851). *Platonis Dialogi...* (i). Teubner.

Irwin, T. (1977). Plato's Heracleiteanism. *Philosophical Quarterly, 27*, 1–13.

Joly, H. (1986). Platon entre le maître d'école et le fabriquant de mots. Remarques sur les *grammata*. In *Philosophie du langage et grammaire dans l'antiquité* (pp. 105–136). OUSIA.

Kenig Curd, P. (1990). *Parmenides* 142b5–144e7: The "unity is many" arguments. *Southern Journal of Philosophy, 28*, 19–35.

Mann, W.-R. (2000). *The discovery of things. Aristotle's* Categories *and their context*. Princeton University Press.

McDowell, J. (Trans. & Comm.). (1973). *Plato, Theaetetus*. Clarendon Press.

Miller, M. H., Jr. (1980). *The philosopher in Plato's* Statesman (2nd ed.). Parmenides Publishing.

Oberhammer, A. A. (2016). *Buchstaben als* paradeigma *in Platons Spätdialogen*. de Gruyter.

Owen, G. E. L. (1970). Notes on Ryle's Plato. In O. P. Wood & G. Pitcher (Eds.), *Ryle* (pp. 341–372). Macmillan.

Owen, G. E. L. (1973). Plato on the undepictable. In E. N. Lee, A. P. D. Mourelatos, & R. M. Rorty (Eds.), *Exegesis and argument: Studies in Greek philosophy presented to Gregory Vlastos* (pp. 349–361). Van Gorcum.

Palumbo, L. (1995). Realtà e apparenza nel *Sofista* e nel *Politico*. In C. Rowe (Ed.), *Reading the* Statesman. *Proceedings of the III Symposium Platonicum* (pp. 175–183). Academia.

Panaccio, C. (1999). *Le discours intérieur. De Platon à Guillaume d'Ockham*. Seuil.

Prauss, G. (1968). Ding und Eigenschaft bei Platon und Aristoteles. *Kant-Studien, 59*, 98–117.

Price, A. W. (1989). *Love and friendship in Plato and Aristotle*. Clarendon Press.

Reshotko, N. (2010). Restoring coherence to the gods' gift to men: *Philebus* 16c9–18b7 and 23e3–27b8. In J. Dillon & L. Brisson (Eds.), *Plato's Philebus. Selected papers from the Eighth Symposium Platonicum* (pp. 92–97). Academia.

Robinson, R. (1953). *Plato's earlier dialectic* (2nd ed.). Clarendon Press.

Rowe, C. J. (Ed., Trans., and Comm.). (1995). *Plato, Statesman*. Oxbow.

Schofield, M. (1973). A neglected regress argument in the *Parmenides*. *Classical Quarterly, 23*, 29–44.

Sedley, D. (1982). The Stoic criterion of identity. *Phronesis, 27*, 255–275.

Sedley, D. (2004). *The midwife of Platonism. Text and subtext in Plato's* Theaetetus. Clarendon Press.
Silverman, A. (2002). *The dialectic of essence. A study of Plato's metaphysics.* Princeton University Press.
Skemp, J. B. (Trans. & Comm.). (1952). *Plato's Statesman.* Routledge & Kegan Paul.
Stallbaum, G. (ed. & comm.) (1841). *Platonis Opera Omnia* (ix.i). Hennings.
Striker, G. (1970). *Peras und Apeiron. Das Problem der Formen in Platons Philebos.* Vandenhoeck & Ruprecht.
Van Cleve, J. (1985). Three versions of the bundle theory. *Philosophical Studies, 47*, 95–107.
White, F. C. (1977). Plato's middle dialogues and the independence of particulars. *Philosophical Quarterly, 27*, 193–213.
White, N. P. (1976). *Plato on knowledge and reality.* Hackett.
Woolf, R. (2013). Plato and the norms of thought. *Mind, 122*, 171–216.

Anti-Platonism in Aristotle's *Categories*

Francesco Ademollo

My aim in this paper is to examine some aspects of Aristotle's ontology in the *Categories* that involve disagreement with Plato's views. Along the way we shall receive help and clarification from other Aristotelian works. While some of the cases I shall discuss are well known to scholars, others are less familiar; I hope that the whole resulting picture may be of some interest to readers—and especially to Massimo Mugnai, whose lectures, thirty years ago, initiated me into the Organon and the history of logic.

1 Introduction: Aristotle's 'Meta-Ontology'

As is well known, *Categories* ch. 2 sets out two main distinctions.[1] The first ($1^{a}16$–19) is between two classes 'of things said' (τῶν λεγομένων): those which are said 'with interweaving' (κατὰ συμπλοκήν) and those which are said 'without interweaving' (ἄνευ συμπλοκῆς).[2] Aristotle puts forward two complete declarative sentences ('(A) human runs', '(A) human wins') as instances of the former class and four

[1] For a good introduction to *Cat.* 2, and to the treatise in general, see Matthews (2009). Some preliminary clarifications may be useful. (i) I use the terms 'predicate' and 'attribute' more or less interchangeably and in a very generic way, to refer to any entity that holds of something, either essentially or accidentally. (ii) I use 'accident' to refer to any non-essential predicate or attribute of some substance. (iii) I treat 'individual' and 'particular' as equivalent. (iv) On a number of occasions I replace humans with cats and Socrates with Jeoffry.

[2] The term 'interweaving' here is an allusion to Plato, *Sophist* 261d–263d, where it is a metaphor for the combination of names and verbs in a sentence.

F. Ademollo (✉)
Dipartimento di Lettere e Filosofia, Università degli Studi di Firenze, Firenze, Italy
e-mail: francesco.ademollo@unifi.it

Scuola Normale Superiore, Pisa, Italy

F. Ademollo et al. (eds.), *Thinking and Calculating*, Logic, Epistemology, and the Unity of Science 54, https://doi.org/10.1007/978-3-030-97303-2_2

individual terms ('human', 'ox', runs', 'wins') as instances of the latter.[3] As he explains in ch. 4, 2^a4–10, the distinction is based on the fact that the elements of the former class are either true or false, whereas those of the latter are neither true nor false.

The second distinction (1^a20–$^b9$) is between four classes of entities (τῶν ὄντων), according to whether or not each entity '*is said of some subject*' (καθ' ὑποκειμένου τινὸς λέγεται) and '*is in a subject*' (ἐν ὑποκειμένῳ ἐστίν). I shall assume that, as most commentators believe, 'being said of a subject' consists in being an *essential* predicate of something—i.e. a predicate that could figure in the definition of something—whereas 'being in a subject' consists in being a *non-essential* or *accidental* predicate of something.[4]

The four classes of entities identified by the second distinction constitute what has been called the 'meta-ontology' of the *Categories*,[5] as distinct from the ontology set out from ch. 4 onwards. In what follows I list the four classes referring to each in the 'meta-ontological' terms of ch. 2 and then adding, between parentheses and in italics, the 'ontological' information we can gather from the sequel of the treatise, especially ch. 5.

(I) Entities which are essential predicates of something but not accidental predicates of anything (= *universal, 'secondary' substances and their constitutive differentiae*).

(II) Entities which are not essential predicates of anything but are accidental predicates of something (= *non-substantial individuals, i.e. individual accidents*).

(III) Entities which are both essential predicates of something and accidental predicates of something else (= *non-substantial universals, i.e. universal accidents*).

(IV) Entities which are neither essential nor accidental predicates of anything (= *individual, 'primary' substances*).

Entities 'individual and one in number' (τὰ ἄτομα καὶ ἓν ἀριθμῷ) are those which fail to be 'said of any subject' (1^b6–7). If every entity is either universal or individual, it follows that universal entities are those which, unlike individuals, are 'said of some subject'. So the difference between universals and individuals is a matter of *essential* predication: universals are essentially predicated of something, whereas individuals are not essentially predicated of anything.

The most controversial aspect of this theory is the nature of the individuals in class (II). According to a traditional, widespread interpretation—which I take to be correct—these are unrepeatable property-instances similar to the 'tropes' of contemporary metaphysics: e.g. the crimson of this cloak, as distinct from the crimson of that

[3] On the problems posed by this distinction and by Aristotle's examples see Ackrill (1963, 73–74).

[4] This standard interpretation is based on 5. 3^a15–28 and on all the examples Aristotle gives in chs. 1–5. Crivelli (2017) advances a non-standard interpretation based on a couple of recalcitrant passages in ch. 10; for an alternative construal of those passages, according to which they do not threaten the standard interpretation, see Rapp (in preparation).

[5] See Wedin (2000, 16 etc.).

cloak, even though the two cloaks are exactly the same hue.[6] Consider the famous definition given in 2. 1ᵃ24–5:

> I call 'in a subject' [ἐν ὑποκειμένῳ] that which, being in something not as a part, cannot possibly exist separately from that in which it is [ὃ ἔν τινι μὴ ὡς μέρος ὑπάρχον ἀδύνατον χωρὶς εἶναι τοῦ ἐν ᾧ ἐστίν].

On the interpretation at stake, this definition strictly speaking implies that, if X 'is in' Y 'as in a subject', then X cannot exist without Y, i.e. X cannot exist if Y does not exist.[7] It is, however, important to point out that Aristotle seems to be much less interested in these entities than many commentators assume:in the *Categories* they are barely mentioned outside this chapter[8] and do not seem to be assigned any significant role.

Now—it can be asked—if this is a distinction among *entities*, then why does Aristotle refer to essential predication as 'being *said* of a subject'? The question is all the more relevant because in the first distinction of ch. 2, just a few lines before 'being said of a subject' is introduced, the expression 'things said' is clearly used to refer to *linguistic* items, i.e. terms and sentences.

John Ackrill (1963, 75–76) suggested that the answer can be found in a passage from ch. 5, 2ᵃ19–34, where Aristotle contrasts items which 'are said of some subject' and items which 'are in a subject' as follows. If X is 'said of' Y 'as of a subject', then both the name of X and its definition (say, 'ABC') are true of Y—i.e. both 'Y is X' ('Jeoffry is a cat') and 'Y is ABC' ('Jeoffry is such-and-such an animal') are true. By contrast, if X 'is in' Y 'as in a subject', it is not the case that both the name of X and its definition are true of Y: e.g. tenacity 'is in' Jeoffry, but 'Jeoffry is tenacity' and 'Jeoffry is such-and-such a virtue' are false. So, Ackrill supposes, Aristotle adopted the phrase 'said of' to express the former kind of predication, in which it necessarily follows that 'Y is X' is true.

Ackrill's remark is a step in the right direction. The linguistic difference on which he focused, however, must presumably be the surface manifestation of some deeper and more substantive difference. The crucial point, I think, is that Aristotle aims to distinguish the essential from the non-essential attributes of something. Borrowing a helpful terminology introduced by Alan Code, we could say that the former attributes constitute what Jeoffry *Is*, as opposed to something Jeoffry merely *Has*.[9] Now I find almost irresistible the further conjecture (advanced by Christof Rapp) that, correspondingly, Aristotle also aims to distinguish two different interpretations of the

---

[6] See Ackrill (1963, 74–75) and Heinaman (1981) (who points out that Aristotle elsewhere commits himself to the existence of such entities anyway); the extended discussion in Wedin (2000, 38–66); and Ademollo (in preparation). I shall occasionally speak of 'inherence' and 'inhering' in place of 'being in a subject'.

[7] This is problematic in the light of 5. 2ᵃ34–ᵇ5, where Aristotle claims that universal accidents (e.g. colour) 'are in' individual substances as subjects; see n. 14.

[8] See 5. 4ᵃ14–15 (individual colours and actions) and 8. 10ᵇ30–2 (where the 'justices' mentioned are presumably individual tokens).

[9] See Code (1986), and cf. Matthews (2009, 148–149) and Ademollo (2021, part 1).

generic scheme 'X is F'[10]: one on which 'F' signifies an essential attribute of X and another one on which 'F' instead signifies a non-essential attribute of X. In a certain strong sense, only in the former case can we say that X *is* really F (i.e. that X *Is* F), and hence that the F is really 'said' or predicated of X. In the latter case, instead, 'X is F' is only a linguistic device through which we express the fact that X *Has* a non-essential attribute, F-ness (or, equivalently, that F-ness 'is in' X 'as in a subject').

2 Fundamental and Non-fundamental Entities

From the distinction drawn in *Categories* 2 between four classes of entities it does not follow that the four classes are meant to be all on a par and equally fundamental. The distinction is actually compatible with the possibility that Aristotle may believe that something may count as an entity either in a broader, weaker sense or in a narrower, stricter sense and therefore regard some classes of entities as more fundamental than others.

One way of developing this basic idea has been proposed by Michael Wedin. His idea is that the members of classes (I) and (III) of the 'meta-ontology', i.e. universals, are such that *their existence consists just in their being essential predicates* of something and that they are in some sense *dependent* upon, or *grounded* by, their subjects—i.e., in the final analysis, individuals, which are no longer essentially predicated of anything else. According to this picture, we can distinguish between a more fundamental level of reality consisting only of individual entities (falling under different categories) and a less fundamental level consisting of universal entities which are ultimately 'said of' the individuals.[11] As a particular case, universal, 'secondary' substances (i.e. the genera and species under which primary, individual substances fall) are called 'substances' only in a derivative and somewhat improper way.[12]

As we are going to see in the next sections of this paper, Wedin's thesis is borne out by some evidence as far as the relation between universal, 'secondary' substances and

---

[10] See Rapp (in preparation). We might suppose that this is also, at the same time, a distinction between two cases covered by Plato's generic notion of 'participation': see Duerlinger (1970, 179–181), Striker (2011, 143). We shall get back to this idea in Sect. 11.

[11] See Wedin (2000, 70) ('the rock-bottom entities of the *Categories* are individuals ... to be is, fundamentally, to be a non-recurrent particular, whether substantial or nonsubstantial'), 86–92, 111–20. A forerunner of this interpretation is Sellars (1957, 690–691) and n. 7. Wedin also remarks that ch. 2's inclusion of universals among 'entities' does not require that they are entities in any robust sense, although Aristotle 'does not address head-on the issue of what, if any, ontological status is to be accorded species and genera' (see 115, 120). The first commentator who held that the *Categories* does not grant full ontological status to universals seems to have been Boethus of Sidon in the first century BC: see frr. 12–13 in Chiaradonna and Rashed (2020).

[12] 'The appellation does not simply indicate a second kind of substance but has obvious demotional force ... to say that species and genera are secondary substances is to say that they are substances in a secondary way only, and this invites a deflationary reading of their ontological status' (Wedin 2000, 96–97).

individual, 'primary' substances is concerned. There is, however, no clear evidence that the same sort of relation holds also between universals and individuals in the other, non-substantial categories.[13] This is in keeping with the limited interest Aristotle seems to take (as I pointed out in Sect. 1) in individual accidents. As we shall see, Aristotle seems to take a simpler, more economical view instead: one kind of individuals—i.e. 'primary' substances—are directly fundamental with respect to *all* the other entities—not only universal substances but also universal and individual accidents (i.e. to all members of classes (I), (II), and (III) of the 'meta-ontology').

3 Primary Substances as Ultimate Subjects

We can start with some internal evidence from the *Categories* itself. At 5. 2^{a}34–5 Aristotle claims that

> All the other items either are said of the primary substances as subjects or are in them as subjects [τὰ δ' ἄλλα πάντα ἤτοι καθ' ὑποκειμένων λέγεται τῶν πρώτων οὐσιῶν ἢ ἐν ὑποκειμέναις αὐταῖς ἐστίν].

He immediately proceeds to explain this claim as follows (2^{a}35–^{b}3):

> This is evident if we consider the matter on the basis of the individual cases. For example, animal is said of human, therefore also of the particular human; *for it were said of none of the particular humans, neither would it be said of human at all* [εἰ γὰρ κατὰ μηδενὸς τῶν τινῶν ἀνθρώπων, οὐδὲ κατὰ ἀνθρώπου ὅλως]. Again, colour is in body, therefore also in a particular body; *for if it were not in some of the singular instances, neither would it be in body at all* [εἰ γὰρ μὴ ἐν τινὶ τῶν καθ' ἕκαστα, οὐδὲ ἐν σώματι ὅλως].

Here Aristotle is inferring, from the fact that (a) universal substances are subjects for essential and accidental predicates, that (b) individual substances are subjects for those same predicates. Actually, on a natural way of reading the two clauses which I have italicized in the translation, he is saying that (b) is the ground for (a): it is precisely *because* F is 'said of'/'is in' individual Gs that F is 'said of'/'is in' G.[14] Admittedly, this construal is not strictly mandatory in the case of F being 'said of' G, i.e. essential predication. For in 3. 1^{a}10–15 Aristotle has already told us that essential predication is transitive (if X is 'said of' Y, and Y is 'said of' Z, then X is 'said of' Z), and this might, in principle, seem to be sufficient to prove that (as Aristotle says

[13] On Wedin's interpretation see further n. 33.

[14] See Wedin (2000, 90–91). Strictly speaking, the claim that universal accidents 'are in' primary substances seems to be inconsistent with the definition of entities 'in a subject' stated in ch. 2, if that definition entails (as it would seem to do—see Sect. 1) that, if X 'is in' Y, then X cannot exist without Y. Aristotle should rather say something different: e.g. that colour (i.e. the universal colour) 'is in' body, a secondary substance, which in its turn 'is said of' individual bodies; or, alternatively, that colour is 'said of' an individual colour, which in its turn 'is in' an individual body, a primary substance; therefore, either way, *in the final analysis*, 'All the other items either are said of the primary substances as subjects or are in them as subjects'. See e.g. Ackrill (1963, 83) and Duerlinger (1970, 183–186).

here) if F is not 'said of' any individual G, then it cannot be 'said of' G either. In the case of F 'being in' G, however, the situation is different: the claim that, if F 'is' not 'in' some individual G, then it 'is' not 'in' the universal G at all does not seem to be derived from anything that has been said so far in the treatise. Therefore it must rest upon some underlying assumption. And the assumption, I submit, is likely to be that *facts about universals are explained or grounded by facts about individuals*—at least as far as universal and individual substances are concerned, as here.[15]

Later on in ch. 5, Aristotle argues that the genera and species of primary substances are the only secondary substances. He gives two arguments.

The first (2^b30–7) is that only their genera and species represent *what* primary substances *are*, their essence or τί ἐστιν. This confirms that secondary substances *derive* their substantial status from primary substances, as the very adjective 'secondary' suggested from the start.

In the second argument (2^b37–3^a6) Aristotle starts by picking up his previous characterization of primary substances as ultimate subjects:

> Furthermore, primary substances are said to be substances most properly [κυριώτατα] because they underlie all other things as subjects [διὰ τὸ τοῖς ἄλλοις ἅπασιν ὑποκεῖσθαι]. But as primary substances are to all the other items, so the species and genera of primary substances are to all the remaining items [ὡς δέ γε αἱ πρῶται οὐσίαι πρὸς τὰ ἄλλα πάντα ἔχουσιν, οὕτω τὰ εἴδη καὶ τὰ γένη τῶν πρώτων οὐσιῶν πρὸς τὰ λοιπὰ πάντα ἔχει]; for all the remaining ones are predicated of these [κατὰ τούτων γὰρ πάντα τὰ λοιπὰ κατηγορεῖται].

So secondary substances are subjects for all the non-substantial items. Surely in order to fulfil this function they must have a robust ontological status? Yet Aristotle's words here should be taken with a grain of salt. To start with, 'as ..., so ...' need not mean that secondary substances are subjects for their predicates *in exactly the same way* as primary substances are; the point may just be the more modest one that secondary substances, like primary ones, are (in some sense or other) subjects for the non-substantial items. Such caution is indeed justified in the light of the example Aristotle proceeds to offer:

> For you will call the particular human literate [τὸν γὰρ τινὰ ἄνθρωπον ἐρεῖς γραμματικόν]; therefore you will also call both a human and an animal literate [οὐκοῦν καὶ ἄνθρωπον καὶ ζῷον γραμματικὸν ἐρεῖς].

The example suggests that the genera and species of primary substances are subjects only in a *derivative* way, i.e. parasitically upon primary substances: only in so far as some individual cats are wise can the universal cat, the species, be said to be itself wise.[16] This obviously harmonizes with the construal of 2^a35–$^b3$ which I recommended above.

---

[15] Duerlinger (1970, 199–201) seems to reconstruct another argument based on the transitivity of essential predication: universal accidents 'are said of' individual accidents, each of which in its turn 'is in' an individual substance; therefore universal accidents 'are in' individual substances. This, however, is unlikely to be what Aristotle has in mind: individual accidents are not even mentioned in the text.

[16] Cf. Rapp (in preparation). The notion that secondary substances are subjects of predication only parasitically upon primary substances might perhaps yield the solution to a thorny problem. As I

4 Self-Contradictory Universals?

The passage we have just read may also seem to have certain other implications, even more radical and more problematic. It will prove helpful to make a detour to see what these implications are and why the passage is actually innocent of them.

Suppose someone advances the following argument. Aristotle is saying that from the fact that some particular humans are literate it follows that 'a human' (or 'an animal') is literate; and he is taking this to mean that the kind human (or animal) is literate. By the same token, from the fact that some particular humans are white it follows that the kind human is white; and from the fact that some other humans are black it also follows that the kind human is black. Therefore the kind human is both white and black—and both good and bad, both tall and short, etc. Surely therefore *universals admit of contrary predicates*, and hence in the final analysis *fail to obey the Principle of Non-Contradiction*? And surely this in turn suggests that in some sense they are not genuine entities?[17]

This argument would, I think, be unsound. The passage does, indeed, imply that universals admit of contrary predicates; it does not, however, follow from this that universals fail to comply with the Principle of Non-Contradiction. It is instructive to investigate why.

As we know from ch. 2, the only entities 'numerically one' are individuals, which are not 'said of' any subject (see Sect. 1). Aristotle will pick up that point and apply it to substances later on in our very chapter, 5. 3^b10–21. There he draws the following distinction between primary and secondary substances. Each primary substance—or rather each singular term for a primary substance—'signifies a particular *this*' (τόδε τι σημαίνει), 'for the thing indicated is individual and numerically one' (ἄτομον γὰρ καὶ ἓν ἀριθμῷ τὸ δηλούμενόν ἐστιν). Secondary substances may sometimes mislead us into thinking that the same holds of them, because descriptions such as 'the cat' seem syntactically akin to proper names such as 'Jeoffry' and are even intersubstitutable with them in certain contexts ('The cat is on the mat'). In fact, however, primary and secondary substances are different. For in a sentence such as 'The cat is an animal', where 'the cat' refers to a kind, 'the subject is not one as the primary substance is, but the human and the animal are said of many things' (οὐ γὰρ ἕν ἐστι τὸ ὑποκείμενον ὥσπερ ἡ πρώτη οὐσία, ἀλλὰ κατὰ πολλῶν ὁ ἄνθρωπος λέγεται καὶ τὸ ζῷον). That is to say, a secondary substance (more generally, a universal) is not one in the way in which a primary substance (more

recalled above, in ch. 3 Aristotle claims that 'being said of a subject' holds both between universals (the animal 'is said of' the cat) and between universals and individuals (both animal and cat 'are said of' Jeoffry). In fact, however, these are two completely different kinds of predication—and Aristotle should be in a position to appreciate the difference in the light of his insistence, against Plato, on the distinction between universals and individuals (see Sect. 4 and n. 19 below). So, if his view turned out to be that F's being 'said of' G can be analysed away in terms of F's being 'said of' the individual Gs, this might allow him to defuse the problem. I cannot pursue this issue any further here; see Duerlinger (1970, 195–197) and Ademollo (in preparation).

[17] Some worries along the same lines (though not fully explicit) are pondered by Wedin (2000, 98–101) and Perin (2007, 142–143).

generally, an individual) is one, i.e. numerically, because that is incompatible with its being said of many things. Presumably, it will be one in some other, weaker way; as is well known, Aristotle elsewhere distinguishes between different ways of being one, i.e. in number, in species and in genus.[18] A secondary substance—or rather a general term for a secondary substance—instead 'signifies a certain qualification' (ποιόν τι σημαίνει), i.e. a certain *sort* of substance.[19]

Now, it is reasonable to suppose that, in so far as a universal is not one, it can be regarded as composed of *parts*, and that, more precisely, individuals are the ultimate parts of which universals are composed, and into which they can be divided. This hypothesis is confirmed by a familiar piece of evidence: the term 'individual' (ἄτομον, literally 'indivisible') itself suggests that there is a sense in which universals are ultimately divided into individuals, whereas individuals are the entities which cannot be divided further. And since universals 'are said of' both more specific universals—if there be any—and individuals, whereas individuals are no longer 'said of any subject' (2. 1^b6–7, see Sect. 1 above), the obvious conjecture is that the relevant sense of 'division' coincides with 'being said of a subject': something can be said to be 'divided' into the entities which it 'is said of' as its 'subjects', i.e. is essentially predicated of.[20]

Now, these things being so, nothing prevents a universal from being and not being F at the same time, if some individuals falling under it are F and some others are not F. The situation is (in the relevant respect) analogous to that of a material object being and not being F at the same time in so far as some material parts of it are F and some others are not F. Therefore it does not follow that universals violate the Principle of Non-Contradiction.

This conclusion is consistent with what we get if we consider[21] that in our passage Aristotle seems to be thinking that literacy's being accidentally predicated of the universal human by 'being in' it 'as in a subject' can be expressed by a sentence such as '(A) human is literate'.[22] Now, sentences of the form '(A/The) F is G', '(A/The) F is not G'—i.e. sentences where the subject term, whether or not it is accompanied by a definite article, lacks a quantifier—are technically what Aristotle in the

---

[18] See *Metaph.* Δ6. 1016^b31–1017^a3 and Frede (1978, 52).

[19] Aristotle's claim that the universal cat is not 'a particular this' and is not 'one in number' has in itself an anti-Platonic ring to it. Elsewhere Aristotle takes Plato to be committed to the deeply problematic view that forms, besides being universals, are also, at the same time, individuals of a special sort (*Metaph.* Z15. 1040^a8–9) and 'particular thises'—an assumption that plays an important role in triggering the Third Man regress (*Soph. El.* 7. 169^a33–6; 22. 178^b36–179^a10). On the connection between Aristotle's criticism of the Third Man and the theory of the categories see Kahn (1978, 244–247).

[20] See again Frede (1978, 52). He helpfully compares the scholastic distinction between the *partes integrales* of a *totum integrale* and the *partes subiectivae* of a *totum universale*: see e.g. Abelard, *Dialectica* 546. 21–547. 5 De Rijk, and Walter Burley, *De toto et parte* (text of the latter in Shapiro and Scott [1966]; trans. of both in Henry [1991, 66–67, 411–412]). The distinction actually stems from Aristotle himself, *Metaph.* Δ26.

[21] As also Rapp (in preparation) does.

[22] '(*A*) human' and '(*An*) animal', because there is no definite article before 'human' and 'animal' at 3^a5. The matter would, however, be unaffected if the article were instead present: see what follows.

Prior Analytics (1.1. 24^a16–22; 1.4. 26^a29–39; 1.7. 29^a27–29) calls 'indeterminate' sentences. There and in the *De interpretatione* (7. 17^b6–12, 26–37) he appears to think that, although such sentences *can* be used as equivalent to universal sentences ('All F is G', 'All F is not G'), generally speaking they should be given a weaker interpretation and should be taken to be equivalent to *particular* sentences ('Some F is G', 'Some F is not G'). But of course such sentences are not contradictories to each other and therefore can be true together unproblematically.[23]

5 Identity through Time

Let us get back to our main line of enquiry. We encounter another interesting piece of evidence at the end of ch. 5, 4^a10–21, where Aristotle states the 'proprium' of substance, namely being capable of changing from one contrary predicate to another while remaining numerically one and the same:

> It seems to be most proper to substance that it is something which, while being numerically one and the same, is capable of receiving contraries [μάλιστα δὲ ἴδιον τῆς οὐσίας δοκεῖ εἶναι τὸ ταὐτὸν καὶ ἓν ἀριθμῷ ὂν τῶν ἐναντίων εἶναι δεκτικόν].

Let us refer to the feature which Aristotle identifies in this passage, and which he puts forward as the 'proprium' of substance, as 'Σ'. Now, as we know from *Topics* 1.5, the proprium of X should be a feature which holds of *all and only* the things that are X. Does Σ meet this condition? In the sequel Aristotle raises the question whether Σ really holds *only* of substances. He discusses some possible counterexamples, i.e. some items (most notably, sentences and beliefs) which are not substances but which might nevertheless seem (falsely, as he argues) to be such that Σ holds also of them. But Aristotle does not also raise the question whether Σ holds of *all* substances. If we do so on his behalf, our answer has to be 'No'; for Σ can only hold of primary, individual substances and cannot hold also of secondary, universal ones, as already the ancient commentators saw.[24] The reason is that the only entities 'numerically one' are individuals, as we saw in Sects. 1 and 4.

So at the end of *Categories* 5 Aristotle is aware that Σ holds of primary substances but not of secondary substances. And yet he puts forward Σ as the proprium of substance. Why?

One possible answer is that Aristotle is actually stating just the proprium of *primary* substance and leaving it to us to generalize this to a more comprehensive

---

[23] Actually, in the *De interpretatione* Aristotle makes matters needlessly confused by claiming that pairs of indefinite sentences of the form '(A/The) F is G' and '(A/The) F is not G' *are* contradictory to each other—but are exceptional in that in their case it is possible for both contradictories to be true. See Weidemann (2014, 206–207).

The material discussed in this section could be profitably compared and contrasted (as I cannot do here for reasons of space) with the Stoic views which Caston (1999, 187–192) reconstructs on the basis of a report in Sextus Empiricus, *Adv. Math.* 7. 246.

[24] Ammonius, 51.13–52.10 Busse; Philoponus, 77.25–79.6, 80.12–14 Busse; Simplicius, 113.13–31 Kalbfleisch. Cf. Ackrill (1963, 89) and Kohl (2008, 154–155).

account: 'Something is a substance iff it *either* is Σ *or* is a genus or species of something that is Σ' (cf. Ammonius 52.11–14 Busse).[25] This answer is fine as far as it goes, but it does not correspond to what Aristotle says: he presents Σ as the proprium of substance in general, not of primary substance. He expresses himself as if primary substances were the only substances there are. Moreover, this answer proposes to identify the real proprium of substance as a *disjunctive property*—a notion which Aristotle would probably not countenance, and of which there is, to my knowledge, no instance in his writings.

An interesting alternative answer is that Aristotle expresses himself as he does precisely because he regards secondary substances as entities which are less fundamental than, or grounded by, or parasitical upon, primary substances. If this is so, then he does after all believe that—in a sense—primary substances are the only substances there are.

6 Substances and Accidents

In ch. 4 and in the following chapters the *Categories* distinguishes ten different kinds of entities, i.e. the categories. This distinction is independent of (and orthogonal to) the one between universal and individual entities. It is also different from (and encompassed by) the 'meta-ontology' set out in ch. 2, as can already be seen from the outline I provided in Sect. 1.

Among the ten kinds of entities one is called οὐσία, literally 'entity'—or 'substance' according to our customary, misleading translation. As Michael Frede pointed out, this term by itself (partly because of it previous use in Plato—we shall come back to this in Sect. 9) is sufficient to make it clear that these are the entities par excellence, the basic or fundamental entities, in relation to which the other entities have to be conceived of and on which they depend for their status as entities.[26]

Furthermore, *Categories* 5 distinguishes between primary and secondary substances. As we know, primary substances are individual, particular substances. They notably include *sensible* particulars; thus Aristotle gives as instances particular humans, horses, trees, and their material parts (head, hands etc.). He also mentions the soul and the body of a primary substance as subjects of the inherence of accidents (ch. 2), and explicitly describes the body of a primary substance as being itself a substance (5. 2^b1–3); the same will presumably hold also of the soul.[27]

25 I myself sympathized with this solution before: see Ademollo (2010).

26 See Frede (1985, 73; 1992). I elaborate on this notion in Ademollo (2021, part 1).

27 The reason why the bodily parts of a primary substance are regarded as substances is, arguably, that they are able to serve as ultimate *subjects* (or, more precisely, as ultimate subjects not themselves predicated of any other subject) for the inherence of certain predicates which via them hold indirectly of the whole substance. Thus e.g. a cat may be white in respect of his teeth and black overall (cf. *Soph. El.* 5. 167^a7–10). We can suppose that, analogously, the reason why also soul and body are taken to be substances is that they are conceived of as *parts* of the whole compound substance and as the ultimate subjects (not themselves predicated of any other subject) of certain predicates which

The primacy of substance among all categories, and of individual substances among substances, means that individual substances are the most fundamental among the fundamental entities. The meaning of this, however, has to be understood more precisely; we do so in the next section.

7 Dependence

Let us return to Aristotle's claim—which we encountered in Sect. 3—that 'all the other items either are said of the primary substances as subjects or are in them as subjects' (5. 2^{a}34–^{b}5). As we also saw in Sect. 3, later in the chapter Aristotle invokes this as the reason why 'primary substances are said to be substances most properly' (2^{b}37–3^{a}1). But before that further development, at 5. 2^{b}5–6, immediately after making the former claim, Aristotle draws a momentous inference from it:

> Therefore, if the primary substances did not exist, it would be impossible for any of the other things to exist [μὴ οὐσῶν οὖν τῶν πρώτων οὐσιῶν ἀδύνατον τῶν ἄλλων τι εἶναι].

So there is a kind of ontological dependence of all other entities on primary substances. We can construe it as follows:

(1) Necessarily (if primary substances do not exist, then all the other entities do not exist).

In other words, the other entities—i.e. secondary substances and universal and individual accidents—exhibit what is sometimes called 'modal-existential' dependence on individual substances.[28]

Let us now consider the converse of (1), which is not stated in the passage:

(2) Necessarily (if all the other entities do not exist, then primary substances do not exist).

Would Aristotle accept this? Yes. For both in the *Categories* and elsewhere he clearly assumes that every substance is necessarily endowed with both essential and accidental attributes.

---

via them hold indirectly of the whole substance. This hypothesis is proved correct by a passage in the *Topics*, 4.5. 126^{a}17–29. See further Ademollo (2021, part 2).

[28] The *Categories* sets forth several other cases of modal-existential dependence: thus e.g. in ch. 7, 7^{b}15–8^{a}12, Aristotle discusses whether or not one of two correlative kinds (e.g. master and slave, or knowledge and the knowable) 'co-destroys' (συναναιρεῖ) the other—that is to say, whether or not it is the case that, if no X existed, then no Y would exist either. There are other related discussions in chs. 12 and 13, which I am going to mention later on, and in various passages from other works (*Top.* 4.2. 123^{a}14–15, 6.4. 141^{b}28–9; *Phys.* Δ1. 208^{b}35–209^{a}2; *Metaph.* Δ11. 1019^{a}1–4). In Sect. 1 we encountered another, notorious case: if X 'is in' Y 'as in a subject', then X cannot exist without Y. Therefore an individual accident X, which 'is in' an individual substance Y, depends on Y (at least) in this way. There are, however, several differences between this case of modal-existential dependence and that of all the other entities on primary substances: see n. 33.

So Aristotle is actually committed to the truth of both (1) and (2). That is to say, the modal-existential dependence between primary substances and all the other entities is in fact *symmetric*. This assumption is crucial to the argument against Parmenides' monism in *Physics* 1.2, 185^{a}20–32: it is impossible that only substances—or one substance—exist, and it is impossible that only accidental attributes—or one accidental attribute—exist, because the existence of substances requires that of accidental attributes and vice versa.[29]

Nevertheless, we expect some sort of *asymmetric* dependence of the other entities on primary substances. For the relation of subjecthood between primary substances and the other entities is asymmetric; and this relation—as we saw above and in Sect. 3—is cited as the basis both for (1) and, later on, for the primacy of primary substances among substances. Hence if Aristotle here only states (1) and does not also state (2), the reason is likely to be that he treats his statement of (1) as conveying an *implicature* of some other kind of dependence which is instead asymmetric.

On the most straightforward account, the required kind of asymmetric dependence of all other entities on primary substances is *explanatory* dependence or 'grounding'. When in our passage Aristotle claims that 'If the primary substances did not exist, it would be impossible for any of the other things to exist', one reason why he does not hasten to add 'and vice versa' may be that he is at the same time implicating that all the other things exist *because* primary substances exist (and are the way they are). As Simplicius says in his commentary, 'the individual substance is for the other things the cause of their being' (αἰτία ... ἡ ἄτομος οὐσία ἐστὶν τοῖς ἄλλοις τοῦ εἶναι, 87.25–6 Kalbfleisch).[30]

This interpretation fits well with Aristotle's use of the very term οὐσία. As Frede (1992) pointed out, already in Plato it is constitutive of the meaning of this term that it refers to something which is both what other things fundamentally are and a fundamental entity—which involves the capacity to explain the existence of the other entities.[31]

The hypothesis of an explanatory role for primary substances accords also with the view, which Aristotle certainly holds in other works, that it is in virtue of having some relation to substances that all other kinds of entities can be said to exist. That view is asserted in the *Posterior Analytics*, where Aristotle claims that only substances are 'in themselves' or 'in their own right' (καθ' αὑτά), whereas non-substances are 'accidental' or 'coincidental' entities (συμβεβηκότα), existing in virtue of being predicated of (individual) substances, and that whenever a non-substance is F, it is F only 'in virtue of being something other' (ἕτερόν τι ὄν), i.e. of being predicated of an underlying substance which is the real subject of F (*An. Post.* 1.4. 73^{b}5–10, 19.

[29] For an analysis of the argument (which is not fully explicit in the text) see Clarke (2019, 20–29). On Aristotle's commitment, in the *Categories*, not only to (1) but also to (2) see Duerlinger (1970, 181–182, 187, 197–198), Sirkel (forthcoming), Edelhoff (2020, 51–66).

[30] Among modern interpreters see especially Frede (1985, 72–74; 1992); cf. among others Schaffer (2009, 351) and Sirkel (forthcoming). This view is also broadly in the spirit of Katz (2017) (although she does not discuss substance in the *Categories* except marginally: see 33 n. 20, 58 and n. 82).

[31] *Pace* Burnyeat (2001, 107): 'The *Categories* simply lacks the concept of a primary kind of being that explains the remaining (dependent) kinds of being'. On Plato see below, Sect. 9.

$81^b25–9$, 22. $83^a1–35$). The same view returns in the *Metaphysics*, where it becomes part of the doctrine that 'that which is is said in many ways' and that being a substance is the primary way of being (Γ2. $1003^b5–10$, Z1. $1027^b10–1028^a31$).

This last doctrine is not yet present—or at least not stated—in the *Categories*, where the distinction between the categories is not cast as a distinction between different *kinds or ways of being* and substance is not explicitly identified as the primary among such kinds or ways. But it is reasonable to regard it as a final development of notions which are already present in the *Categories* as well as in other 'logical' works.[32]

Is there any further evidence within the treatise for the explanatory account of primary substance? Some scholars believe that there is some in chs. 12–13. I examine these texts in the next section.[33]

8 Explanatory Priority in *Categories* 12–13

Cat. 12 distinguishes several different ways for something to be 'prior' (πρότερον) to something other, including this ($14^b11–13$):

> Among things which reciprocate with respect to the implication of being [τῶν ... ἀντιστρεφόντων κατὰ τὴν τοῦ εἶναι ἀκολούθησιν], what is in any way a cause of being for the other might be reasonably called prior by nature [τὸ αἴτιον ὁπωσοῦν θατέρῳ τοῦ εἶναι πρότερον εἰκότως φύσει λέγοιτ' ἄν].

Aristotle gives an example ($14^b13–22$):

> And that there are some such cases is clear. For a human's being [τὸ ... εἶναι ἄνθρωπον] reciprocates with respect to the implication of being [ἀντιστρέφει κατὰ τὴν τοῦ εἶναι ἀκολούθησιν] with the true sentence about it: if a human is, the sentence whereby we say that a human is is true, and reciprocally – since, if the sentence whereby we say that a human

---

32 See *Top.* 4.5. $127^a26–38$ (being is not a genus), *Soph. El.* 33. $182^b13–27$ etc. (being and one 'are said in many ways'). The view that being is not a genus seems to be tacitly presupposed also in the *Categories*, where Aristotle for each of the categories does not try to provide a fully-fledged definition by genus and differentia, but just a proprium.

33 Wedin (2000, 67–121) offers an alternative account of the asymmetric dependence of all other entities on primary substances. It is a combination of two stages, each of which turns on an asymmetric relation of dependence: (i) universals—both substantial and accidental—explanatorily depend on individuals—both substantial and accidental—in so far as they exist in virtue of being predicated of individuals (cf. Sect. 2 above); (ii) among individuals, each accident enjoys modal-existential dependence on the primary substance which it 'is in': the individual instance of tenacity which 'is in' Jeoffry cannot exist if Jeoffry does not exist, but of course Jeoffry can exist if it does not exist (cf. Sect. 1).

There are two drawbacks to this interpretation. One is its emphasis on the pivotal role of individual accidents, which is highly speculative and unwarranted by the text. Another is that the two stages are heterogeneous and sit uncomfortably with each other: the dependence at issue is explanatory at stage (i), but modal-existential at stage (ii); and it holds between whole classes of entities at stage (i)—as in the passage of the *Categories*—but between *each* individual accident and the individual substance it 'is in' at stage (ii). Therefore Wedin's account is uneconomical.

> is is true, then a human is. But the true sentence is in no way a cause of the object's being, whereas the object appears to be somehow the cause of the sentence's being true [τὸ μέντοι πρᾶγμα φαίνεταί πως αἴτιον τοῦ εἶναι ἀληθῆ τὸν λόγον]: for it is in virtue of the object's being or not being that the sentence is called true or false [τῷ γὰρ εἶναι τὸ πρᾶγμα ἢ μὴ ἀληθὴς ὁ λόγος ἢ ψευδὴς λέγεται].

The example is not very clear and its interpretation is controversial.[34] Fortunately, however, Aristotle refers to the same notion again in the next chapter, 13. 14^b27–33, to describe a case to which it does *not* apply. He identifies a way in which things can be said to be 'together' (ἅμα) in so far as neither is 'prior' to the other in the specific way we have just encountered:

> Those things are called 'together by nature' which are such that they reciprocate with respect to the implication of being, but neither is in any way a cause of being for the other, as e.g. in the case of the double and the half. These do reciprocate, since if there is a double there is a half and if there is a half there is a double, but neither is the cause of being for the other.

So it becomes clear that the kind of priority at issue can take the following form:

(3) (Necessarily, Xs exist iff Ys exist) and (Xs exist because Ys exist).

And this might be obviously relevant to our concerns.[35] Still, Aristotle makes no attempt whatsoever to connect this discussion with the theory of the categories and apply these conceptual tools to the issue of the priority of primary substances over all other entities, when it would be very easy to do so. So I am sceptical that *Cat.* 12–13 can be helpful for our purposes.

---

[34] It is unclear how we should construe the 'being' whose 'reciprocal implication' is at issue here, which seems to cover both the *existence* of something (the most natural interpretation of the Greek ἔστιν ἄνθρωπος) and the *truth* of a sentence asserting the existence of that thing. See Edelhoff (2020, 27–31). Actually, it is tempting to surmise that in this specific passage the fact that the example concerns something's existence is purely accidental and that Aristotle is rather interested in something whose general form is this: '("P" is true iff P) and ("P" is true because P)'—or, even more generally, '(α iff β) and (α because β)'. If this is so, then the 'being' in question here is rather likely to be the *obtaining* or being-the-case of a proposition or state of affairs (which is perhaps what the term πρᾶγμα, 'object', refers to at 14^b19–21). See John Buridan, *Sophismata*, c. 8, second sophism, and among modern interpreters Nuchelmans (1973, 33–34), Katz (2017, 39 n. 41), Caston (2018, 42 and n. 15). See Cavini (2011, 392–394) for a helpful discussion and a different suggestion.

[35] Cf. 14^a29–32, earlier in ch. 12, where another kind of thing which is 'prior' to something other is referred to as 'what does not reciprocate as to the implication of being, e.g. one is prior to two: for if there are two things it immediately follows that there is one, whereas if there is one it is not necessary that there be two'.

NB: (3) would have been closer to the actual formulation of Aristotle's claim about double and half, if I had not added 'Necessarily' at the beginning and after '(Xs exist because Ys exist)' had instead added 'and not (Ys exist because Xs exist)'. Both of these departures from the letter of the text are substantially innocuous and faithful to its spirit.

9 Anti-Platonism

The picture we have been drawing so far amounts, in several respects, to a reversal of the Platonic priority of universals over individuals. This claim is frequently made about the *Categories*, and I myself have already partly anticipated it in Sect. 6 above. It is, however, important to understand its actual purport as precisely as possible.

Plato has no trouble acknowledging that there is a sense in which intelligible, changeless universal forms and sensible, changing particulars constitute two kinds of entities (δύο εἴδη τῶν ὄντων, *Phaedo* 79a). However, he also believes that among all such entities forms are the fundamental ones, the entities par excellence, whereas sensible particulars are entities only in a weak, derived way. Therefore he refers to the realm of forms collectively as 'the being which really is' (ἡ ... οὐσία ὄντως οὖσα, *Phaedrus* 247c) and on one occasion says, or rather implies, that a member of that realm 'is perfectly a being' (τελέως ... εἶναι ὄν, *Rep.* 10. 597a). Therefore Plato can also refer to the realm as a whole more simply as 'being' (οὐσία, *Rep.* 7. 525bc, 534a, etc.), and either to the realm or to each of its member forms as 'that which is' (τὸ ὄν, *Rep.* 5. 477b, 478a; 6. 484*cd*; 7. 518c; *Phaedrus* 247de; etc.). Sensible particulars, on the other hand, hover between what is and what is not (*Rep.* 5. 479ae) and constitute the realm of 'coming-to-be' (γένεσις, *Rep.* 7. 525bc, 534a; τὸ γιγνόμενον, *Rep.* 7. 518c, 521d, etc.). So for Plato universal forms are more fundamental than at least a specific subclass of individuals, namely sensible particulars. For Aristotle, instead—as we have seen—sensible particulars are the paradigm case of 'primary substances', i.e. of the most fundamental instances of the most fundamental kind of entities.[36]

As far as sensible particulars are concerned, Aristotle disagrees with Plato not only about whether or not they should be regarded as οὐσίαι or entities par excellence, but also about other (connected) issues. Here is one: at least in some contexts (*Symp.* 207d–208b; *Phaedo* 87de, 91d), Plato seems to hold that sensible particulars, e.g. our own bodies, are, strictly speaking, *not identical through time*. What we usually see as the persistence of an identical sensible object is actually a succession of distinct but similar objects. Aristotle, by contrast, as we saw in Sect. 5, maintains that it is distinctive of (primary) substances that they remain identical through time and change from one contrary predicate to another.[37]

It is interesting to ask how Plato would react on opening the *Categories* and encountering Aristotle's claim—which we discussed in Sect. 7—that, necessarily, if primary substances did not exist, then all the other entities would not exist. That claim agrees with what Plato would say about the items to which *he* applies the term 'substance' (οὐσία), i.e. the forms. It is, however, incompatible with what Plato would say about the items *Aristotle* himself calls '(primary) substances'.[38] Consider

---

[36] The anti-Platonic significance of Aristotle's use of the term οὐσία in the *Categories* is especially emphasized by Frede (1985, 73; cf. 1992). See also Kahn (1978, 247).

[37] On Plato's conception of the change undergone by sensible particulars, on its relation to Plato's claim that sensible particulars 'are' not and merely 'come to be', and on Aristotle's reaction in the *Categories*, see Ademollo (2018) and Ademollo (2021, part 2).

[38] Cf. Duerlinger (1970, 181–182, 198–201) and Edelhoff (2020, 62–63).

in particular the *Timaeus*. According to the story told there, universal forms already existed before God decided, out of his goodness, to take them—or some of them—as his models in bringing order to a primordial chaos and thus creating the universe we now live in, including sensible particulars and particular souls (27d–31b etc.). This of course entails that forms would exist even if sensible particulars (and particular souls) did not exist. So Plato would presumably read Aristotle's claim (1) as contradicting him.

10 Universals in the *Posterior Analytics*

At this point it is helpful to take into account a couple of passages from another of the 'logical' works, the *Posterior Analytics*. The first passage is at 1.11. 77^{a}5–9:

> There need not necessarily be forms, or some one item in addition to the many [εἴδη μὲν οὖν εἶναι ἢ ἕν τι παρὰ τὰ πολλὰ οὐκ ἀνάγκη], in order for there to be demonstrations. *Necessarily, however, it must be true to say that one thing holds of many* [εἶναι μέντοι ἓν κατὰ πολλῶν ἀληθὲς εἰπεῖν ἀνάγκη]. For there will not be what is universal if this is not the case [οὐ γὰρ ἔσται τὸ καθόλου, ἂν μὴ τοῦτο ᾖ]; and if there is not what is universal, then there will be no middle term, and hence no demostration. There must, therefore, be something – one and the same item – which holds of several cases non-homonymously [δεῖ ἄρα τι ἓν καὶ τὸ αὐτὸ ἐπὶ πλειόνων εἶναι μὴ ὁμώνυμον].[39]

For demonstration and science to be possible, argues Aristotle here, it is not necessary to assume that there are Platonic forms or, more generally, universals existing 'besides' or 'in addition to' (παρά) their individual instantiations.[40] The universality we need rather consists in the truth of a universal *claim*: 'it must be true to say that one thing holds of many'. This, says the final sentence, is the sense in which there is one and the same item holding of many, i.e. in which there are universals.

Now it might seem, on the face of it, that thereby Aristotle is reducing the existence of universals to the truth of certain *sentences* in which a universal is predicated of some individual. This, however, would be problematic; for Aristotle often insists that our sentences are true because of the way things are in the world, not the other way round.[41] Perhaps, therefore, Aristotle's reference to the truth of our claims is actually

---

[39] Trans. after Barnes (1993), both here and in what follows.

[40] *Pace* several commentators, the preposition need mean no more than that, according to the view in question, a universal is an entity *fundamentally distinct from* its individual instantiations. Cf. Pl. *Phaedo* 74a11, and in Aristotle, among the other passages cited by Bonitz (1870, 562), *An. Post.* 2.19. 100^{a}7; *Metaph.* B3. 999^{a}7–23 (especially ^{a}17–19), H6. 1045^{a}10, M1. 1076^{a}11; *Phys.* 4.3. 210^{a}17. See also *Cat.* 14. 15^{a}25; *De gen. et corr.* 1.1. 314^{b}11, 2. 316^{b}7; and the *De ideis* as reported by Alex. Aphr. *in Metaph.* 79.16–19, 84.4–7 Hayduck. The view that there are universals distinct from particulars is meant to be *weaker* than the view that there are Platonic forms. Cf. the first of the two *De ideis* passages cited above (where Aristotle—at least as Alexander understands and paraphrases him—actually *endorses* the existence of 'common' entities in addition to the many individuals).

[41] See *Cat.* 5. 4^{a}36–^{b}1, ^{b}8–10; 12. 14^{b}15–22 (discussed in Sect. 8); *Metaph.* Θ10. 1051^{b}6–9.

meant as an indirect reference to the way in which things are in the world and in virtue of which our claims are true. Then the idea is that *for a universal F to exist is just for some individual to be F*: the existence of the universal cat, or wisdom, consists in, or is grounded by, the fact that some individuals are cats, or are wise.[42] In the present context it is reasonable to suppose that the individuals in question are in all cases meant to be individual *substances*. This further supposition is recommended by the absence of any reference to individual accidents in this context and by the fundamental role assigned to substances in the *Posterior Analytics*, as we saw in Sect. 7. And this obviously harmonizes with the interpretation of the *Categories* which I am proposing.

The second passage from the *Posterior Analytics* is in 1.24. Aristotle there discusses whether universal or particular demonstrations are better. According to one argument he considers (85^a31–7), universal demonstrations are worse,

> if a universal is not something in addition to the singulars [τὸ μὲν καθόλου μὴ ἔστι τι παρὰ τὰ καθ' ἕκαστα], and if a demonstration instils an opinion that the item with regard to which it demonstrates is something [εἶναί τι τοῦτο καθ' ὃ ἀποδείκνυσι], i.e. that this is some sort of nature among the things that are [τινα φύσιν ὑπάρχειν ἐν τοῖς οὖσι ταύτην] (e.g. a triangle in addition to the individual triangles, a figure in addition to the individual figures, a number in addition to the individual numbers); and if a demonstration about something that is is better than one about something that is not, and a demonstration by which we will not be led into error is better than one which will lead us into error; and if universal demonstrations are of this latter type.

Thus universal demostrations mislead us into believing falsely that there universal entities 'in addition to' particulars or individuals, when in fact there are no such entities. More precisely, since every demonstration presupposes the existence of universals, the more universal a demonstration is, the worse it is.

In the sequel Aristotle defends the merits of universal demonstrations. In particular, at 85^b15-22 he turns the tables on the argument we have just read:

> If there is a single account of a universal and it is not homonymous, then it will be no less than some of the particulars but even more so [εἴη τ' ἂν οὐδὲν ἧττον ἐνίων τῶν κατὰ μέρος, ἀλλὰ καὶ μᾶλλον], in so far as what is imperishable is found among the universals and it is rather the particulars that are perishable [τὰ ἄφθαρτα ἐν ἐκείνοις ἐστί, τὰ δὲ κατὰ

---

[42] Barnes (1993, 144–145) writes that 'The answer Aristotle should have given is perfectly clear: knowledge is of universal *propositions*; only particular *objects* are real: universal propositions do not require universal objects as their subject-matter. Aristotle is groping for this answer in our passage; but he does not grasp it properly'. On the present suggestion, Aristotle's point is more sophisticated: the existence of universal objects is grounded by the truth of universal propositions about particular objects. See Wedin (2000) mentioned in Sect. 2 and n. 11 above; Wedin quotes the *An. Post.* 1.11 passage and makes a few remarks about it at 119 and n. 82. See also Mignucci (2007, 249): 'gli universali per Aristotele non hanno esistenza autonoma. In realtà quello che si domanda quando si chiede "esiste X?" è se esistano le cose che sono X'. As far as Wedin is concerned, however, remember that he takes universals to be grounded by individuals *both substantial and accidental*, whereas I see no clear evidence for this either in the *Cat.* or in the *An. Post.*: cf. Sect. 2 above and n. 33. The simpler picture I favour seems more similar to the view of Loux (2015, 38): Aristotelian universals 'are suches; that is, they are *ways* self-standing thises are ... essentially predicative entities; they are things that are, so to speak, adjectival and, hence, dependent on their subjects'.

μέρος φθαρτὰ μᾶλλον]. Again, there is no need to believe that a universal is something in addition to the particulars on the grounds that it indicates one thing [οὐδεμία ἀνάγκη ὑπολαμβάνειν τι εἶναι τοῦτο παρὰ ταῦτα, ὅτι ἓν δηλοῖ], any more than in the case of those other items which signify not a substance but either a quality or a quantity or a relation or a doing [οὐδὲν μᾶλλον ἢ ἐπὶ τῶν ἄλλων ὅσα μὴ τὶ σημαίνει ἀλλ' ἢ ποιὸν ἢ πρός τι ἢ ποιεῖν]. And if the belief is instilled after all, then the explanation lies not with the demonstration but with the audience.

Aristotle's reply is actually twofold. In the first part (^{b}15–18) he makes an interestingly provocative move: he insists that there is indeed a sense in which universals 'are' even more than particulars, in so far as at least some of them are 'imperishable', whereas particulars are 'perishable'. Thereby he is probably flirting with Platonic terminology, only to give it a twist by applying it to universal and singular *propositions*: universal propositions cannot change their truth value, whereas singular ones can.[43] The implication is that some of Plato's general views on epistemology can be preserved in a different framework: Plato was after all right to claim that knowledge has its own special objects—universal, necessary, eternal—although his specific conception of those objects was wrong.

The second part of Aristotle's reply, from 85^{b}18 'Again, there is no need ...' onwards, responds more directly to the rival argument and is more immediately in keeping with 1.11. Aristotle *acknowledges* that, although a universal does 'indicate one thing', this is no reason to regard it as an entity existing in addition to its particular instantiations. He also seems to add that universals in the category of substance are no different from universals in the other categories in this respect. Thereby he seems to be inviting us to extend to the former class of universals a treatment we are already familiar with in the case of the latter class. Now, as we saw in Sect. 7 and I recalled above, in the *Posterior Analytics* he does indeed hold (and has repeated in the previous chapter, 1.22. 83^{a}1–35) that non-substances exist and possess their attributes parasitically upon (individual) substances.[44] Therefore his point here in 1.24 is again that the existence of *any* F (including cases in which this is in the category of substance), consists in, or is grounded by, some individual substance's being F.[45]

---

[43] See Barnes (1993, 185), referring to *An. Post.* 1.6. 74^{b}32–9 and 1.8.

[44] It is exactly on these grounds that Aristotle, in the *An. Post.* 1.22 passage, makes the famous claim that we can 'say farewell' to Platonic forms, because they are empty words and have nothing to do with actual demonstrations (83^{a}32–5). There he is not saying farewell to Plato's forms because they are universals; his point there is the more specific one that those Platonic forms which from his own perspective correspond to non-substantial universals (the just, the good, the beautiful, etc.) cannot be self-standing subjects of attributes as Plato takes them to be.

[45] The picture I have been drawing so far fits well also—at least according to one prominent line of interpretation—with some aspects of Aristotle's views in the *Metaphysics*: his denial that universals can be substances (Z13) and his commitment to individual substantial forms which are qualitatively identical for all members of the same species and thus 'identical in virtue of the universal account' (τῷ καθόλου λόγῳ ... ταὐτά, Λ5. 1071^{a}29). All this, however, is the subject of complex controversies which I cannot enter here. For the interpretation I am endorsing see Frede and Patzig (1988, 1. 48–57) and e.g. Mignucci (1993).

11 Essentialism and Anti-essentialism

Some scholars have suggested that there is a further difference between the *Categories* and Plato: 'Aristotle holds, in opposition to Plato, that sensible particulars are endowed with essential natures'.[46] This suggestion has to be examined carefully. Let us call 'essentialism' the view that, for any sensible particular *x*, there is some F such that *x* is essentially F. The claim then is that Plato rejects essentialism whereas Aristotle endorses it.

Now, that Aristotle in the *Categories* endorses essentialism is beyond dispute: every individual, in any category, is the subject of essential predicates which are 'said of' it.[47] As for Plato, he does not discuss essentialism explicitly; so we have to search for some evidence that he is committed to rejecting it.

The evidence is indeed irresistible in the *Timaeus*, where sensible particulars are conceived of as regions of a spatial, propertiless 'Receptacle' temporarily characterized by forms (48e–53c). In other dialogues, however, and in particular in those which are usually regarded as belonging to Plato's maturity, the issue is much less clear cut.

As is well known, Plato holds that any situation which we would ordinarily describe as one in which a sensible particular X *is* F should be described as a situation in which X bears a certain *relation* (which he variously characterizes as 'participation', 'imitation', 'being called after', etc.—let us call it 'Π' for short) to the form of F. Now, this by itself need not be incompatible with essentialism. For, generally speaking, X's bearing a certain relation to Y may be part of X's essence.[48] Hence in the specific case of sensible particulars bearing Π to Platonic forms we should ask whether, in Plato's view, they can do so essentially.

A passage in the *Phaedo*, 102bc, might seem to suggest a negative answer to the last question. There Socrates discusses a situation in which Simmias is taller than Socrates but smaller than Phaedo in virtue of the presence in him both of largeness and of smallness. Socrates makes the following claim: strictly speaking it is not true to say that 'Simmias overtops Socrates'. For this formulation would imply that Simmias is such as to overtop Socrates 'in virtue of his being Simmias' (τῷ Σιμμίαν εἶναι), whereas in fact it is 'in virtue of the largeness he happens to have' (τῷ μεγέθει ὃ τυγχάνει ἔχων). This might seem to suggest that there is a contrast between two mutually exclusive cases: (a) X is *intrinsically* or *essentially* F—the only case in which, strictly speaking, it is appropriate to say that X 'is F' or 'Fs'; (b) X bears Π to the form of F. And since on Plato's view the latter sort of formulation is always

[46] Code (1986, 430, cf. 426–427). Cf. Frede (1988, 31): Plato 'seems to think that, though there is such a thing as the nature or essence of an F, no particular F, no particular object of experience which is an F, is or has the nature or the essence of an F. And hence he thinks that he has to set up natures or essences as a second set of objects, in addition to the ordinary objects of experience, existing independently and separately from them'.

[47] See Sect. 7 above and Matthews (2009, 148–149) on what he calls 'Aristotle's Principle'.

[48] Think of the relation which the set {Jeoffry} bears to Jeoffry: the essence of the set (if there be any such thing) seems to consist precisely in this relation. Cf. Fine (1994, 4–5).

correct to describe sensible particulars, it would then follow that there is nothing that a sensible particular X is intrinsically or essentially; X merely bears Π to a number of forms.[49]

This interpretation would not be compelling, however. We can also understand the contrast drawn by Socrates in a partly different way, namely by supposing that there are *two ways* in which X may bear Π to the form of F: either (a') in such a way that X is intrinsically or essentially F—the only case in which, strictly speaking, it is appropriate to say that X 'is F' or 'Fs'—or (b') in such a way that X is not F, but merely bears Π to the form. In other words, bearing Π to the form is a necessary but not sufficient condition for being F.

This alternative interpretation is actually supported by the sequel of the *Phaedo* passage (103e–107a). There Socrates points out that in some cases

> it is not only the form itself that is entitled to its own name for all time [μὴ μόνον αὐτὸ τὸ εἶδος ἀξιοῦσθαι τοῦ αὑτοῦ ὀνόματος εἰς τὸν ἀεὶ χρόνον], but also something other which is not that form, but always has its character whenever it exists [ἀλλὰ καὶ ἄλλο τι ὃ ἔστι μὲν οὐκ ἐκεῖνο, ἔχει δὲ τὴν ἐκείνου μορφὴν ἀεί, ὅτανπερ ᾖ]'. (103e)[50]

That is to say, in some cases X bears Π to the form of F in such a way that 'F' is always true of X whenever X exists. Socrates offers the following examples: (i) X = snow, F = cold; (ii) X = fire, F = hot; (iii) X = 3, F = odd; (iv) X = 2, F = even; (v) X = soul, F = alive. Now, in all these five cases Socrates not only emphasizes that 'F' holds of X whenever X exists, but does not hesitate to express this in terms of X *being* F (103d, 104b, 106cd). This is especially interesting because examples (i) and (ii) clearly concern also *particular* fires and snowballs. Indeed, in the case of example (iii) Socrates even claims that such is the 'nature' (οὕτω πεφυκέναι, οὕτω πέφυκε, 104a) of the number 3—and this example is not flagged as different in any way from (i) and (ii) in this respect. So it is unclear why we should refrain from saying that F is an *essential* predicate of X in all the examples considered, including those which involve a reference to sensible particulars.[51]

So much for the *Phaedo*. Another text where scholars have claimed to detect Platonic anti-essentialism is *Republic* book 5, 476a–480a, where sensible particulars are described as possessing opposite properties—on the face of it, both simultaneously, in different respects, and successively, at different times—and therefore being 'tumbled about' in the mid-region between what is and what is not. It is, however, very unclear whether the opposite properties at stake in that passage are really meant to include *essential* properties.[52] Unfortunately I have no space to examine the passage

---

[49] Code (1986, 427 nn. 30–3) invokes the *Phaedo* passage as support for his interpretation. Notice that, according to him (1986, 426), only the F itself is essentially F: it is not just that forms cannot be essentially predicated of sensible particulars; a form cannot be essentially predicated of another form either. This strikes me as both intrinsically hard to accept and inconsistent with the way in which Socrates expresses himself in the passage.

[50] Trans. after Gallop (1975).

[51] See White (1978, 142–143).

[52] For a balanced discussion of this and other aspects of the passage, see White (1978), although I am sceptical about his conclusion that in the end the anti-essentialist interpretation is preferable.

in any detail.So I shall content myself with a cautious, generic conclusion: it is far from obvious that a *general*, unqualified rejection of essentialism can be safely attributed to Plato. The attribution, however, remains sound at least with regard to the *Timaeus*; and at least to that extent it is correct to claim that the essentialism of the *Categories* is anti-Platonic.

Acknowledgements I am very grateful to Paolo Fait and to audiences in Paris, Turin, Utrecht and New York—especially Riccardo Chiaradonna, Marko Malink and Jessica Moss—for extremely helpful comments and suggestions, both oral and written. Thanks also to all participants (in particular Matteo Caparrini, Wolfgang Carl and Luca Castagnoli) in seminars in Florence, Bologna and at the Scuola Normale Superiore in Pisa over the years.

References

Ackrill, J. L. (1963). *Aristotle: Categories and De Interpretatione*. Clarendon Press.

Ademollo, F. (2010). The principle of bivalence in Aristotle, *De interpretatione* 4. *Oxford Studies in Ancient Philosophy, 38*, 97–113.

Ademollo, F. (2018). On Plato's conception of change. *Oxford Studies in Ancient Philosophy, 55*, 35–83.

Ademollo, F. (2021). The anatomy of primary substance in Aristotle's *Categories*. *Oxford Studies in Ancient Philosophy*, 60.

Ademollo, F. (in preparation). *Categories* 1–3. Paper to be presented at the 22nd Symposium Aristotelicum, Oslo 2022.

Barnes, J. (1993). *Aristotle: Posterior Analytics* (2nd ed.). Clarendon Press.

Bonitz, H. (1870). *Index Aristotelicus* (2nd ed.). Akademische Druck- und Verlagsanstalt.

Burnyeat, M. F. (2001). *A map of* Metaphysics *Zeta*. Mathesis Publications.

Caston, V. (1999). Something and nothing: The Stoics on concepts and universals. *Oxford Studies in Ancient Philosophy, 17*, 145–213.

Caston, V. (2018). Aristotle on the reality of colors and other perceptible qualities. *Res Philosophica, 95*, 35–68.

Cavini, W. (2011). Vero e falso nelle *Categorie*. In M. Bonelli & F. G. Masi (Eds.), *Studi sulle* Categorie *di Aristotele* (pp. 371–406). Hakkert.

Chiaradonna, R., & Rashed, M. (2020). *Boéthos de Sidon: Exégète d'Aristote et philosophe*. De Gruyter.

Clarke, T. (2019). *Aristotle and the Eleatic one*. Oxford University Press.

Code, A. (1986). Aristotle: essence and accident. In R. E. Grandy & R. Warner (Eds.), *Philosophical grounds of rationality* (pp. 411–439). Clarendon Press.

Crivelli, P. (2017). Being-said-of in Aristotle's *Categories*. *Rivista di filosofia neo-scolastica, 3*, 531–556.

Duerlinger, J. (1970). Predication and inherence in Aristotle's *Categories*. *Phronesis, 15*, 179–203.

Edelhoff, A. L. (2020). *Aristotle on ontological priority in the* Categories. Cambridge University Press.

Fine, K. (1994). Essence and modality. *Philosophical Perspectives, 8*, 1–16.

Frede, M. (1978). Individuals in Aristotle. In Frede (1987, 49–71).

Frede, M. (1985). Substance in Aristotle's *Metaphysics*. In Frede (1987, 72–80).

Frede, M. (1987). *Essays in ancient philosophy*. University of Minnesota Press.

Frede, M. (1988). Being and becoming in Plato. *Oxford Studies in Ancient Philosophy* (Suppl.), 37–52.

Frede, M. (1992). Acerca de la noción de sustancia en Aristóteles, otra vez. *Méthexis, 5*, 79–98.

Frede, M., & Patzig, G. 1988. *Aristoteles "Metaphysik Z"* (2 Vols.). C.H. Beck.
Gallop, D. (1975). *Plato: Phaedo.* Clarendon Press.
Heinaman, R. (1981). Non-substantial individuals in the *Categories. Phronesis, 26*, 295–307.
Henry, D. P. (1991). *Medieval mereology*. Grüner.
Kahn, C. H. (1978). Questions and categories. In H. Hiż (Ed.), *Questions* (pp. 227–278). Reidel.
Katz, E. (2017). Ontological separation in Aristotle's *Metaphysics. Phronesis, 62*, 26–68.
Kohl, M. (2008). Substancehood and subjecthood in Aristotle's *Categories. Phronesis, 53*, 152–179.
Loux, M. J. (2015). An exercise in constituent ontology. In G. Galluzzo & M. J. Loux (Eds.), *The problem of universals in contemporary philosophy* (pp. 9–45). Cambridge University Press.
Matthews, G. (2009). Aristotelian categories. In G. Anagnostopoulos (Ed.), *A companion to Aristotle* (pp. 144–161). Blackwell.
Mignucci, M. (1993). La sémantique des termes généraux chez Aristote. *Revue Philosophique de la France et de L'etranger, 183*, 355–373.
Mignucci, M. (2007). *Aristotele: Analitici secondi. Organon IV*. Laterza.
Nuchelmans, G. (1973). *Theories of the proposition. Ancient and medieval conceptions of the bearers of truth and falsity*. North-Holland.
Perin, C. (2007). Substantial universals in Aristotle's *Categories. Oxford Studies in Ancient Philosophy, 33*, 125–144.
Rapp, C. (in preparation). Essentialism in Aristotle's *Categories*: Some queries and suggestions. https://www.academia.edu/40326251/Essentialism_in_Aristotles_Categories_Some_Queries_and_Suggestions
Schaffer, J. (2009). On what grounds what. In D. J. Chalmers, D. Manley, & R. Wasserman (Eds.), *Metametaphysics. New essays on the foundations of ontology* (pp. 347–383). Oxford University Press.
Sellars, W. (1957). Substance and form in Aristotle. *Journal of Philosophy, 54*, 688–699. Reprinted in Sellars W. (1967). *Philosophical perspectives: History of philosophy* (4th ed., pp. 108–118). C.C. Thomas.
Shapiro, H., & Scott, F. (1966). Walter Burley's *De Toto et Parte. Archives d'histoire doctrinale et littéraire du Moyen Age, 41*, 299–303.
Sirkel, R. (forthcoming). Ontological priority and grounding in Aristotle's *Categories*. In M. Roques (Ed.), *Grounding in medieval philosophy*. Brill. https://www.academia.edu/472499/Ontological_Priority_and_Grounding_in_Aristotles_Categories_
Striker, G. (2011). A note on the ontology of Aristotle's *Categories*, Chapter 2. In B. Morison & K. Ierodiakonou (Eds.), *Episteme, etc.: Essays in honour of Jonathan Barnes* (pp. 141–150). Oxford University Press.
Wedin, M. V. (2000). *Aristotle's theory of substance*. Oxford University Press.
Weidemann, H. (2014). *Aristoteles: Peri hermeneias* (3rd edn.). De Gruyter.
White, F. C. (1978). The *Phaedo* and *Republic* V on essences. *Journal of Hellenic Studies, 98*, 142–156.

Aristotle on Common Axioms

Vincenzo De Risi

1 Introduction: Common Axioms and Universal Science

In the *Posterior Analytics*, with the intention of laying the foundations to a general account of scientific demonstration, Aristotle presented a rich and important theory of the first principles (ἀρχαί) of science. Aristotle's main polemical target was Plato, who famously argued time and again that only philosophy (or dialectics) could provide the foundations for all other disciplines.[1] Aristotle, by contrast, advocated the autonomy of non-philosophical sciences and intended to secure *specific* foundations for each of them individually, on account of the fact that each science investigates a specific *genus* of things (τὸ γένος ὑποκείμενον), and that there is no genus common to them all. For Aristotle, the worst methodological mistake in science consisted in mixing up different genera (μετάβασις εἰς ἄλλο γένος), as in employing the results established in one science to draw conclusions in the domain of another.

Most of Aristotle's efforts in his epistemological works were accordingly devoted to discussing and expanding the *proper* principles of science, which he understood as the veritable foundations of each discipline. Aristotle called them *theses* (θέσεις), and further divided them into *definitions*, *hypotheses*, and possibly *postulates* and yet other principles. The *Posterior Analytics* are mostly devoted to explaining the epistemic role of these proper principles. What comes as striking is rather to find out that, in addition to the above-mentioned principles proper to each science, Aristotle also posited some principles *common* to a plurality of disciplines. He called these principles *axioms*, or *common axioms* (κοινὰ ἀξιώματα).

Common axioms seem to have a twofold origin in Aristotle's system. On the one hand, they clearly stem from the Academic discussions on dialectics as the

[1] See for instance Plato, *Resp.* H, 533b–534e.

V. De Risi (✉)
Laboratoire SPHère, Université de Paris Cité, Paris, France
e-mail: vincenzo.derisi@gmail.com

F. Ademollo et al. (eds.), *Thinking and Calculating*, Logic, Epistemology, and the Unity of Science 54, https://doi.org/10.1007/978-3-030-97303-2_3

foundation of all sciences. Aristotle, to be sure, wanted to oppose these Platonic claims, but agreed that there was some truth in them, and himself developed a theory of first philosophy as a grounding discipline *of sorts*. Aristotle's first philosophy deals indeed with very general principles, such as the Principle of Contradiction or the Principle of Excluded Middle, which apply to every object and must be accepted in every science. These logical and ontological principles are arranged by Aristotle among the ἀξιώματα.

On the other hand, the theory of common principles had a mathematical source. In Plato's time, mathematics was not considered to be a unitary discipline, and it was rather a name encompassing the different sciences of plane geometry, stereometry, arithmetic, and sometimes also optics, astronomy, harmonics, and others.[2] Similarly, Greek mathematicians considered solid bodies, plane figures, lines, points, and numbers, as objects of different sorts, with few or no connections to one another—objects belonging to different *genera*, as Aristotle would have said.[3] This general picture of the mathematical sciences and their plural objects was however challenged during Aristotle's lifetime. Aristotle states, in fact, that some mathematicians had begun to prove theorems that could be universally applied to different kinds of mathematical objects, and in this way they had generalized previous theories that proved in different ways the same theorem for lines, plane figures and numbers. Aristotle's claim is confirmed by Proclus' testimony that Eudoxus had "increased the number of general theorems", thereby unifying mathematical sciences that had been, up to that point, separate disciplines.[4] Aristotle's explicit example of these general theorems comes from the theory of proportions (ἀναλογίαι), and while it is uncertain which theory of proportions Aristotle had in mind, one can make the educated guess that this was indeed Eudoxus' theory.[5] The same theory, or a close version of it, was the one later expounded in Euclid's Fifth Book of the *Elements*. Eudoxus' theory of proportions dealt with a novel kind of mathematical object, called *magnitude* (μέγεθος), which encompassed lines, plane figures, solid bodies, and possibly angles and other

[2] Plato still distinguished, for instance, plane geometry and stereometry as two different disciplines. See *Resp.* H, 528b-c.

[3] See De Risi (2021a) as an example of a multi-sorted (as opposed to set-theoretic) approach to Greek mathematics.

[4] See Proclus, *In primum Euclidis* 67, which seems to be based on a book by Eudemus on the history of ancient mathematics. The fact that Plato did not recognize common axioms as principles of mathematics (they do not fit at all with his description of ὑποθέσεις in the important passages of *Meno*, 86e and *Resp.* Z, 510d; H, 533b-c), may confirm that they were first thematized in Eudoxus' time, shortly after the composition of the *Republic*. Some principles similar to Euclid's common notions are however (very obliquely) mentioned in *Parm.* 154b and *Theaet.* 155a. These passages may offer some evidence that these principles were already being discussed by the mathematical community.

[5] In the early *Top.* Θ 3, $158^{b}29$–35, Aristotle mentions an older (pre-Eudoxian) anthyphairetic theory of proportions; for a similar passage in the *Metaphysics*, see Pritchard (1997). On Aristotle and the theory of proportions, see Mendell (2007) and Rabouin (2016). On Eudoxus' theory of proportions, see the classic Knorr (1975). For a discussion of Eudoxus' dating and involvement with the Academy, see for instance Zhmud (1998).

mathematical items. As a consequence, a Eudoxian "general" theorem on proportions may be employed to find the solution of an array of related problems in different mathematical sciences. This new theory was extremely powerful and represented an important turning point in the history of Greek mathematics.[6]

The formulation of general theorems reverberated in the theory of principles, since such results should be themselves grounded on some basic assumptions. Euclid's *Elements*, indeed, are prefaced by a set of principles labeled "common notions" (κοιναὶ ἔννοιαι) that seem to refer to any mathematical objects whatsoever and are applied, in the course of the proofs, to lines, surfaces, solids, angles, magnitudes in general, as well as numbers. The third of these common notions, for instance, states that *equal things subtracted from equal things result in equal things*. This principle (henceforth CN3) is mentioned by Aristotle several times as the main example of a "common axiom", and he explicitly says that this kind of general principles were called "axioms" (ἀξιώματα) *by the mathematicians*.[7] We may easily surmise, then, that some common axioms, or common notions, had been spelled out in Aristotle's time (possibly by Eudoxus himself) in order to ground general theorems. This new kind of principle was discussed by Aristotle and was later incorporated into Euclid's *Elements*.[8]

In an isolated passage, Aristotle also refers to a "universal mathematics", i.e. a καθόλου μαθηματική (later to be called *mathesis universalis*), that was *common* (κοινή) to several genera of objects.[9] We have no independent testimony attributing

[6] In *An. post.* A 5, 74^{a}17–25, Aristotle says that a theorem of the theory of proportions used to be proven by ancient mathematicians through different demonstrations, bearing on numbers, lines, solids and time intervals, respectively, but more recent mathematicians had unified all these demonstrations into one. The theorem that Aristotle has in mind may well be the one of the alternation of proportion (cf. also *An. post.* B 17, 99^{a}8–11, discussed later at *note* 36). In Euclid's *Elements*, such a theorem has *two* demonstrations: a first which is valid for lines, surfaces and solids (i.e. magnitudes in general), given in *Elements* V, 16; and a second, valid for numbers, given in *Elements* VII, 13. The *Elements*, therefore seem to represent a stage of generalization predating the proof hinted at by Aristotle. But also more in general, Book X of Euclid's *Elements*, which is often attributed to Theaetetus, still classifies ratios according to the dimensionality of the objects involved (lines vs. surfaces). This is not the case with the theory of proportions expounded in Book V of the *Elements* (commonly attributed to Eudoxus) which offers a unified treatment of ratios between "magnitudes", be they lines, surfaces, solid bodies, or other geometrical objects. On the other hand, *Elements* X, 5 and 6 mix together magnitudes and numbers. It is clear that all these theorems express different conceptions of the generality and scope of the proofs.

[7] *Metaph.* Γ 3, 1005^{a}18–20. It is not entirely clear why Euclid called κοιναὶ ἔννοιαι what Aristotle called κοινὰ ἀξιώματα, but the term may have enjoyed some flexibility, and Aristotle himself calls these principles κοιναὶ δόξαι in *Metaph.* B 2, 996^{b}28. It has been suggested that the term κοινὴ ἔννοια is Stoic, and that the label of these principles at the beginning of the *Elements* was first given, or modified, in post-Euclidean times. See Todd (1973).

[8] In De Risi (2021b), I have claimed that Euclid's first three common notions were originally devised in order to ground a theory of equivalence for plane figures, which is expounded in *Elements* I, 35–45. They were later generalized so as to encompass other kinds of mathematical objects. I have also claimed that at the times of Aristotle and Euclid it is likely that only the first three common notions of the *Elements* had been spelled out, while the others were later additions to the text.

[9] The main passages in which Aristotle mentions a universal mathematics are E 1, 1026^{a}25–27, and K 7, 1064^{b}7–9, the second of which may be spurious and depending on the first. A possible further

to Eudoxus or other mathematicians of the fourth century the idea of a universal mathematics, and it is debated among the interpreters whether the general theorems mentioned by Aristotle were ever put together into a "universal mathematics" as an independent science.[10]

Be this as it may, Aristotle clearly envisaged the possibility of conceiving of a unitary science of mathematics (encompassing both arithmetic and geometry) and discussed it in several passages of his work. Once again, Aristotle's main targets seem to be various theories developed in the Academy. On the one hand, universal mathematics may have appeared to Aristotle as an analogue of Platonic dialectics for mathematics, that is to say a fabricated super-science intended to provide a foundation for mathematical disciplines. On the other hand, it seems that universal mathematics had suggested several ontologies of mathematical objects. Aristotle opposed some reductionists ideas aimed to derive geometrical magnitudes from numbers, for instance, that may have been advanced by Pythagorizing Platonists, and that were probably connected with a parallel reduction of geometry to arithmetic—considered itself as the *mathesis universalis*.[11] It seems, also, that some other philosophers suggested that universal mathematics was not to be identified with arithmetic or geometry, and was rather a different science having *objects of its own*, that were neither numbers nor magnitudes but higher-level entities. Aristotle strongly opposed this view as well, and denied that there might be a mathematical science dealing with

> no number, no line, no surface, no solid, but something else.[12]

These criticisms notwithstanding, Aristotle clearly praised the mathematical power of Eudoxus' theory of proportions, and explicitly thematized the notion of

---

reference may be *Metaph.* Γ 2, 1004^a2–9, in which Aristotle mentions a "first part of mathematics". On the development of this notion from Aristotle to the early modern age, see Rabouin (2009).

[10] In particular, the important Rabouin (2009) sheds doubts on the hasty conclusion that Aristotle's discussion of general theorems was connected with his references to a *mathesis universalis*, as well as on its identification with any extant mathematical theory. The Eudoxian theory of proportions expounded in Book V of the *Elements*, insofar as it deals with any kind of magnitudes (lines, surfaces, solid bodies) has been pointed out as a specimen of Aristotle's *mathesis universalis*. Yet, it does not apply to numbers, and in this respect it appears more as a fragment of a "universal geometry" of sorts than as a universal mathematics in Aristotle's sense—and in fact, in the passage in which Aristotle mentions universal mathematics, he explicitly *contrasts* this latter with geometry.

[11] A large part of the discussion of *Metaph.* M 9, in fact, revolves around Speusippus' views, and in general on whether the number may be considered a principle for lines, surfaces and solids (and so, possibly, arithmetic being the first mathematical science), or whether these three geometrical genera may be reducible to one another. A passage from Diog. Laert. IV, 2 (frag. 70 in Tarán, 1981) states that Speusippus was the first who considered what is *common* (κοινόν) to several μαθήματα, which might be a reference to a *mathesis universalis* (but also a tamer reference to μαθήματα as sciences in general).

[12] See *An. post.* A 24, 85^a37–85^b1. A similar statement is read in *Metaph.* M 2, 1077^a9–12, in which Aristotle says that the mathematicians formulate some universal axioms or theorems (the Greek only has ἔνια καθόλου), that should apply to substances that are "neither number, nor point, nor magnitude, nor time" (but, Aristotle adds, such substances do not exist). Cf. also *Metaph.* M 3, 1077^b17–22: "Just as universal propositions in mathematics are not about separate objects over and above magnitudes and numbers, but are about these...".

μέγεθος in his philosophy of mathematics.[13] He also praised general theorems applying transversally to several kinds of objects, and intended to elaborate a theory of science that could accommodate these important results.

Aristotle needed a theory of mathematics that could oppose the Platonic "univocation" of this science and its ontology, and at the same time account for the new discoveries pushing towards a *mathesis universalis.* It is clear, therefore, that Aristotle perceived similar philosophical issues in general mathematics and first philosophy. It comes therefore as no surprise that he conceived of a theory of the "common" principles of science (axioms) that could account for both these hyper-generic disciplines.

This meant that common axioms such as CN3, that apply to numbers, lines, surfaces, and solids, do not apply to "something else" over and above the previous items. It also meant that such axioms may not ground a science of their own, the καθόλου μαθηματική, as if they were the *proper principles* of a higher genus. These common principles, therefore, had to be sharply distinguished from the *theses*. Similarly, the logical and ontological principles of dialectics, such as the Principle of Contradiction, cannot be said to apply to a definite genus of things nor to ground a single science, and are therefore axioms as well. Aristotle's *Posterior Analytics* attempted to provide a unified theory of these κοινὰ ἀξιώματα in mathematics, first philosophy and science in general. This theory had the aim, on the one hand, of explaining the universal applicability of axioms; and, on the other, of reassessing the claim that each science is nonetheless grounded on its own proper foundations.

In what follows, I will sketch Aristotle's theory of axioms and the way in which it may accomplish his epistemological goals. We will first look at the main texts in which the theory of common axioms is presented (Sect. 2). Then we will explore Aristotle's original solution to the problem of reconciling the existence of common principles with the idea that each science should be grounded on proper principles (Sect. 3). We will further look at the role of common axioms in Aristotle's theory of demonstration, and offer a standard reading of it: the "schematic" interpretation of axioms (Sect. 4). This latter reading, however, presents some serious exegetical issues, and I will defend an alternative interpretation of Aristotle's theory (the "inferential" interpretation of common axioms), that may solve these problems (Sect. 5). Finally, I expound some more general, but also more conjectural consequences of the inferential interpretation for Aristotle's theory of demonstration (Sect. 6).[14]

---

[13] An important question is whether, according to Aristotle, the notion of a *magnitude* constitutes a veritable genus or a weaker form of unity among the objects of geometry, and therefore whether geometry should be considered to be a unitary discipline. In several passages, such as the one in *Metaph.* Δ 13, Aristotle seems to incline towards the idea that lines, surfaces and solid bodies represent three different *summa genera* which are irreducible to one another, even though they may all be labeled as *magnitudes* (μεγέθη). This is the opinion, for instance, of Apostle (1952, 105–106). A contrary stance seems however to be taken in the array of objections in *Metaph.* M 9, $1085^{a}9$–23, mentioned above. I will not discuss this difficult issue here, since it does not affect the interpretation of Aristotle's common axioms.

[14] The most extensive treatments of common axioms in recent literature are to be found in Mueller (1991) and McKirahan (1992). My own interpretation of axioms is completely at odds with McKirahan's—who distinguishes between logical and mathematical axioms; accepts a schematic reading

2 Logical and Mathematical Axioms

Aristotle's main discussion of axioms takes place in two passages in the *Posterior Analytics* and one passage in the *Metaphysics*.[15]

In *An. post.* A 2, Aristotle offers a first classification of the principles of science, and distinguishes between theses and axioms (ἀξιώματα), by saying that both kinds of principles are unprovable, but axioms have the further feature that they must be grasped by anyone who is going to learn anything whatever. He does not provide any examples in this connection, but a few lines above he explicitly spells out the Principle of Excluded Middle.[16] The characterization of axioms as principles needed to learn anything at all fits well with the status of the highest logical principles such as the Excluded Middle or the Principle of Contradiction. In this passage, axioms are not qualified as "common", nor are theses ascribed to a single genus. The whole context seems much more philosophical than mathematical.[17]

Aristotle offers a classification of principles again in *An. post.* A 10. Here the context is more explicitly mathematical, and the divide between theses and *common* axioms (κοινὰ ἀξιώματα) falls on their being either genus-specific or common to several sciences. The discussion of theses is expanded into an array of cases that had not been considered in the previous chapter, and new terms as "postulates" (αἰτήματα) or specific "hypotheses" (ὑποθέσεις, in a different sense from in the previous chapter) are introduced. The philosophical "definitions" (ὁρισμοί)

of them; denies their primary inferential role; takes categories to be genera; and thinks that Aristotle embraced a fully-fledged *mathesis universalis*. Mueller's essay has the great advantage of providing some of the mathematical background of Aristotle's theory of axioms, but still rejects the possibility of an inferential interpretation of axioms. I will deal with all these points in the course of the present essay.

[15] The term ἀξίωμα only rarely appears in Aristotle's works, the main occurrences being limited to the technical notion employed in the *Analytics* and the *Metaphysics*. It is sometimes also employed, however, in the general, non-technical sense of a premise in *Top.* Θ 1, 156^{a}23, or *Soph. El.* 24, 179^{b}14. In *An. pr.* B 11, 62^{a}12–13 an ἀξίωμα is an ἔνδοξον, this term possibly betraying the dialectical origin of these principles. It seems, therefore, that Aristotle gave a more technical meaning to the term only in his later works. This may draw on the constitution of a mathematical theory of common axioms (possibly devised by Eudoxus) during Aristotle's lifetime. On Aristotle's mathematical terminology, see Einarson (1936).

[16] In *An. post.* A 1, 71^{a}11.

[17] "An immediate deductive principle I call a thesis if it cannot be proven but need not be grasped by anyone who is to learn anything. If it must be grasped by anyone who is going to learn anything whatever, I call it an axiom (there are principles of this kind); for it is of this sort of principle in particular that we normally use this name. A thesis which assumes either of the parts of a contradictory pair – what I mean is that something is or that something is not – I call a hypothesis. A thesis which does not I call a definition" (*An. post.* A 2, 72^{a}14–24; transl. Barnes modified). Note, however, that the Greek uses the indefinite ὅστις where Barnes (and many others) translate "anything whatever". The meaning might also be "some (unspecified) thing", referred to a particular science, and in the latter interpretation a principle would be an axiom if there is a science for which it is necessary to learn it. Given Aristotle's example in this connection (the Excluded Middle), as well as the parallel passages in the *Metaphysics* referring to the Principle of Contradiction, I endorse Barnes' stronger translation.

mentioned in A 2 are here substituted by a reference to the "terms" (ὅροι) employed in a demonstration, and ὅροι is indeed the word commonly found at the beginning of the lists of definitions in Greek *mathematical* texts. The examples are mathematical throughout, and as an instance of a *common axiom* we do not find the Excluded Middle but rather the statement on the subtraction of equals, CN3.[18]

The reason for this double presentation in the *Posterior Analytics* is unclear. It is possible that Aristotle was first introducing the general ideas of his theory of science (A 2), and then returning in greater detail to the same topic (A 10), but there are too many small discrepancies between the two chapters to suggest that we are in fact confronted with two different versions of the same discussion, one philosophical and one mathematical, that were put together into a single treatise. Indeed in the two chapters we seem to witness the twofold origin of Aristotle's interest on the topic.

This does not mean, however, that the *Posterior Analytics* offer two different theories of axioms, such that the logical "axioms" needed to be grasped in order to learn anything whatever are different principles from the "common axioms" employed in the mathematical sciences. On the contrary, the double discussion of logical and mathematical axioms is intertwined in the whole text of the *Posterior Analytics*, showing that we are confronted with different aspects of the same kind of principles rather than with different sorts of axioms. The two examples of CN3 and the Excluded Middle are very recurrent in Aristotle's works, and they represent the standard instances of axioms in several contexts.[19] A passage in *An. post.* A 11 explicitly mentions both the Excluded Middle and CN3 as falling under the same label:

> All the sciences associate with one another in respect of the common principles [κοινά] (I call common the principles which they use as the basis from which to demonstrate—not those about which they prove nor what they prove); and dialectic associates with them all, and so would any science which attempted to give universal proofs of the common principles (e.g. that everything is asserted or denied, or that equals from equals leave equals, or any principles of this sort).[20]

---

[18] Here is the main passage on common axioms: "Of the (principles) used in the demonstrative sciences some are proper to each science and others common—but common by analogy, since something is only useful insofar as it is in the genus under the science. Proper: e.g. that a line is such-and-such, and straight so-and-so. Common: e.g. that if equals are removed from equals, the remainders are equal" (*An. post.* A 10, $76^{a}36$–41; transl. Barnes modified). The term "common axioms" does not appear as such in this passage, and Aristotle simply adds a few lines below that the "common things" from which all demonstrations begin are called axioms (τὰ κοινὰ λεγόμενα ἀξιώματα).

[19] The Principle of the Excluded Middle appears in *An. post.* A 1, $71^{a}11$; A 11, $77^{a}22$ and $77^{a}30$; A 32, $88^{a}36$-$^{b}1$. The passage in *An Post.* A 11, $77^{a}22$ has a clear correspondence in *An Pr.* B 11, $62^{a}13$, which is the only reference to axioms in the latter work. The common axiom CN3 is mentioned in *An. post.* A 10, $76^{b}21$ and $76^{a}41$–42; A 11, $77^{a}26$–31; *An. pr.* A 24, $41^{b}22$–23; *Metaph.* K 4, $1061^{b}20$.

[20] *An. post.* A 11, $77^{a}26$–31. I use and modify Barnes' translation, who renders "κοινά" as "common items" rather than as "common principles" or "common axioms". Note that the passage is arguing about Platonists, and it may be read as stating that *they* should think that logical and mathematical axioms fall under the same label. I do not think that this is the case, and rather believe that Aristotle himself endorsed such a view. But the above dialectical reading may account for some different interpretations of Aristotle's theory (such as Theophrastus').

The passage is also relevant insofar as it explicitly connects both *logical and mathematical* common axioms to "all the sciences" (πᾶσαι αἱ ἐπιστῆμαι), and therefore seems to envisage an unrestricted notion of their "commonality" (which is probably in line with *An. post.* A 2, and its claim on the need of axioms in order to learn anything whatever).

The convergence of the two characterizations of axioms is further developed in the third important passage dealing with this kind of principles, in *Metaph.* Γ 3. Here the axioms at stake are the logical and ontological principles, such as the Excluded Middle and the Principle of Contradiction. Aristotle famously says that the Principle of Contradiction is the best known principle, the one on which no one can err, the principle that must be known in order to learn anything else, and is a perfectly unhypothetical principle.[21] These (and other) features of this axiom dovetail with the characterization of axioms in *An. post.* A 2 (first principles needed in order to learn anything whatever). Nonetheless, a large part of this chapter of the *Metaphysics* is in fact devoted to belaboring the point that such logical principles are investigated by the philosopher insofar as they pertain to being *qua* being and, therefore, hold for all genera whatsoever:

> It is obvious that the investigation of these axioms pertains to one science, namely the science of the philosopher, for they apply to all existing things, and not to a particular class separate and distinct from the rest. Moreover all thinkers employ them – because they are axioms of being *qua* being, and every genus possesses being – but employ them only in so far as their purpose require; i.e., so far as the genus extends about which they are carrying their proofs. Hence these axioms apply to all things *qua* being (for this is what is common [κοινόν] to them), it is the function of him who studies being *qua* being to investigate them as well.[22]

The main argument in the passage is therefore precisely that such logical axioms are *common* and hyper-generic, just as was the case with *An. post.* A 10.

In *Metaph.* Γ, mathematical principles such as CN3 are not mentioned, but Aristotle claims that *equality* is a notion common to several genera, and therefore it is investigated by first philosophy. In this respect CN3, articulating a fundamental property of equality, seems to be regarded by Aristotle as being on the same footing of the other logical-ontological principles.[23]

Further on, the long discussion about the peculiar "demonstration" by ἔλεγχος of the Principle of Contradiction carried out in *Metaph.* Γ 3 is applied (with important modifications) to the Principle of the Excluded Middle as well (in *Metaph.* Γ 7), and we have seen above that in *An. post.* A 11 Aristotle explicitly mentioned the issue of the demonstrability of the Excluded Middle together with that of CN3,

---

[21] *Metaph.* Γ 3, 1005^b12–19.

[22] *Metaph.* Γ 3, 1005^a21–29 (transl. Tredennik). In a similar vein, Aristotle begins the *Rhetoric* (A 1, 1354^a1–4) by stating that there are principles for the particular sciences and principles that are common (κοινά) to all. Just as *dialectic* deals with the latter (here the reference seems to be the discussion of common axioms in the *Metaphysics*), rhetoric itself, conceived of as the "countermelody" (ἀντίστροφος) of dialectic, is the art of expressing any science and any argument, irrespective of their particular genus. Aristotle does not mention explicitly axioms in this passage, and the κοινά may refer to either notions or principles.

[23] *Metaph.* Γ 2, 1004^a18.

and indeed of all common axioms (δεικνύναι τὰ κοινά). Whatever we may think of these dialectical arguments on principles, the very fact that they seem to concern both logical and mathematical axioms shows that Aristotle envisaged a uniform treatment of ἀξιώματα in general.

Based on these considerations, it seems reasonable to claim that the *Posterior Analytics* and the *Metaphysics* expound the same theory of axioms, just as it seems safe to conclude that Aristotle regarded logical axioms, such as the Excluded Middle, and mathematical axioms, such as CN3, as instances of the same kind of principles.[24]

At first glance, this claim might come as a surprise, since the two kinds of axioms seem to possess very different epistemic features. The Principles of Contradiction and the Excluded Middle seem to be *much more general* than an axiom on the subtraction of equals, that only applies to mathematical objects. More than this: in the passage in *Metaph.* Γ, Aristotle can be seen to be arguing that a principle needed to know anything whatever (i.e. an axiom according to *An. post.* A 2) must be a principle common to all sciences (i.e. an axiom according to *An. post.* A 10). The inverse implication however is far more problematic to justify, and it is hard to see how CN3 could possibly be a principle, without which one could not learn anything whatever. As a matter of fact, in the same *Metaph.* Γ Aristotle states that the Principle of Contradiction is *the principle of all other axioms* (ἀρχὴ τῶν ἄλλων ἀξιωμάτων πάντων), therefore explicitly stressing its preeminence over all common principles such as CN3.[25]

Given the apparent heterogeneity of the logical and mathematical principles, and their double origins in dialectical discussions and Eudoxian theories, both ancient and modern interpreters have detected a tension in Aristotle's treatment of axioms and attempted to separate once more the pieces of a theory that Aristotle had attempted to join together. Since the generation after Aristotle, a difference has been established between the two kinds of common axioms, and we have a reference by Themistius on the fact that Aristotle's student Theophrastus had himself already distinguished between common axioms that are truly universal, such as the logical-ontological principles, and common axioms that apply to mathematics only.[26] Theophrastus' reading was generally accepted by ancient, medieval and early modern commentators, and it is still endorsed by many modern interpreters.[27]

---

[24] I may add that in a (probably spurious) passage in *Metaph.* K 4, 1061^b20, the theory of logical axioms expounded in *Metaph.* Γ is explicitly connected with CN3, this being the only reference to this axiom in the whole *Metaphysics*.

[25] *Metaph.* Γ, 1005^b34–35.

[26] The passage is in Them. *In an. post.* 10: "Theophrastus defines an axiom as follows. An axiom is a statement either concerning things of the same genus, such as 'equals subtracted from equals', or concerning absolutely everything, such as 'either the affirmation or the negation'".

[27] The only Aristotelian passage that may help such an interpretation, as far as I know, is in *An. post.* A 32, 88^a36–88^b3, where Aristotle distinguishes common principles such as the Principle of Contradiction, and principles pertaining to a genus, such as "quantity". The principles pertaining to a genus are, I would say, hypotheses and definitions (theses), and Aristotle is using "quantity" in an equivocal sense to mean "either numbers or magnitudes". But the text is open to other readings, such as understanding that both kinds of principles are common axioms, the first being logical

Modern scholars have reinforced Theophrastus' interpretation by noting that in various contexts Aristotle employed the word "common" (κοινόν), in the double sense of "common to all" and "common to some", without bothering too much to distinguish between the two meanings. This happens, for instance, in Aristotle's famous distinction between proper sensibles and common sensibles. The latter are sometimes regarded as common to *all* senses, and sometimes simply as sensibles that can be perceived by *more than one sense.*[28] It has been surmised, therefore, that in the case of common axioms mentioned in *Analytics* a similar meaning of commonality is at play, and that Aristotle did not distinguish between principles that are common to all sciences (the logical principles) and principles common to more than one science (the mathematical principles). He may have intended κοινόν in the weaker sense of "common to several items, and possibly to all of them".

It must be stressed, in any case, that the (alleged) difference between the two kinds of common axioms rests entirely on conceptual interpretation, and it is not grounded on any explicit Aristotelian claim. On the contrary, it seems to go directly against Aristotle's express intention to equate the epistemic role of logical and mathematical axioms. Moreover, we have seen that in *An. post.* A 11 Aristotle also explicitly refers to CN3 as a mathematical example of a common axiom applying to *all* sciences (πᾶσαι αἱ ἐπιστῆμαι). In the final sections of this paper, therefore, I propose a unitary interpretation of the role of κοινὰ ἀξιώματα in Aristotle's epistemology, that takes "commonality" in the stronger meaning of (formal) generality.

Before offering such a reading, however, we must look at Aristotle's arguments for accepting common axioms among principles in the first place.

3 Generic Unity and Unity by Analogy

Aristotle's theory of common axioms was devised in order to account for universal logical and mathematical principles in the context of a theory of science with the aim of grounding individual (i.e. genus-specific) disciplines in proper principles. To this effect, Aristotle developed some important epistemological strategies, which were especially deployed in the case of the *mathematical* common axioms.

Mathematics, for Aristotle, deals with quantity (ποσόν), and a principle such as CN3 does indeed express a general claim about equality, sums and differences, which are the main notions that Aristotle connects with quantity as such.[29] In several

---

common axioms, and the second being mathematical common axioms (for further interpretations see Mignucci [1975, 627–629] and Barnes [1993, 196]).

[28] Aristotle's examples of common sensibles are often sensibles perceived by some senses (e.g. *De sensu* 4, 442^{b}7, where he says that common sensibles are common "at least to touch and sight"). He explicitly characterizes common sensibles as sensibles perceived by all senses in *De an.* B 6, 418^{a}19 (cf. also *De an.* Γ 1, 425^{a}15). The neat comparison between this use of κοινόν in psychology and the one concerning axioms is in Barnes (1993, 100).

[29] See for instance *Cat.* 6, 6^{a}26–27: "Most distinctive of a quantity is its being called both equal and unequal" (transl. Ackrill).

passages, Aristotle explains that everything *having parts* may be considered to be a quantity, and further distinguishes between *continuous* and *discrete* parts of a whole. A quantity, the parts of which are continuous, is called a *magnitude* (μέγεθος); a quantity with discrete parts is called a *number* (ἀριθμός).[30] This important distinction between quantities demarcates two different mathematical sciences: geometry, dealing with magnitudes, and arithmetic, dealing with numbers.[31] A common axiom such as CN3 should apply to the whole mathematical domain of geometry and arithmetic, and to the mixed, subordinate sciences such as astronomy or music.

In spelling out this classification of the mathematical disciplines, Aristotle deals with quantity *as if* it were a genus, with specific differences (continuity and discreteness) and subordinate species (magnitudes and numbers). This would open the perspective that a single science (universal mathematics) could investigate the single genus of quantity, and principles such as CN3 would be *theses* (i.e. *proper* principles) of a *mathesis universalis* rather than common axioms. Geometry and arithmetic would be downplayed to the role of subordinate sciences of the καθόλου μαθηματική.

Aristotle, however, resisted this idea. He claimed that quantity is a *category* rather than a genus, and therefore it does not produce the kind of unity that a genus imposes on its subordinate species. According to Aristotle, since there is no *genus* of quantity, there is not, and cannot be, a science of mathematics in general. Instead, magnitudes and numbers are themselves *summa genera*, that do *not* communicate with one another thanks to superordinate kinds of objects (alleged universal quantities, which would be "no number, no line, no surface, no solid, but something else"). It can therefore be concluded that, for Aristotle, mathematical sciences (in the plural) deal with numbers and magnitudes (and other items); but there is no *one* science (mathematics) that deals with a single genus of objects (quantities).

An insightful connection may be made with Aristotle's extended discussion, in the *Posterior Analytics*, on the epistemic need to find the *primary subject* of a property. The property of having an interior angle sum of two right angles, for instance, belongs to triangles as its primary subjects. Aristotle insists that *isosceles* triangles cannot be said to have this property per se, insofar as they have this certain angle sum *qua* triangles and not *qua* isosceles. In this case, the genus (triangle) is the primary subject of the property, even though, of course, its species (such as the isosceles triangle) also share this property.[32] Should one ask, now, which are the primary subjects of CN3, it would be clear that they cannot be numbers, nor lines, nor surfaces, nor solids, since none of them may encompass the others. The common axiom applies indeed to all these objects, which are not, however, species of any common genus. But since the

---

[30] The main passages on the classification of quantities are *Cat.* 6 and *Metaph.* Δ 13.

[31] See for instance *An. Post.* A 7, 75^a38–39 on the separation between arithmetic and geometry, and the mistake of μετάβασις εἰς ἄλλο γένος which is made by proving a geometrical theorem by arithmetical means.

[32] See for instance *An. Post.* A 4, 73^a34–73b3, but the example of the isosceles triangle is recurrent in this and other works.

common genus is missing (the alleged "something else" mentioned by Aristotle) the κοινὰ ἀξιώματα (contrary to proper principles) *do not have any primary subject*.

Still, according to Aristotle, the generalization of magnitudes and numbers into "quantities" is not *entirely* nominal. Rather, magnitudes and numbers *do have* a sort of intrinsic commonality, which depends on the unity conferred to them by their belonging to the same category. The term "quantity", in short, is not an *equivocal* name, arbitrarily (νόμῳ) encompassing two altogether different kinds of objects. It is not univocal either, however, as if it expressed a veritable genus. In Aristotle's view, the category of quantity indeed offers a unity, but a *unity by analogy* (that is to say, *by proportion*) of numbers and magnitudes:

> Of the (principles) used in the demonstrative sciences some are proper to each science and others common—but common by analogy [κατ᾽ἀναλογίαν], since something is only useful insofar as it is in the genus under the science.[33]

This is Aristotle's breakthrough in the epistemology of common axioms. Geometry and arithmetic are different sciences, and they cannot be fused together into a universal mathematics. As such, they are actually grounded on *their own* specific principles (*theses*) that are *not* common to one another (e.g. the definition of a line, or the definition of a number). Nonetheless, there are relations among numbers and relations among magnitudes that are similar to one another. These relations can be captured by some principles expressing their common behavior. Such principles are common axioms, showing the structural similarities between different domains of objects, by analogy.[34]

Thus, for instance, "equal" means different things when referred to numbers or geometrical magnitudes. Numbers are generally conceived of, in Greek mathematics, as collections of units: and to be equal, in numbers, may therefore mean that two collections of units may be put in biunivocal correspondence. Geometrical magnitudes, on the other hand, have quite different conditions of equality: for instance, they may be dissected into equal collections of superimposable parts.[35] Yet, no matter how different the notions of equality among numbers and magnitudes may be, it is still true that equal parts, subtracted from equal wholes, make the remainders equal. This is a structural similarity between numbers and magnitudes concerning mereology and equality. As such, it may be expressed by a "common axiom" capturing the "analogy" among these two different domains. Other principles may be stated as well, and the collection of these principles suggests a kind of "unity" of the domain

---

[33] *An. post.* A 10, 76^{a}37–40 (transl. Barnes modified). The Greek word ἀναλογία is generally translated as "analogy" in philosophical contexts and as "proportion" in mathematical ones. It is the same notion, however, employed by Eudoxus and Aristotle for similar aims. Aristotle explicitly says that an analogical unity does not require a genus-specific unity in *Metaph.* Δ 6, 1017^{a}1–3.

[34] See for instance *Metaph.* N 6, 1093^{b}16–21: "For all things have connections and have unity by analogy. There is something analogous, indeed, in all categories of being. As the straight is to the line, so the plane to the surface, and, possibly, the odd to the number and the white to the color".

[35] *Cat.* 6, 6^{a}27–30. On numbers, see for instance the passage on the equal number of sheep and dogs in *Phys.* Δ 14, 224^{a}2–3. See again my De Risi (2021b) for a fuller treatment of the matter in Euclid's geometry.

of numbers and magnitudes, that may be formulated as the unity κατ'ἀναλογίαν of a category—the category of quantity. Aristotle himself makes the further example of the rule of alternation in proportions (his basic example of a theorem of universal mathematics), which is, and yet is not, the same for the two different genera of numbers and geometrical lines:

> E.g. why do proportionals alternate? The explanation in the cases of lines and of numbers is different—and also the same: as lines it is different, as having such-and-such a ratio it is the same. And so in all cases.[36]

This being and not-being the same is expressed in the formula κατ'ἀναλογίαν.

The treatment of *logical* axioms, such as the Principle of Contradiction and the Excluded Middle, is also based on the same idea. While, according to Aristotle's *Metaphysics*, there is no science common to being *qua* being, there are universal relations among beings that are the same in all genera. The fact that it is impossible for the same thing to hold good and not to hold good simultaneously for the same thing and in the same respect (i.e. the Principle of Contradiction in Aristotle's famous formulation)[37] expresses a relation among things that is structurally similar in all genera of objects.

We should be careful, of course, not to over-modernize Aristotle: the notions of structural identity and isomorphism (or the notion of a structure *tout court*) do not belong as such to Aristotle's epistemology. Nonetheless, the idea of a similarity of relations, which is the starting point of the modern notion of isomorphism and structure, is certainly to be found in Eudoxus' mathematical theory of proportions (ἀναλογίαι) as it is expounded in Euclid's *Elements*.[38] In such a theory, indeed, ratios

---

[36] *An. post.* B 17, 99^{a}8–11 (transl. Barnes). The example on alternation (ἐναλλάξ) is the same as Aristotle had mentioned in his description of the mathematical breakthrough of the invention of universal theorems in *An. post.* A 5, 74^{a}17–25 (see above, *note* 6). The passage in *An. post.* B 17 continues by remarking that not every case in which we use an identical term to express a relation (such as "proportional") is *analogous* to any other. There are more frequent cases of mere homonymy. Thus, when we say that two geometrical figures are "similar" and that two colors are "similar", here similarity is not predicated κατ'ἀναλογίαν in the two cases, for there is not any identical structure shared in the two cases. Indeed, colors and figures do not belong to the same category. Here is the passage: "The explanation of a color's being similar to a color and a figure to a figure is different for the different cases. Here similarity is homonymous: in the latter case it is presumably having proportional sides and equal angles; in the case of colors it is the fact that perception of them is one and the same (or something else of this sort)". A more general discussion on the analogy of principles (ἀρχαί), this time conceived of as physical and metaphysical causes, is to be found in *Metaph.* Λ 4. On the notion of analogy in Aristotle, see at least Courtine (2005) and Aubenque (2009).

[37] *Metaph.* Γ 3, 1005^{b}19–20.

[38] An important breakthrough towards modern structuralism took place in Leibniz' theory of relations and theory of "expression" (isomorphism), and it is likely that Dedekind's (and others') discussions on structuralism in the nineteenth century drew in fact on Leibniz' ideas. On the other hand, Leibniz' route towards his own proto-structuralism and theory of relations is likely to have drawn precisely on Eudoxus' theory of proportions (as hinted at, for instance, in Sect. 47 of Leibniz' *Fifth Paper to Clarke*, in GP vii, 400–402). On Leibniz' theory of relations and expression, and its sources, see Mugnai (1992, 2012). On the birth of modern structuralism, see Reck and Schiemer (2020).

are conceived of as very abstract relations between magnitudes, and "analogies" as identities among these relations. Aristotle's philosophical theory of analogy is likely to depend on, and generalize over, the basic ideas of this new theory of proportions. It is remarkable that Eudoxus' very refined theory of mathematical analogies may have brought this mathematician to develop the idea of a universal mathematics as a general science of magnitudes. Aristotle's "meta-mathematical" generalization of the notion of analogy transferred the idea of a structural identity from mathematics to epistemology and applied it to a plurality of "analogous" scientific theories. In this way, Aristotle employed Eudoxus' conceptual breakthrough against the very idea of a universal mathematics as a unified superordinate science, and showed that a general theory of magnitudes and ratios still allows for a plurality of structurally similar, and yet different, mathematical sciences.

Thus, even if Aristotle cannot be credited for a fully-fledged theory of isomorphism between theories, he saw the possibility of spelling out *structurally identical principles* ruling different theories. In this sense, Eudoxus may have first envisaged the idea of "common notions" of the kind we find in Euclid, and Aristotle developed this view in his epistemology by thematizing the need of κοινὰ ἀξιώματα alongside the *proper* principles of each science. Thanks to this idea, Aristotle attempted to solve the mathematical and dialectical puzzles concerning the universality of science.

According to Aristotle, then, each science is based on some principles that are proper to it (definitions, hypotheses, and possibly other kinds of "theses"). Nonetheless, these principles are complemented by others that are not proper to the science in question but are in common with other *analogous* disciplines.

We must now look at how these common principles may be applied to the individual sciences.

4 Specialized Common Axioms: The Schematic Interpretation

A κοινὸν ἀξίωμα states some very general relations between all *beings* whatsoever. The Principle of Contradiction or CN3 are indeed formulated so as to apply to things or beings in general (the Greek makes use of the neuter pronouns expressing the maximum possible generality). Aristotle, however, still wants every science to be *only* grounded on principles referring to the genus that is investigated by it. *All* principles of arithmetic, say, must be about *numbers*.

Aristotle is explicit in stating that axioms, which are hyper-generic principles, are concretely *applied* in science in some specialized version. They are only employed in a demonstration to the extent in which they deal with the genus under consideration. We have already mentioned the most important passages to this effect:

> Of the principles used in the demonstrative sciences some are proper to each science and others common—but common by analogy, since they are only useful in so far as they bear on the genus under the science ... It is sufficient to assume each of these in so far as it bears

> on the genus; for it will produce the same results even if it is assumed as holding not of everything but only for magnitudes (or, for arithmeticians, for numbers).
>
> ...
>
> It is obvious that the investigation of these axioms pertains to one science, namely the science of the philosopher, for they apply to all existing things, and not to a particular class separate and distinct from the rest. Moreover all thinkers employ them – because they are axioms of being *qua* being, and every genus possesses being – but employ them only in so far as their purpose require; i.e., so far as the genus extends about which they are carrying their proofs.[39]

Ancient commentators offered the first and most influential interpretation of these passages. It was first expounded (in the texts we have) by Alexander of Aphrodisias, and later almost unfailingly repeated by other Greek, Latin and modern interpreters. I will call it the *schematic interpretation* of Aristotle's common axioms.

According to it, common axioms may be conceived of, in modern parlance, as "axiom schemata" of sorts. Rather than encompassing a number of related principles of one theory, though, these schemata would be meta-theoretical: each "analogous" theory would be grounded on one instance of the axiom schema. A statement such as CN3 ("*If equal things be subtracted from equal things, the remainders are equal things*"), thus, would be a general principle about quantity that should be specialized as a proper principle of arithmetic or geometry in order to be actually employed: "*If equal numbers be subtracted from equal numbers, the remainders are equal numbers*", "*If equal magnitudes be subtracted from equal magnitudes, the remainders are equal magnitudes*". The same happens with equal lines, equal surfaces, equal times or motions, and so forth. The structural similarity between numbers and other magnitudes (i.e. their categorial unity κατ'ἀναλογίαν) would allow for stating the above principles as the common schema CN3, dealing with all "things" in general.[40]

According to the schematic interpretation, the specialized principles on numbers, lines, etc., are *assertions* that may be assumed as premises of syllogistic reasoning, and therefore may play the role of principles and starting points in a demonstration. For instance, if we want to formalize in Aristotle's logic a geometrical demonstration concerning angles, we should construct a "syllogism" having as a premise a specialized version of CN3, such as "*If equal angles be subtracted from equal angles, the remainders are equal angles*", and as another premise that "angle a is equal to angle b, and angle c is equal to angle d", in order to conclude that "angle $(a-c)$ is equal to angle $(b-d)$". This is not a syllogism according to any standard definition, of course,

[39] *An. post.* A 10, 76ᵃ37–76b2; *Metaph.* Γ 3, 1005ᵃ21–29.

[40] See Alex. *In metaph.* 265: "For each of the genera with which the sciences are concerned is a kind of being; this is what these people cut off from what is common and on it they use the axioms, so far as it is also useful for them to do so: the geometer applies the axioms to magnitudes, and uses them to that extent; the arithmetician applies them to numbers. They do so because the axioms belong to all things and are not proper to any one kind of thing. For example, the geometer assumes as an axiom that 'things equal to the same thing are equal to one another' and substitutes 'magnitudes', for his treatise is about magnitudes. The arithmetician, in his turn, transfers the axioms to numbers" (transl. Madigan). Among ancient commentators, this view is restated in Them. *In an. post.* 29–30, and Philop. *In an. post.* 123 and 141–142.

and we see here at work the main problem of adapting Aristotle's logic to mathematical reasoning. The general idea should however be clear: a common axiom is not itself a premise of a reasoning since it is not even an assertion, but its specialized instances may work as premises and therefore enter into the Aristotelian model of a deductive science.

It is important to stress that this interpretation is entirely conjectural. Nowhere in his writings does Aristotle offer an explicit example of a specialized principle, or the way in which an axiom is actually employed in a proof, and the above references to the genus-specific instances of CN3 (on numbers and magnitudes, or on angles) were made up and suggested by Alexander and the other commentators.[41]

This schematic interpretation agrees with Theophrastus' general idea that the axioms of quantity and the axioms of logic are two different kinds of principles. The schematic interpretation, indeed, takes very seriously the idea that categories are *genera of sorts*, which are further divided into *species of sorts*. Thus, since quantity would be a kind of "quasi-genus" divided into the "quasi-species" of numbers and magnitudes, a general statement about quantity (such as CN3) may be specialized into particular statements about numbers or about magnitudes.

This seems to be the most important but also the most problematic feature of the schematic interpretation. A statement such as "*If equal angles be subtracted from equal angles, the remainders are equal angles*" would be, according to Aristotle's classification of principles, a *thesis* (being a claim about one particular genus). In the schematic interpretation, therefore, a common axiom would be nothing but a general expression for a collection of theses sharing a similar structure; it would be a *schema of theses*. As a matter of fact, this is explicitly stated by ancient commentators, who regarded common axioms as collections of proper principles.[42] But nothing similar is *ever* suggested by Aristotle, who always stresses the *irreducibility* of these different kinds of principles: common axioms are considered by him to be totally different

[41] The passage mentioning CN3 in *Metaph.* K 4, 1061^{b}17–27 (see above, *note* 19) is the most liable to be interpreted along the lines of Alexander's reading. The Eleventh Book of the *Metaphysics*, however, is most certainly spurious, and probably written in the early Peripatetic school. Here is the passage: "Since even the mathematician uses the common axioms only in a particular application, it will be the province of first philosophy to study the principles of these as well. That when equals are taken from equals the remainders are equal is an axiom common to all quantities; but mathematics isolates a particular part of its proper subject matter and studies it separately; e.g. lines or angles or numbers or some other kind of quantity, but not *qua* being, but only in so far as each of them is continuous in one, two or three dimensions" (transl. Tredennick, modified). It may be noted that this passage does not mention *magnitudes* (μεγέθη) but only specific instances of them (such as lines). This is connected with the problem of understanding whether magnitudes are for Aristotle a veritable genus or not (see above, *note* 13), and therefore whether, according to the schematic interpretation, CN3 should be specialized into a statement on magnitudes or (e.g.) on lines, in order to be employed in a demonstration. Ancient commentators, in any case, showed no doubts that μεγέθη constitute a proper genus.

[42] Philop. *In an. post.* 141–142, even stated this view about the Principle of Contradiction: "So the principle of contradiction without qualification is *common* to every science, but it becomes *proper* insofar as it concerns magnitudes, numbers, or the attributes of these things" (transl. McKirahan).

from theses, to the point that their difference demarcates the first division among ἀρχαί.[43]

One may attempt to defend the schematic interpretation and speculate that Aristotle would have not considered a specialized axiom (such as "*If equal angles be subtracted from equal angles, the remainders are equal angles*") to be a thesis after all. Insofar as this assertion expresses a special case of a general principle, it would just be "a common axiom limited in scope" or "the common axiom itself, *qua* specialized" rather than a proper principle in the Aristotelian sense. This latter reading would at least account for Aristotle's distinction between axioms and theses. The problem remains that a principle such as the one on "equal angles" above would be labeled as a specialized version of the common axiom CN3 (rather than a thesis) *only insofar as it happens* that another genus of things (say, numbers) also obeys an analogous principle. The state of affairs that both angles and numbers fall under the scope of a common principle is to be taken at face-value as a brute fact, since no common genus of them may be found (as an alleged *primary subject* of predication) that might explain their similar structure. Being common, according to this reading, is not an intrinsic feature of axioms as such—which distinguishes them from theses—but is derived from their scope of application. In a way, according to the schematic reading, mathematical axioms are theses (accidentally) gone wider.

In conclusion, the schematic interpretation succeeds in accounting for Aristotle's claim that each science only proceeds by and through principles that are referring to its subject-matter, accepting at the same time the existence of some common axioms. I still see three main, related problems stemming from such an interpretation. (#1): The idea of separating mathematical and logical axioms introduces a division among axioms where Aristotle appears to have seen none. The schematic interpretation fails to provide a unitary account of the role of axioms. It rather assumes that Aristotle was not able to reconcile the different sources from which his theory originated, namely, the mathematical and dialectical claims on the univocation of being. (#2): The schematic interpretation reduces axioms to collections of proper principles, and is incapable of tracing a real divide between axioms and theses. In this sense, it offers a weak explanation of Aristotle's main epistemological claim that principles are first and foremost distinguished into proper and common ἀρχαί. (#3): The schematic interpretation also has the additional issue (which is a standard problem of many readings of Aristotle's theory of demonstration) that common axioms are not expressed in a subject-predicate form that would make them eligible to be assumed as syllogistic premises in a demonstration. Nonetheless, according to the schematic interpretation, specialized common axioms (on angles, on numbers, etc.) should be assumed as premises of syllogistic reasoning. The three claims above are to some

[43] I may add that in *Metaph.* Γ 3, $1005^{b}14$, Aristotle says that the Principle of Contradiction (a common axiom) is *not hypothetical* (ἀνυπόθετον). While the notion of hypothesis hinted at in this passage may be that of an assumption that is provable (i.e. following one of the looser definitions of ὑπόθεσις given in *An. post.* A 10, rather than the metaphysical one offered in *An. post.* A 2; such seems to be Ross' suggestion), I would still find it odd if Aristotle could call ἀνυπόθετον a principle that he had conceived of as a collection of ὑποθέσεις. On the relation between Aristotelian hypotheses and the unhypothetical first principle, see Upton (1985).

extent independent from one another (claim #2 being the core of the schematic interpretation), but they are usually found together in ancient and modern interpreters, and indeed all together they make a very consistent interpretation of Aristotle's theory of axioms.

In the next sections of this essay, I advance a different reading of axioms, with the aim of solving these exegetical problems.

5 The Inferential Interpretation of Common Axioms

In his influential commentary on the *Analytics*, David Ross briefly suggested an interpretation of axioms which is completely at odds with the standard reading. Logical principles such as the Principle of Contradiction or the Excluded Middle, Ross claimed, cannot be thought of as syllogistic premises of a demonstration. Rather, they shape and constrain the logical form of the demonstration itself. Axioms, therefore, are not to be interpreted as schemata of assertions, but rather as *formal* principles of sorts.[44]

Ross also pointed to a couple of passages in the *Posterior Analytics* that should substantiate this reading. In them, Aristotle states, indeed, that a demonstration proves its conclusion *through* (διά) the axioms, rather than *from* (ἔκ) them as premises.[45] In yet another passage, Aristotle explicitly claims that, in general, demonstrations do not assume the Principle of Contradiction as a premise, even though he seems to leave open the possibility that particular demonstrations may do so.[46]

Recent interpreters have generally disregarded the textual evidence presented by Ross as inconclusive, and rejected his general suggestion that axioms may be formal principles of sorts. I think we have to agree with them on the poverty of the textual evidence substantiating Ross' claim. I have already remarked, however, that Theophrastus' and Alexander's readings, as well as the schematic interpretation as a whole, do not score any better. The schematic interpretation is generally accepted

---

[44] Ross (1949, 531–532): "even if we insert the law of contradiction as a premiss, we shall still have to use it as a principle in order to justify our advance from that and any other premiss to a conclusion". In the course of the years, several other interpreters have endorsed weaker forms of this "inferential" interpretations. Something to this effect, for instance, may be found in Granger (1976, 81–82), where axioms are called "outils de raisonnement" (but later on also "schémas de régles", 93).

[45] The passages mentioned by Ross, in which Aristotle employs the term διά, are *An. post.* A 10, 76^b10 and *An. post.* A 32, 88^a36–88^b3. Ross' argument has been rebutted by Mignucci (1975, 141–143), who showed that Aristotle uses of διά and ἔκ to mean the starting points from which, and the means through which, a proof is performed, are quite inconsistent. A few lines below one of the passages just quoted, in *An. post.* A 10, 76^b15 and again 22, for instance, Aristotle mentions common axioms as principles ἐξ ὧν a demonstration is carried out. Also in the other passage quoted by Ross, Aristotle employs the expression ἐξ ὧν (*An. post.* A 32, 88^a37). The same is stated in *An. post.* A 7, 75^a42 and in *Metaph.* B 2, 997^a8–9.

[46] *An. post.* A 11, 77^a5–25. The passage is, however, quite obscure, and has been the subject of divergent readings. For a recent interpretation, see Harari (2004, 51–55).

thanks to a certain intrinsic conceptual plausibility, joined with an ancient lineage, but it is scarcely based on Aristotle's actual statements. In sum, any interpretation of Aristotle's conception of axioms must complement textual evidence with some conceptual guidance, and in this respect Ross' general remarks on the role of the Principle of Contradiction in logical deduction seem to advance a sound point about the role of logical principles. I may add that Ross' interpretation is not unprecedented. In the Middle Ages, an inferential interpretation of common axioms (or some of them) was held by a minority of authors, such as Albert the Great and Giles of Rome. They claimed that axioms are principles *secundum virtutem* (i.e. formally) rather that *secundum substantiam* (i.e. semantically), and that they are only causes of the conclusion through the mediation of the proper principles assumed as premises.[47]

I do believe, in fact, that there are good reasons to follow Ross' suggestion and develop a fully-fledged *inferential interpretation* of Aristotle's axioms. This should be extended well beyond Ross' initial idea. Ross, indeed, confined his interpretative suggestion to the *logical* axioms themselves, and claimed that *mathematical* common axioms were regarded by Aristotle as assertions and syllogistic premises.[48] Ross was thus preserving the most important (and, I think, critical) aspect of Theophrastus' reading, namely, the distinction between two sorts of axioms. By contrast, I advance the hypothesis that *all* axioms (logical and mathematical alike) were primarily conceived of by Aristotle as formal principles.

According to this interpretation, a mathematical inference on angles employing CN3 would be construed as: "angle a is equal to angle b", and "angle c is equal to angle d"; therefore: "angle $(a - c)$ is equal to angle $(b - d)$". In this inference CN3 is employed as a kind of *rule* allowing the conclusion to be drawn from the two assertions used as premises.

In this interpretation, no application is made of the specialized principle: "*If equal angles be subtracted from equal angles, the remainders are equal angles*". The latter does not appear in the proof either as a premise or a rule. It is true that CN3, according to this reading, only applies to angles in the present context; and its scope when it applies to other objects (such as numbers) is strictly speaking different, and only *analogous* to the scope that we use in the deduction of angles. Nonetheless, CN3 is applied to angles (or numbers, or magnitudes, etc.) as a formal principle; and its use is restricted not by itself (i.e. being specialized), but by the subject matter to which it applies. Common axiom CN3 *per se*, in fact, applies to the *category* of quantity as such, and not to a plurality of genera. In this sense, it would be misleading to lay down a specialized form of CN3 such as the one on angles that we have mentioned above (which is a thesis, in Aristotle's sense). In fact, we have

[47] For a few passages in Albert, see his commentary on the *Posterior Analytics* I, 5, 4 and 8 (Borgnet, 1890, 138 and 148–149) and the *Ethics*, II, VI, 5, q. 5 (Kübel, 1987, 426). I take the references from the discussion in Corbini (2006, 72–74, who also mentions Giles, offering further references, in 83–85).

[48] The same was the case for the above-mentioned medieval authors. It is remarkable that the main passage used by Ross to foster his own inferential interpretation of logical axioms is *An. post.* A 10, 76^b2–12, which gives the example of a mathematical demonstration and is probably referring to *mathematical* common axioms.

seen that Aristotle did *not* make any reference to specialized forms of CN3. I would surmise that this missing reference is not to be accounted for as an accident of the text of the *Posterior Analytics*, that was later corrected and integrated in the ancient commentaries mentioning such specialized principles. Specialized principles simply *do not exist* in Aristotle's epistemology.[49]

I should stress that in order to endorse the inferential interpretation, we do not need to claim that axioms *cannot be* syllogistic premises and assertions, or that they are indeed devoid of any semantic content. We may indeed modify Ross' view that *some* common axioms are used inferentially (the logical ones), whereas *some other* common axioms are used as premises (the mathematical ones). Aristotle's texts may support the more refined claim that *all* axioms are employed as formal principles, and *all* of them may *also* be formulated in assertoric form and be assumed to be syllogistic premises. The inferential interpretation only maintains that axioms are employed in demonstrations as formal principles, since no deduction can happen without them; but it allows axioms to be also assertions, and to be employed as premises in special cases.[50] This seems to be important for a historically sound attempt at reconstructing Aristotle's theory of principles. We cannot credit Aristotle with a neat distinction between our notions of a semantic axiom vs. a rule of deduction. According to him, axioms (like any other principles) are *about something* insofar as they are principles, and indeed are *true* statements about their subject matter. They *are*, therefore, assertions; and Aristotle often presents them in this form. An axiom such as CN3 is a true statement about quantity, and the Principle of Contradiction expresses a fundamental truth about being *qua* being. Nonetheless, quantity and being are not genera, and much less the *primary subjects* of these statements. Therefore, I would claim, these axioms *cannot be employed as assertions* and syllogistic premises in any standard scientific demonstration—since scientific demonstrations only make use of genus-specific premises. The only way in which axioms may enter a scientific demonstration is by shaping the deduction itself, and the inferential interpretation claims that, according to Aristotle, they may indeed play this role.[51]

---

[49] Aristotle applies CN3 to angles in his example of *An. pr.* A 24. The issue is further complicated by the fact that Aristotle seems to deny that angles are *quantities* at all, and instead categorizes them among *qualities* (see *De caelo* B 14, 296^b20; B 14, 297^b19; Δ 4, 311^b34, where *similar* (ὅμοιαι) angles are mentioned; cf. Simplicius, *In de caelo* 538, for a similar observation). I think that the schematic interpretation cannot really account for the fact that CN3, dealing with quantities in general, may be specialized into a statement on angles. On the contrary, the inferential interpretation, by applying CN3 in its full generality to angles, seems to fit with Aristotle's usage. According to such an interpretation, angles are simply *employed as quantities* in the demonstration. On some notions of angles in Greek geometry, see my De Risi (forthcoming).

[50] This seems to be a viable interpretation of the above-mentioned passage in *An. post.* A 11, 77^a5–25.

[51] Indeed, in *An. post.* A 2, 72^a7–8 Aristotle introduces for the first time the notion of a principle in science, by stating that *any* principle whatsoever (therefore also an axiom) is an assertion and a premise. Some interpreters have employed this passage to discredit Ross' idea that an axiom may be a formal principle (a rule of sorts); see an argument to this effect in Mueller (1991, 68, fn. 11). I think, on the contrary, that the passage is consistent with the fact that for Aristotle the Principle of Contradiction is an assertion, but it is nonetheless employed in demonstration as a rule of sorts.

I would like to venture here a further conjecture. The dual character of axioms, counting both as assertions and formal principles, may perhaps be better explained if we conceive of axioms as assertions *grounding* and validating inferential rules. We know, after all, that axioms such as the Principle of Contradiction and the Excluded Middle (or the *dictum de omni*) are used by Aristotle in order to ground the validity of the moods of syllogistic figures.[52] We can surmise, therefore, that a principle such as CN3 should not be employed *itself* as a rule of inference (as I did in the above-mentioned example on angle equality), but should rather be regarded as a statement grounding (in some way to be better explained) a valid deductive rule. According to this latter interpretation, we would have a uniform reading of common axioms as assertions, while at the same time their role in demonstration would be totally different from that of proper principles insofar as axioms would still only shape the *form* of valid reasoning. In the absence of any textual evidence to this effect, I would not belabor this idea any further and in the rest of the paper I will not distinguish between formal rules and assertions grounding such rules. I can add, however, that Galen may have suggested something along these lines when advancing his theory of "relational syllogisms".[53]

The extensive discussion on first philosophy carried out in *Metaph.* Γ seems to take axioms as assertions and true statements on a wide domain of objects (being *qua* being in the case of the Principle of Contradiction, for instance). Nonetheless, I would say that the very significance of the whole discussion in *Metaph.* Γ shows that axioms cannot possibly be considered to be on a par with proper principles, nor collections of theses, and that their epistemic role was conceived of by Aristotle to be completely different from those. Aristotle was here launching his strongest attack against Plato's conception of dialectics as a hyper-science of sorts, dealing with all genera of things. Aristotle's conception of first philosophy is modeled in opposition to this idea, and he did not see philosophy as a discipline dealing with a hyper-genus or a quasi-genus (being *qua* being), but rather as a *formal* science with different aims and goals with respect to the genus-specific sciences. First philosophy is, so to speak, orthogonal to

---

[52] The debate on the principles grounding Aristotle's syllogistics is ample and difficult, and it is not entirely clear the role that the Principle of Contradiction and the Excluded Middle play in this connection. I will not enter this topic: for a recent assessment, see the important Malink (2013).

[53] Galen's treatment of relational syllogisms is in *Inst. log.* XVI–XVIII, where he also mentions several Euclidean common notions, and CN3 in particular (in XVI, 8), but also non-mathematical axioms (such as "Every man is the son of his father", XVI, 11) that validate other relational inferences. The details of Galen's theory are notoriously opaque. At times, he clearly took axioms to be premises of inferences; at other times, he seems to have regarded them as formal rules of sorts that do not enter among the premises of an inference. Morison (2008) aptly signals all passages in which Galen takes axioms to be syllogistic premises (*Inst. log.* I, 3; XVI, 12; XVI, 13) and in which he denies that they are premises (XVI, 1; XVI, 2; XVI, 3; XVI, 4; XVIII, 4; XVIII, 5). Among the various explanations of this confusing behavior, it has been suggested that Galen may have considered mathematical axioms, such as CN3, precisely as assertions validating rules, and therefore as *sui generis* premises (say, "formal premises") of syllogistic inferences. For a solution along these lines, see Barnes (2007, 437–438), where he calls "Galen's metatheorem" the idea that syllogisms and relational syllogisms are validated by common axioms. On the legacy of Galen's theory of demonstration in the works of the Aristotelian commentators, see Chiaradonna (2009).

all other sciences, rather than above them. Consequently, it cannot prove (contrary to Plato's opinion) the proper principles of each particular science (the theses), and cannot be regarded as their foundation. First philosophy, nonetheless, still deals with the common principles, the ἀξιώματα. These axioms, like the science to which they belong, are not hyper-generic or quasi-generic principles (i.e. principles modeled after the theses, and sharing a similar, but wider, demonstrative role), but *formal principles* shaping the structure of being *qua* being. Thus, notwithstanding the fact that the Principle of Contradiction and the Excluded Middle are generally presented in *Metaph.* Γ as true assertions (for they indeed *are* true assertions), I would say that the theory expounded throughout the Book clearly shows that their primary function and meaning must be itself orthogonal to that of the proper principles: formal, rather than semantic.

In the *Posterior Analytics*, the formal role of common axioms is more explicitly stated (and Ross, indeed, took his lead from this work). In particular, the inferential interpretation of axioms may explain some debated passages of the *Posterior Analytics* in which Aristotle mentions axioms as the *only* principles employed in a demonstration.[54] Hypotheses and definitions (i.e. proper principles) do not appear at all in these passages, and they are only indirectly hinted at through the remark that the demonstration is *about* something (the γένος ὑποκείμενον, which is probably represented in the proof by genus-specific definitions and hypotheses). We have already seen a passage from *An. post.* A 11 to this effect, to which another from *An. post.* A 7 may be added:

> There are three things involved in demonstrations: first, what is being demonstrated, or the conclusion (this is what holds of some genus in itself); second, the axioms (*axioms are the principles from which the demonstrations proceed*); third, the underlying kind whose attributes—i.e. the accidents *per se* that the demonstrations make plain.
>
> ...
>
> I call common the principles which they use as the basis from which to demonstrate—not those about which they prove nor what they prove.[55]

[54] This is especially important since in the passages quoted in the *note* 45 above Aristotle states that common axioms are the principles ἐξ ὧν the proof is carried out, and in *An. post.* A 32, 88^b28–29, he says, conversely, that the principles ἐξ ὧν are common axioms. There is thus an identity between principles ἐξ ὧν and axioms. It is ironic that Ross considered these passages as hindrances for his own interpretation, insofar as they do not mention axioms as principles *through* (διά) which something is proven.

[55] The first quotation is from *An. post.* A 7, 75^a39–75^b2 (cf. an analogous passage in A 10, 76^b11–16); the second is the already quoted *An. post.* A 11, 77^a27–28 (transl. Barnes modified). Mignucci (1975, 141–144) attempts to solve the problem raised by the first passage by claiming that there Aristotle was using a broader notion of "axiom" as to encompass proper principles as well. This strikes me as highly implausible. On the contrary, Mueller (1991) straightforwardly accepts the conclusion that (1) common axioms are *premises* of syllogistic reasoning, and (2) Aristotle took them to be the only premises of a mathematical demonstration. Mueller attempts to explain this latter preposterous claim by pointing to the fact that Aristotle could mean by "demonstration" a specific part of the Euclidean proof, i.e. the part that Proclus later called precisely the ἀπόδειξις of the proof (Proclus, *In Euclidis* 203–207). Yet, there are no hints that Proclus' classification of the parts of a proof predated Proclus' late commentary on the *Elements* (see Netz, 1999), and in any case Aristotle's references to mathematical proofs often take into consideration other parts of the

If we take common axioms to be *premises* of syllogisms, this seems to imply that definitions and hypotheses do not appear as premises at all, since the whole demonstration is carried out from (ἔκ) and through (διά) common axioms. This is a quite implausible claim to ascribe to Aristotle, who elsewhere stresses the importance of *theses* in demonstrations and indeed has the aim of grounding science on proper principles. On the contrary, if we accept the idea that common axioms make the formal part of a demonstration, Aristotle's statements become more palatable: definitions and hypotheses, in this reading, may well be premises of syllogistic chains, but the proof is still said to be carried out through and from common axioms in the sense that these are the actual *inferential means* that allow a conclusion to be drawn from some premises. More generally, in the first book of the *Posterior Analytics*, Aristotle seems to attribute great weight to common axioms in the making of science, and his statements to this effect conflict with the many other passages (especially in other works) in which he stresses the role of definitions and hypotheses as principles of knowledge. This apparent oddity can be smoothed out if we think of axioms as formal principles, in which case in different passages of different works Aristotle would be highlighting the different, and yet necessary, roles of the formal and the semantic principles of demonstration.

Finally, it is important to acknowledge that there is a passage in Aristotle's works that appears to militate against an inferential interpretation of common axioms. In *An. pr.* A 24, Aristotle argues that at least one of the premises of a syllogism must be universal. He explains several issues that may arise in science from syllogisms with only particular premises and gives some non-mathematical examples. Then, he says that "this is even better seen in geometrical demonstrations" and offers a rather obscure example of a pre-Euclidean geometrical theorem that Aristotle seems to articulate in three "syllogisms" needing universal premises. The third of these inferences, Aristotle says, cannot employ a statement on the equality of differences between angles but instead needs CN3 in its full strength in order to be universal and therefore properly deductive.[56] Ancient commentators such as Alexander and Philoponus read this passage as a confirmation of their schematic interpretation, according to which axioms are assumed to be universal premises of syllogisms (rather than rules of inference). These commentators even exploited this passage to claim that, for Aristotle, mathematics is entirely reducible to chains of syllogisms.[57] Modern interpreters generally also agree that in this example CN3 is taken as an assertion and a syllogistic

demonstration as well (see Mendell, 1998). Mueller seems to have been aware of some problems in his own interpretation, but still did not accept an inferential interpretation of axioms (see above, *note* 51).

[56] *An. pr.* A 24, 41^b13–28. The mathematical example has been the subject of much debate among interpreters, and we possess several reconstructions of it. See at least McKirahan (1992), Mendell (1998), and Malink (2015).

[57] Alex. *In an. pr.* 268–269; Philop. *In an. pr.* 253–254. From these ancient commentaries, a long and important tradition of attempts at *barbarizing* mathematics (put it into syllogisms in *barbara*) took the lead, with significant developments in the middle ages and the early modern age.

premise, thus blocking an inferential reading of axioms.[58] I note, however, that the schematic interpretation does not fare any better. On the one hand, interpreting that CN3 is here used as a syllogistic premise would mean to assume a common axiom in its full generality rather than in a specialized form (this is Aristotle's main contention in this passage). But we have seen that Aristotle is very explicit elsewhere that this should not be done. The need of restricting the axioms to a genus, as soon as they are employed as syllogistic premises, is in fact the core of the schematic interpretation. Such an interpretation is therefore more damaged than supported by this passage. On the other hand, there is no reasonable way to frame CN3 as a syllogistic premise, since it is not a judgment in subject-predicate form. It is highly unlikely, I think, that Aristotle would have considered this statement as a viable syllogistic premise for a mathematical theorem in *barbara*.[59] Finally, it should be noted that the passage in *An. pr.* A 24 seems to contradict Aristotle's explicit claim in the above-mentioned *An. post.* A 11, according to which a common axiom (the Principle of Contradiction) is not generally employed as a syllogistic premise. I can suggest two solutions in order to deal with the delicate passage in the *Prior Analytics*. It is simply possible that Aristotle had not yet developed a theory of common axioms at the time in which he wrote the *Prior Analytics*. In this work, after all, the word ἀξίωμα rarely appears, and it has a non-technical meaning. In this hypothesis, the passage would hinder neither the schematic nor the inferential interpretation of axioms, since the theory of axioms would only appear later in the *Posterior Analytics*.[60] Alternatively, it is also possible that Aristotle, contrary to the obvious interpretation of the passage ("this is even better seen in geometrical demonstrations"), was giving the mathematical example to make a general point: even geometrical demonstrations, that appear to deal with individual angles and lines (Aristotle mentions particular angles identified by letters) need universal assumptions in order to be deductions. *A fortiori*, syllogisms would also need universal premises. This reading would be compatible with an inferential interpretation of CN3, employed in the passage as a general example

---

58 Note that this specific problem did not arise in Ross' interpretation, since for him *mathematical* axioms are assertions and not rules.

59 A syllogistic reconstruction of this proof, along the lines hinted at by the ancient commentators, is given in the above-mentioned (Malink, 2015; McKirahan, 1992; Mendell, 1998). The main logical issue in any syllogistic reconstruction of the argument is that it has to assume premises of the form "A and B are equal", thus treating equality as a standard monadic predicate with plural subjects. But whereas "A and B are red" may be analyzed into "A is red" and "B is red", this cannot be done with "equal", which is a relation and therefore behaves in a completely different way. It is possible that Aristotle did not realize this fact, but I note that in *An. pr.* A 36, 49ᵃ2–3, Aristotle explicitly takes "equal" to imply an *oblique* predication in the dative (to be equal *to* something), and therefore it is difficult to think that he may have conceived of a syllogistic premise with the predicate "…are equal" as ancient and modern commentators claim.

60 We have already mentioned that in *An. pr.* B 11, 62ᵃ12–13 the word ἀξίωμα means ἔνδοξον. On the meaning of πρότασις in the *Prior Analytics*, see Crivelli Charles (2011). On the relative chronology of the *Prior* and *Posterior Analytics*, see some conjectures in Barnes (1981). I am not entirely fond of this solution, since in Sect. 6 below I would like to connect Aristotle's views on common axioms with the theory of oblique inferences expounded in *An. pr.* A 36.

of a universal statement (or rule) used in science rather than as a specific instance of a universal syllogistic premise.

In conclusion, I think that this isolated and rather obscure example cannot be taken as strong evidence against an inferential interpretation of axioms. On the contrary, the latter interpretation may solve some important exegetical issues that affect more standard readings. On the one hand (issue #1 in Sect. 4 above), such an interpretation clearly puts together logical and mathematical axioms, thus providing grounds for Aristotle's unitary understanding of all sorts of axioms as formal principles. On the other hand (issue #2), it also distinguishes in a very sharp way common axioms (as formal principles) from proper (semantic) principles, thus explaining Aristotle's main divide in the classification of ἀρχαί. In the next section, we discuss in which sense the inferential interpretation may also contribute to explain non-syllogistic inferences in Aristotle's theory and the fact that axioms are not presented in the subject-predicate form (issue #3, further reinforced by the above example).

6 Axioms, Logic and the Categories

Further developments of the inferential theory of axioms below presuppose a general interpretation of the meaning of Aristotle's logic and theory of categories, that I am unable to fully provide in the present paper.[61] These developments are also grounded on a thinner textual basis than the previous discussion. As such, they should be especially regarded as interpretative proposals and conjectures.

I surmise that the inferential interpretation of common axioms may explain why they are not stated in the usual subject-predicate form, but rather in the form of relations. According to this interpretation, axioms do not express a property of anything in particular, and not even the properties of several things in different genera (as in the schematic interpretation). On the contrary, they express formal relations among things and properties in general. The main difficulty is of course that the kind of inferences licensed by common axioms (such as CN3) are not syllogistic, and it is not entirely clear whether Aristotle's theory of science may accommodate for "asyllogistic" inferences (i.e. inferences that are not ruled by the moods expounded in the *Prior Analytics*).

As is well known, there are passages in which Aristotle seems to hint at a broader understanding of deduction, in which the notion of predication (the κατηγορεῖσθαι of a predicate to a subject) is extended to a more general notion of belonging (ὑπάρχειν) that may encompass at least some kinds of relations. Aristotle gives examples of deductions in which one premise, or both premises, allow for oblique predication, i.e. they are not such that in them a predicate "is said of" a subject: if wisdom is a science (standard direct predication) and wisdom is of the good (oblique ὑπάρχειν), the conclusion is that a science is of the good. These examples have been discussed

[61] In particular, I follow here a *logical* interpretation of the categories that is similar to that advanced by Apelt (1891), and which has been later debated by several interpreters.

at length by the interpreters, who in different degrees of generosity have attributed to Aristotle a fully-fledged theory of relations, or various confusions on the latter. It seems clear that Aristotle did not regard his extended logic as a treatment of relations in the strict sense, since he does not give any special relevance to the category of πρός τι in this connection. He seems rather to have in mind that relations are treated as a special sort of oblique predication.[62]

I would venture the hypothesis, then, that the principles ruling these kinds of oblique inferences may be *common axioms* interpreted along the lines of the inferential reading.

Aristotle's very scanty references to axioms and oblique ὑπάρχειν make it difficult to provide textual evidence for this interpretative hypothesis. It may be noted, however, that as an example of oblique predication Aristotle mentions the "being equal to" or the "being double of". These relations are the main content of the mathematical *common notions* that we find in Euclid, and that were possibly formulated by Eudoxus.[63] The very plausible identification of Euclid's κοιναὶ ἔννοιαι with Aristotle's κοινὰ ἀξιώματα supports the idea that Aristotle's cases of oblique predication *in quantity* may have been ruled by common axioms such as CN3.

This interpretation does not need to assume that a general logic of relations would be needed to implement these inferences. In Aristotle's theory of deduction no *general* rules for oblique and relational inferences are given beforehand. On the contrary, in the absence of a general logical framework that allows for a treatment of relations, there are a few cases of relations and oblique predications that have their own particular rules, which are encoded in common axioms. This might explain, at least, why Aristotle seems to have never envisaged or treated at any length a logic of relations, and rather conceived of these occurrences as particular cases. I think, therefore, that the inferential interpretation of common axioms may contribute to solving the difficult question of the existence of asyllogistic inferences in Aristotle's works. If, on the contrary, we interpret common axioms as syllogistic premises, we would have to explain how they may be assumed in a syllogism, since they are not given in subject-predicate form. We would have either to admit that Aristotle simply could not fathom how to do this, or to imagine that Aristotle had conceived a logic of relations of sorts in order to draw conclusions from premises such as CN3.[64] The lack of evidence for

---

[62] The main text is *An. pr.* A 36, in which a broader notion of ὑπάρχειν is introduced. The meaning of this whole chapter is very controversial. Aristotle offers a treatment of oblique inferences mainly based on some linguistic features of the Greek, and it is hard to evaluate its scope and significance. Many interpreters interested in Aristotle's philosophy of mathematics have employed this passage to argue that Aristotle had a conception of logic that was broad enough to accommodate at least some mathematical asyllogistic inferences. The most significant attempt in this respect is the important paper by Mendell (1998). See also, from the same year, Kouremenos (1998), who however has less bold claims about the reconstruction of Aristotle's theory of relational syllogisms. I follow their lead here, but the textual basis is thin.

[63] Examples of oblique inferences involving quantity are to be found in *An. pr.* A 36, 48^b40–49^a5. Aristotle makes examples of predications in which the copula must be interpreted as "to be equal to" also in other passages, such as in *An. pr.* B 25, 69^a30–34, or *An. post.* B 11, 94^a27–34.

[64] This seems to be the path taken by Mendell (1998), who basically admits the possibility of inferring "A is equal to B" from "A and B are equal" (p. 206), thus resolving a theory of relations

the latter seems to me a further argument to conceive of common axioms as formal principles complementing standard syllogistic inferences.

In a further passage in the *Prior Analytics*, Aristotle states indeed that ὑπάρχειν is said *according to all the categories*.[65] This important statement seems to imply (should we further extend the present interpretation) that different common axioms rule different categorial contexts. The relations expressed by a common axiom such as CN3 are not properties intrinsic (say) to triangles or lines, but relations that such objects acquire in so far as they are *predicated as quantities* (ὑπάρχειν "according to quantity") in a given propositional context, such as a geometrical theorem. We can apply CN3 to triangles not because triangles are quantities *per se* (they are rather *magnitudes*, and quantity is not a genus), but because sometimes, in doing geometry, we do employ triangles as quantities. A common axiom, in sum, does not express a collection of genus-specific properties of an object, but the *formal behavior of any thing whatsoever* in so far as it plays a certain role in a proposition and in an inference, and therefore it is predicated according to a given category.

In this respect, the domain of application of CN3 is no more restricted than is the domain of the Principle of Contradiction: in so far as they are both formal (ontological) principles, they apply to all beings. We may admit, of course, degrees of *formal* generality, and state that the Principle of Contradiction, in so far as it applies to *all* categories (not just quantity) rules every kind of predication (every kind of ὑπάρχειν), and therefore it may be said to be the ἀρχὴ τῶν ἄλλων ἀξιωμάτων πάντων. But this greater generality does not concern the genera of things falling under the axiom, but the different ways of predication to which it applies. In this respect, we may fully account for Aristotle's claim (in *An. Post.* A 11) that axioms are common to *all* sciences, without having to accept the suggestion of some interpreters that commonality would only mean "common to *some* things". The meaning of "being common" (κοινόν) is rather to be equated with "being formal" or "being categorial".

This interpretation also allows for a uniform reading of Aristotle's statement in *An. Post.* A 2 to the effect that axioms are needed to learn *anything whatever*. We do not have to restrict the latter claim to the "logical" axioms only, such as the Principle of Contradiction: common axioms in general (also CN3) are rather to be understood as

---

into a theory of inferences between monadic predications. I am aware that there are further difficulties in attributing a theory of asyllogistic inferences to Aristotle, and for instance a referee of the present essay rightly remarked that such a theory might become an insurmountable obstacle for Aristotle's important claim (based on syllogistic inferences) that every demonstration is finite (*An. post.* A 19–22).

[65] *An. pr.* A 37, 49^{a}6–8. This chapter of the *Prior Analytics* is short and controversial, and seems not to connect smoothly with the previous discussion in *An. pr.* A 36. The significance of the term κατηγορίαι has been deflated by some interpreters as if Aristotle here meant just "predications" in a grammatical sense. I do not see much room for sharply distinguishing between grammatical and philosophical meanings of the term. For an interpretation of κατηγορίαι as categories in these passages, see Alex. *In metaph.* 366; and very recently Striker (2009, 226).

formal principles, and therefore as principles employed in any reasoning whatsoever (as long as predication according to the relevant categories is employed).[66]

Even more importantly, an inferential reading of axioms may explain the analogical unity of things falling into different genera. The inferential interpretation agrees with the schematic interpretation on the fact that objects falling under different genera may have a unity κατ'ἀναλογίαν, which is captured by the axioms. This hyper-generic unity expressed by the axioms, however, is no longer regarded as an extrinsic, brute fact in the inferential interpretation. Numbers and geometrical magnitudes (two genera of things) do not simply *happen* to have relations and structures in common (as it is the case in the schematic interpretation). On the contrary, numbers and geometrical magnitudes (and times, motions, etc.) have analogous relations and structures *insofar as* they are predicated according to the same category in a propositional context. Their similar propositional roles make them fall under the same relations. The functional unity of the category explains the unity κατ'ἀναλογίαν expressed by κοινὰ ἀξιώματα.

This interpretation may therefore fully explain the sense in which the demonstration on the alternation of proportions "is different, and also the same" for lines and numbers.[67] The *deductive structure* of the two proofs is the same, and it employs the same common axioms to draw analogous conclusions. But the subject matter of the two proofs is different, as well as the definitions which explain the nature of the objects involved. This structural identity of several demonstrations dealing with different objects, which is a common feature of mathematical practice, is explained by Aristotle (I surmise) with a reference to the identical role (κατ'ἀναλογίαν) that different objects play in the proofs, insofar as they are employed as quantities.[68]

We also see why Aristotle could, even if he consistently rejected the idea of a *mathesis universalis* over and above the individual mathematical disciplines, nonetheless refer to a universal mathematics as an established doctrine. This was certainly for historical reasons—Eudoxus and others had proven hyper-generic results in mathematics. But we see that there is a sense in which Aristotle could accept a *sui generis* science that did not revolve around any particular genus: this was the case of first philosophy, conceived of as a formal discipline. Universal mathematics, like first philosophy restricted to the *category* of quantity (as opposed to Platonic dialectic restricted to the *genera* of magnitudes and numbers) had the same status.[69]

In sum, I have shown that an inferential interpretation of common axioms may solve several important exegetical issues. It puts logical and mathematical axioms on

---

[66] For the sentence on the Principle of Contradiction as the principle of all other axioms, see above *note* 25. For the passage in *An. Post.* A 11, see *note* 20. For the passage in *An. Post.* A 2, see *note* 16.

[67] Following the above-mentioned passage in *An. post.* B 17, 99^{a}8–11 (cf. *note* 36).

[68] See for instance the appendix to Rabouin (2016), which offers parallel proofs of the same result about magnitudes and numbers in Euclid's *Elements* (X, 3 and VII, 2). I would say that Aristotle regarded these proofs as identical κατ'ἀναλογίαν.

[69] In this way, I endorse David Rabouin's idea of a *mathesis universalis* in Aristotle as a science of the "equal and unequal", according to the category of quantity (and grounded on the common axioms on equality). See Rabouin (2009, 58–67).

the same footing, as Aristotle himself seems to have done. It neatly separate axioms from proper principles by granting them a different role in demonstrations. It explains their not being formulated in a subject-predicate form, and gives some hints towards the solution of the difficult problem of the existence of non-syllogistic inferences in Aristotle's epistemology. Finally, it provides a better grounding for the whole theory of analogy by showing functional reasons for the commonality of axioms. In addition, Galen's reading of common axioms as rules of inferences (or grounds for these rules) shows that in antiquity such a theory was at least conceivable.[70]

I am aware that in the lack of explicit textual evidence such an inferential interpretation of common axioms is destined to remain conjectural, but I think that this alternative reading is well worth entertaining.

Acknowledgements I thank Francesco Ademollo, Orna Harari, Marko Malink, Mattia Mantovani, Marco Panza, and David Rabouin for their brilliant remarks and careful corrections of previous drafts of this essay. I have learned a lot from their comments, and the paper has substantially improved thanks to their help.

References

Apelt, O. (1891). *Die Kategorienlehre des Aristoteles.* In *Beiträge zur Geschichte der griechischen Philosophie* (pp. 101–216). Teubner.

Apostle, H. G. (1952). *Aristotle's philosophy of mathematics.* Chicago University Press.

Aubenque, P. (2009). *Problèmes aristotéliciens. Philosophie théorique.* Vrin.

Barnes, J. (1981). *Proof and the syllogism.* In E. Berti (Ed.), *Aristotle on science: The posterior analytics* (pp. 17–59). Antenore.

Barnes, J. (1993). *Aristotle. Posterior Analytics.* Clarendon.

Barnes, J. (2007). *Truth, etc.* Clarendon.

Borgnet, A. (Ed.). (1890). *Albertus Magnus. In libris Posteriorum Analyticorum* (*Opera Omnia*, Vol. 2). Vivès.

Chiaradonna, R. (2009). *Le traité de Galien "Sur la Démonstration" et sa posterité tardo-antique.* In R. Chiaradonna & F. Trabattoni (Eds.), *Physics and philosophy of nature in Greek Neoplatonism* (pp. 43–78). Brill.

Corbini, A. (2006). *La teoria della scienza nel XIII secolo. I commenti agli Analitici Secondi.* Galluzzo.

Courtine, J.-F. (2005). *Inventio analogiae. Métaphysique et ontothéologie.* Vrin.

Crivelli, P., & Charles, D. (2011). 'ΠΡΟΤΑΣΙΣ' in Aristotle's Prior Analytics. *Phronesis, 56*, 193–203.

De Risi, V. (2021a). Gapless lines and gapless proofs. Intersections and continuity in Euclid's Elements. *Apeiron, 54*, 233–259.

De Risi, V. (2021b). Euclid's common notions and the theory of equivalence. *Foundations of Science, 26*, 301–324.

De Risi, V. (forthcoming). Euclid's fourth postulate. *Science in Context.*

Einarson, B. (1936). On certain mathematical terms in Aristotle's logic. *The American Journal of Philology, 57*, 33–54, 151–172.

[70] See *Inst. log.* XVIII, 8. Galen even claimed that an interpretation of axioms as principles licensing inferences was older than him, and had already been endorsed, possibly, by Posidonius.

Granger, G. G. (1976). *La théorie aristotélicienne de la science.* Aubier Montaigne.
Harari, O. (2004). *Knowledge and demonstration: Aristotle's Posterior Analytics.* Kluwer.
Knorr, W. (1975). *The evolution of the Euclidean Elements.* Reidel.
Kouremenos, T. (1998). Aristotle on syllogistic and mathematics. *Philologus, 142*, 220–240.
Kübel, W. (Ed.). (1987). *Albertus Magnus. Super Ethica commentum et questiones* (*Opera Omnia*, Vol. XIV 2). Aschendorff.
Malink, M. (2013). *Aristotle's modal syllogistic.* Harvard University Press.
Malink, M. (2015). The beginnings of formal logic: Deduction in Aristotle's Topics vs. Prior Analytics. *Phronesis, 60*, 267–309.
McKirahan, R. D., Jr. (1992). *Principles and proofs. Aristotle's theory of demonstrative science.* Princeton University Press.
Mendell, H. (1998). Making sense of Aristotelian demonstration. *Oxford Studies in Ancient Philosophy, 16*, 161–225.
Mendell, H. (2007). Two traces of two-step Eudoxan proportion theory in Aristotle: A tale of definitions in Aristotle, with a moral. *Archive for History of Exact Sciences, 61*, 3–37.
Mignucci, M. (1975). *L'argomentazione dimostrativa in Aristotele. Commento agli Analitici Secondi.* Antenore.
Morison, B. (2008). Logic. In R. J. Hankinson (Ed.), *The Cambridge companion to Galen* (pp. 66–115). Cambridge University Prss.
Mueller, I. (1991). *On the notion of a mathematical starting point in Plato, Aristotle, and Euclid.* In A. C. Bowen (Ed.), *Science and philosophy in classical Greece* (pp. 59–97). Garland.
Mugnai, M. (1992). Leibniz's theory of relations. *Studia Leibnitiana Supplementa, 28.*
Mugnai, M. (2012). Leibniz's ontology of relations: A last word? *Oxford Studies in Early Modern Philosophy, 6*, 171–208.
Netz, R. (1999). Proclus' division of the mathematical proposition into parts: How and why was it formulated? *The Classical Quarterly, 9*, 282–303.
Pritchard, P. (1997). Metaphysics Δ 15 and pre-Euclidean mathematics. *Apeiron, 30*, 49–62.
Rabouin, D. (2009). *Mathesis Universalis. L'idée de "mathématique universelle" d'Aristote à Descartes.* Presses Universitaires de France.
Rabouin, D. (2016). The problem of a "general" theory in "mathematics": Aristotle and Euclid. In K. Chemla, R. Chorlay, & D. Rabouin (Eds.), *The Oxford Handbook of generality in mathematics and the sciences* (pp. 113–134). Oxford University Press.
Reck, E. H., & Schiemer, G. (Eds.). (2020). *The prehistory of mathematical structuralism.* Oxford University Press.
Ross, D. (1949). *Aristotle's Prior and Posterior Analytics. A revised text with introduction and commentary.* Clarendon.
Striker, G. (2009). *Aristotle. Prior Analytics. Book I.* Clarendon Press.
Tarán, L. (1981). *Speusippus of Athens. A critical text with a collection of the related texts and commentary.* Brill.
Todd, R. B. (1973). The stoic common notions: A re-examination and reinterpretation. *Symbola Osloenses, 48*, 47–75.
Upton, T. V. (1985). Aristotle on hypothesis and the unhypothesized first principle. *The Review of Metaphysics, 39*, 283–301.
Zhmud, L. (1998). Plato as "architect of science". *Phronesis, 43*, 211–244.

Chrysippus' Logic in a Natural Deduction Setting

Marcello D'Agostino and Mario Piazza

1 Introduction

It is well known that the Stoic theory of deduction is the largest contribution to a system of propositional logic in antiquity. From the 1950s it has been at the center of a considerable body of historical and theoretical research (cf. Becker, 1957; Bobzien, 1996, 1997; Casari, 2017; Frede, 1974; Hülser, 1987-1988; Mates, 1953; Mignucci, 1993; Milne, 1955; Mueller, 1978; O'Toole & Jennings, 2004). Despite many general accounts of Stoic logic, however, there is little consensus among scholars about motivations and details of some of its most controversial theses. These are traditionally associated with Chrysippus of Soli (*c.* 280–206 BCE), third head of the Stoa and its leading logician. The aim of this paper is to frame in a contemporary context some salient but overlooked aspects of Chrysippus' system of deduction, which speak eloquently to our own logical concerns.

We subscribe to Susanne Bobzien's recent claim that Stoic logic "deserves more attention from contemporary logicians" (Bobzien, 2019, 234). According to her, one of the reasons why Stoic logic should catch our interest is that Stoic analysis is "closest to methods of backward proof search for Gentzen-inspired substructural sequent logics" (Bobzien, 2019, 234). Of course, a word of methodological caution is in order at the outset. A logical system of antiquity brings together logical commitments with distinctive metaphysical and epistemological outlooks, so that any attempt to link that system up to a contemporary proof-theoretic setting demands a considerable level of idealisation. Modulo this *caveat*, anyway, we are going to suggest that Chrysippus' system of deduction is instructive and attractive for a contemporary logician not only as anticipation of some proof-theoretic frameworks of the twentieth century, but also, more subtly and surprisingly, by virtue of the (peaceful) coexistence of what in the eyes of modern logicians are *classical* and *non-classical* patterns of inference.

M. D'Agostino (✉)
Department of Philosophy, University of Milan, Milan, Italy
e-mail: marcello.dagostino@unimi.it

M. Piazza
Classe di Lettere e Filosofia, Scuola Normale Superiore di Pisa, Pisa, Italy
e-mail: mario.piazza@sns.it

F. Ademollo et al. (eds.), *Thinking and Calculating*, Logic, Epistemology, and the Unity of Science 54, https://doi.org/10.1007/978-3-030-97303-2_4

In particular, we propose to implement Stoic logic within a *relevant natural deduction* setting as an alternative to other reconstructions relying on sequent calculus like the one offered by Bobzien herself (Bobzien, 2019; see also Bobzien & Dyckhoff, 2019). Moreover, we will show how adding suitable elimination rules to Chrysippus' natural deduction system, one automatically obtains classical logic as a kind of deductive upper bound.

The plan is as follows. In Sect. 2, we sketch Chrysippus' account of connectives. In Sect. 3 we frame the package of Chrysippus' five indemonstrable within a relevant natural deduction system and we argue that the Stoic notion of proof also involves the commitment to the *reductio ad absurdum* rule. Section 4 traces out the distinction between "shallow" and "deeper" proofs. In Sect. 5, we show how the underlying natural deduction system can be extended to bridge the gap with classical propositional logic. Section 6 contains some concluding remarks.

2 Chrysippus' Connectives

Conjunction

Chrysippus' *conjunction* (συμπεπλεγμένον ἀξίωμα) is defined as one proposition "which is put together by the connective particle 'and' as for example, 'It is day and it is light'" (DL, VII, 72, Diogenes Laertius, 2013). Conjunction may be regarded as truth-functional as attested in this passage from Sextus Empiricus:

> [Stoics] say, just as in life we do not say that the piece of clothing that is sound in most parts, but torn in a small part, is sound (on the basis of its sound parts, which is most of them), but torn (on the basis of its small torn part), so too the conjunction, even if it has only one false component and majority of true ones, will as a whole be called false on the basis of that one. (Sextus Empiricus, 2005, 114)

No matter how big the number of true conjuncts is, one false conjunct is enough to nullify all these truths in a conjunction.

Negation

Negation (ἀποφατικὸν ἀξίωμα) is truth-functional: if added to true propositions, it makes them false; if added to false ones it makes them true (AM, VIII 103, Sextus Empiricus, 2005). Negation is involutive, i.e. $\neg\neg A = A$ (DL, VII, 69, Diogenes Laertius, 2013) (see *infra*).

Conditional

Characteristically, Chrysippus' analysis of the conditional (συνημμένον ἀξίωμα) relies on the relation of *incompatibility*. In Sextus Empiricus's presentation of four different accounts of the conditional from the Hellenistic period, the third one is commonly attributed to Chrysippus although Sextus does not mention any particular philosopher, while ascribing the "material" conditional to Philo of Megara and its omnitemporal interpretation to Diodorus Cronus. So Sextus:

> Those who introduce "connectedness" (συνάρτησις) say that a conditional is sound (ὑγιὲς) when the contradictory of its consequent conflicts (μάχηται) with its antecedent; according to them, the conditionals mentioned above [*sc. If it is day, I converse* and *If there are not indivisible elements of the things, there are indivisible elements of the things*] will be unsound, while *If it is day, it is day* will be true. (PH, II, 111, Sextus Empiricus, 1996)

And Diogenes Laertius:

> A conditional is true in which the contradictory of the consequent conflicts with the antecedent (τὸ ἀντικείμενον τοῦ λήγοντος μάχηται τῷ ἡγουμένῳ), for example *If it is day, it is light*. This is true, for *Not it is light*, the contradictory of the consequent, conflicts with *It is day*. A conditional is false in which the contradictory of the consequent does not conflict with the antecedent, for example *If it is day, Dion is walking*. For *Not Dion is walking* does not conflict with *It is day*. (DL VII, 73, Diogenes Laertius, 2013)

The primitive notion of incompatibility or conflict (μάχη) marks off a sort of negative connectedness between propositions. Conflict embraces not only a logical incompatibility but also an empirical one, insofar as the experience of the world provides constraints on the circumstances in which incompatibility can occur. In short, connectedness requires that the reasoner inspects and matches the specific contents of the propositions involved in the conditional statement. From the Chrysippean "connexive" conditional two consequences immediately follow:

1. no conditional of the form $A \to \neg A$ is true, insofar as the contradictory of $\neg A$ (i.e. A) is never incompatible with A.
2. the principle *ex falso sequitur quodlibet* —which allows one to infer anything from inconsistent premisses (i.e., $(A, \neg A) \vdash B$)— doesn't hold. For instance, from *Dion is walking; not Dion is walking* it doesn't follow that *It is day* inasmuch as *Not it is day* doesn't conflict with the premisses.

Disjunction

The notion of conflict regulates also the behaviour of the Chrysippean disjunction (διεζευγμένον), making it essentially *exclusive* in the binary case: it is true only when exactly one disjunct is true and the disjuncts are in conflict with one another (PH, II,

191, Sextus Empiricus, 1996). In other words, disjunction is not truth-functional, but rather equivalent to the conjunction of the two "connexive" conditionals $A \rightarrow \neg B$ and $\neg A \rightarrow B$. This reading of disjunction is entrenched in Greek mathematical practice. Clear cases of disjunctive conflicts are given in Euclid's *Elements* by the exclusive disjunctions encapsulating conflicting properties of actual mathematical objects; for instance, "if a straight line set up on a straight line make angles, it will make <them> either two rights or equal to two rights" (*Elements* I, 13) (Acerbi, 2008).

However, according to some ancient sources (e.g. Gellius, Galen), the Stoics considered disjunctions of arity of three or greater as true when only one of the disjuncts is true. But on the exclusive reading this becomes obviously problematic since a ternary exclusive disjunction is true also when its three disjuncts are all true. We have sympathy for Jennings' hypothesis that this divergence between modern logicians and the Stoics might reside on the modern use of brackets to separate the disjuncts; for example, let $\oplus$ denote the exclusive disjunction, the ternary disjunction $(A \oplus B \oplus C)$ could also be expressed as $(A \oplus (B \oplus C))$: assume then that A is true, then $B \oplus C$ must be false and this happens either when both B and C are false or when both B and C are *true*. Instead, according to Jennings, Stoic logicians conceived of the n-ary disjunction in terms of a prefix notation of the kind $\oplus(A, B, C)$. For instance: our disjunction $(2 + 2 = 4) \oplus ((2 + 3 = 5) \oplus (2 + 4 = 6))$ is *true* because its second disjunct is false (both of its disjuncts being true), whereas the Stoic disjunction $\oplus((2 + 2 = 4), (2 + 3 = 5), (2 + 4 = 6))$ is *false* because more than one of its disjuncts is true (Jennings, 1994, 256–257).

Note that, due to the exclusive character of disjunction, Chrysippus' rejection of *ex falso* is compatible with disjunctive syllogism (DS), alias the fifth indemonstrable argument (ἀναπόδεικτος λόγος) ascribed to Chrysippus by Diogenes Laertius and Sextus (DL, VII, 80–81, Diogenes Laertius, 2013; PH, II, 157–158, Sextus Empiricus, 1996) (see *infra*). In the twentieth century, logicians of relevantist persuasions have instead renounced *ex falso* by claiming the invalidity of DS associated with an *inclusive* reading of disjunction. This is the moral they draw from a venerable argument *promoting ex falso*, which goes back at least to Alexander Neckam's *De naturis rerum* (c. 1200) (Read, 1989) and revived in C. I. Lewis and C. H. Langford's *Symbolic Logic* (1932, 250–252). Namely,

1. A Assumption.
2. $\neg A$ Assumption.
3. $A \vee B$ 1, $\vee$-intro.
4. B 2,3, DS.

Relevantists tend to block the use of DS in order to short-circuit the above argument, instead of disputing either the disjunction-introduction, or the very transitivity of implication which the argument incorporates (Anderson & Belnap, 1975).

3 Chrysippus' Logic as a Natural Deduction System

In this section, we provide a proof-theoretic reconstruction of Stoic logic in terms of a kind of natural deduction system and compare it with the reconstruction given by Bobzien (2019) in terms of Gentzen-style sequent calculus. In most cases proofs are omitted or informal and sketchy arguments are given instead.

Here are Chrysippus's five "indemonstrables" in schematic form, as reported by Sextus [PH, II, 157f, Sextus Empiricus, 1996; AM, VIII, 227f, Sextus Empiricus, 2005]:

$$If\ A,\ then\ B;\ but\ A;\ therefore\ B. \tag{1}$$

$$If\ A, then\ B;\ but\ not\ B;\ therefore\, not\ A. \tag{2}$$

$$Not\, both\ A\, and\ B;\, but\ A;\ therefore\, not\ B. \tag{3a}$$

$$Not\, both\ A\, and\ B;\ but\ B;\ therefore\, not\ A. \tag{3b}$$

$$Either\ A\, or\ B;\ but\ A;\ therefore\, not\ B. \tag{4a}$$

$$Either\ A\, or\ B;\ but\ B;\ therefore\, not\ A. \tag{4b}$$

$$Either\ A\, or\ B;\ but\, not\ A;\ therefore\ B. \tag{5a}$$

$$Either\ A\, or\ B;\ but\, not\ B;\ therefore\ A. \tag{5b}$$

These inference schemes can be seen as *elimination rules* of a kind of natural deduction system:

$$\frac{A \to B \quad A}{B}\ (1) \qquad \frac{A \to B \quad \neg B}{\neg A}\ (2)$$

$$\frac{\neg(A \wedge B) \quad A}{\neg B}\ (3a) \qquad \frac{\neg(A \wedge B) \quad B}{\neg A}\ (3b)$$

$$\frac{A \oplus B \quad A}{\neg B}\ (4a) \qquad \frac{A \oplus B \quad B}{\neg A}\ (4b)$$

$$\frac{A \oplus B \quad \neg A}{B}\ (5a) \qquad \frac{A \oplus B \quad \neg B}{A}\ (5b)$$

Observe that: (i) all the rules have two premisses, (ii) there are no "discharge" rules, (iii) there are elimination rules only for certain types of "non-simple" propositions, that the Stoics called "mode-forming" (τροπικόν).

Moreover, the Stoics stated at least four metatheorems (θέματα), of which only the first and the third are extant.

> There is also another proof common to all syllogisms, even the indemonstrable, called "(reduction) to the impossible" and by the Stoics termed "first metatheorem" or "first exposition". It is formulated thus: "If some third is deduced from two, one of the two together with the opposite of the conclusion yields the opposite of the other". (Bocheński, 1970, 22.12)

So, the first metatheorem corresponds to a contraposition principle for proofs. In modern notation:

$$\frac{A, B \vdash C}{A, C^* \vdash B^*} \tag{T1}$$

where A^* is the complement of A. The complement of A is B if $A = \neg B$ and $\neg A$ otherwise. The notion of complement is necessary to account for the double negation rule that does not appear explicitly among the Stoic indemonstrables, although, as we said, they certainly accepted the idea that a double negation of A is equivalent to A: "The supernegative is the negation of the negative; e.g., 'not – it is not day'. This posits 'it is day'" (DL, VII, 69f, Diogenes Laertius, 2013).

The other extant metatheorem corresponds to what is probably the most characteristic principle of logical deduction, namely the principle of *transitivity* (or "Cut" in the terminology of Gentzen's sequent calculus).

> The essentials of the so-called third metatheorem look like this: if some third is deduced from two and one (of the two) can be deduced syllogistically from others, the third is yielded by the rest and those others. (Bocheński, 1970, 22.13)

This third metatheorem then amounts to the following proof rule[1]:

$$\frac{A, B \vdash C \quad \Gamma \vdash A(B)}{\Gamma, B(A) \vdash C} \tag{T3}$$

the peculiar form of the first premiss sequent being probably due to the fact that all the indemonstrables have exactly two premisses. According to many commentators, the stoics also used other variants of this metatheorem to be able to express full transitivity, but this is a point on which historians can only conjecture, since there is no explicit description of the other metatheorems in the ancient sources.[2] However, (T3) in itself is sufficient to justify the following notion of *direct proof*:

---

[1] By "proof rule" in the context of this paper we mean a rule that takes proofs as premisses and conclusion, as opposed to an "inference rule" that takes propositions as premisses and conclusion.

[2] On this point see Bobzien (2019).

Definition 1 A *direct proof* of A depending on Γ is a tree such that (i) all its nodes are labelled with formulae (ii) its root is labelled with A, (ii) its leaves are labelled with the formulae in Γ, (iii) each formula in Γ labels at least one of the leaves, (iv) each node other than the leaves is labelled with a formula that results from the two formulae labelling the two nodes above by means of an application of one of the indemonstrables (1)–(5b).

Observe that in such direct proofs all formulae occurring in the leaves are actually *used* in order to obtain the conclusion.

Example 3.1 "If the first, then if the first then the second; but the first, therefore the second" [AM, VII, 230–233, Sextus Empiricus, 2005].

This theorem in modern logic corresponds to what in the context of structural proof theory is known as the principle of *contraction* and it amounts to allowing multiple uses of the same proposition in the proof. A direct proof is as follows:

$$\cfrac{\cfrac{A \to (A \to B) \quad A}{A \to B}\,(1) \qquad A}{B}\,(1)$$

We mention in passing that this kind of proof was called by the Stoics "homogeneous", because it makes use of only one indemonstrable. An example of a non-homogeneous proof is the following argument:

Example 3.2 "If the first and the second, then the third; not the third; the first; therefore not the second" [AM, VIII, 234–236, Sextus Empiricus, 2005].

It is not homogeneous because it makes use of two distinct indemonstrables. A direct proof is as follows:

$$\cfrac{\cfrac{A \wedge B \to C \quad \neg C}{\neg(A \wedge B)}\,(2) \qquad A}{\neg B}\,(3a)$$

Example 3.3 "If the first, then the first; but the first; therefore the first".

This is ascribed to Chrysippus by Alexander (see Kneale & Kneale, 1962, 167) and its direct proof is as follows:

$$\cfrac{A \to A \quad A}{A}\,(1)$$

The interest of this example lies is in the fact that, taken for granted that the Stoics were interested in proofs satisfying a relevance criterion (Bobzien, 2019), they adopted a rather sophisticated one. The above proof is in fact relevant, because both its premisses are *used* in the deduction, although its assumptions are redundant, since obviously the conclusion follows immediately from the first premiss. Therefore, they seem to adopt a relevance-as-use criterion similar to that underlying modern systems of relevance logic. On the other hand, it seems that the Stoics did not admit of a proper inference with only one premiss (Bobzien, 2019), so that, from a formal viewpoint, this might have appeared to be the only way of deducing the conclusion A from the assumption A (in accordance with the formal ban on one-premiss inferences) in a relevant, albeit redundant, fashion.

It can be argued that the Stoics endorsed a notion of proof that included those by *reductio ad absurdum* as well as direct proofs. The validity of a proof is construed by them in terms of a conflict between the denied conclusion and (the conjunction of) the premisses, just as the truth of a conditional is construed in terms of a similar conflict between the denied consequent and the antecedent. This is in accordance with the underlying idea that in order for an argument to be valid, the (connexive) conditional whose antecedent is the conjunction of the premisses and whose consequent is the conclusion of the argument, must be true.

> Some arguments are conclusive [περαντικοί], others not conclusive [ἀπέραντοι]. They are conclusive when a connected proposition [i.e. a conditional], beginning with the conjunction of the premisses of the argument and ending with the conclusion, is true. (PH, II, 137, Sextus Empiricus, 1996)

> Of arguments, some are not conclusive, others conclusive. Not conclusive are those in which the contradictory opposite of the conclusion is not incompatible with the conjunction of the premisses. (DL, VII, 77, Diogenes Laertius, 2013)

Moreover, simple arguments by *reductio ad absurdum* can be justified by means of rule (1), and the metatheorems (T1) and (T3), as follows[3]:

(1) $A \rightarrow \neg B, A \vdash \neg B$ by (1)
(2) $A, B \vdash \neg(A \rightarrow \neg B)$ from 1 by (T1)
(3) $A \rightarrow B, A \vdash B$ by (1)
(4) $A \rightarrow B, A \vdash \neg(A \rightarrow \neg B)$ from 2,3 by (T3)
(5) $A \rightarrow B, A \rightarrow \neg B \vdash \neg A$, from 4 by (T1)

Hence, a notion of *indirect proof* seems to be fully justified in Stoic terms. In order to accommodate such indirect proofs in our natural deduction system, let us introduce a basic *conflict rule* that corresponds to the Aristotelian *Principle of Non-Contradiction*:

$$\frac{A \quad \neg A}{\bot} \qquad \text{(PNC)}$$

where $\bot$ is not a logical constant, as in intuitionistic logic, but simply a marker that a conflict has been revealed.

[3] What is essentially the same argument in a slightly different setting presented in Hitchcock (2006).

Definition 2 A *refutation* of Γ is a tree such that (i) all its nodes except the root are labelled with formulae (ii) its root is labelled with the special symbol $\bot$, (ii) its leaves are labelled with the formulae in Γ, (iii) each formula in Γ labels at least one of the leaves, (iv) each node other than the leaves is labelled with a formula that results from the two formulae labelling the two nodes above by means of an application of one of the indemonstrables (1)–(5b).

Definition 3 An *indirect proof* of A depending on Γ is a refutation of $\Gamma \cup \{A^*\}$.

In the sequel we shall call NS_0 the proof system consisting of the inference rules (1)–(5b) together with the associated notions of direct and indirect proofs. (The reason for the subscript will be apparent in Sect. 4 below.)

Example 3.4 "If the first, then the second; if the first, then not the second; therefore not the first".

This is clearly related to the *reductio ad absurdum* pattern that was extensively used by the Greeks. According to Kneale and Kneale (1962, 172) the Stoics probably held this theorem in high esteem, and it is mentioned by several authoritative sources, under the name of "theorem with two mode-forming premisses". Indeed, there is no indemostrable that can conclude from two mode-forming premisses, so a direct proof—as we have called it—cannot even start. Hence, its proof must be achieved by indirect means. A simple way of doing it would consist in accepting what we have called "indirect proofs" in Definition 3 as first-class citizens in the realm of proofs.

$$\dfrac{\dfrac{A \to B \quad A(= (\neg A)^*)}{B}\,(1) \qquad \dfrac{A \to \neg B \quad A(= (\neg A)^*)}{\neg B}\,(1)}{\bot}\,(\text{PNC})$$

Another related way, illustrated in Bobzien (2019, 251), consists in making explicit use of (T1) as a rule of proof. In this case the proof takes the form of a *sequent proof*, namely a tree whose nodes are labelled with sequents, the leaves with sequents corresponding to the five indemonstrable, and all nodes except the leaves with sequents obtained from the sequents above by means of a metatheorem: T1 or T3 or a variant of T3. For a careful reconstruction of the variants of T3 the reader is referred to Bobzien (2019), where the author endorses the view that these variants are nothing but the missing methatheorems of the Stoics. Here we shall loosely refer to all the variants of T3 as "Cut".

$$\dfrac{\dfrac{\dfrac{A \to B, \neg B \vdash \neg A}{A, \neg B \vdash \neg(A \to B)}\,(1) \qquad A \to \neg B, A \vdash \neg B}{A \to \neg B, A \vdash \neg(A \to B)}\,\text{Cut}}{A \to B, A \to \neg B \vdash \neg A}\,(1)$$

A third kind of proof is illustrated in Kneale and Kneale (1962, 171–172) via "conditionalization" (loosely speaking a Stoic version of the deduction theorem), which might be taken as an auxiliary metatheorem, using previously proven theorems as lemmas.

Indeed, there is evidence that the Stoics considered an inference valid whenever the conditional whose antecedent is the conjunction of the premisses and whose consequent is the conclusion is true. This suggests that they might have had no qualms in exploiting conditionalization in their deductive practice in the following sense: if an argument was found valid, the corresponding conditional was regarded as true and could be freely assumed in another proof without counting as a genuine assumption or, if you wish, representing a sort of "unassailable" assumption. More precisely, we can distinguish two forms of conditionalization:

Weak Conditionalization. Every conditional corresponding to an indemostrable can be freely assumed in a proof.

Strong Conditionalization. Every conditional corresponding to a valid inference can be freely assumed in a proof.

Neither of them appears to be equivalent, given (T1) and Cut, to the full version of the deduction theorem, even with the restriction that the cardinality of Γ is at least 2[4]:

$$\frac{\Gamma, A \vdash B}{\Gamma \vdash A \to B} \quad \text{(DT)}$$

There seems to be no extant example of a Stoic proof of a sequent whose conclusion is a conditional, and we found no reference in the literature about how such a proof could be achieved by means of Stoic analysis.

Let us call *LS* the sequent calculus consisting of (i) all the sequents corresponding to the five indemonstrables as basic axioms, (ii) Cut and T1 as sequent proof rules, (iii) Weak Conditionalization as a source of auxiliary "unassailable" premisses, i.e., conditionals corresponding to indemonstrables. Let us also call LS^+ the similar system in which Weak Conditionalization is replaced by Strong Conditionalization, allowing the free introduction of conditionals corresponding to previously proven inferences.

The discussion of the last example suggests that the two approaches—the natural deduction system with direct and indirect proofs on the one hand, and the sequent calculus with T1, Cut, and Conditionalization on the other—might be equivalent. Indeed, it can be shown that LS^+ can simulate NS_0. Suppose there is a refutation of Γ, A^*; then there must be direct proofs of $\Delta_1 \vdash B$ and $\Delta_2 \vdash B$ for some B and some Δ_1, Δ_2 such that $\Delta_1 \cup \Delta_2 = \Gamma \cup \{A^*\}$. Thus, $\bigwedge \Delta_1 \to B$ and $\bigwedge \Delta_2 \to \neg B$ are true conditionals by Strong Conditionalization. Then:

---

4 According to most scholars, the Stoics did no allow inferences with less than two premisses; see Bobzien (2019).

$$\frac{\bigwedge\Delta_1 \to B, \neg B \vdash \neg\bigwedge\Delta_1 \quad \bigwedge\Delta_2 \to \neg B, \bigwedge\Delta_2 \vdash \neg B}{\bigwedge\Delta_1 \to B, \bigwedge\Delta_2 \to \neg B, \bigwedge\Delta_2 \vdash \neg\bigwedge\Delta_1}\text{Cut} \tag{3}$$

and exploiting the equivalence between B and its "supernegative" $\neg\neg B$,

$$\frac{\bigwedge\Delta_1 \to B, \bigwedge\Delta_1 \vdash B \quad \bigwedge\Delta_2 \to \neg B, \neg\neg B \vdash \neg\bigwedge\Delta_2}{\bigwedge\Delta_1 \to B, \bigwedge\Delta_2 \to \neg B, \bigwedge\Delta_1 \vdash \neg\bigwedge\Delta_2}\text{Cut} \tag{4}$$

Now, if $A^* \in \Delta_1$, then

$$\neg\bigwedge\Delta_1, \Delta_1 \setminus \{A^*\} \vdash A^{**} = A \tag{5}$$

is an axiomatic sequent by a plausible generalization of the second indemonstrable; and if $A^* \in \Delta_2$, then

$$\neg\bigwedge\Delta_2, A^* \vdash \neg\bigwedge\Delta_2 \setminus \{A^*\} \tag{6}$$

is also an axiomatic sequent by the same argument. So, by an application of Cut to (3) and (5) and omitting the "unassailable" conditionals corresponding to the indemonstrables

$$\Delta_1 \setminus \{A^*\}, \bigwedge\Delta_2 \vdash A. \tag{7}$$

Moreover, given that

$$\Delta_2 \vdash \bigwedge\Delta_2 \tag{8}$$

is a theorem of Stoic logic, which can be derived from the second indemonstrable by (T1), we obtain, by means of an application of cut to (7) and (8):

$$\Delta_1 \setminus \{A^*\}, \Delta_2 \vdash A. \tag{9}$$

If $A^* \notin \Delta_2$ the above sequent is the endsequent and a similar argument can be developed starting from (4) for the case in which $A^* \in \Delta_2$ but $A^* \notin \Delta_1$. Otherwise, by an application of (T1) to (7),

$$\Delta_1 \setminus \{A^*\}, A^* \vdash \neg\bigwedge\Delta_2 \tag{10}$$

and then

$$\Delta_1 \setminus \{A^*\}, A^* \vdash \neg \bigwedge \Delta_2 \setminus A^* \text{ from (6) and (10) by Cut} \tag{11}$$

$$\Delta_1 \setminus \{A^*\}, \bigwedge \Delta_2 \setminus \{A^*\} \vdash A \text{ from (11) by (T1)} \tag{12}$$

$$\Delta_1 \setminus \{A^*\}, \Delta_2 \setminus \{A^*\} \vdash A \text{ from (12) and (8) by Cut} \tag{13}$$

and hence $\Gamma \vdash A$. The above argument establishes the following:

Proposition 3.1 *If* $\Gamma \vdash_{NS_0} A$, *then* $\Gamma \vdash_{LS^+} A$.

In the next section we shall show that the converse holds only for a special case of *LS*-proofs and shall define a stronger notion of *NS*-proof that is indeed equivalent to that of LS^+-proof.

4 Shallow and Deep Arguments

The proof rule expressed by (T1) can be easily simulated by the notion of indirect NS_0-proof. Suppose there is an indirect NS_0-proof of C depending on $\{A, B\}$, namely a refutation of $\{A, B, C^*\}$. Then such a refutation can immediately be read as a proof of B^* depending on $\{A, C^*\}$. Is this the whole story? Not quite.

First, observe that in NS_0 Cut is satisfied only by direct proofs and not by indirect proofs.

Example 4.1 As shown in Example 3.4,

$$A \to B, A \to \neg B \vdash_{NS_0} \neg A$$

can be established only by means of an indirect proof. Similarly,

$$C \to D, C \to \neg D \vdash_{NS_0} \neg C.$$

Moreover, it is easy to verify that

$$\neg A, \neg C \vdash_{NS_0} \neg(A \oplus C)$$

can be established by means of an indirect proof. However,

$$A \to B, A \to \neg B, C \to D, C \to \neg D \nvdash_{NS_0} \neg(A \oplus C),$$

because if the premisses are all "mode-forming", the proof (whether direct or indirect) cannot even start.

As a consequence, indirect NS_0-proofs can simulate (T1) under the restriction that the latter is used *only as the last step in the proof*. Let LS^- be the restriction of *LS* allowing only proofs such that (T1) is applied only as the last step in a sequent proof.

It can be shown that Weak Conditionalization does not increase the expressive power of NS_0 and so:

Proposition 4.1 *If* $\Gamma \vdash_{LS^-} A$, *then* $\Gamma \vdash_{NS_0} A$.

In order to achieve full equivalence between *NS*-proofs and *LS*-proofs we need to introduce a discharge rule, which might well have been acceptable by Stoic standards, although it appeals to what they would have called "hypothetical reasoning". The rule takes the form of a non-constructive dilemma rule:

$$\dfrac{\begin{array}{cc} \Gamma, [A] & \Gamma, [\neg A] \\ \vdots & \vdots \\ \bot/B & \bot/B \end{array}}{\bot/B} \tag{14}$$

where the slash notation is to be understood as follows: if at least one of the conclusions of the subordinate arguments is B, then the conclusion is B, otherwise it is $\bot$. In either case, the application of the rule discharges the assumptions A and $\neg A$, so that the conclusion of the main proof no longer depends on them.

With this addition, Cut is satisfied for indirect as well as direct proofs and the proof of Example 4.1 is accomplished as follows. Let T_1 be the following proof tree:

$$\dfrac{\dfrac{\dfrac{A \oplus C \quad [\neg A]}{C}\,(5b) \qquad \neg C}{\bot}\,\text{PNC} \qquad \dfrac{\dfrac{A \to B \quad [A]}{B}\,(1) \qquad \dfrac{A \to \neg B \quad [A]}{\neg B}\,(1)}{\bot}\,\text{PNC}}{\bot}\,\text{PNC}$$

and T_2 be the following proof tree:

$$\dfrac{\dfrac{C \to D \quad C}{D}\,(1) \qquad \dfrac{C \to \neg D \quad C}{\neg D}\,(1)}{\bot}\,\text{PNC}$$

Then,

$$\dfrac{\begin{array}{cc} [C] & [\neg C] \\ T_2 & T_1 \\ \bot & \bot \end{array}}{\bot}$$

is a refutation of

$$\{A \oplus C, A \to B, A \to \neg B, C \to D, C \to \neg D\}$$

and therefore an indirect proof of $\neg(A \oplus C)$ depending on

$$\{A \to B, A \to \neg B, C \to D, C \to \neg D\}.$$

Since the effect of (T1) can be simulated by the notion of indirect proof, it follows that any *LS*-proof can be simulated by the system *NS* obtained by adding to NS_0 the single discharge rule (14).

On the other hand, (14) is a derived rule in LS^+. We have shown in the previous section that if $\Gamma, A^* \vdash_{NS_0} \bot$ then $\Gamma \vdash_{LS^+} A$. So, if $\Delta, A \vdash_{NS_0} B$, by Proposition 3.1, $\Delta, A \vdash_{LS^+} B$ and, by Cut, $\Gamma, \Delta \vdash_{LS^+} B$. The general case for *NS* can be shown by induction on the number of nested applications of the the rule itself in its subordinate proofs. For the case in which both subordinate proofs conclude to B, the simulation is obtained in a similar way.

It can be shown that Strong Conditionalization does not increase the deductive power of *NS*. In general, it holds that:

Proposition 4.2 *If* $\Gamma \vdash_{LS^+} A$, *then* $\Gamma \vdash_{NS} A$.

Moreover, (14) is a derived rule in LS^+, therefore:

Proposition 4.3 *If* $\Gamma \vdash_{NS} A$, *then* $\Gamma \vdash_{LS^+} A$.

5 Bridging the Gap with Classical Logic

The natural deduction system *NS*, as well as its equivalent sequent calculus LS^+, is a relevance logic and therefore falls short of classical propositional logic. What is missing? From this point of view, it is interesting to compare *NS* to the classical refutation system *KE* (D'Agostino & Mondadori, 1994). The *KE* elimination rules for the mode-forming premisses are the same as the *NS*-rules, except that the primitive disjunction in *KE* is inclusive and not exclusive. However, the *NS*-rules for $\oplus$ can be easily derived in *KE* using the translation of exclusive disjunction in terms of inclusive disjunction, negation and conjunction. On the other hand, *KE* includes more elimination rules than *NS*. These are elimination rules for propositions of the form $\neg(A \to B)$, $A \wedge B$, $\neg(A \oplus B)$ plus an explicit double negation rule.

$$\frac{\neg(A \to B)}{A} \qquad \frac{\neg(A \to B)}{\neg B}$$

$$\frac{A \wedge B}{A} \qquad \frac{A \wedge B}{B}$$

$$\frac{\neg(A \oplus B) \quad A}{B} \qquad \frac{\neg(A \oplus B) \quad B}{A}$$

$$\frac{\neg(A \oplus B) \quad \neg A}{\neg B} \qquad \frac{\neg(A \oplus B) \quad \neg B}{\neg A}$$

$$\frac{\neg\neg A}{A}$$

The addition of the above rules to *NS* yields a refutation system that is sound and complete for classical propositional logic.

Interestingly enough, the missing rules are all objectionable from a Chrysippean viewpoint. The additional rules for the denied conditional, via (14) or, equivalently, via (T1), justify the following introduction rules for the conditional:

$$\frac{\neg A}{A \to B} \qquad \frac{B}{A \to B}$$

and so yield Philo's conditional with all the associated paradoxes. From the elimination rules for $A \wedge B$, again via (14) or (T1), one can easily derive the following introduction rules:

$$\frac{\neg A}{\neg(A \wedge B)} \qquad \frac{\neg B}{\neg(A \wedge B)}$$

Either of the them yields a proof of *ex falso quodlibet*. For example:

$$\dfrac{\dfrac{A \qquad \dfrac{\neg A}{\neg(A \wedge \neg B)}}{\neg\neg B}}{B}$$

Finally, the rules for $\neg(A \oplus B)$ turn $\oplus$ into the Boolean exclusive disjunction and justify the following introduction rules:

$$\frac{A \quad \neg B}{A \oplus B} \qquad \frac{\neg A \quad B}{A \oplus B}$$

which are at odds with the intensional load of Chrysippean disjunction.

6 Conclusion

We have argued that Stoic modes of inference can be sensibly reconstructed in modern terms as a natural deduction system with a *single discharge rule* expressing a form of non-constructive dilemma that governs hypothetical reasoning and allows for a natural distinction between "shallow" and "deeper" proofs. Finally, we have shown how this system can be extended to bridge the gap with classical propositional logic. This is a rational reconstruction that does not claim to be historically accurate. However, it can be seen as further and independent support to Bobzien's view that Stoic logic can be seen as a forerunner of modern propositional logic and its proof-theoretic settings (Bobzien, 2003, 2019). At the same time, our reconstruction highlights exactly what is missing in Stoic logic in order to provide a classically complete system.

References

Acerbi, F. (2008). Conjunction and disjunction in Euclid's elements. *Histoire, Épistémologie, Langage, 30*, 21–47.

Anderson, A.R., & Belnap, N. D. (1975). *Entailment: The logic of relevance and necessity*. Princeton University Press.

Becker, O. (1957). *Zwei Untersuchungen zur antiken Logik*. Harrassowitz.

Bobzien, S. (1996). Stoic syllogistic. *Oxford Studies in Ancient Philosophy, 40*, 133–192.

Bobzien, S. (1997). The Stoics on hypotheses and hypothetical arguments. *Phronesis, 42*(3), 299–312.

Bobzien, S. (2003). Stoic logic. In B. Inwood (Ed.), *The Cambridge companion to Stoic philosophy*. Cambridge University Press.

Bobzien, S. (2019). Stoic sequent logic and proof theory. *History and Philosophy of Logic, 40*(3), 234–265.

Bobzien, S., & Dyckhoff, R. (2019). Analyticity, balance and non-admissibility of cut in Stoic logic. *Studia Logica, 107*(2), 375–397.

Bocheński, J. M. (1970). *A history of formal logic*. Chelsea Publishing Company.

Casari, E. (2017). *La logica stoica*. A cura di Enrico Moriconi. ETS.

D'Agostino, M., & Mondadori, M. (1994). The taming of the cut. Classical refutations with analytic cut. *Journal of Logic and Computation, 4*, 285–319.

Diogenes Laertius. (2013). *Lives of eminent philosophers* (T. Dorandi, Ed.). Cambridge University Press.

Frede, M. (1974). Stoic vs. Aristotelian syllogistic. *Archiv für Geschichte der Philosophie, 56*(1), 1–32.

Hitchcock, D. (2006). The peculiarities of Stoic logic. In A. D. Irvine & K. A. Peacock (Eds.), *Mistakes of reason: Essays in honour of John Woods* (pp. 224–242). University of Toronto Press.
Hülser, K. (1987–1988). *Die Fragmente zur Dialektik der Stoiker: Neue Sammlung der Texte mit deutscher Übersetzung und Kommentaren* (4 vols.). Frommann-Holzboog.
Jennings, R. E. (1994). *The genealogy of disjunction*. Oxford University Press.
Kneale, W. C., & Kneale, M. (1962). *The development of logic*. Clarendon Press.
Lewis, C. I., & Langford, C. H. (1932). *Symbolic logic*. Dover.
Mates, B. (1953). *Stoic logic*. University of California Press.
Mignucci, M. (1993). The Stoic themata. In K. Döring & T. Ebert (Eds.), *Dialektiker und Stoiker: Zur Logik der Stoa und ihrer Vorläufer* (pp. 217–238). Steiner.
Milne, P. (1995). On the completeness of non-philonian Stoic logic. *History and Philosophy of Logic, 16*(1), 39–64.
Mueller, I. (1978). An introduction to Stoic logic. In J. M. Rist (Ed.), *The Stoics*. University of California Press.
O'Toole, R. R., & Jennings R. E. (2004). The Megarians and the Stoics. In M. D. Gabbay & J. Woods (Eds.), *Handbook of the history of logic, Volume I. Greek, Indian and Arabic logic* (pp. 397–522). Elsevier.
Read, S. (1989). *Relevant logic: A philosophical examination of inference*. Blackwell.
Sextus Empiricus. (1996). *The skeptic way: Sextus Empiricus' outlines of Pyrrhonism* (Translated, with Introduction and Commentary, by B. Mates). Oxford University Press.
Sextus Empiricus (2005). *Against the logicians* (R. Bett, Ed.). Cambridge University Press.

The Middle Ages and the Scholastic Tradition

"*Generaliter De Nullo Enuntiabili Aliquid Scio*": Meaning and Propositional Content in the *Ars Meliduna*

Christopher J. Martin

1 Introduction

From their close reading of the texts bequeathed to them by antiquity philosophers of the eleventh and twelfth centuries acquired a limited terminology for talking about propositional meaning and a few hints of the metaphysical problems that arise in considering it. In particular they found in Boethius a quite indiscriminate use of both '*propositio*' and '*enuntiatio*' to characterise the sentences with which we say something about the world and a number of expressions to be employed in talking about their meaning, '*eventus rei*',[1] '*essentia rei*', '*consequentia rerum*', '*intellectus*', '*significatio*' '*sententia*', and '*sensus*'. The expression '*existentia rei / rerum*' is also employed by early twelfth century writers, and in particular by Abaelard, to locate propositional meaning but does not seem to occur in his authorities where we find only '*essentia rei*'. In particular, we have from Boethius that:

<Aristotle> says that the consequence of things (*consequentia rerum*) is that the truth of the proposition follows the subsisting thing (*res subsistens*) and that the being of the thing (*essentia rei*), about which a proposition speaks, accompanies (*comitatur*) the truth of the proposition.[2]

And from Priscian:

Continuative propositions are those which show the connection and consequence of things (*consequentia rerum*), [...] that is, they signify the ordering and nature of things with some

[1] '*Eventus rei*' is limited by Boethius to the discussion of future contingents and it is retained for this purpose by twelfth century writers.

[2] Boethius (1877, 109) (On *De Interpretatione*, 18a39-b1): "Ait enim hanc esse rerum consequentiam, ut rem subsistentem propositionis ueritas consequatur, ueritatem propositionis rei, de qua loquitur propositio, essentia comitetur."

C. J. Martin (✉)
School of Humanities, Auckland University, Private Bag 92019, Auckland 1142, New Zealand
e-mail: cj.martin@auckland.ac.nz

F. Ademollo et al. (eds.), *Thinking and Calculating*, Logic, Epistemology, and the Unity of Science 54, https://doi.org/10.1007/978-3-030-97303-2_5

doubt about the being of the things (*essentia rerum*); subcontinuatives, on the other hand, show both the consequent cause of connection, and the being of the things.[3]

'*Res*' here is used as *pragma* is in Greek to mean a state-of-affairs and a proposition is true just in case the corresponding thing in this sense exists.[4]

Abaelard adopts this problematic terminology in his *Dialectica* but uses '*essentia rei*' for the signification of the proposition as a whole, insisting that it is something different from the *res* signified by the terms of the proposition. Each *essentia rei*, however, is composed of the *res* signified by those terms. In the case of a true proposition what the proposition as a whole signifies is their standing in the relationship in which they do stand.

This account of propositional meaning is very reminiscent of the theory propositions advocated by Bertrand Russell in the early stages of his development of logical atomism and it suffers from the same difficulties.[5] How to explain the meaning of true negative propositions, false propositions, and compound propositions. The latter is a comparatively simple task for Russell whose logic is extensional, and whose conditional is material implication. It is not so simple for Abaelard whose logic is hyperintensional and whose conditional is relevantist.[6] The meanings of propositions are furthermore leading candidates for the bearers of logical relations and for Abaelard these relations hold independently of the existence of any particular thing in the world.

The risks of using the same terms to talk about both states-of-affairs and their constituents and the problems of formulating a theory of states-of-affairs is presumably what led Abaelard to develop a new terminology for talking about propositional meaning and a new account of it. Propositional meaning is explained in his *Logica Ingredientibus* in terms of *dicta*, what it is that is asserted in using propositions assertively. *Dicta* are, Abaelard insists, entirely nothing, but stand eternally in logical relations quite independently of the existence of anything in the world. The relationship between Abaelard's two terminologies and his move from identifying propositional meaning with states-of-affairs to the theory of *dicta*, and the question, indeed, of whether this is in fact a new theory, has exercised many commentators and

[3] Priscian (1859, 94): "Continuatiuae sunt, quae continuationem et consequentiam rerum significant, ut si, cum εἰ Graecum significant, [...] Proprie autem continuatiuae sunt, quae significant ordinem praecedentis rei ad sequentem, ut 'si stertit, dormit' et 'si aegrotat, pallet' et 'si febri uexatur, calet'. Non enim conuerso ordine in his consequentiam sententiae seruat oratio: non enim qui dormit omnimodo stertit, quomodo qui stertit omnimodo dormit, nec qui pallet omnimodo aegrotat, quomodo qui aegrotat omnimodo pallet, nec qui calet omnimodo et febri uexatur. Et hae quidem [id est continuatiuae] qualis est ordinatio et natura rerum, cum dubitatione aliqua essentiam rerum significant; subcontinuatiuae uero causam continuationis ostendunt consequentem cum essentia rerum, ut 'quoniam', 'quia', ut 'quoniam ambulat, mouetur'; 'quia sol super terram est, dies est': utrumque enim significat fieri ordine consequenti".

[4] We have the corresponding Greek expression in Stephanus of Alexandria's commentary on the same text, Stephanus (1885, 36): "ἔστι δὲ τοῦτο ὅτι τῇ ἀληθείᾳ τῶν λόγων ἕπεται ἡ ὕπαρξις τοῦ πράγματος καὶ τῷ ψεύδει τῶν λόγων ἡ ἀνυπαρξία."

[5] For Russell's early theories of propositions see Klement (2016).

[6] See Martin (2004a).

I do not intend to consider it further here. Rather I want to examine the development of a theory of propositional meaning which recalls Frege rather than Russell, the theory of *enuntiabilia*, or assertables, which we find in the work of the followers of Robert of Melun, the *Melidunenses*.

The major surviving logical work of the *Melidunenses* is the very extensive treatise called by De Rijk the *Ars Meliduna* (*AM*), dating from the third quarter of the twelfth century and preserved in the Bodleian manuscript Digby 174 (ff. 211–241), from which he published substantial extracts in his *Logica Modernorum*[7] and which has been transcribed as a whole by Yukio Iwakuma.[8] Aside from the *Ars Meliduna*, the next most extensive source for the views of the *Melidunenses* is the list of theses insisted on by the school given by the author of the manifesto called by De Rijk the *Secta Meliduna*. Sadly, the explanation and justification of all but the first four of the fifty three theses is lost.[9] We should note in particular numbers eleven and twelve[10]: (11) Nothing follows from a false. (*Nihil sequitur ex falso*), (12) No false is (*Nullum falsum est*).

Let me refer to the author of the *Ars Meliduna* as *AM* and assume that it was a man.

2 The Ars Meliduna on Semantics of Terms

AM distinguishes between '*appellatio*', which I will translate as 'denotation', and '*significatio*', which can be safely left as 'signification'. Words (*voces*) are introduced in acts of imposition, or institution, to enable us to make our thoughts manifest to one another.[11] These acts are performed according to *AM* not to establish a signification for a word but rather solely with the goal of securing its denotation. Like Kripke in the twentieth century, *AM* takes the baptism of a child with a proper name to show that initially, at least, we only need to refer to things:

> That words were instituted to denote can be credibly enough concluded from the imposition of a word which is employed when a name is imposed on a child. One does not ask here

[7] De Rijk (1967, ch. VI–X). De Rijk treats the four parts of *AM* in four chapters 1 = ch. VII, 2, = ch. VII, 3, = ch. IX, 4 = ch. X, he divides each book into sections, in what follows the references of the form 'x.y' are to book x, section y of this division.

[8] The *Ars Meliduna* and the *Melidunenses* have so far been very little studied. In addition to De Rijk see Nuchelmans (1973, ch. 10), Iwakuma (1997), Biard (1987), Biard (1999), De Libera (1996, 1999), and Martin (2004b).

[9] Given the size of the surviving fragment if as much attention was devoted to the missing theses as to those which remain, the resulting work would have been very large.

[10] For the full list see De Rijk (1967, pp. 283–286). Many of these theses are discussed in the *Ars Meliduna*.

[11] *AM* 1.6, fol. 213ra-b: "Causa institutionis vocum fuit manifestatio intellectus, id est ut haberet quis quo alii intellectum suum manifestaret. Ideoque sicut intellectu duo principaliter comprehendimus, suppositum scilicet et quod de eo dicitur, ita quoque inventa sunt duo genera dictionum, nomina scilicet et verba, haec ad supponendum, illa ad apponendum."

what the name will signify, or with what name the child will be signified, but rather what it will be called (i.e. denoted).[12]

Unlike Kripke, however, *AM* is not committed to a causal theory in which the meaning of a name is simply, and always, its denotation while everything else is a matter of arbitrary subjective association. Rather the signification of a term follows upon its denotation and it is neither arbitrary nor subjective.

To establish the nature of the signification of terms *AM* examines and rejects a number of old and new theories. Against the old theories he argues that the signification of a term is not an image in a human mind—agreeing here with Kripke's rejection of subjective descriptivism. Nor is it an idea in the divine mind, though this suggestion, taken from Priscian, has the advantage that the signification remains when the word ceases to denote[13] and, as we will see, does play a role in the *AM*'s theory of meaning. Finally a group of old theories is rejected all of which hold that what is signified by a word is an understanding (*intellectus*). There is, however, no sense of understanding, including Abaelard's theory of the act of intellectual attention, *AM* argues, which can guarantee the objectivity of signification required to ground our ability to communicate the contents of our minds to one another.

AM next considers two theories proposed by the *moderni* to explain the meaning of terms. The first, again recalling Kripke, identifies the signification and denotation of a proper name but distinguishes them in the case of common names, maintaining that such terms denote determinately individuals of the kinds of things (*maneries rerum*) which they signify indeterminately. A barrage of arguments is mounted against this theory but, as with so much in the *Ars Meliduna*, it is rather difficult to follow them.

The second modern theory and apparently the one which *AM* accepts characterises the signification of words with the term of art, *status,* introduced by Abaelard to locate the common cause of imposition of the general names which for him are species and genera:

> The second opinion, which we have made one of our theses, maintains that words (*dictiones*) signify common or private *status*, that is *status* participable by one alone or several. So that the name 'man' signifies a special *status*, that is, one participated in only by things of one species, 'animal' signifies a general *status*, that is, one participated in by thing of opposed species, 'Socrates', on the other hand, <signifies> a private *status*, that is, one participable by only one thing.[14]

---

[12] *AM* 1.6, fol. 213rb: "Quod autem ad appellandum fuerint voces institutae, satis probabiliter coniectari potest ex illa impositione vocis quae fit cum puero nomen imponitur; ibi enim non quaeritur quid significabit illud nomen vel quo nomine puer significabitur, sed potius quis appellabitur."

[13] Though, as we will see below, *AM* argues that a failure of denotation, such as that of 'rose' when roses no longer exist, renders incongruous a present tense sentence with that term for its subject.

[14] *AM* 1.8, fol. 213rb-va: "Secunda opinio, quam nos in positionem traduximus, fatetur dictiones significare communes status vel privatos, id est participabiles ab uno solo vel a pluribus, ut hoc nomen 'homo' significat specialem statum, id est participatum tantum a rebus unius speciei; 'animal' vero generalem, id est participatum a rebus oppositarum specierum; 'Socrates' vero privatum statum, id est ab uno solo partici<pa>bilem." It is not clear to me how best to understand '*quam nos in positionem traduximus*', here translated as 'which we have made one of our theses'. It apparently

Proper names thus signify private *status*, and common names common *status*, which may be either special or general, while verbs signify only common *status*.[15] Support for this theory is found by *AM* in Priscian whose references to general and special substance he construes as claims about the general and special *status* signified by the names of genera and species. From Priscian too he takes the claim that a species name is the common, or appellative, name for all of the members of that species but also a proper name for the corresponding incorporeal species.[16] This second theory of the *moderni*, and that adopted by *AM*, thus holds that kind terms construed as common names denote each and every one of the individuals of that kind but construed as proper names denote the incorporeal *status* which those individuals participate in.

AM returns to this question again later and asks directly whether, given that the name 'human being' signifies the species, it is a proper or an common name for it. Priscian's authority is invoked once more in favour of its being a proper name and indeed for the claim that every name is a proper name of what it signifies.[17] In addition we learn something more about the signification of proper names such as Socrates:

has something of a technical sense for *AM* since shortly after this he devotes a long discussion to the different *causae positionis*, reasons that one might have for upholding theses. The *Ars Meliduna* opens (De Rijk, 1967, 264, fol. 211ra) with the remark that: "Propositum quidem negotii est circa opinionis nostrae positiones singula diligenter inquirere, ut sic et nobis ipsis iucundum comparemus exercitium et sociis progressum; facile namque fieri poterit ut in quo scriptor sumet exercitationem, lectoris etiam diligentia suam instruat ruditatem."

[15] *AM* 1.8, fol. 213va: "Verba quoque communes status significant, ut hoc verbum "legit" quoddam accidens participatum ab omni re legente; neque enim dicimus verba significare actionem vel passionem, sed copulare; significant autem accidentia quorum praedicatione ostenditur actio vel passio inesse, ut dicto quoniam legens legit." This is qualified in *AM* 2.13, fol. 221vb: "De verbis dicimus quod fere omnia universalium sunt significativa, praeter substantivum, quod omnia sequitur generalissima, et ideo non potest sub aliquo contineri; impersonalia quoque excipiuntur, ut 'paenitet', 'taedet'; et quae solis vocibus conveniunt, ut 'declinari', 'derivari', 'proferri'; aut solis significatis, ut 'praedicari', 'dici', 'subici'; aut quae actiones vel passiones artificiales, ut ita dicam, copulant, ut 'aedificari, 'fieri', et similia."

[16] *AM* 1.8, fol. 213va: "In libro etiam *Constructionum* frequenter dicit nomina significare substantiam generalem vel specialem; generalem substantiam vocans substantialem statum significatum nomine generis, specialem speciei. Et iterum hoc nomen 'homo' esse proprium ipsius speciei incorporalis, licet omnium hominum communem." See Priscian (1859, 130): "Cum dico uero quid est animal rationale mortale? speciem mihi uolo manifestari, id est hominem, quae quamuis uideatur esse communis omnium hominum, tamen est etiam propria ipsius speciei incorporalis."

[17] *AM* 1.11, fol. 213vb: "Dicimus quoniam omne nomen est proprium nomen sui significati. Quod manifeste Priscianus vult in libro *Constructionum*, ubi dicit 'quamvis sit omnium hominum commune, tamen est etiam proprium ipsius speciei incorporalis.' Paulo etiam inferius dicit quod 'quantum ad generales et speciales formas rerum, haec quoque nomina propria possunt esse quibus genera et species naturae rerum demonstrantur.'" See Priscian (1859, 135): "[...] quantum ad generales et speciales formas rerum, quae in mente diuina intellegibiliter constiterunt antequam in corpora prodirent, haec quoque propria possint esse, quibus genera et species naturae rerum demonstrantur."

> The name 'Socrates' is both the proper name of *this individual* and the proper name of *this human* but in different ways, for it is said to be the proper name of the individual because it signifies it, but of this human, because it denotes him.[18]

It follows, although *AM* does not draw the conclusion here that private *status* of Socrates is *this individual.*[19]

The distinction between signification and denotation had been used by Abaelard to explain how present tense sentences such as '*rosa est*' and '*rosa non est*' may be true and false, and *a fortiori* meaningful, even though their subject term is empty. Meaningfulness for him is guaranteed by the subject term continuing to constitute the common understanding of roses in the mind of someone who hears it.[20] *AM* disagrees and argues that where denotation is lacking sentences containing such common names are incongruous, by which he means that they are neither true nor false, and so not propositions.[21] This holds, he claims, for indefinite, particular, and universal sentences with empty subject terms.[22] On the other hand, if the common name is used as a predicate the resulting sentence is, according to *AM*, congruous and

[18] The objection to which this is a reply is that if a name is a proper name for its significatum, then a proper name will denote what it signifies and so be equivocal, *AM* 1.10, fol. 213vb: "Eadem ratione dicetur hoc nomen 'Socrates' esse proprium nomen sui significati, scilicet individui; sed ipsum etiam est proprium nomen eius quod appellat, nihil autem significat quod appellet, ergo est proprium nomen duorum, unde videbitur esse aequivocum." The reply: "Ad tertiam rationem respondemus quoniam hoc nomen 'Socrates' est proprium nomen huius individui et est proprium nomen huius hominis, sed aliter et aliter; nam individui dicitur esse nomen proprium, quia illud significat, hominis vero quia illum appellat. Et fortasse gratia significati dicitur etiam proprium appellati." See also below, n. 31.

[19] Unfortunately *AM* seems to be rather careless with the term 'individual' sometimes using it for the *status* and sometimes for the sensible substance.

[20] Abaelard (1919, 30): "[...] uniuersalia nomina nullo modo, uolumus esse, cum rebus eorum peremptis iam de pluribus praedicabilia non sint, quippe nec ullis rebus communia, ut rosae nomen <non> iam permanentibus rosis, quod tamen tunc quoque ex intellectu significatiuum est, licet nominatione careat, alioquin propositio non esset 'nulla rosa est'".

[21] *AM* 1.17, fol. 218va: "Hiis ita determinatis hoc solum circa terminorum appellationem restat inquirendum utrum terminus nulli conveniens per appellationem possit congrue sumi ad supponendum verbo praesenti affirmato vel negato, ut dicatur bene 'rosa est' vel 'non est'. Quod non patitur ratio. Nam terminus in subiecto vult poni pro aliquo suorum appellatorum, unde cum nihil subest eius copulationi, non habet ibi locum velut non habens quid comprehendat; cum verbo enim praesenti non pertinet ad praeterita vel futura, ut ex supradictis palam."

[22] *AM* 1.17, fol. 218va: "Sunt quidam qui indefinitam recipiunt 'rosa non est' et 'rosae non sunt', et etiam universalem negativam 'nulla rosa est'; sed alia signa negant posse apponi. [...] Verius autem solvetur dicto tam indefinitam quam particularem quam universalem incongrue dictam esse, audita quippe huiusmodi voce deficit animus, non inveniens quid comprehendat."

See Nuchelmans (1973, 174), referring to a diferent point made in the *Ars Meliduna* and the characterisation of a sentence as '*nugatoria*': '"*Socratem diligere filium suum*", the dictum belonging to the containing sentence "*Socrates diligit filium suum*" ('Socrates loves his son'), becomes nugatory when Socrates ceases to have a son. It is clear from the context that 'nugatory' means the same as 'neither true-nor-false'. If the sentence 'Socrates loves his son' is uttered at a moment when Socrates has no son, the *dictum* is neither true nor false, but simply incongruous.' Nuchelmans here apparently assumes that incongruity and nugatoriness are the same but they are perhaps different for *AM* with the first characterising propositions and the second assertables. He may be making just this distinction in the passage to which Nuchelmans refers, *AM* 4.8, fol. 236vb: "Nam quemadmodum a proposition because as predicates "words are used indefinitely and confusedly".[23] Thus when no roses exist "*nulla rosa est*" is meaningless while "*nihil est rosa*" is true, as indeed is "*nihil est aliqua rosa*" but "*nihil est omnis rosa*" is meaningless since there are no roses for the quantifier to distribute over.[24] *AM*, indeed, seems committed to the position that any sentence of the form "*nullus A est*", where '*A*' is a common name, is either meaningless or false—meaningless if there are no *A*s and false if there are some. Rather curiously, however, he also maintains that sentences such as "*Caesar est albus*" in which the subject terms are proper names whose denotation has ceased to exist continue to be propositions because they 'cause something to be determinately understood'.[25]

Although *AM* accepts that names are introduced in order to communicate our thoughts to others he rejects the claim that a word signifies a proper or common *status* by causing it to be conceived by the mind.[26] More probable he suggests is that a word is said to signify a *status* because its apposition to the subject in a predicative proposition 'causes' (*facit*) the *status* to be predicated. This is an obscure observation but the point seems to be that the predicative structure of categorical sentence reflects a fundamental distinction in reality corresponding to subject and predicate in virtue of which, according to *AM*, in predication the predicate shows the *status* of the subject.[27] In the most basic form of categorical propositions this relationship between subject and predicate is that between universal and singular explored in Part 2 of the *Ars Meliduna* devoted to the *significata* of terms.

AM begins by quickly dispatching those theories which hold that universals are words. Noting next that there are many theories that assert universals to be things he

---

locutio quae congrua est fit ex rei mutatione incongrua, ita *et ipsa enuntiabilia videntur fieri nugatoria*, ut Socratem esse album quod ipse est vel diligere filium suum, quando desinit esse albus aut habere filium; neque enim intelligibile est qualiter haec falsa esse possit, sed erit intellectus talis vanus et cassus."

[23] *AM* 1.17, fol. 218vb: "Licet autem in subiecto praedicti termini vitium faciant, tamen non aequum est ut similiter in praedicato, quoniam ibi habent poni dictiones infinite et confuse. Sit itaque congrue dictum 'nihil est rosa'."

[24] *AM* 1.17, fol. 218vb: "Et poterit quidem addi signum particulare, ut "nihil est aliqua rosa"; at universale additum vitium faciet, quia nulla sunt appellata huius nominis "rosa" quorum vellet facere distributionem."

[25] *AM* 1.17, fol. 218vb: "Potest autem quaeri utrum eadem in propriis habenda sit consideratio. Quod plerique autumant, iudicantes incongrue dici 'Caesar, sive Antichristus, est' vel 'non est'. Nobis videtur quod, quia semper tam in subiecto quam in praedicato aut etiam per se prolata aliquid determinate faciunt intelligi, possunt cum quolibet verbo sumi ad illud supponendum."

[26] *AM* 1.11, fol. 213vb: "Deinde quaeritur quid sit vocem significare communem statum vel proprium. Ad quod dicunt quidam quoniam est ipsum menti repraesentare, audita quippe hac voce 'homo' statim mens concipit talem statum in quo conveniunt omnes homines ex eo quod sunt homines, deinde etiam intelligit rem illius. Sed quia hoc forte difficile videbitur intelligere in propriis, dicunt alii quoniam vox dicitur ideo significare statum quia appositione sua facit illum praedicari (ut praedicationem hic large intelligas), voces quippe instrumenta sunt praedicandi."

[27] *AM* cites Aristotle in support of this, 1.11, fol. 213vb: "Dicit enim Aristoteles in *Praedicamentis*, loquens de incomplexe significatis, quod quaedam eorum significant substantiam, quaedam qualitatem, quaedam quantitatem, etc.; id est quaedam praedicant <substantiam, quaedam> qualitatem, etc.; hoc est quaedam praedicatione sua ostendunt quid aliquid sit, quaedam quale aliquid sit, etc."

first argues against the view that universals are themselves substances or accidents and then devotes a considerable amount of space to refuting the form of collective realism that holds that the universal human is the collection of all individual humannesses. His own view is that[28] "every universal is an intelligible thing participable by many, that is, something which can be perceived only by the understanding". A universal is, of course, what is signified by a common term, and this we know is a *status*. Bringing these ideas together *AM* notes that universals are often referred to as the being of things (*esse rerum*) and characterised as 'the indifferent substance of different things'.[29] Following those who hold rather that no universal or singular is the being of a thing he notes that this latter is to be understood as the principle, for example, that a genus is "an indifferent substance, or *status*". It follows, he claims, that:

> Universals are neither substances nor properties but have their own form of being, as do assertables, times, words, and voids. Whence they exist apart from sensibles and are understood apart, as, for example, the species human, or the individual Socrates, <exists> apart from Socrates, but is understood with respect to (*circa*) him. For if it were in him, it would have to be in him as a part or as a property [...] and further what is more unbelievable than that a universal is in a finger or a nose or in the rear end of a donkey.[30]

'Voids' here translates '*inania*' which as Sten Ebbesen has pointed out makes better sense than '*fama*' in the manuscript.

The claim here that the individual Socrates, like the universal human, is intelligible, and inaccessible to sense, corresponds to the semantical thesis that common and proper names signify common and private *status* and is confirmed by the discussion of the nature of universals and particulars which follows. What we have in the *Ars Meliduna* is thus much more than simply a theory of universals. From the structure of categorical sentences and their semantics *AM* argues for a distinction between two kinds of being, one sensible, consisting of substances such as Socrates with his various properties, and other intelligible, which includes singulars such as the private *status* signified by 'Socrates' and universals corresponding to his species, genus, and

---

[28] *AM* 2.A4, fol. 219rb: "Nos dicimus quoniam omne universale est res quaedam intelligibilis et a pluribus participabilis, id est quiddam quod solo intellectu habet percipi."

[29] *AM* 2.A5, fol. 219rb: "Solent etiam universalia esse rerum appellari, ut hoc universale animal indifferens esse omnium hominum, hoc individuum Socrates proprium esse Socratis. Unde et aliae solent dari descriptiones, hoc modo: 'genus est indifferens substantia rerum differentium <specie, species> numero'. 'Substantia', id est substantiale esse, et praedicabile in substantiam, id est in quid; 'indifferens', id est convenire faciens utpote ea a quibus participatur."; *AM* 2.A6, fol. 219rb: "Alii dicunt quod nullum universale vel singulare est esse rei, sed esse Socratis est constitutio eius ex partibus illis suis, ne necesse sit confiteri plurium quodlibet esse Socratis esse. Unde et praedictas descriptiones sic intelligunt: 'genus est indifferens substantia etc.', id est 'indifferens substantia vel status'".

[30] *AM* 2.A6, fol. 219rb: "Non sunt ergo universalia substantiae nec proprietates, sed habent suum esse per se, sicut enuntiabilia, tempora, et voces, et inania [*ms.* fama]. Quare sunt extra sensibilia et extra intelliguntur, ut haec species homo vel hoc individuum Socrates extra Socratem, sed tamen circa ipsum intelligitur. Si enim esset in eo, oporteret ut in eo esset tanquam pars vel tanquam proprietas. Nec esset intelligibile qualiter aliquid quod partes non habet esset in pluribus. Adhuc autem quid improbabilius quam universalia esse in digito vel in naso aut in posteriore asinae?"

accidents.[31] By participating in various universals, i.e. common *status*, Socrates is the kind of thing that he is and has the features that he does, by participating alone in the proper *status*, he is who he is. The intelligible nature of the singular, as distinct from the sensible Socrates, is made particularly clear by *AM* in his answer to the question "What is a singular?":

> A singular is something subjectable, for example with the name 'Socrates', which is in no way predicable, which is an intelligible thing at which sense is not directed. For it is the singular and private *status* whereby Socrates has his being Socrates, and similarly the singular this animal, is that whereby something is this animal, just as the species human is that whereby something has its being a human [...].[32]

This derivation of ontology from the structure of categorical sentences with the significations of subject and predicate distinguished from their denotations and located in the realm of intelligible being cannot fail to recall both Frege's procedure in developing the distinctions between concepts, i.e. predicates, and objects and, like the *Ars Meliduna,* his insistence that their significations are something which exist apart from the sensible world.

Frege famously held that the distinction between sense and denotation applied to sentences as well as to their component words. The sense of an indicative sentence is the content that remains the same if the sentences is uttered assertively, or about which a question is asked, or something commanded, if the sentence is formulated interrogatively or imperatively. Frege called such content a *Thought* and held that it occupies a third realm of intelligible being, distinct from that inhabited by external objects and internal ideas. Thoughts exist objectively and eternally independently of their expressions in token sentences of our various languages. Part 4 of the *Ars Meliduna,* contains what is probably the most extensive anticipation of these Fregean ideas on propositional meaning to be found in mediaeval philosophy.

3 The Ars Meliduna on Assertables

The question of what it is that a proposition signifies has been settled, at least in part, by characterising it as (a) true or (a) false, i.e. something of which 'true' or 'false'

---

[31] *AM* 2.A6, fol. 219va: "Nos dicimus quod hoc individuum non est idem specie huic asino, nec differens specie. Sed tamen recipimus quod et sensibilia differunt specie et individua, sed aliter et aliter; nam differre specie est esse res oppositarum specierum vel sub oppositis speciebus contineri, quorum primum dicitur propter sensibilia, reliquum propter individua. Quare non erit recipienda coniunctio 'individua et res differunt specie', modus enim diversus prohibet id esse verum, immo forte incongrue dicitur, quoniam significata non habent appellatis connumerari, sicut nec voces."

[32] *AM* 2.E1, fol. 223vb: 'Singulare est subicibile quod nullo modo est praedicabile, velut hoc nomine 'Socrates', quod est res quaedam intelligibilis. Ad quam sensus non conatur, status enim quidam singularis et privatus unde Socrates habet esse Socrates; similiter hoc singulare hoc animal <est> illud unde aliquid est hoc animal sicut haec species homo est id unde aliquid habet esse homo, quaedam scilicet unitiva communio in qua Socrates et Plato ostenduntur uniri et convenire, dicto eos esse homines, participatione enim speciei plures homines unus."

is predicable which *AM* also refers to using 'true' and 'false' substantively so the definite or indefinite article may be needed. This division is of what *AM* usually calls '*enuntiabilia*' but sometimes refers to as '*dicta*' and which I translate as 'assertables':

> Assertables are the *significata* of propositions. They are so called from assertion or from suitability for assertion for they are suited to assert, that is, to be said (*dici*) in assertion.[33]

Once we have a general name for what is signified we may ask of what kind of item that name denotes; whether something reasonably familiar such as an idea, or something *sui generis* like Abaelard's *dicta*. To form names for particular assertables *AM* employs the Latin infinitive accusative construction. The usual way of translating this into English is with 'that' followed by the corresponding sentence. But, as we will see below, *AM* argues that the corresponding Latin construction cannot be used to do this and so I will leave the relevant expressions in Latin. Thus, for example, to the sentence '*Socrates legit*' there corresponds the name '*Socratem legere*'.

The Nature of Assertables

AM considers four earlier theories of the nature of assertables.[34] According to the first they are 'understandings conceived through words':

> [...] thus the true *Socratem legere* is the understanding constituted by the proposition '*Socrates legit*', and so a property since every understanding is a property.[35]

'Understandings' are taken by *AM* to be to purely mental items and most of the criticisms of the theory are standard anti-psychologistic objections, some of which will be repeated by Frege in his criticism of the view that the senses of propositions are ideas. Since mental events are tied to individual minds as the properties of a thinking subject there will be as many meanings as there are individuals conceiving an understanding for a given utterance. Likewise should everyone be asleep, there would be no assertables and all sentences would be meaningless. Furthermore there will be no contradiction between '*Socratem legere est verum*' and '*Socratem legere non est verum*'[36] since such propositions are indefinite according to this account of

[33] *AM* 4.1, fol. 236ra: "Enuntiabilia sunt propositionum significata. Dicta sic ab enuntiando sive ab aptitudine enuntiandi, quia apta sunt enuntiari, id est enuntiatione dici."

[34] See Iwakuma (1997), also Nuchelmans (1973, ch. 10).

[35] *AM* 4.1, fol. 236ra: "Et fuerunt qui enuntiabilia dixerunt esse intellectus per voces conceptos, ut hoc verum *Socratem legere* intellectum constitutum hac propositione 'Socrates legit'; et ita proprietatem, cum omnis intellectus sit proprietas."

[36] *AM* 4.1, fol. 236ra: "Multorum etiam quodlibet erit hoc enuntiabile *Socratem esse hominem*, ut intellectus huius, id est quem iste in mente concipit ex prolatione propositionis, et similiter intellectus illius; sed iste alius est ab isto, quia uterque proprietas est quae est in anima; nulla autem proprietas in pluribus esse potest ita quod in quolibet eorum; erit ergo infinitorum intellectuum quilibet illud enuntiabile; ex quo accidet hanc propositionem indefinitam esse 'Socratem esse hominem est verum', quia appellatio enuntiabilis nullum illorum intellectuum supponit determinate, sed potest

meaning. Their subject terms do not denote determinately a single 'thing' but rather many distinct understandings.

AM's first objection to this theory is rather disturbing: according to it every false will something (*secundum hanc opinionem erit omne falsum aliquid*) since a false is a property.[37] In objecting to this *AM* is apparently agreeing with the twelfth thesis of the *Secta Meliduna*: "*Nullum falsum est*". We saw above, however, that for him this proposition must be meaningless or false. If it is to be meaningful and either true or false then '*nullum*' must have the sense of '*nihil*' and the thesis "*nihil est falsum*". It will turn out that *AM*'s own view is apparently that both a true and a false are something.

The second theory of assertables seems to be a version of the account of propositional meaning, mentioned above, which was advocated by Russell in his early formulation of logical atomism and for which propositions, corresponding to *AM*'s assertables, are facts, and their constituents are things. *AM* tells us that[38]:

> Others held that a true is the composition of a predicate with a subject, a false the division of them from one another, saying that the true *Socratem esse album* is nothing other than the composition or coherence of the predicable white with Socrates. For that is said to be true because whiteness coheres with Socrates, and *Socratem esse asinum* is false because the species ass is not participated in by Socrates. As in the case of the first theory *AM* objects that it follows that a false is something (*Secundum hanc opinionem, sicut et secundum primam, oportebit falsum esse aliquid*), since division is a property.[39]

Among his contemporaries we may note that the *Porretani* identified the senses of assertables as compositions and divisions but while they accepted that a true is something they denied that a false is something.[40]

---

pro quolibet poni; et has non esse contradictorias 'Socratem esse hominem est verum' 'Socratem esse hominem non est verum', velut nec <in> indefinitis 'homo est iustus', 'homo non est iustus'."

[37] *AM* 4.1, fol. 236ra: "At vero secundum hanc opinionem erit omne falsum aliquid, quia intellectus; omnis autem intellectus proprietas est animae; quare aliquid."

[38] *AM* 4.1, fol. 236ra: "Alii posuerunt verum esse compositionem praedicati ad subiectum, falsum vero divisionem eorum ab invicem, dicentes hoc verum *Socratem esse album* nihil aliud esse quam compositionem sive cohaerentiam huius praedicabilis album cum Socrate. Ideo enim dicitur illud esse verum, quia albedo cohaeret Socrati, et *Socratem esse asinum* falsum esse, ideo quoniam haec species *asinus* non participatur a Socrate."

[39] *AM* 4.1, fol. 236ra: "Secundum hanc opinionem, sicut et secundum primam, oportebit falsum esse aliquid, quia omnis divisio proprietas."

[40] See Anonymous (1983). *Compendium Logicae Porretanum*, 63–64: "Ratio quare dicatur *verum omne esse aliquid, sed non quicquid est verum est aliquid*: [...] Consistit autem veritas circa duo, scilicet circa formam substantie ut 'Socrates est albus', et circa substantiam forme ut 'albedo est color'; nam et hac et illa verum dicitur: est enim quasi compositio coloris ad albedinem; itaque quamdiu forma in subiecto est, et eius compositio est. Ergo sicut vere forme sunt, sic et earum compositiones sunt; immo quia † potius † compositiones et forme. Ergo et omnia vera sunt. Sed cum veritas et falsitas sint opposita ut privatio et habitus, et omnis veritas sit habitus (id est compositio), cum privatio nulla sit <et privatio veri sit> falsitas sive divisio forme a subiecto, *falsitas sive falsum non est aliquid* [...]."

A third theory of assertables was that nothing is (a) true and nothing (a) false (*nihil est verum, nihil est falsum*).[41] It maintained, recalling Abaelard's reduction of nominal to adverbial modes, that the adjective 'true' is is used to form propositions equivalent to those formed with the adverb 'truly'. To say '*Socratem legere est verum*', is thus just to say '*Socrates vere legit*' and to say this is to say no more than '*ita in re est quod Socrates legit*'. This is the formulation which Abaelard uses to explicate truth and falsity but presumably every theory of truth as correspondence has to say something like it. *AM* adds only, apparently in support of the theory, and correctly, that nowhere in Aristotle or Boethius will one find the identification of true and false with assertable; as I noted above, neither of them uses the term.

The fourth theory according to *AM* was that of his own teacher and with it, if '*nullum*' is understood to mean '*nihil*', we have an explicit contradiction of the thesis '*Nullum falsum est*' included in the *Secta Meliduna*'s 'exhaustive account of the Meldunian faith':

> None of the theories mentioned pleased our teacher, and setting out his own he said that some things are true(s) and some are false(s) (*aliqua esse vera et aliqua esse falsa*), and that both this and that are assertables, which are neither substances nor properties, but rather have their own mode of being, like time and *word*, and they are grasped only by reason and understanding, for sense, for example vision and hearing, cannot presume to reach them.[42]

Above we were told that universals belong with assertables in having a mode of being accessible only to understanding and here we are told again that neither of them are substances or properties. Properties seems to mean accidents in this context and so the claim is that universals, assertables, and the other items given in the lists do not fall into one of the ten predicaments. As Peter King has pointed out, the longer list given in the first characterisation of this mode of being is practically the Stoic list of incorporeals,[43] sayable (*lekton*), void, place, and time.[44] Perhaps then '*vox*' translates '*lekton*', though it is singularly inappropriate to do so, and its inclusion

---

[41] *AM* 4.1, fol. 236ra-rb: "Tertia opinio fuit quod nihil est verum, nihil est falsum, sed sunt huiusmodi nomina determinationes quaedam quorundam modorum loquendi, ut sit quidam tropus loquendi, dicto Socratem legere esse verum vel Socratem esse asinum esse falsum, id est Socrates vere legit, quod dicit 'ita est in re quod legit', similiter Socrates falso est asinus, id est non est ita in re. Legendo namque et relegendo Aristotelem et Boethium, nunquam scriptum invenies aliquod verum vel falsum esse enuntiabile vel e converso, sed sumit semper Aristoteles enuntiabile pro 'praedicabile', ut enuntiabile de aliquo, id est praedicabile de aliquo, et enuntiari pro praedicari, unde propositio est enuntiatio alicuius de aliquo."

[42] *AM* 4.2, fol. 236rb: "Nostro vero praeceptori nulla praedictarum placuit opinionum, sed suam proferens dixit aliqua esse vera et aliqua esse falsa, et tam haec quam illa esse enuntiabilia; quae nec substantiae sunt nec proprietates, sed habent suum esse per se, similiter tempori aut voci; et comprehenduntur sola ratione et intellectu, nec enim contingit ad ea sensum conari ut nec visum nec auditum."

[43] See King (1982, 110).

[44] The problem with appealing to the Stoic incorporeals is that the only source for the list, Sextus Empiricus, *Adversos Mathematicos*, was not known in the twelth century. Time and the vacuum as special forms of being were, however, available in Seneca, *Epistola*, 58, though the characterisation of them as such is attributed there to Plato.

immediately at odds with the observation at the end of the last quotation that these are items which sight and hearing cannot touch.

Concluding the discussion, *AM* complains that none of these theories succeeds in telling us what an assertable is since to know that we would have to be directly acquainted with them through our senses, and that is impossible. Even if *per impossibile* the assertable '*Deum esse*' and others were made visible we would not be able to tell which one it is. This is nice rhetorical touch but quite misses the point which *AM* like Frege relies upon in explicating their theories of assertables and Thoughts. Our access to these items is through our use of our languages to express them. It is only by the examination of properties of all the different kinds of propositions we may employ that we can say anything about their meanings and *AM* succeeds in what follows in saying a great deal.[45]

The Semantics of Propositions and Assertables

The semantical picture for propositions and assertables proposed by *AM* looks like this[46]:

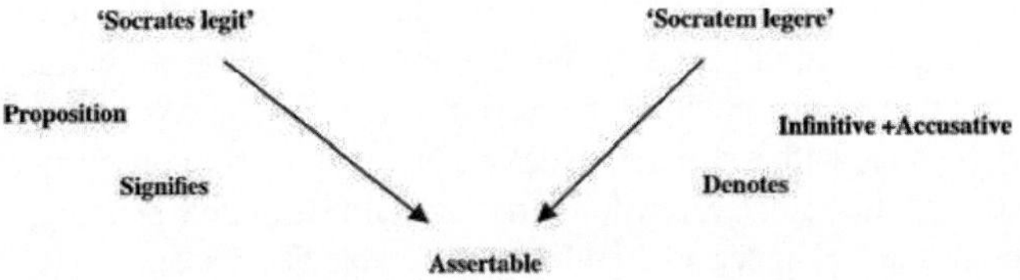

Because the infinitive accusative construction forms a name, and indeed a proper name, like any other name it both denotes and signifies. What it signifies is, according

45 *AM* 4.3, fol. 236rb: "Nondum tamen ex praedictis satis innotuit quid sint enuntiabilia, sed magis quid non sint. Quippe cum de his solis quae sensui subiacent sciri possit plene quid ipsa sint; sensus siquidem vias praeparat cognitioni; ergo ubi sensus deficit, cognitio non ascendit. Et praeterea cum diversi diversas assignaverint enuntiabilibus essentias, non est facile dinoscere qui verum attingerint; potuerunt autem bene dicere tam hii quam illi, cum viri essent magnae auctoritatis; itaque non novi utrum *deum esse* sit verum, aut utrum *Socratem esse hominem* sit *ipsum esse asinum.* Et generaliter de nullo enuntiabili aliquid scio; si enim, quod impossibile est, contingere<t> enuntiabilia fieri visibilia, et demonstratis omnibus, quaereret quispiam istorum esset *deum esse*, non esset qui interrogationem sciret certificare; qua inspectione nec novi aliquid de aliqua anima."

46 *AM* 4.3, fol. 236rb: "Habent autem enuntiabilia proprias appellationes quae sumuntur nominativo propositionis flexo in accusativum et verbo in infinitivum, ut *hominem legere* <est> appellatio huius enuntiabilis quod significat haec propositio 'Socrates legit'; cui similiter in aliis. Videtur etiam huiusmodi imperfecta oratio enuntiabile significare, non solum appellare. Nihil enim facit intelligi propositio quin detur intelligi per orationem imperfectam, quia utraque tantum enuntiabile facit intelligi. Unde idem videntur significare, quod plerique recipiunt; sed differ<un>t in modo, quia haec appellando, illa enuntiando. Quibus obviat descriptio assignata propositionis a Boethio; quae si convenienter et convertibiliter assignata est, erit propositio omne suscipiens ipsam, ergo vel oratio imperfecta non est oratio verum vel falsum significans aut est propositio."

to *AM*, not the assertable itself but, corresponding to the singulars signified by proper names, 'something like a singular' (*quiddam simile singulari*), *AM* gives no name to this item but let us call it a '*quasi singular*'. The quasi singular functions with respect to an assertable in a way parallel to that in which private and common *status* functions with respect to the things and kinds of things which participate in those *status*.[47] *AM* holds furthermore that synonymous terms agree both in signification and denotation and so the substitution of one for another does not affect the signification of proposition into which such a substitution is made.[48] 'Marcus is white' and 'Tully is candid' thus signify the same assertable and one and the same assertable may be denoted by many different accusative infinitive constructions. Cicero's various names are often used in modern discussions to illustrate the failure of identity of denotation to preserve meaning in certain contexts and Frege's proposed solution to the problem. Synonymy, however, allows no difference of descriptive content and unfortunately *AM* does not discuss the properties of proper names which agree in denotation but differ in signification, if, indeed, he allows that there are any. Nor does he raise the converse question of whether propositional tokens of the same type, for example those containing indexicals, signify different assertables.

An accusative infinitive construction functions like a proper name in denoting an assertable though it is not imposed to do this but rather constructed from the corresponding proposition. The question thus arises for *AM* of whether a proper name could be instituted to *signify* an assertable.[49] Given what we have seen of his account of the meaning of proper names the answer is, as we we would expect, no, though proper names can, he holds, be instituted to *denote* an assertable. This would presumably be done with a stipulative definition invoking the appropriate accusative infinitive construction since ostension is not available for intelligible items. A proper name cannot remain proper and signify an assertable because:

> [...] the name would have to <constitute> the understanding of the three expressions, 'Socrates', 'is', <and> 'white', and so would signify an adjective quality <i.e. whiteness> which would be contrary to the nature and property of proper names because all are proper in quality. A proper name cannot, therefore, be instituted to signify an assertable but only to denote <one> [...].[50]

---

[47] *AM* 4.3, fol. 236rb: "Ideo non dabimus quod significet enuntiabile cuius est appellatio, sed potius quiddam simile singulari cuius res est illud enuntiabile. Maxime tamen enuntiabile facit intelligi, quippe quaelibet vox magis facit intelligi appellatum quam significatum, velut nec iste terminus 'haec species' speciem significat quam demonstrat, quia potius individuum significaret quam speciem, si alterutrum; nec hoc nomen 'mille' significat mille, quae appellat; aut iste terminus 'hic homo' hunc hominem; aut 'quaelibet res' quamlibet rem; eodem modo de his orationibus sentiendum <est:> 'hoc enuntiabile', 'hoc verum', et de hoc nomine 'verum', non enim significant enuntiabile sed appellant."

[48] *AM* 4.2, fol. 236rb: "Ex hoc palam quod non secundum pluralitatem vel singularitatem enuntiabilium attenditur singularitas auditorum vel pluralitas; siquidem illi quorum unus audit hanc propositionem 'Marcus est albus', reliquus istam 'Tullius est candidus', non unum dicuntur audisse sed tantum diversa, quia diversas propositiones."

[49] On various discussions of this problem in the twelfth century see Kneepkens (1997).

[50] *AM* 4.4, fol. 236rb: "Illud tamen rationabiliter quaeri potest utrum possit aliquod nomen proprium esse significans hoc enuntiabile *Socratem esse album* vel aliquod aliud. Quod inde videtur, quia

A proper name thus lacks the internal propositional structure of an assertable but that structure is retained in the accusative infinitive construction which has both a nominal and a verbal component. We know from *De Interpretatione* that the characteristic feature of a verb is that it consignifies time and so the question arises of how these various components function in larger propositional contexts. In propositions, that is, which are used to make claims about assertables, for example about their modality. And in particular how does the tense of the infinitive in the expression denoting the assertable interacts with the tense of main verb of the proposition in which it appears as the subject.

Assertables and Tense

Explaining how tense functions in accusative infinitive constructions became more and more important as theologians in the twelfth and thirteenth century developed and opposed the theory that articles of faith are assertables. Their positions evolved as responses to Peter Lombard's appeal in his *Sentences* (I.41.3) to assertables as the unchangeable objects of God's knowledge. Later theologians associated the theory of the assertable employed by Lombard with the *Nominales* but, so far at least, no *Nominalist* account of assertables has been found. What we do have in the *Ars Meliduna* is the treatment by a contemporary school of the same problem.

Unfortunately *AM* limits his consideration to the behaviour of present tense infinitives. In his discussion of the effect of tense on the truth conditions of categorical propositions he maintains that if the verb is in the present tense, the denotation of the subject is all and only the presently existing things to which the name applies. With future and past tenses, however, there is an ampliation of its denotation to all the past and future things which fall under the name. The infinitive accusative construction functions, as we have seen as a proper name for an assertable so we might expect a similar account of its behaviour and, indeed, *AM* first argues that the time consignified by the embedded verb varies with the tense of the principal verb. The meaning of a proposition such as '*Antichristum legere erit verum*' on this account of it is thus that at some time, t_1, in the future it will then, at t_1, be true that the Antichrist is reading, the embedded temporal reference having narrow scope. Despite a series of arguments for this position, however, it is finally rejected by *AM* in favour of taking the embedded temporal reference with wide scope so that the claim '*Antichristum legere erit verum*' is true now, at t_0, if and only at a later time t_1 it will be true that the Antichrist is reading now, at t_0. This has the striking consequence that:

---

voces ad placitum sunt. Sed in contrarium validior est ratio. Haberet enim nomen illud intellectum istarum trium dictionum 'Socrates' 'est' 'albus', et ita significaret qualitatem adiectivam, quod esset contra naturam et proprietatem propriorum nominum, quae omnia in propria sunt qualitate; non potest ergo nomen proprium ad significandum enuntiabile institui, sed tantum ad appellandum, ne peccaret contra artem recte loquendi accidat."

> You will be guilty of a solecism if you say [...] '*Socratem diligere filium suum*' will be true or will not be true if he does not yet have a son ... because nothing is denoted by such a word, and so there is nothing which it subjects <to the predicate>.[51]

It thus seems that according to *AM* while we could before the birth of Christ have congruously and truly asserted '*Christus nasciturus est*' we could not then have congruously asserted '*Christum esse natum erit verum*', lacking a present denotation for 'Christ', the sentence would have been neither true nor false.

The Meaning of Quod *Constructions*

As I noted above the ugly mixture of English and Latin here is due to the necessity in English of translating accusative infinitive constructions with 'that' clauses. *AM*, however, rejects the proposal that '*quod*,' followed by a proposition may, like the accusative infinitive construction be used to denote the assertable signified by that proposition. His treatment of the question is very brief but it interestingly recalls twentieth century discussions of *oratio obliqua* in which 'that' is taken with the preceding verb rather than as an operator on the proposition which follows.[52] '*Quod Socrates legit*', for example, is not a well formed expression, and so certainly not a name. 'It seems', *AM* says:

> [...] that such an expression does not denote an assertable but rather that the verb is used impersonally, nor is a significant expression formed from these words. Similarly if you were to say, '*ab aliquo dicitur quod Socrates legit*' [...], the words placed after the verb do not form a significant expression, just as also happens in all of these cases: '*desidero quod Socrates currat*', '*servo praecipitur ut faciat ignem*', '*video hominem currere*' [...].[53]

The inclusion of '*video hominem currere*' is not explained by *AM* but is presumably intended to show that in some contexts, here with a verb of perception, infinitive accusative constructions, like '*quod*' clauses generally, do not denote assertables. We

[51] *AM* 4.4, fol. 236rb: "Unde nulla appellatio cum extrinsecis temporibus commode sumetur nisi et cum praesenti congrue fuerit sumpta, velut neque demonstratio. Et ideo soloecismum facis, si dicas omnem hominem diligere filium suum erit vel non erit verum aut Socratem diligere filium suum posito quod nondum habeat filium, aut Antichristum esse id quod ipse est, vel Socratem esse album quod ipse est, si nondum sit albus, aut angelos canere cras audietur ab aliquo, quia nihil subest appellationi talis vocis, et ideo non habet quid supponat."

[52] See Davidson (1968, 130–146) and Dummett (1973, 370–382).

[53] *AM* 4.5, fol. 236rb: "Sumitur autem et aliter appellatio enuntiabilis, ut plerique volunt, praeposita coniunctione 'quod' ipsi propositioni, ut 'quod Socrates legit est verum', nec aliud quam *Socratem legere*. Verius tamen videtur esse ut talis vox non sit appellatio enuntiabilis, sed ponitur impersonaliter verbum, nec fit oratio ex illis dictionibus. Similiter si dicas 'ab aliquo dicitur quod Socrates legit' vel 'aliqua propositione significatur'. In hac etiam oratione 'iste dicit quod Socrates legit' non faciunt orationem dictiones positae post verbum, velut et in his omnibus accidit 'desidero quod Socrates currat', 'servo praecipitur ut faciat ignem', 'video hominem currere'. Etenim si dicas 'aliquod quod est verum est quod Socrates legit' aut 'quod Socrates currit non est lapis' vel 'quod Socrates currit et quod Plato disputat sunt diversa', minus congrua apparebit locutio."

are surely intended to read '*video hominem currere*' *de re* as about the subject of the embedded clause. Likewise with '*desidero quod Socrates currat*' what I express is my desire for Socrates to run rather than for the assertable supposedly denoted by '*quod Socrates currat*'. Here, then, we have the beginnings of a theory of intensional verbs.

The remainder, and the bulk, of Part 4 of the Ars Meliduna is devoted to the division of assertables into trues and falses and the consideration of each of these in turn.

Force and Content

Frege has been credited by Peter Geach with the discovery of the 'Frege Point', the distinction between force and content which enables the same Thought to be asserted, inquired about with a question, commanded with an imperative, connected without being asserted with another Thought in a conditional proposition, and so on. Frege was certainly not the first to make the distinction, however, and it plays a crucial role in Abaelard's account of propositionality and propositional operations. Abaelard notes the distinction in explaining the relationship between different kinds of speech act to one another and relies on it in his account of the difference between separative and extinctive, or propositional negation.[54] The former signifying the separation of predicate from subject, the latter operating on a proposition to produce another defined to have the opposite truth value. I have argued in an earlier paper that *AM* more than any other twelfth century author understands that a propositional component of a compound proposition retains its signification in the compound but loses is assertoric force.[55] He is likewise aware of the distinction between force and content as it is applied to speech acts though his treatment of it is rather obscure.

The issue arises in the discussion of the adequacy of the division of *assertables* into trues and falses. Arguments that it is not turn on putative counter examples which are said to be true or false but not assertables. Thus someone may be in doubt about what Socrates is (*quid sit Socrates*), ask what he is, and thereby come to know what he is. Since only what is true is knowable it seems that *what-Socrates-is* (*quid sit Socrates*) is (a) true, and so something other than an assertable is true.

AM replies first by insisting, as he did when discussing '*quod*' constructions, that '*quid sit Socrates*' is not the name of an item which is said to be known when we assert '*scitur quid sit Socrates*'. Rather the proposition is to be read *de re* as a claim about (*de*) Socrates, saying of him that it is known what he is.[56] More generally he argues

---

[54] See Martin (2004a).

[55] See Martin (2004a).

[56] *AM* 4.6, fol. 236va: "Sed nos in hoc solvemus ut in priori, quia non fit oratio ex his dictionibus 'quid', 'sit', 'Socrates', dicto aliquem scire vel dicere quid sit Socrates; sed ponitur verbum absolute; quod manifestum fit per hoc verbum absolutum 'dubito', quod ibi poni potest, non enim dicitur 'dubito hoc', sed 'de hoc.'"

that question and assertion have the same content, asserted by uttering a sentence with assertive force and asked about by employing an interrogative construction. In the terminology perhaps introduced by Abaelard[57] such a construction is called a '*quaestio quaerens*', the content asked about and also assertable, is the '*quaestio quaesita*':

> [...] every assertable is an interrogable. For example someone who asks 'are the stars even in number or not?' asks about *astra esse paria*, and he only asks one thing. This, however, is not *astra esse paria*, so that one may legitimately ask whether the stars are even in number [...] because there is a difference between asking *about* something and asking *that*, and what interrogatively proposes the former is a *quaestio quaerens*, the later a *quaestio quaesita* [...].[58]

An assertable is thus the common propositional content of an assertion and the corresponding question and as we have seen, according to *AM*, it inhabits the realm of intelligible being. It may be expressed in different ways, and though our access to it is only through its various expressions it exists quite independently of any language. Assertables, like Frege's Thoughts, are the bearers of truth and falsity. Unlike Thoughts, however, which correspond to sentences from which all indexicality has been removed and so bear their truth value immutably,[59] assertables retain the context dependence of the sentences which express them and, like Stoic 'changing *axiomata*',[60] change their truth value and fail to have a truth depending on the state of the world:

> [...] some assertable often begins and ceases to be true as things vary, and of some contradictory opposites it is not necessary that one be true and the other false.[61]

Note that *AM*'s remark here on contradictory opposites is in line with his general understanding of negation as separative. While he does, in his discussion of compound propositions, allow that negation may be preposed to a proposition he nevertheless always reads it separatively.[62]

---

[57] See Abaelard (1969, 256) and Jolivet (1982, 191–193).

[58] *AM* 4.6, fol. 236va: "Neque recipimus praedictam distinctionem, immo omne enuntiabile est interrogabile, ut *astra esse paria* interrogat qui sic quaerit 'sunt astra paria necne?', et unum solum interrogat; tamen astra esse paria non est, ut licite dicatur; nec *omnem animam esse immortalem* 'utrum omnis etc.', quia aliter dicitur hoc 'interrogare de quo' et aliter 'illud', ut quod interrogative proponit istud est quaestio quaerens, illud quaesita, velut et per imperativum ut 'impero tibi ut legas' <est> 'impero etiam legere', tamen 'legere' non est 'ut legas' [...]."

[59] For a critical discussions of Frege's account of Thoughts as eternal see Carruthers (1984); also Dummett (1973, ch 11).

[60] The *metapiptonta*. See Ierodiakonou (2006). Ierodiakonou translates '*axioma*' as 'assertible' to distinguish the notion from that of a proposition—the term which has replaced 'Thought' in modern discussions of eternal bearers of truth and falsity.

[61] Continuing the text quoted in n. 21, *AM* 4.8, fol. 236vb: "Ponendum itaque talia enuntiabilia nugatoria fieri posse. Ex quo accidit aliquod enuntiabile pluries incipere et desinere esse verum vel falsum iuxta rei variationem et aliquorum contradictoriae oppositorum non esse necesse unum esse verum et alterum falsum, licet hoc contrarium Aristoteli inveniatur, nisi quia dictum est illud <de> eo genere loquendi quo et hominem esse et animal esse necessarium aut contradictoriarum unam esse veram et alteram esse falsam."

[62] See Martin (2004b).

AM's general discussion of trues and falses continues with a discussion of the relationship between (a) true (*verum*) and truth (*veritas*) which seems to conclude that there is no difference between them, or between (a) false and falsity.[63] After considering in great detail and for various kinds of proposition what is properly said to be true or false of what, and concluding that it is assertables which are said to be true of things,[64] *AM* finally moves on to discuss the proper characterisation of true and false.

Trues

After examining the properties of the assertables corresponding to various tenses of categorical propositions, including the future tense and discussing the problem of future events,[65] *AM* eventually comes to the question of the role of assertables in argument. Abaelard had effectively distinguished the theory of argument from that of the conditional. He claimed that so far as a dialectician is concerned, the purpose of argument is to convince an opponent of some claim about which he is initially in doubt. The means for doing this is the *argumentum*, defined as a reason bringing conviction in some doubtful matter. In the *Dialectica* Abaelard argues that the *argumentum* must be what it is that is said to be so by the premise and glosses Boethius as characterising an *argumentum* as necessary if it is impossible for it to be true and the conclusion false at the same time. In his later *Glosses on the Topics,* however, he argues against the claim that *argumenta* are the *dicta* of premises and

[63] The discussion seems incomplete. *AM* first presents the case for truth and falsity as the properties which make trues true and falses false, as whiteness makes white things white. His only challenge to this position is that someone who holds it must tell us whether the property in question can be signifed with some expression. If so that expression must be a proposition or some other kind of expression. It cannot be a proposition because a proposition would then signify both an assertable and a property of assertables. And that's all he has to say, *AM* 4.9, fol. 237ra: "Sed ab eo illud quaeremus utrum veritatem aliqua voce dici contingat vel significari. Quo sumpto erit vox illa propositio vel alia vox; et si propositio, accidet qualibet propositione significari duo. Similiter utrum mendacium sit idem falso aut idem falsitati vel aliud ab utroque."

[64] *AM* 4.11, fol. 237rb: "His itaque nihil impedientibus dicimus enuntiabilia esse vera de aliquibus." Again we have a apparent disagreement with *Secta Meliduna*, Thesis 15: "<N>ulla propositione dicitur eius dictum de aliquo."

[65] *AM* 4.19, fol. 238vb: "De eventibus tamen futuris non minima inter veteres habita est dissentio." Beginning his discussion of assertables about the past *AM* strikingly remarks "In his primum notabis omne dictum singularis et affirmativae de praeterito cuius veritas non pendet ex eventu praesenti vel futuro, necessarium esse si fuerit verum, non tamen impossibile si falsum." (*AM* 4.19, fol. 238vb) This suggests that he holds the the reference to the past infects such propostions and renders them contingent, the position rejected by Thomas Aquinas in his treatment of divine foreknowledge and human freedom. See e.g. Thomas Aquinas, *Summa theologiae*, I^{a}, q. 14, a. 13, ad 2. *AM*'s discussion of the necessity of future events includes a refutation of the 'Consequence Argument' by distinguishing between *de dicto* and *de re* readings of 'possibile est aliquam rem aliter evenire quam eveniet'.

insists that Boethius' account of them requires instead that they are the propositional tokens employed in presenting the argument.

AM canvases four theories of the nature of *argumenta* before giving his own. The first is the theory of Abaelard's *Glosses* and *AM*'s objection is that the theory make the existence of reasons and proof entirely contingent upon the existence of conventional spoken language. It is clear from Abaelard's discussion that there are two related questions about argument which have to be settled in formulating the theory of argument. Firstly what kind of thing is the *argumentum* and secondly how is it guaranteed to play the role assigned to it of providing conviction where something is in doubt, how, that is, does the *argumentum* prove the conclusion? The second and third theories propose that it does the latter by guaranteeing the connection between premises and conclusion. According to the second it does this because it is the *dictum* of the general conditional proposition formed from the premises and conclusion of the argument, for example the *dictum* of 'if something's a human then it's an animal' for the enthymeme 'Socrates is a human, therefore he's an animal'. According to the third it is simply the *dictum* of the conditional formed from the premises and conclusion. For the fourth theory the *argumentum* is the topical relationship between the terms appearing in the argument.

AM rejects all of these theories because they fail to explain how it is that the *argumentum* proves the conclusion. His concern is that the *argumentum* justify confidence in the truth of the conclusion and it can only do this, he holds, by being true, that is to say by being a true assertable from which the truth of the assertable corresponding to the conclusion follows. So for the relation consequence to exist *AM* requires not simply that the connection between assertables be necessary but also that the assertables both be true.[66] So, finally we come to the theory of the false.

Falses

The final chapters of the *Ars Meliduna* are devoted to false *assertables*[67] and the first question is whether they exist (*utrum ipsa sint*). We have seen that according to the *Secta Meliduna,* provided we understand '*nihil*' for '*nullum*' they do not. This is a rather attractive position since it might yield Thesis 11, *Nihil sequitur ex falso,* the thesis associated with the *Melidunenses* by their contemporaries, by appeal to the principle that nothing comes form nothing. After presenting a large number of direct and indirect arguments including one which considers whether the proposition '*nihil*

[66] *AM* 4.39, fol. 240va: "Itaque nulla naturalis hypothetica est vera nisi cuius utraque pars est vera; ceterae autem falsae simpliciter aut, quod verius, incongruae, quia haec coniunctio 'si', quotiens indicativo proprie iungitur, confirmative ponitur et certitudinem notat." Thesis 10 of the *Secta Meliduna* is: '<N>ulla consequentia naturalis affirmativa vera est, nisi et antecedens et consequens ipsius verum sit.'

[67] *AM* 4.34–37. The text is perhaps incomplete since it finishes without an '*explicit*' though at the end of a long series of objections and responses to the thesis that nothing follows from a false.

est falsum' or the contradictory opposite is true, but nowhere mentioning the claim '*nullum falsum est*', *AM* concludes that:

> Persuaded by all these arguments we say that just as (a) good is something so also (a) bad, just as (a) true, so also (a) false.[68]

This is not the only point on which *AM* disagrees with the *Secta Meliduna* so we may wonder which of them was the true follower of Robert of Melun. They do agree, however, on many claims and in particular on Thesis 11. *AM* characterises this thesis as the second thesis about the false[69] sadly without telling us what the first is but presumably it was a claim about the existence of the false.

Frege, too, notoriously insisted on a version of Thesis 11:

> From false premises nothing at all can be concluded. A mere Thought, which is not recognised as true, cannot be a premise. Only after a Thought has been recognised by me as true, can it be a premise for me. Mere hypotheses cannot be used as premises.[70]

Frege's concern here is with justification and proof, just as *AM*'s was in its treatment of *argumenta* and it is to the theory of *argumenta* that he appeals to support it[71]:

An *argumentum* is a reason bringing conviction to something in doubt. But nothing lacking conviction may provide conviction for another; therefore nothing false, for how could it give to another what it cannot give to itself. Therefore it cannot be an *argumentum* for something. So nothing comes from a false.

Unlike Frege, however, who accepted the use in argument of conditionals with false antecedents, *AM* insists that the thesis applies to natural conditionals, as well as to the corresponding enthymemes.[72] This provides him with a solution to Alberic's embarrassing argument against Abaelardian connexive logic,[73] but the price for his own logic is very high. Contraposition can never hold and transitivity only when all of the conditionals in the chain have true antecedents. *AM* apparently accepts at least the first of these consequences but the discussion which follows the statement of Thesis 12 is very hard to follow, abounding in the *instantiae* which are defining feature of the *Ars Meliduna* and I have nothing useful to say about them yet. So let me conclude.

---

[68] *AM* 4.35, fol. 240rb: "His rationibus persuasi dicimus quod sicut bonum est aliquid, ita et malum, sicut verum, ita et falsum."

[69] *AM* 4.37, fol. 240va: "Secunda positio de falsis est quod ex nullo falso aliquid sequitur."

[70] Frege (1980, 182).

[71] *AM* 4.39, fol. 240va: "Argumentum est ratio rei dubiae faciens fidem. Sed nullum carens fide poterit alii praestare fidem; ergo nullum falsum (nam quomodo dabit alii quod non potest sibi ipsi?). Ergo non potest esse argumentum ad aliud. Quare ex falso nihil."

[72] See n. 63.

[73] See Martin (2004a).

4 Conclusion

The *Ars Meliduna* is an enormously complex and sophisticated work. It contains a theory of propositional meaning which very much anticipates Frege's. The great difference, of course, is that Frege takes propositions to be a variety of name and so concludes that truth values are objects. There can be nothing like this in the *Ars Meliduna* since the distinction between names and propositions must be strictly observed. Perhaps this is not such a bad thing, however, since Michael Dummett, for one, holds that Frege's theory of the denotation of propositions had disastrous consequences for his semantical project.[74] Disaster for the *Meliduneses* perhaps lies elsewhere, in the attempt to make logical relations vary contingently with truth-value. Whether this disaster can be avoided can only be settled by further study of the *Ars Meliduna* and so let me finish by encouraging this.

References

Primary Texts

Abaelard, P. (1919). *Logica Ingredientibus (Isagoge): Peter Abaelards Philosophische Schriften,* I.1 (B. Geyer, Ed.) (*Beiträge zur Geschichte der Philosophie des Mittelalters,* XXI.1). Aschendorf.

Abaelard, P. (1969). Super Topica Glossae. In M. Dal Pra (Ed.), *Scritti di Logica* (2nd ed., pp. 206–330). La Nuova Italia.

Anonymous. 1983. *Compendium Logicae Porretanum, CIMAGL* 46.

Boethius. (1877). *In Librum Peri Hermenias. Editio Prima* (C. Meiser, Ed.). Teubner.

Priscian. (1859). *Institutiones Grammaticae* (H. Keil, Ed.). Teubner (*Grammatici Latini* III).

Stephanus. (1885). *Stephani in Librum Aristotelis De Interpretatione Commentarium* (M. Haydruck, Ed.). Reimer (Commentaria in Aristotelem Graeca XVIII. 3).

Secondary Literature

Biard, J. (1987). Semantique et ontologie dans l'Ars Meliduna. In J. Jolivet & A. de Libera (Eds.), *Gilbert de Poitiers et ses contemporains* (pp. 121–144). Bibliopolis.

Biard, J. (1999). Le langage et l'incorporel. Quelques réflexions à partir de l'Ars Meliduna. In J. Biard (Ed.), *Langage, sciences, philosophie au XIIe siècle* (pp. 217–234). Vrin.

[74] Dummett (1973, 196): "Unhappily, the doctrine [...] which regards truth-values as objects and hence assimilates sentences to complex proper names, undermined the sharpness of the original perception. If sentences are merely a special case of complex proper names, if the True and the False are merely two particular objects amid a universe of objects, then, after all, there is nothing unique about sentences [...] This was the most disastrous of the effects of the misbegotten doctrine that sentences are a species of complex name, which dominated Frege's later period: to rob him of the insight that sentences play a unique role, and that the role of almost every other linguistic expression (every expression whose contribution to meaning falls within the division of sense) consists in its part in forming sentences [...]."

Carruthers, P. (1984). Eternal thoughts. *The Philosophical Quarterly, 34*, 186–204.
Davidson, D. (1968). On saying that. *Synthese, 19*, 130–146.
De Libera, A. (1996). *La Querelle des universaux*. Éditions du Seuil.
De Libera, A. (1999). *L'Art des généralités: Théories de L'Abstraction*. Aubier.
De Rijk, Lambert M. (1967). *Logica Modernorum* (Vol. II.2). Van Gorcum.
Dummett, M. (1973). *Frege: Philosophy of language*. Harper & Rowe.
Frege, G. (1980). In B. F. McGuinness (Ed.), *Philosophical and mathematical correspondence* (Hans Kaal, Trans.). Blackwell.
Ierodiakonou, K. (2006). Stoic logic. In M. L. Gill & P. Pellegrin (Eds.), *A companion to ancient philosophy* (pp. 505–529). Blackwell.
Iwakuma, Y. (1997). Enuntiabilia in XIIth Century Logic and Theology. In C. Marmo (Ed.), *Vestigia, imagines, verba semiotics and logic in medieval theological texts (XIIth–XIVth century)* (pp. 19–35). Brepols.
Jolivet, J. (1982). *Arts du langage et théologie chez Abélard*. Vrin.
Klement, K. (2016). Russell's logical atomism. In E. N. Zalta (Ed.), *The stanford encyclopedia of philosophy* (Spring 2020 Edition). http://plato.stanford.edu/archives/spr2016/entries/logical-atomism/. Accessed 31 January 2021.
King, P. (1982). *Abelard and the problem of universals in the twelfth century*. Princeton University.
Kneepkens, C. H. (1997). Please, don't call me Peter: I am an enuntiabile, not a thing. A note on the enuntiabile and the proper noun. In C. Marmo (Ed.), *Vestigia, imagines, verba semiotics and logic in medieval theological texts (XIIth–XIVth century)*, ed. (pp. 83–98). Brepols.
Martin, C. J. (2004a). Logic. In J. E. Brower & K. Guilfoy (Eds.), *The Cambridge companion to Abelard* (pp. 158–199). Cambridge University Press.
Martin, C. J. (2004b). Propositionality and logic in the *Ars Meliduna*. In A. Maierù & L. Valente (Eds.), *Medieval theories on assertive and non-assertive language* (pp. 111–128). Olschki.
Nuchelmans, G. (1973). *Theories of the proposition, ancient and medieval conceptions of the bearers of truth and falsity*. North Holland.

Complete Forms, Individuals and Alternate World Histories: Gilbert of Poitiers

Graziana Ciola

1 Introduction

In his seminal contribution on the emergence of a synchronic conception of alethic modalities in Gilbert of Poitiers' commentaries on Boethius' *Opuscula sacra*,[1] Simo Knuuttila sketched, in passing, an analogy between Gilbert's account of individuality and Leibniz's complete concept.[2] He did so without omitting a poignant remark on the most evident difference between Leibniz's view and what he rightfully took to be Gilbert's: for Gilbert, the complete form of an individual contains not only those predicates that are actualized at a given point of an individual's history, but possible predicates as well.

Comparisons of this sort, connecting disparate points across time as well as across philosophical traditions and discussions, are unavoidably farfetched. If framed in the light of a search for the forerunners of any later renowned theories or philosophical approaches—the quest for the infamous *first-one-to-claim-this-or-that*, who can always be found at an earlier point in time than previously thought—such comparisons would be of little interest and even less import, especially if unsupported by a provably continuous textual or intellectual tradition. Neither such a continuity of textual transmission nor of a more or less direct intellectual influence of Gilbert's authentic doctrines can be tracked up to Leibniz with any certainty.

However, similarly far-fetched comparisons do highlight conceptual similarities between theories dealing with somewhat similar questions and reaching similar conclusions without any confirmed connections to each other. In doing so, these

[1] Edited in Häring (1966). All quotations of Gilbert's works are from this edition.

[2] Knuuttila (1987, 217).

G. Ciola (✉)
Center for the History of Philosophy and Science Faculty of Philosophy, Theology and Religious Studies, Radboud University, Erasmusplein 1, 6525 HT Nijmegen, The Netherlands
e-mail: graziana.ciola@ru.nl

F. Ademollo et al. (eds.), *Thinking and Calculating*, Logic, Epistemology, and the Unity of Science 54, https://doi.org/10.1007/978-3-030-97303-2_6

comparisons help in tracking the history of the risings and resurgences of genuine philosophical problems, even or especially in widely different intellectual contexts and conceptual frameworks. From this angle, such comparisons not only are informative on a given philosophical problem's conceptual and historical articulation, but they are just as helpful to interrogate earlier texts formulating innovative interpretative hypotheses, asking new questions, and coming to conceptually relevant answers.

In the aforementioned study, Simo Knuuttila formulated another suggestive hypothesis. Because of his metaphysics of individuality, his reframing of modal notions within a synchronic rather than diachronic framework, and his theological account of God's alternative providential plans, Gilbert of Poitiers had all the conceptual instruments to conceive the question—which Leibniz himself asked and which still is one of the core points of dispute in modal metaphysics—about the possibility of an individual's counterfactual identity. In the present study I rely on Knuuttila's hypothesis, addressing it "from within" Gilbert's philosophy and theology in order to test the idea that Gilbert can indeed conceive, entertain and treat the notion of counterfactual identity. I do so by bringing together several relevant features of Gilbert's ontology, his innovative take on the problem of individuation and his account of creation.

Knuuttila's reading is quite strong and far from being uncontroversial, mainly insofar as two general theses are concerned: (i) that Gilbert's account of modal notions is properly synchronic and is grounded on an embryonal conception of genuinely alternative possible worlds; and (ii) that modal properties—"modal" in this sense—are coherently included within Gilbert's theory of individuation. The present study aims to propose a textually based presentation and an analysis of Gilbert's own account that for the most part supports—and in some case refines—both (i) and (ii).

Over the course of this study, I will not overtly engage with any opposing readings addressing Knuuttila's.[3] However this study intends to show that, on the one hand, (i) and (ii) are neither obviously incoherent with Gilbert's text nor particularly problematic; and, on the other hand, that Gilbert's theory can indeed treat many of the potential issues stemming from the modal definition of individual—be they Gilbert's own problems or Knuuttila's.

I will get there by paving the way as follows. In the first place, I briefly outline those features of Gilbert's own well-known developments on Boethius' ontology that are most relevant for the matter at hand (1). I then stress the modal features of Gilbert's account of individual substances, thus informally defining the modal conception that such features imply and the type of modal space within which they should be situated (2). I consequently examine Gilbert's take on the Augustinian doctrine of God's alternative providential plans or alternative histories of the world in God's mind, underlying the modal implications of such an account (3). And finally I address the issue of whether Gilbert of Poitiers has the conceptual framework to conceive of a form of counterfactual identity or of cross-worlds counterpart relations (4).

---

[3] See in particular Marenbon (1998, 171–172), and Valente (2011, 412–413).

2 Porretan Ontology—An Outline

Gilbert's ontology certainly is his most renowned philosophical contribution[4]—especially insofar as the metaphysics of individuality is concerned—both as a predominantly philosophical matter and in connection to his theology.

Still, several major interpretative disagreements split the scholarship. Some concern Gilbert's collocation on the realist-nominalist spectrum in the context of the infamous problem of universals. Is Gilbert a moderate realist? Is he a realist insofar as immanent forms are involved but not about universals?[5] Or is he some sort of trope theorist in the same vein as Abelard and holding fewer ontological commitments than normally expected?[6] Other open questions address the impact, in Gilbert's thought, of Neoplatonic influences, along with the conceptual relationship between Gilbert's theological concerns and his seemingly more strictly philosophical theses focusing on the ontology of ordinary beings and on matters of language.[7] While these exegetical issues are crucial for a proper framing of Gilbert's intellectual production and its impact, I do not touch on these matters here, instead limiting my focus to those features of Porretan ontology that have a direct relevance for the task at hand—i.e. assessing whether Gilbert's ontology could admit any form of counterfactual identity.

Therefore, in this section, I introduce the relevant broad strokes of Gilbert's ontology and, given Gilbert's notoriously obscure style, I do so by examining a few central terms and distinctions, beginning with the notorious Porretan distinction between *subsistens* and *subsistentia.*

"Subsistens" and "Subsistentia"

Gilbert uses the term *subsistens* for what Boethius called *id quod est*[8]—an expression which Gilbert and some of his followers still employ interchangeably.[9] *Subsistens*, "the subsisting thing", or *id quod est*, "that which is", is the determinate entity, that

[4] Despite none of his strictly logical works having survived, Gilbert was also a celebrated logician and the head of one of the major logical trends of the 12th-century, that of the Porretani. See e.g. Ebbesen et al. (1983), Ebbesen (1992), and Ebbesen and Iwakuma (1992). On the Porretan school and tradition beyond their logical positions see in particular Catalani (2008).

[5] Most interpretations are actually on a fairly moderate spectrum. See e.g. Forest (1934a, 1934b), Hayen (1935–1936), Vicaire (1937), Williams (1951), Schmidt (1956), Vanni-Rovighi (1956), Westley (1959–1960), van Elswijk (1966), Maioli (1979), Nielsen (1982), de Rijk (1988a, 1989), and Valente (2008).

[6] For a particularly deflationist assessment see e.g. Erismann (2014). A fairly balanced interpretation is e.g. in Jolivet (1992).

[7] Besides de Rijk (1988a, 1989), for synthetic references on these matters see e.g. Marenbon (1988).

[8] Cfr. Maioli (1978, 13–31), and de Rijk (1981, 1988b, 1–29).

[9] Valente (2008).

thing which is what it is due to the inherence of several forms in a substratum.[10] In other words, a "subsistent" is any concrete and singular thing, considered "as one self-contained entity, [...] whose identity and ontological unity are due to the singularity of what is proper to it (*sue proprietatis singularitas*)".[11] A subsistent, then, is a structured concretion (*concretio*) of those articulated forms and properties inhering (*inesse*) to the subsistent itself. As such, a "subsistent" amounts to what, in a more widespread philosophical jargon, is commonly called "substance" in the sense of a determinate entity and the substrate of inhering properties.

Subsistentia, on the other hand, is equivalent to Boethius' *esse* (being) or *id quo est* ("that in virtue of which [something] is"), i.e. the formal principle due to which an *id quod est* is what it is and has its essential features.

In this sense, a *subsistentia*, then, is equivalent to a property inherent in a substratum. However, the concept of *subsistentia* is neither unitary nor is it univocally characterised, its use being at least twofold. Namely, "subsistence" is both a singular essential property pertaining to a determinate subject (e.g. mortality for an animal) and an *id quod est's* form as a whole, i.e. that form in virtue of which an *id quod est* is that particular *id* and not another. This subsistence of a subsistent as a whole is also called its "complete form" or "proper form" (*tota forma/subsistentia, propria forma/subsistentia*). The complete form is composed by the subsistent's singular and simple subsistences (*subsistentie*) or forms or essences (*essentie*)—Gilbert's terminology is notoriously inconsistent—which are the properties that make an *id quod est* have the essential features it has and, together, make it into that *quod est.* A subsistent's complete form, then, includes a set of core essential properties pertaining to that subsistent and constituting it. For the complete form to be "complete", there should presumably be a sense in which this set includes all the subsistent's relevant properties and some of them might as well be accidental.[12] For it to be a "form" or a "subsistence" itself, there has to also be some sort of unity involved—be it the unity of a mereological compound, or that of a bundle of forms articulated hierarchically, or a supervening property. It is nonetheless clear that the complete form has an internal articulation of its properties and that it can either be taken as the whole set (*tota proprietate sua*) or considered with respect to a restricted group of those properties (*aliqua sue proprietatis parte*)", i.e. a proper subset.[13] A similar internal articulation is not a prerogative of the *tota forma*, but it applies to any subsistence that is not simple. For example, the form of humanity (*humanitas*) is not a simple form, but is articulated in other subsistences making it up, e.g. corporeity (*corporeitas*), animality (*animalitas*), rationality (*rationalitas*), etc. Not all of these forms stand in the same relation to each other or to the form of humanity since, for example, to be an animal requires to be corporeal; therefore corporeity is a simpler subsistence than animality and is subsumed under it. Plato's humanity, with an internal structure of this sort, is in turn encompassed under Plato's proper complete form, i.e. that

[10] Valente (2008, 194).

[11] de Rijk (1988a, 74).

[12] See sections '"Accidens" and "Status"' and '"Participatio" and "Habitus"'.

[13] Valente (2008, 195).

subsistence that is only Plato's and is characterising and individual, in other words his *Platonitas*. Plato's *Platonitas* is a set of properties of the kind described. Plato's complete form—i.e. what makes Plato into Plato and not Socrates or Aristotle—is the principle that individuates Plato. Those features typical of Porretan ontology, in this framework, appear quite clear and are justified. Humanity is what, in the Porretan jargon, is called a "dividual" (*dividua*) property, i.e. a property that is not individual. Dividual properties are intensionally the same for all individuals instantiating them. This means, in this case, that humans are human because they have the subsistence of humanity, which is defined in the exact same way in Plato, Socrates, Aristotle, and so on for all humans. However, humanity is—or, better, the "humanities" are –numerically distinct from one subsistent to the other. While Plato and Socrates are human in the exact same way—and this is the way in which humanity is definitionally identical and indiscernible in Plato and in Socrates—Socrates' own humanity is not Plato's: they are not numerically the very same humanity, but Plato has his own subsistence of humanity within his complete form and Socrates has his own. There is a relationship of "conformity" (*conformitas*) or "exact resemblance" (*similitudo*)[14] between Plato's and Socrates's humanities, but they remain metaphysically and actually distinct. This type of account grounds the dominant interpretation of Gilbert's thought as an instance of moderate realism, i.e. a realism of the dividual forms, as concreted subsistences (*concretio*) within the subsistents, but not of separate universals.[15] The relationship of conformity or exact similitude between subsistences of the same kind could be interpreted either as a properly realist element of Gilbert's metaphysics, if one, for example appeals to some further (and rather universal) form doing for dividual subsistences what these do for the subsistents instantiating them. A deflationist reading going in the nominalist direction—e.g. by pointing at the abstractive nature of such similarity—is just as possible.[16] Either interpretation can find support in theoretical and textual arguments. Overall, for my purposes here what counts is not so much whether these relations require some degree of reification: it

[14] On this, see Erismann (2014).

[15] E.g. Gilbertus Porreta, *DTrin.* I, 1, 27, p. 76,68–71: "Dicuntur etiam multa subsistencia unum et idem non nature unius singularitate sed multarum, que ratione similitudinis fit, unione. Hac enim plures homines 'unus vel idem homo' et plura animalia 'unum vel idem animal' dicuntur". See also Valente (2008, 214 ff.), Maioli (1979), Maioli (1974), Gracia (1984a) and especially Jolivet (1992, 148–149): "Comparé à Roscelin Abélard apparaît réaliste; on parvient évidemment à un résultat différent quand on le compare à Gilbert. Cette comparaison peut se résumer ainsi: tous deux ont une ferme doctrine de l'individu irréductible et séparé; mais si Abélard ne peut se défère d'une tendance platonisante, Gilbert est clairement réaliste, bien que son réalisme porte sur la forme et non sur l'universel; cette différence entre leurs ontologie respectives s'exprime aussi dans leur relation à Boèce: Abélard n'en retient que les commentaires de l'*Organon*, Gilbert s'attache à commenter les opuscules théologiques. Il y trouve une métaphysique de l'*esse* et du flux ainsi qu'une épistémologie [...] il puise l'essentiel de son platonisme chez l'auteur qui en fut pour les médiévaux une des sources principales alors que celui d'Abélard ne se règle guère que sur le schème impliqué dans l'arbre de Porphyre."

[16] Some scholars (e.g. Valente, 2008, 215), count the *conformitas* or *similitudo* between different subsistences of the same "kind"—e.g. between Plato's and Socrates's humanity—as a properly realist feature of Gilbert's thought. Others have argued for a radically opposite position—see e.g. Erismann (2014).

is instead the clear and crucial theorization of the notion of conformity or exact similitude between subsistences instantiated by different subsistents.

"Accidens" and "Status"

As we have seen, the "complete form" constitutes, for Gilbert, the formal cause and the metaphysical principle making each subsistent that particular subsistent instead of another. This represents one of Gilbert's major contributions to the history of the metaphysics of individuality and constitutes a development of and a shift from the account that normally goes by the label of Standard Theory of Individuality (STI).[17] STI locates the principle of individuation in "one or more accidents (usually place) of a substance" or in "the collection of all features (including non-accidental ones) belonging to the substance".[18] While Gilbert still operates within the STI framework (i.e. individuation is the result of a collection of properties belonging to the substance), his approach implies a reconsideration of what metaphysical individuality ultimately amounts to.[19]

For Gilbert, the complete form encompasses some essential properties—i.e. those subsistences that are forms of being (*forme essendi)*—seemingly along with properties that are accidental. The *subsistentie* as *forme essendi* inhere (*inesse*) to the subsistent and are, in a sense, the very core of the complete form. However, there is a degree of ambiguity about what, for Gilbert, constitutes an accidental property and about which accidental properties are determinant for individuation.[20]

In discussing the ten Aristotelian categories, Gilbert seems to count all the qualifications included under the nine categories other than substance as *accidentia*. Nothing particularly unusual there, at least at first sight:

> Accidentia vero de illis quidem substantiis que ex esse sunt aliquid, dicuntur – sive in eis creata sive extrinsecus affixa sint – sed eis tantum qua esse sunt accidunt. Quare illis recte non "inesse" sed "adesse" dicuntur.[21]

While the unqualified "*accidens*" applies to the usual extent of Aristotle's *symbebêkota,*[22] an important distinction between two sets of accidents is immediately evident. On the one hand, some classes of accidents are intrinsic to the subsistent to which they pertain (*adsunt*). Accidents of this sort are those within the category of quality and quantity. Their *adesse* to the subsistent amounts to a pertinence relationship which, despite being a properly distinct type of relation, is somewhat comparable to the *inesse* proper of substantial forms, in the sense that—just like *subsistentie*—they are naturally connoting the subsistent to which they pertain. No subsistent can

---

[17] Gracia (1984a, 1984b).

[18] Gracia (1994, esp. 21–36).

[19] On Gilbert's shift from STI, see Valente (2011, 412–413).

[20] For a systematic exposition on this subject, see de Rijk (1988a, 101 ff).

[21] Gilbertus Porreta, *Dtrin.* I, 4, 19, p. 118,8–11.

[22] See also, e.g., *CEut.* I, 84, p. 260,91–97.

be unqualified or unquantified; all the quantities and qualities that a subsistent holds (or can hold) are *in eis creata*—i.e. any subsistent is naturally apt to hold the quantities and qualities it holds. On the other hand, other classes of accidents are extrinsic (*extrinsicus affixa*), i.e. they are extrinsically attached to the subsistent. As de Rijk had already noted, Gilbert tends to consider as proper accidents only those accidents that are intrinsic: they are concomitant with subsistences and, as such, they are predicable of and ascribable to the subsistent itself.[23] "Nature" (*natura*) is one of those words that find several uses in Gilbert's prose.[24] Namely, "nature" is often employed to refer to formal *subsistentie,* but it sometimes encompasses intrinsic accidents as well in cases when *natura* is used as an exact equivalent of complete form.[25] It is then the case that the complete form includes not only formal subsistences but intrinsic accidents as well: along with formal subsistences, the intrinsic accidents pertaining to them make their subsistent be that *quod est* and not another.

As for extrinsic accidents (*accidentia extrinsecus affixa*), they pertain to what Gilbert calls *status*. A subsistent's *status*—which is sharply distinct both from formal subsistences and from intrinsic accidents—includes some entity's extrinsic collocation and transitory condition,[26] i.e. all those purely contingent attributes whose presence or absence, for Gilbert, does not affect what a *quod est* is. Whereas intrinsic accidents are limited to quality and quantity, those which make up the status comprehend the other seven "little" categories: relatives; somewhere; sometime; being in a position; having; acting; and being acted upon.[27]

---

[23] *DTrin.* I, 4, 21, p. 119,21–25: "Pars vero predicamentorum loco rationis est in numero accidencium: scilicet cum de subsistentibus dicantur, tamen eorum subsistencias comitantur. Et sic quidem in naturalium genere quecumque predicantur, naturalium propriis rationibus 'substancie' vel 'accidentia' nominantur". See de Rijk (1988a, 101).

[24] Cf. e.g. Valente (2008, 195–196) and de Rijk (1988a, 106–111).

[25] For examples, *CEut.* 2, 11, pp. 266–267,67–78: "Naturam autem alie sunt substantie alie accidentes"; 5, 25, p. 319,59–62: "Natura enim subsistentis est qua ipsum subsistens aliquid est. He vero sunt substantiales forme et, que illis in ipso subsistente adsunt, qualitates et intervallares mensure." de Rijk includes also *DTrin.* I, 2, 28, p. 84 and I, 2, 53, p. 89 in the list of relevant passages in this sense. However, the inclusion of these passages may be unwarranted, since there is not any explicit element uncontroversially referring to the inclusion of intrinsic accidents along with formal subsistences.

[26] *CEut.* 5, 25, p. 319,58–59; 5, 26, p. 319,63–66; and 5, 27, pp. 319–320,67–71: "[...] de humane nature statu clarius intelligi poterit quod videlicet cuiuslibet subsistentis aliud est natura, aliud status. [...] Cetera vero, que de ipso naturaliter dicuntur, quidam eius 'status' vocantur eo quod nunc vero aliter – retinens has, quibus aliquid est, mensuras et qualitates et maxime subsistentias – statuantur. Nam – sepe manente colore et trium vel quatuor vel quolibet cubitorum lineis, semper autem veri nominis subsistentiis manentibus – homo nunc illo situ vel loco vel habitu vel relatione vel tempore vel actione vel passione statuitur et – idem permanens – secundum extrinsecus sibi accidentia variantur."

[27] *CEut.* 5, 28, p. 320,72–74: "Idem enim est homo salendo quod stando: et extra domum quod intra: et inermis quod armatus: et dominus quod servus: et mane quod vespere: et quiescendo quod agendo: et letus quod tristis."; 8, 16, pp. 357–358,19–24: "Superius dictum fuisse recordor quod cuiuslibet subsistentis aliud est natura, aliud status: et quod natura sit id quo ipsum subsistens est aliquid. Cetera vero, de ipso extrinsecus illi affixa dicuntur, eiusdem 'status' vocantur eo quod

"Participatio" and "Habitus"

Both the *inesse* and the *adesse* relations—of, respectively, formal subsistences and accidents that are intrinsic to the subsistent—are subsumed under the umbrella of participation (*participatio*).[28] The fundamental characterisation of participation, as resumed by Gilbert, is thematised in Boethius' sixth theorem of the *De hebdomadibus*, addressing the basic relation between being (*esse*) and being something (*esse aliquid*).[29] Against a Neoplatonic background and on a theological grounding of metaphysics, for Gilbert the primary and paradigmatic kind of participation is that of being itself, i.e. the most general subsistence (*generalissima subsistentia*): by participating in this most general subsistence, everything that is *is*.[30] The participation of all other subsistences—be they generic or specific—is subordinated to the participation of this most general subsistence: for anything to be able to be something, it has to first be, i.e. to have some sort of being. Since only God necessarily is, then the participation of the most general subsistence in all created beings is contingent; all the other subsistences and intrinsic accidents that could belong to anything

nunc sic nunc vero aliter retinens ea, quibus est aliquid, et maxime perpetua subsistentias, divina voluntate statuatur."

[28] On the Porretan notion of *participatio* and on its both Aritsotelian Neoplatonic features in particular see: Garin (1958), Gregory (1958), van Elswijk (1966), Southern (1970), Häring (1974), Maioli (1979), Nielsen (1982), and de Rijk (1988a).

[29] *DHeb.* 1, 53–56, pp. 198,96 – 199,16 and 1, 98–100, pp. 208,64 – 209,81: "Supra in regula tercia qua dictum est: 'quod est, participare aliquo potest' participationem dicebat id, quod quod est, cum suo esse aliud habere quiddam. Unde in quarta aperte dicebat: 'Id, quod est, habere aliud preter quam quod ipsum est potest'. In quo etiam et in eo quod in tercie – clausula ponebat dicens: 'est autem aliud cum esse susceperit', et in fine secunde subiungens: 'Quod est accepta essendi forma est', patenter ostendit quoniam habere ipsum esse participatio est. Utrumque – videlicet et quo habetur ipsum esse et quo aliud aliquid cum ipso ab uno solo i.e. ab eo quod est – et in hac sexta regula manifeste 'participationem' appellat et ait: OMNE QUOD EST scilicet omne subsistens PARTICIPAT EO QUOD EST eius ESSE non quidem ut eo sit aliquid sed ad hoc tantum UT eo SIT. Cum eodem VERO idem subsistens quodam ALIO PARTICIPAT UT eo SIT ALIQUID. Sed illa participatio, quae eo quod est participat, natura prior est: altera vero posterior. Unde inferit: AC PER HOC. Quasi: quia videlicet non potest esse aliquid nisi prius naturaliter sit, ID QUOD EST, sicut dictum est, PARTICIPAT EO QUOD EST ESSE UT SIT. EST VERO naturaliter prius UT deinde PARTICIPET ALIO QUOLIBET quo aliquid sit. [...] Ad quod dicimus quod participatio, sicut et in his que premisse sunt regulis significatum est, pluribus dicitur modis. Cum enim subsistens in se aliquid – ut naturam qua sit vel aliquid sit – habet, dicitur quod ipsum ea natura participat. Natura vero que, quoniam inest subsistenti, dicitur ab eo participari, alia ita prima est, ut nullam pre se, quam sequatur, nisi primordialem habeat causam: ut ea, que omni subsistenti inest, generalissima subsistentia. Alia huius prime quodam modo comes est et, post causam primordialem, illam quoque ita causam habet ut ad potentiam eius ipsa pertineat et proprietate, qua sine ea esse non possit, adhereat. Tales sunt omnes differentie ille quecumque vel huic generalissimo proxime cum ipso quedam contractioris similitudinis constituunt genera – qua a logicis sub naturali, que ab ipsis est, subsistentium appellatione 'subalterna' vocantur – vel subalternis similiter adherentes quamlibet sub ipsis subsistentiam specialem componunt. He omnes non modo habitu illo quo inherent subsistenti verum etiam illo, quo generibus eius predicta potestate atque proprietate adherent, dicuntur haberi. Ac per hoc duplici ratione participantur."

[30] Cf. Catalani (2008, 159–323), in particular 187 ff.

created are in a sense potential. A sort of extrinsicality sneaks into the very notion of participation, and Gilbert does indeed speak sometimes of "extrinsic participation" (*participatio extrinseca*) both pertaining to the subsistent of the *subsistentia generalissima* and for that of its substantial and intrinsic accidental properties constituting its complete form.[31]

For those properties that are instead extrinsically affixed, i.e. those properties pertaining to the status of a subsistent, Gilbert never speaks of "participation" but instead of some sort of having: the habit (*habitus*).[32]

"Modi Coniungendi"

As we have seen, the relations between a subsistent, its being, its subsistences and its accidental properties can be treated in general as types of participation. The dividual properties of which the subsistent participates are singularised within the scope of the subsistent itself, rather than separated and independent universals. A short overview of the modes of relation between subsistents, formal subsistences, and intrinsic and the extrinsic accidents—in Gilbert's jargon: *modi coniungendi*—follows.[33]

Apposition (*appositio*) is the weakest of these modes of conjunction and amounts to a kind of juxtaposition and concerns extrinsically attached accidents. It is a concomitance relation between entities and it does not have any substantial or qualitative effect on the entities themselves. Apposition can be exemplified as the type of relation between two items being in proximity to each other or as the relation between a body and the clothes it wears. In this case, neither an actual unity nor a true composition results from the juxtaposition of the entities involved; what is predicable of one does not become predicable of the other. Overall, this is the relation that a subsistent entertains with all those e*xtrinsecus affixa* forming its *status*.[34]

---

[31] E.g. *DHeb.* 1, 43, p. 196,36–43: "ID QUOD EST HABERE ALIQUID PRETER QUAM QUOD IPSUM EST POTEST. IPSUM VERO ESSE NICHIL ALIUD PRETER SE HABET ADMIXTUM. Hec regula quodam modo precedentis sensum. Ideo namque id quod est participare aliquo dictum est quoniam ID ipsum, QUOD EST POTEST HABERE ALIQUID PRETER QUAM illud sit QUOD IPSUM, quod est, EST, i.e. preter quam sit esse quo ipsum est: ut corpus preter corporalitatem cum ipsa, qua est, corporalitate habet colorem." For the notion of "participatio extrinseca" see e.g. *DTrin.* I, 45, p. 88,68–69; I, 44, p. 123,57–59. See de Rijk (1989, 19, n. 18).

[32] See e.g.: *DTrin.* I, 70, 12–15, p. 129: "Cetera vero que quolibet modo sibi invicem adunantur – scilicet vel intrinseco concretionis vel extrinseco cuiuslibet appositionis habitu [...]." Clearly, this use of *habitus* is not to be confused with the Arestotelian *echein* aristotelico (occurring, for example, in *DTrin.* II, 3, p. 163,18), nor with the *habitus* opposed to *dispositio* (as in *DTrin.* I, 3, p. 57,22).

[33] Here, I introduce exclusively the relevant *modi coniungendi in naturalibus*. For a more detailed exam of Gilbert's treatment of such relations *in divinis*, see de Rijk (1989, 21 ff.).

[34] *CEut.* 4, 14–17, pp. 290,82 – 291,97: "Diligenter attende quod his verbis breviter et obscure significatum est: diversos scilicet esse coniungendi quelibet modos. Ait enim quod duo corpora ita sibi coniunguntur quod in alterum nichil ex alterius pervenit qualitate. In quo innuit quod etiam ita sibi invicem aliqua coniunguntur ut in alterum ex alterius qualitate aliquid perveniat. Niger enim lapis, albo lapidi appositus, loco quidem alter alteri iuxta est. Seque neque, qui niger est, albi

The relation of composition (*compositio*) implies instead a real unity between the related entities, while at the same time they remain distinct. This is typically exemplified by the case of the composition between soul and body—which, for Gilbert, is paradigmatic.[35] Composition is the relation, within a subsistent, between the subsistences that express and enact different modes of being within the same compound entity they constitute.[36] For Gilbert, as de Rijk had already noted, composition is the mode of conjunction captured by logical predication.[37]

Finally, the strongest among the modes of conjunction is the one that Gilbert calls "mixture" (*commixtio*)[38] or "confusion per composition" (*per compositionem confusio*).[39] Any two or more mixed entities are entangled in such a way that either one of them or both are transferred into each other or lose their proper form. In other words, the components involved in a *commixtio* lose their separate properties: their

qualitate dicitur 'albus': neque, qui albus est, nigri qualitate dicitur 'niger'. Ligno autem ferrum vel aurum apponitur. Et dicitur quidem appositionis habitu lignum ipsum 'ferratum' vel 'auratum'. Sed nondum ferri vel auri qualitas predicatur de ligno. Contingit tamen in huiusmodi appositionibus etiam alterius sed alterum quadam denominatione qualitatis nomen transsumi, ipsam vero qualitatem nequaquam in altero fieri: vel in vestium seu armorum appositionibus."

[35] See e.g. *CEut.* 4, 22–23, p. 292,20–32: "Unde manifestum est unum esse aliquid in quo diversa sibi invicem coniuncta dicuntur. Cui uni sunt esse omnes speciales et he, ex quibus speciales constant, subsistentie illorum que in ipso sibi invicem coniunguntur: et preter has ille etiam que in eodem creantur ex habitu coniunctorum: ut homini, qui ex corpore et spiritu sibi coniunctis unus est, sunt esse omnes corporis atque spiritus subsistentie et alie quedam que in ipso ex eorum fiunt concursu. Idem vero homo ex his que subsistentiis adsunt, qualitatibus et mensuris intervallaribus aliquid est. Et quoniam hominis ex corpore et spiritu compositio ita fit quod nec utrumque nec alterum in eo confunditur, omnes ille, quas modo diximus, subsistentie et qualitates et intervallares mensure immo etiam intervallarium termini de ipso homine recte dicuntur." For an outline of the distinction between *coniunctio* and *commixtio* see *DTrin* I, 79, pp. 95–96,87–99: "Putant quidam imperiti ex hoc quod ait: ' non vel corpus vel anima' quod nec etiam dici horum alterum sine altero liceat i.e. quod non sit vera dictio si quis dicat 'homo est corpus' non addens 'et anima' aut si dicat 'homo est anima' non addens 'et corpus' opinantes quod – ex quo diveras, ut unum componant, coniuncta sunt – esse utriusque adeo sit ex illa coniunctione confusum ut sicut cum album et nigrum permiscentur, quod ex illis fit, nec 'album' nec 'nigrum' dicitur sed cuiusdam alterius coloris ex illa permixtione provenientis ita, quod ex diversis constat, neutrius deinceps nomen suscipiat sed sit aliquid ex eo quod ex permixtione provenit: et ex hoc sensu dictum esse 'homo est corpus et anima' non quod ipse sit corpus vel anima sed quod ipse sit quiddam quod provenit ex permixtione que ex corporis et anime coniunctione contigit."

[36] See de Rijk (1989, 25).

[37] See de Rijk (1989, 25). See *CEut.* 4, 27–29, p. 293,46–56: "In quo diligenter est attendendum quod, etsi quandoque non eiusdem sint generis que sibi in compositionibus coniunguntur, semper tamen in aliquo sunt eiusdem rationis. Quamvis enim corpus et spiritus diversi generis sint, in hoc tamen eiusdem rationisquod utraque his, que predicatur, supposita sunt.ipsa vero impossibile est predicari. Numquam enim id, quod est, predicatur. Sed esse et quod illi adest predicabile est: et sine tropo non nisi de eo quod est. Simplices quoque subsistentie diversorum sunt generum: ut rationalitas animatio. Una tamen earum est ratio qua eorum, que sunt, 'esse' dicuntur."

[38] See e.g. *CEut.* 4, 24, p. 292,33–36: "Hec enim spiritus corporisque coniunctio compositio est, non commixtio. Non enim omnis compositio commixtio est: sicut non omnis coniunctio est compositio. Omnis vero commixtio compositio est. Unum enim aliquid in sese mixta componunt."

[39] *CEut.* 5, 5, p. 327,20: "Commixtio namque est per compositionem confusio[…]."

respective properties are not maintained in the resulting mixture, where a different property emerges instead.[40]

3 "Individuum"

The dividual subsistences and intrinsic accidents, as we have seen, are singular as singularly instantiated in the subsistent, constituting its complete form. This complete form itself is what determines an individual being.[41] Then, not everything that is singular is also individual.[42] Individual is only that entity whose complete form—resulting from a collection of singular properties constituting a new unity (unum)—is not attributable to anything else:

> Alia vero ab aliis omnibus aliqua sue proprietatis parte dissimilia. Que sola et omnia sunt huius dissimilitudinis ratione individua, ut hic lapis, hoc lignum, hic homo.[43]

Any determinate and distinct item is an individual—this stone and not that one, this piece of wood and not that one, this person and not that one.

Given Gilbert's characterisation of the singular properties whose combination determines an individual's *tota forma*, there is no need for any extrinsic accidents beyond substantial properties and intrinsic accidents concurring in the process of individuation. Let's take two stones of the exact same shape, colour, weight—and so on for all the proper characteristics of a stone. For Gilbert, the criterion of distinction between the two stones is still the singularity, in each of them, of their substantial properties and intrinsic accidents: these features, while being exactly similar qua dividual, are numerically distinct. It is then not necessary to appeal to any extrinsic features distinguishing them, such as the fact that they do not occupy the exact same space at the exact same time. Of course, epistemically, one can immediately know

---

[40] *CEut.* 4, 25, p. 292,37–40: "Sed vel alterius vel utrisque qualitates aliquas mixtura confundit: ut cum album nigrumque miscentur, neque componentia neque compositum albi et nigri retinet qualitates sed alterius specieis afficiuntur colore."; *CEut.* 6, 7, p. 327,27–33: "Hic dicendum videtur quod eorum, que vere miscentur, corporum nature non nisi per denominationem dicuntur 'misceri': per subiectorum tamen corporum mixturam recte ab asque denominationis tropo dicuntur 'confundi': ut albedo et nigredo nequaquam miscentur quoniam incorporales sunt, albi tamen atque nigri permixtione confunduntur. Igitur sola illa, que sunt, misceri: illa vero, quibus sunt, confundi contingit."

[41] Another important aspect of Gilbert's ontology (and of his ontological terminology) is the distinction between "individuum", "singular" (which has been outlined here) and "persona". For an outline of the relevance of this discussion, in particular insofar as Gilbert's use of this notion shifts from Boethius', see: Valente (2008, 200).

[42] Nielsen (1982, 61, n. 104), de Rijk (1989, 78). See e.g. *DTrin.* I, 2, prol. II, 6, p. 58,45–47: "[...] Platonis et Ciceronis non solum accidentales proprietates verum etiam substantiales, quibus ipsi sunt verbi gratia vel diversa corpora vel diversi homines, diverse sunt."; *CEut.* 2, 29, p. 270,73–77: "Quidquid enim est, singulare est. Sed non: quidquid est, individuum est. Singularium namque alia aliis sunt tota proprietate sua inter se similia. Que simul omnia conformitatis huius ratione dicuntur 'unum dividuum', ut diversorum corporum diverse qualitates tota sui specie equales."

[43] *CEut.* 2, 30, p. 270,78–80.

that the two stones are not the same stone since they do not occupy the same place at the same time, but this is not the reason for their distinction.

The principle of individuation, for each individual, is its complete form. Now, a legitimate question to ask would be whether the collection of properties within the complete form is sufficient to ground the peculiar individuality of a subsistent, or if something else is required—be it the relations subsisting among some or all of the subsistences and properties in the *tota forma,* or even some distinct and unique individual property, that cannot be predicated of any other individual.

Gilbert is consistent in repeating that no part of an individual is itself an individual[44]; while each formal and accidental property is singularly instantiated in each individual, it is the unity of all forms that makes the individual.[45] Plato's "*Platonitas*", in other words, is not one of the properties within the collection of features constituting Plato's *tota forma*, but is the unity of the *tota forma* itself. This unity of the complete form implies that the relations between its parts are structural and characterising of the subsistent's individuality. But since Gilbert defines the unity of the complete form in temporal and modal terms, it follows that potential properties and the possible relations between them are included in what individuates one individual, differentiating it from any other.

> Illa vero cuiuslibet proprietas, que naturali dissimilitudine ab omnibus – que actu vel potestate fuerunt vel sunt vel futura sunt – differt, non modo 'singularis' aut 'particularis' sed etiam 'individua' vere et vocatur et est. [...] Hac igitur ratione Platonis tota forma – nulli neque actu neque natura conformis – vere est individua.[46]

It is evident that some modal characterisation is essential to Gilbert's definition of *individual.* And, overall, the conception of modality that Gilbert has in mind does not seem to be properly potentialist.[47] Gilbert's use of *potestate* here does not seem to be equivalent to *(in) potentia,* which would maintain a temporal dimension of the *tota forma*'s properties and internal relations, but would effectively get rid of their properly modal dimension—since all properties would have to be actualised at some point in time, be it past, present or future. On the contrary, along with the concomitant

[44] See, for example, *CEut.* 3, 12, p. 274,75–80: "Unde Platonis ex omnibus, que illi conveniunt, collecta proprietas nulli neque actu neque natura conformis est: nec Plato per illam. Albedo vero ipsius et quecumque pars proprietatis eius aut natura et actu aut saltem natura intelligitur esse conformis. Ideoque nulla pars proprietatis cuiuslibet creature naturaliter est individua quamvis ratione singularitatis 'individua' sepe vocetur."

[45] *DTrin.* I, 25, p. 144,69–78: "Attendendum vero quod ea, quibus id quod est est aliquid, aut simplicia sunt ut rationalitas aut composita ut humanitas. Simplicia omnia vel actu vel natura conformia sunt. Ideoque nulla eorum vera dissimilitudinis ratione sunt individua. Composita vero alia ex aliquibus tantum, alia ex omnibus. Que non ex omnibus, similiter sicut et simplicia vel actu vel natura conformia sunt. Ac per hoc nulla eorum sunt individua. Restat igitur ut illa tantum sint individua que, ex omnibus composita, nullis aliis in toto possunt esse conformia: ut ex omnibus, que et actu et natura fuerunt vel sunt vel futura sunt Platonis, collecta platonitas."

[46] *CEut.* 3, 13–15, p. 274,81–90. It should be noted that this modal characterisation pertains to the notion of individual and not only to that of a "person" (*persona*), as believed by Nielsen (Nielsen, 1982, 58–69). On this, see also Knuuttila (1993, 79).

[47] See below, Sect. 4.

occurrence of *natura* contrasted with *actu*, there seems to be an array of dispositional features in Gilbert's account[48] along with the notion of pure possibility, i.e. the idea that possibilities do not need to be actualised at some point in time in order for them to be possible in a pregnant sense.

That Gilbert admits an infinite number of pure *possibilia*, carrying no requirement of actualisation, is evident from other texts, as for example in *Contra Euticen*:

> Fuerunt enim qui iam non sunt: et erunt qui nondum sunt vel fuerunt: et nunc sunt tam actu quam natura homines infiniti. Ideoque ipsorum forme multe similiter natura et actu et fuerunt et sunt et erunt a quibus hoc ipsarum plena inter se conformitate vere dividuum nonem hominibus ipsis inditum est. Unus vero actu solus est sol preter quem nullus actu vel fuit vel est vel erit quamvis natura et fuerunt et sunt et futuri sunt infiniti: ideoque infinite sola natura subsistentie inter se sola natura conformes a quibus hoc vere dividuum et universale nomen est. Sicut enim veri individui plena proprietate nulla neque actu neque natura esse potest ita secundum plene proprietatis quamplibet partem naturalis saltem similitudo.[49]

"Sun", for Gilbert, is a common name, because the concept of sun is a dividual concept. While, for Gilbert, the only existing sun is the actual and individual Sun, the concept of sun includes also an infinite number of suns that could have been—in the past, in the present or in the future—but that never actually are. There is in Gilbert's ontology the idea of authentic possibility that does not require to be actualised in order to remain a possibility in a proper sense. But in order for this to be the case and for such a conception to take shape, the frame of reference cannot be limited to the actual world and what can be instantiated there. In other words, in order to conceive non-actualised and yet authentic possibilities, these possibilities need to be framed in terms of conceivability against a framework of alternative possible worlds. Gilbert has indeed such a framework at his disposal, in the notion of conceivability in the mind of God—who operates under the constraints of the Principle of Non-Contradiction—and in an account of Creation appealing to God's alternative providential plans.

4 On Creation, Necessity and Possible Worlds

Gilbert of Poitiers's account of creation and of the providential plan underlying it rests on a long and complex tradition. Some of its philosophical and theological forerunners and sources are particularly important and evident—e.g. the Augustinian account of the seminal ideas (*rationes seminales*) in the mind of God[50] or

[48] For reasons of space I will not be able to expand on this albeit important matter here.

[49] *CEut*. 3, 10–11, p. 273, 63–74. In this particular passage, Gilbert is making a grammatical argument on the distinction between appellative and proper names, for which I refer to Nielsen (1976). This is not an isolated occurrence – see, e.g., also *DTrin*. I, 4, 72, p. 129,25–28.

[50] In particular, Knuuttila proposed a reading of Augustine's account of Creation and of the seminal reasons in the mind of God as an embryonic account of unrealised *possibilia*—see e.g. Knuuttila (2001).

the Neoplatonic metaphysics of participation mentioned above. Yet, these conceptual materials are reassembled into an innovative configuration and shifted into a profoundly original framework. Along with Gilbert's own ontological conclusions, they yield interesting results, particularly insofar as modal notions are concerned.

Theological accounts of Creation are one of the usual contexts in which to reflect on the notions of necessity and contingency, allowing one to do so from the privileged point of view of God.[51] Gilbert's own account of creation and of God's alternative providential plans is of historical importance for the articulation of modal notions at least under two respects. On the one hand, it relies on a distinction between absolute and relative necessity which—in concept if not in terminology—foreshadows the crucial Scholastic distinction between *de potentia Dei absoluta* and d*e potentia Dei ordinata*. On the other hand, side by side with Gilbert's metaphysics of individual beings, it opens the door to a synchronic account of *possibilia*. Overall, Gilbert's treatment of modal notions exemplifies, in a nutshell, a conception of possibility and necessity whose paternity is usually attributed to Duns Scotus and seems to even presuppose a version of possible worlds semantics at the embryonic stage.

Nothing besides God exists necessarily.[52] In a sense, the necessity of God's existence is the paradigmatic case of absolute necessity. God's creation, then, is a genuinely free act and, as such, it is absolutely contingent: God could have chosen not to create or not to create what he actually created in the way he created it; he could have created something else or could have made it in a different fashion.

Gilbert explicitly and repeatedly commits to the view that God could have actualized a different world with a different history, with different natural laws,[53] or even with different individuals.[54]

---

[51] Modal notions framed in this context are normally grouped under the label of "theological modalities". See e.g. Knuuttila (2008, 517–521).

[52] *DTrin.* II, 1, 7–8, p. 164,34–41: "In ceteris facultatibus, in quibus semper consuetudini regule generalitas atque necessitas accomodatur, non ratio fidem sed fides sequitur rationem. Et quoniam in temporalibus nichil est quod mutabilitati non sit onoxium, tota illorum consuetudini accomodata necessitas nutat. Nam in eis quicquid predicatur necessarium vel esse vel non esse, quodam modo nec esse nec non esse necesse est. Non enim absolute necessarium est cui nomen 'necessitatis' sola consuetudo accomodat."

[53] *CEut.* 5, 41–43, pp. 322–323,36–43: "Sed 'mortalis' dicitur homo quia potestate divina dissolvi potest: ante peccatum quidem absque ulla necessitate, post peccato vero necessario. 'Immortalis' quoque dicitur quoniam eadem divina potestas eum, dum sine peccato fuit, ita conservavit ut non dissolveretur – post reformationem vero ita conservabit ut numquam exinde dissolvatur. Et ideo dicitur non posse dissolvi. Non enim iccirco dicitur non posse vel dissolvi vel non dissolvi quod Deus hec facere non possit sed quod, ut ita se haberet vel ante peccatum vel post peccatum vel post resurrectionem homo, divina voluntas statuit. Secundum hoc dicitur sol non posse non moveri cum tamen divina potestas eum, ut non moveatur, sistere possit. Et huiusmodi sunt infinita. Sunt ergo, ut de ceteris taceamus, hominis mortalitas et immortalitas – sive necessaria sive non necessaria – non subsistentie, quibus ipse sit aliquid, sed eius in diversis temporibus – secundum divinam potentiam sive voluntatem suam – status."

[54] *DTrin.* I, 4, 72, p. 129,25–28: "Eque etenim universa eius subiecta sunt potestati ut scilicet sicut, quecumque non fuerunt, possunt fuisse et, quecumque non sunt vel non erunt, possunt esse ita etiam, quecumque fuerunt, possunt non fuisse et, quecumque non sunt vel erunt, possunt non esse."

In any conception of the actual world as contingently created, there is always an underlying sense in which natural necessities, i.e. those facts that cannot but hold in the actual world (as e.g. the standing laws of nature), are indeed relative to something contingent and therefore are ultimately contingent themselves. However, Gilbert, developing a traditional notion already found Jerome of Stridon and famously embraced by Peter Damian in his *De divina omnipotentia*,[55] endorses a view that is even stronger. Not only are the natural necessities in the actual world contingent upon God's will and power, but they are simple regularities (*necessitas consuetudini accommodata*), holding no intrinsic real necessitating force beyond the modelling of recurring patterns.[56] This is a genuinely philosophical position and represents, in the first place, a claim about the relative and epistemic nature of necessity within the natural world, independently of God's intervention and absolute power.

Gilbert, however, is not a 12th-century forerunner of Hume and this account is still to be ultimately situated within a framework of theological overdetermination, given the assumption that creation is contingent upon God's omnipotence and on the freedom of his creative act.

Nonetheless, this does not mean that, for Gilbert, God is unbound by any kind of necessitation.[57] On the contrary, in contrast with the view that Peter Damian held and that later authors such as Descartes will hold, Gilbert maintains that there are conceptual necessities which are binding even for God.[58] These necessities seem to be of two kinds. On the one hand, God's action does conform to metaphysical

---

[55] See Peter Damian, *Briefe* 199. Peter Damian examines several relevant examples defining God's omnipotence in relation to the laws and facts of nature. Some of these examples, such as God's power of making it so that a virgin gives birth or that a prostitute's virginity is restored, are traditional and Peter Damian himself refers to Jerome's account. For Gilbert's reprise of the example of the virgin giving birth, see the following note. For a general overview of Peter Damian's thought see Holopainen (2016).

[56] *Commentarius in Epistulas Sancti Pauli,* quoted in Nielsen (1982, 136): "Philosophiam vocat scientiam naturalium deductam ex principiis rationum, quorum tamen est consuetudini accommodata necessitas. Sed divina ex una potestate omnium motus est atque substantia et earum, que ad se dicuntur, causarum conexio. Quidam autem imperiti ea, que dicuntur, ex dicendi rationibus minime iudicantes cum audiunt quedam esse necessaria, divine derogant potestati putantes id, quod iuxta nature consuetudinem dicitur, necessarium absolute non posse non esse. Ideoque negant virginem peperisse et huiusmodi alia, que predicto modo dicuntur impossibilia. [...] Sapientia enim mundi huius, cui ratio illa, que sunt creata, non prevenit, sed ex eorum potius cursu atque nativa consuetudine universalitatem regule atque eam, que dicitur, necessitatem non in eis facit, sed ex eis ad eadem humane supponenda intelligentie accipit." On these matters see also Knuuttila, "Possibility and Necessity in Gilbert of Poitiers", 213, upon which this section builds and expands.

[57] On this see also Knuuttila (1987, 213).

[58] See for example *CEut.* 4, 108–109, p. 310,56–63: "Nam unus et idem est Christus: videlicet qui et Deus erat et homo est. Qui se ipsum nulla ratione assumere potuit quia non nisi diversorum ulla potest esse assumptio. Quod si in Christo unam putat esse persona illum qui Deus erat, alterum vero esse personam illum qui homo est, atque illum, qui Deus erat, eum qui homo est assumpsisse, hoc quoque impossibile est. Nam sicut omnino idem ita omnino diversum assumi non potest."; 3, 5–6, p. 272,27–44: "Hoc tamen impossibile esse per hoc intelligitur quod nulla persona pars potest esse persone. Omnis enim persona adeo est per se una quod cuiuslibet plena ex omnibus, que illi conveniunt, collecta proprietas cum alterius persone similiter plena et ex omnibus collecta proprietate de uno vere individuo predicari non potest: ut Platonis et Ciceronis personales proprietates de uno individuo

necessities regulating the possible relations between the constituents of entities, the determination of individuation and the interactions between what is material and what is immaterial. On the other hand, for Gilbert even God's omnipotence is bound by some logical necessities and, namely, by the Law of Non-Contradiction in the first place. Leaving aside the absolute necessity of God's being and existence, since natural necessity is fundamentally voided of any binding force, overall the model of actual necessitation in the Porretan picture seems to be predominantly that of the conditional necessity warranting the following of a logical consequence.[59] It is under these logical constraints and within these parameters that God exercises his omnipotence and his creative action.

Within these borders, God could have still created the world otherwise—with different individuals, different recurring patterns, different histories, and overall according to a different providential plan.[60] All these non-implemented providential plans or unrealized worlds need to be possible in a logical sense (i.e. non-contradictory, since God is bound by the Law of Non-Contradiction) and they are eternally in the mind of God, who is outside of time.

The picture emerging from the concurrence of these views that Gilbert holds consistently is a version of possible worlds semantics grounded in the mind of God, in which these possible worlds are synchronously possible in an authentically logical sense.

These alternative providential plans are eternally and synchronically possible because they are eternally conceived in the mind of God; they are therefore possible in a sense that is neither temporal nor potentialist, since God has not actualized them but might have. In a way, with all the due differences, they are functionally

---

dici non possunt. Tota vero anime Platonis proprietas – i.e. quicquid de ipsa naturaliter affirmatur – de ipso Platone predicatur. 'Naturaliter' dicimus quoniam, quod non naturaliter de anima dicitur, non necesse est de Platone predicari: ut topica ratio, quam Platonis anima 'pars' eius vocatur, de ipso Platone minime dicitur. Dicimus etiam 'affirmatur' quia, quod ab anima Platonis negatur, non necesse est ab ipso negari: ut si dicatur anima esse incorporea quo privatorio nomine corporum subsistentia, que est corporalitas, removetur ab ea, non ideo Plato incorporeus esse dicitur. Et sic quidem humana anima secundum predictam diffinitionem videtur esse: secundum expositam vero rationem videtur non esse persona."; 5, 21, p. 318,32–36: "Tunc enim non modo 'quod' eveniat verum 'ut' eveniat, promitti aliquid dicitur cum promissionem ipsam rei eventus necessario sequitur. Non dico 'necessarius' sequitur sed 'necessario' sequitur. Quod utique fit cum ab eo qui mentiri non potest i.e. a Deo promittitur."; 6, 52–53, pp. 337,95 – 338,3: "Non autem potest esse communis que omnino non est. Certum vero est quoniam nulla omnino, sicut predictum est, est in corporalibus rebus materia. Manifestum est igitur quod non poterunt in se invicem permutari. Recole – si didicisti – argumentationis necessarie artem. Et vide cuius generis atque figure sillogismo propositum probat. Nunc enim – per propositionem maximam que est: ' quorum nulla est communis materia, non poterunt in se invicem permutari' certum reddidit quod nulla incorporea in sese permutantur."; 7, 55, p. 353,41–42: "De utrisque quidem partibus idonee, ut arbitror, disputatum est et quod hec adunatio nullo tempore, nulla omnino ratione fieri potuit, demonstratum."

59 *CEut.* 5, 21, p. 318,32–36 (see text in previous footnote).

60 See *CEut.* 5, 41–43, pp. 322–323,36–43; *DTrin.* I, 4, 72, p. 129,25–28; *CEut.* 2, 16, pp. 267–268,92–98; *CEut.* 8, 17–24, pp. 358–359,28–62.

similar to Arthur Prior's take on alternative histories of the world in the context of the discussions on counterfactual identity.[61]

Given what Gilbert says about how God could have created the world otherwise or implemented different providential plans, these alternate possible worlds seem to come in degrees of similitude with respect to the actual world through variations on the standing natural laws, the possible individuals included in a world, their properties, their histories, and so on.

An account of this type, then, could enable Gilbert both to dismiss the necessity of the present *qua* factually actual and to claim, without contradiction, that even if Socrates is sitting right now it is still genuinely possible that Socrates is standing right now. Overall, then, a theory of this kind, even contained embryonically in a systematic array of ontological and logical theses, is an original development within the Western tradition up to this point.

5 Identicals and Counterparts

Yet, what such a theory would require to support such a claim is, as Knuuttila puts it, "a theory of individuals in which the relationships between an individual and actual as well as possible states of affairs are discussed".[62] But this means that a theory needs to be able to hold that within another divine providential plan there is some subsistent which is identical or exactly similar to Socrates and that such a subsistent is standing while Socrates is sitting—all other things remaining the same, or most things remaining the same, and so on for increasing degrees of variations. In other words, Gilbert's framework, for such a claim to be treatable, needs to be able to conceive of other-worlds counterparts that are counterfactually identical to Socrates (or conform in such a way that they are functionally identical) and for which the statement "Socrates is standing right now" is true when in the actual world Socrates is sitting right now. Can Gilbert think of individuals like that?

Overall, Gilbert disposes of a full array of philosophical tools and fully developed positions—both in his cosmology and his ontology—which would allow him to conceive of the issue. His cosmology seems to imply an embryonal version of possible worlds that is more developed and systematically supported than any previous theory presenting analogous features. His ontology and his metaphysics of individuality rely on a hierarchical composition of actual and possible singular properties and their respective relations, resulting in an innovative shift from STI.

Gilbert speaks quite liberally of alternate providential plans similar enough to the enacted one that one could find items in those worlds that look very much like items in this one—e.g. a W_1Socrates for our @Socrates, a W_1Plato for our @Plato, and so on. This seems to follow from some passing claims of Gilbert, such as that God could have implemented different natural regularities, with all other things remaining the

[61] Prior (1960). For a contrasting position in this debate see e.g. Chisholm (1967).

[62] Knuuttila (1987, 215).

same—so, among those "all other things" one would count also the collection of individuals belonging to this providential plan. However, the question that should be asked is: how seriously can we take Gilbert on these matters, especially insofar as something very much like counterfactual identity seems at play? Quite seriously, it would seem, at least up to a point and with some limitations.

We have seen how all properties that are in some sense conjoined to entities are fundamentally of three kinds: substantial forms, intrinsic accidents and extrinsic accidents. In Gilbert's account, only substantial forms and intrinsic accidents—both actual or possible—compose the complete form of an individual subsistent, i.e. what makes any individual *this* subsistent rather than *that* subsistent. But this is not the case with any accidents other than qualities and quantities: extrinsic accidents, summed under the label of *status*, are not part of an individual's complete form. Accidents like relations, actions, what a subsistent is subjected to, the place it occupies, the time it occupies, if it is in motion or at rest etc. are just attached to a subsistent in such a way that the subsistent's individuality—i.e. what that subsistent, deep down, *is*—is not changed by their addition or remotion. This being the case, then, any or all accidents within a subsistent's status could be different in an alternate history of the world and that subsistent would still be the very same subsistent, all other substantial and intrinsic features remaining the same. If W_1Socrates is sitting at a given time and @Socrates is standing at the same given time, W_1Socrates and @Socrates would still be the same individual, since they would have the same complete form which is the unique principle of their individuation. Within the framework that Gilbert has outlined, this would still hold true even if major changes were to be implemented on other extrinsic properties, with more or less possibly counterintuitive results—e.g. W_1Socrates would still be the same individual as @Socrates even if he had not begotten Lamprocles, Sophroniscus and Menexenus, or if he had acquiesced to Crito's pleas to attempt escaping from prison, etc., as long as all of W_1Socrates' and @Socrates' substantial and intrinsic features stay the same.

One may ask whether @Socrates and W_1Socrates would still be the same individual if an intrinsic property that is merely potential in @Socrates is actualized in W_1Socrates, or vice versa. Strictly speaking, that would be a change within Socrates' complete form, that per its nature is individually unique: if @Socrates and W_1Socrates have different complete forms, then they are not the same individual. In that case, there would still be a strong similarity between @Socrates and W_1Socrates, to the point that one might ask: (a) whether such similarity could be reduced to a kind of "conformity" (*conformitas*) analogous to the one between dividual substantial forms and intrinsic properties of the same kind instantiated in different individuals; and (b) whether this sort of implemented conformity could constitute a metaphysical ground for a more general account of counterpart relations across alternate providential plans. But these are questions for another study.

6 Some Closing Remarks

In the present study, I have developed some seminal suggestions put forward by Simo Knuuttila. I have presented an overview of Gilbert of Poitiers' most relevant contributions to the history of ontology and of the metaphysics of individuality, examining how they integrate with Gilbert's account of creation and of God's alternate (and unrealized) providential plans. The conceptual framework that I have outlined is fundamentally essentialist and seems able to both accommodate a form of counterfactual identity and preserve the identity of a same individual across alternate providential plans, at least to a degree. Given the features of Gilbert's ontology that I have underlined so far, neither Knuuttila's interpretation of Gilberts' account of alternate providential plans as an embryonal version of possible world semantics nor Gilbert's inclusion of modal properties within the definition of individual immediately gives rise to obvious instances of incoherence.

The crucial element of Gilbert's metaphysical construction, leading to this type of development, is his account of the complete form as the principle of individuation. As Knuuttila had already remarked, Gilbert's *tota forma* has indeed remarkable analogies with Leibniz's notion of a complete concept. However, while Gilbert's inclusion of possible properties within the complete form is a major difference with Leibniz's complete concept, an even more crucial and fruitful Porretan peculiarity lies in the exclusion, from the complete form, of all those extrinsic accidents that are instead part of what makes Leibniz's complete concept *complete*. This exclusion, however, ultimately allows Gilbert a margin to preserve individual identity across alternative histories of the world as long as an individual's substantial and intrinsic properties stay the same.

Of course, it might be the case that "Gilbert did not in fact think out his views on modality" all the way through.[63] But there seems to be enough textual elements to allow for a rational reconstruction[64]—along the lines of Knuuttila's interpretation—bringing forth the kind of modal account that those elements can entail, soundly and without it being unwarrantedly speculative.

Overall, Gilbert frames individuals as bundles of both possible and actual dividual properties; he does so by excluding all *extrinsecus affixa* from the *tota forma* ultimately constituting an individual's individuation.

"[T]hat whole forms are indeed individual" and that, therefore, "there is nothing else to which any of them is, or could possibly be, exactly similar"[65] is not an a priori presumption, inasmuch as a consequence of Gilbert's peculiar brand of "realism of the forms and non-realism of the universals"[66] or, in other words, of his "metaphysics of the concrete."[67] No complete bundle of properties is exactly similar to another

---

[63] I am quoting and paraphrasing Marenbon (1998, 171–172).

[64] I refer to the methodological notion of "rational reconstruction" in the history of philosophy thematised e.g. in Cameron (2011).

[65] Marenbon (1998, 171).

[66] Valente (2008, 214 ff.).

[67] Maioli (1979), cited in Valente (2011).

complete bundle of properties, no matter how similar to each other both bundles are. This is the case since all the dividual properties belonging to the first bundle are singular and numerically distinct from all the dividual and singular properties belonging to the second bundle. Castor is not exactly similar to Pollux, even if they are identical twins, because the complete form of Castor is numerically and metaphysically distinct from the complete form of Pollux.

On the one hand, the exclusion of extrinsic accidents from what is relevant for individuation is a handy tool. For example, take the spatio-temporal dissimilarity principle.[68] This is a fact of nature and therefore, given the way the actual world works, Castor and Pollux have a different spatio-temporal collocation. However, as we have seen, the spatio-temporal collocation of an individual does not belong to the individual's complete form, but pertains to the individual's status. While it is the case that, given the way the world works, Castor and Pollux do not share the same spatio-temporal collocation and, thus, Castor and Pollux are recognizable as different individuals simply by seeing that they do not stand in the same spot at the same time, this is not the reason for their being distinct. In an alternate history of the world, right now Castor could have been standing where Pollux stands, and vice versa. And yet, there would be no contradiction in claiming that Castor would still be Castor and Pollux would still be Pollux: extrinsic accidents do not pertain to the complete form and do not affect an individual's identity to itself.

On the other hand, the inclusion of modal properties within the complete form does not pose any more problems than those raised by any bundle theory. For instance, there is no reason to believe that the singularity and metaphysical distinction of the dividual forms instantiated in each and every individual should apply only to actual properties but not to possible ones. Just like Castor's humanity and all his other intrinsic actual features are metaphysically and numerically distinct from Pollux's humanity and all his other intrinsic features, no matter how alike Castor's and Pollux's actual properties are, so one would expect their possible properties to be metaphysically and numerically distinct in the same way—no matter how alike are all the features that Castor and Pollux could possibly have. Then, it does not follow "that the complete forms of individuals of the same species would all be similar to each other, and so dividuae."[69] Neither does it follow that "Gilbert's modal definition of individuality seems to have the undesirable consequence of dissolving the very individuality he wishes to define".[70] "[I]ndeed, if taken literally, Gilbert's definition" could "imply that the complete form of every individual of a species includes all the possible determinations of every other individual of that species".[71] But this would still not be enough to dissolve the individual distinctions between the members of

[68] The matter with the spatio-temporal differentiation principle is addressed by Marenbon (1998, 171–172). The following lines are in dialogue with Marenbon's and Valente's (see Valente, 2011) critical remarks on Gilbert's modal account of individuality.

[69] Valente (2011), referring also to Marenbon (1998, 171).

[70] Valente (2011).

[71] Valente (2011).

that species if the singularity of possible properties functions like the singularity of actual properties—as we have no reason to believe otherwise.

Bringing possible features into the picture does not per se make room for variation "disturbingly wide"[72]—not any more, at least, than the disturbing wideness intrinsic to any notion of complete form including all and only an individual's actual properties, as Leibniz's complete concept.

However, the fact that Gilbert's complete form is not metaphysically incoherent does not imply that it is epistemically serviceable or that it would be preferable to think about Castor's and Pollux' complete forms instead of looking at them standing next to each other, in two different spots at the same time, to conclude that they are not indeed one and the same individual. Gilbert himself is well aware that the epistemic counterpart to the ontological datum of the complete form is too complex and thus can neither be formed nor fully grasped by a finite mind.[73] But this issue is not properly due to the inclusion of modal elements within the complete form as much as to its completeness: indeed, Leibniz himself has to face a similar problem with his complete concept, even sticking exclusively to actual properties.

References

Primary Texts

Gilbertus Porreta, *Expositio in Boecii librum de Trinitate*, in Häring 1966, 53–181 (=*DTrin.*).

Gilbertus Porreta, *De Bonorum Ebdomade*, in Häring 1966, 183–231 (=*DHeb.*).

Gilbertus Porreta, *Contra Euticen et Nestorium*, in Häring 1966, 233–364 (=*CEut.*).

Peter Damian, *De divina onnipotentia*, ed. Kurt Reindel, in *Monumenta Germaniae Historica. Die Briefe der deutschen Kaiserzeit 4.1–4: Die Briefe des Petrus Damiani*, vol. 3, 341–384. München 1989.

Secondary Literature

Cameron, M. (2011). Methods and methodologies: An introduction. In M. Cameron & J. Marenbon (Eds.), *Methods and methodologies: Aristotelian logic east and west, 500–1500* (pp. 1–26). Brill.

Catalani, L. (2008). *I Porretani. Una scuola di pensiero tra alto e basso Medioevo*. Brepols.

Chisholm, R. M. (1967). Identity through possible worlds: Some questions. *Noûs, 1*, 1–8.

de Rijk, L. M. (1981). Boèce logicien et philosophe: ses positions sémantiques et sa métaphysique de l'être. In L. Obertello (Ed.), *Atti del Congresso internazionale di studi boeziani (Pavia 1980)* (pp. 141–156). Herder.

de Rijk, L. M. (1988a). Semantics and metaphysics in gilbert of Poitiers. A chapter of twelfth century Platonism (1). *Vivarium, 26*(2), 73–112.

[72] Marenbon (1998, 171, n. 31).

[73] See: Valente, "Gilbert of Poitiers", 640a.

de Rijk, L. M. (1988b). On Boethius' notion of being. A chapter of Boethian semantics. In N. Kretzmann (Ed.), *Meaning and inference in medieval philosophy. Studies in memory of Jan Pinborg* (pp. 1–29). Springer.
de Rijk, L. M. (1989). Semantics and metaphysics in Gilbert of Poitiers. A chapter of twelfth century Platonism (2). *Vivarium, 27*(1), 1–35.
Ebbesen, S. (1992). What must one have an opinion about? *Vivarium, 30*(1), 62–79.
Ebbesen, S., & Iwakuma, Y. (1992). Logico-theological schools from the second half of the 12th century: A list of sources. *Vivarium, 30*, 173–210.
Ebbesen, S., Fredborg, K. M., & Nielsen, L. (Eds.). (1983). Compendium Logicae Porretanum ex codice Oxoniensi Collegii Corporis Christi 250: A Manual of Porretan Doctrine by a Pupil of Gilbert. *Cahiers de l'Institut du Moyen-âge grec et latin, 46*, 1–113.
Erismann, C. (2014). Explaining exact resemblance. Gilbert of Poitiers "Conformitas" theory reconsidered. *Oxford Studies in Medieval Philosophy, 2*, 1–24.
Forest, A. (1934a). Gilbert de la Porrée et les écoles du XII siècle. *Revue des cours et des conférences, LXXXV*(II), 410–429 and 640–651.
Forest, A. (1934b). Le réalisme de Gilbert de la Porrée dans le commentaire du De Hebdomadibus. *Revue néoscholastique de Philosophie, 37*(44), 101–110.
Garin, E. (1958). *Studi sul platonismo medieval*. Le Monnier.
Gracia, J. J. E. (1984a). *Introduction to the problem of individuation in the early middle ages*. Philosophia Verlag.
Gracia, J. J. E. (1984b). Thierry of Chartres and the theory of individuation. *New Scholasticism, 58*, 1–23.
Gracia, J. J. E. (1994). *Individuation in scholasticism: The later middle ages and the counter-reformation (1150–1650)*. State University of New York Press.
Gregory, T. (1958). *Platonismo medievale*. Istituto storico per il Medioevo.
Häring, N. M. (1966). *The commentaries on Boethius by Gilbert of Poitiers*. Pontifical Institute of Medieval Studies.
Häring, N. M. (1974). Paris and chartres revisited. In J. R. O'Donnell (Ed.), *Essays in honour of Anton Charles* (pp. 268–329). Pontifical Institute of Medieval Studies.
Hayen, A. (1935–1936). Le concile de Reims et l'erreur théologique de Gilbert de la Porrée. *Archives d'Histoire Doctrinale et Littéraire du Moyen Âge, 10*, 29–102.
Holopainen, T. J. (2016). Peter Damian. In E. N. Zalta (Ed.), *The stanford encyclopedia ofphilosophy* (Winter 2016 Edition). https://plato.stanford.edu/archives/win2016/entries/peter-damian/
Jolivet, J. (1992). Trois versions médiévales sur l'universel et l'individu: Roscelin, Abélard, Gilbert de la Porrée. *Revue de métaphysique et de morale, 97*, 11–155.
Knuuttila, S. (1987). Possibility and necessity in Gilbert of Poitiers. In J. Jolivet & A. de Libera (Eds.), *Gilbert de Poitiers et ses Contemporains. Aux origines de la Logica Modernorum* (pp. 199–218). Bibliopolis.
Knuuttila, S. (1993). *Modalities in medieval philosophy*. Routledge.
Knuuttila, S. (2001). Time and creation in Augustine. In E. Stump & N. Kretzmann (Eds.), *The Cambridge companion to Augustine* (pp. 103–115). Cambridge University Press.
Knuuttila, S. (2008). Medieval modal theories and modal logic. In D. M. Gabbay & J. Woods (Eds.), *Handbook of the history of logic*. II: *Medieval and renaissance logic* (pp. 505–578). Elsevier.
Maioli, B. (1974). *Gli universali. Storia antologica del problema da Socrate al XII secolo*. Bulzoni.
Maioli, B. (1978). *Teoria dell'essere e dell'esistente e classificazione delle scienze in M. S. Boezio, una delucidazione*. Bulzoni.
Maioli, B. (1979). *Gilberto Porretano. Dalla grammatica speculativa alla metafisica del concreto*. Bulzoni.
Marenbon, J. (1988). Gilbert of Poitiers; A note on the Porretani. In P. Dronke (Ed.), *A history of twelfth-century western philosophy* (pp. 328–357). Cambridge University Press.
Marenbon, J. (1998). The twelfth century. In J. Marenbon (Ed.), *Routledge history of philosophy* (Vol. III, pp. 150–187). Routledge.

Nielsen, L. (1976). On the doctrine of logic and language of Gilberto Porreta and his followers. *Cahiers de l'institut du Moyen Age grec et latin, 17*, 40–69.
Nielsen, L. 1982. *Theology and philosophy in the twelfth century: A study on Gilbert Porreta's thinking and theological expositions of the doctrine of the incarnation during the Period 1113–1180*. Brill.
Prior, A. N. (1960). Identifiable individuals. *The Review of Metaphysics, 13*(4), 684–696.
Schmidt, M. A. (1956). *Gottheit und Trinität nach dem Kommentar des Gilbert Porreta zu Boethius de Trinitate*. Verlag für Recht und Gesellschaft.
Southern, R. W. (1970). *Medieval humanism and other studies*. Harper & Row.
Valente, L. (2008). Forme e universali in Gilberto di Poitiers e nella Scuola Porretana. *Documenti e studi sulla tradizione filosofica medievale, 19*, 191–246.
Valente, L. (2011). Gilbert of Poitiers. In H. Lagerlund (Ed.), *Encyclopedia of Medieval philosophy* (pp. 409–417). Springer.
van Elswijk, H. C. (1966). *Gilbert Porreta, sa vie, son œuvre, sa pensée*. Spicilegium sacrum Lovaniense.
Vanni-Rovighi, S. (1956). La filosofia di Gilberto Porretano. *Miscellanea del Centro Studi Medievali, 58*, 1–64.
Vicaire, M.-H. (1937). Les Porrétains et l'avicennisme avant 1215. *Revue des sciences philosophiques et théologique, 26*, 449–482.
Westley, R. J. (1959–1960). A philosophy of the concreted and the concrete. The constitution of creatures according to Gilbert de la Porrée. *Modern Schoolman, 37*, 54–90.
Williams, M. E. (1951). *The teaching of Gilbert Porreta on the trinity as found in his commentaries on Boethius*. Pontificia Università Gregoriana.

Turning Potentialities into Possibilities: Early Medieval Approaches to the Metaphysics of Modality

Irene Binini

1 The "Potency-Based" Account of Possibility

Early medieval reflections on modalities are deeply indebted to the modal theories developed by Aristotle in *De Interpretatione*, with which scholars were acquainted through the mediation of Boethius' translation and commentaries. Late eleventh- and early twelfth-century authors inherited from Aristotle and Boethius a specific syntactical structure, one that they used to construct propositions about possibility, impossibility, necessity, and contingency, as well as a number of rules they applied to describe the logical behavior of such propositions. From the same sources, they also inherited a number of philosophical issues related to modalities, such as the discussion about what the realm of possibility should comprise (in particular, whether this should include possibilities that are never actualized) and the problems concerning future contingents and the existence of free agency. Even though these traditional modal views were considerably enriched in the early Middle Ages with a more sophisticated syntax for modal propositions and a more expressive logic, the general framework in which modal language was analyzed remained essentially Aristotelian in spirit. However, the theories of modalities reported by Aristotle and Boethius were also loaded with certain metaphysical assumptions that some early medieval authors felt uncomfortable adopting. This metaphysical background was mostly connected to what Simo Knuuttila has called the "potency-based" account of possibility, that is, the interpretation of possibilities as ontologically grounded in the *potencies* or *potentialities* of things. According to this modal paradigm, the truth-makers of claims about

I. Binini (✉)
Università Di Parma, Parma, Italy
e-mail: irene.binini@unipr.it

University of Toronto, Toronto, ON, Canada

F. Ademollo et al. (eds.), *Thinking and Calculating*, Logic, Epistemology, and the Unity of Science 54, https://doi.org/10.1007/978-3-030-97303-2_7

possibility are the powers, capacities, or tendencies that certain substances have to be otherwise than they actually are.[1]

The term that Aristotle uses for the word "possible" (the Greek *dynatón*) is said to derive its signification from that of "potency" (*dynamis*; see, for example, *Metaphysics* V 12, 1019b34 ff.), which, according to Aristotle, is the principle of motion or change that allows a substance to be mutable with respect to either its existence or its properties. For Aristotle, possibilities may be ascribed to individuals insofar as they possess such a principle of change, which enables them to activate or undergo a process in which they transition from being in a certain state to being in a different one. For instance, saying that "it is possible for a doctor (or for a patient) to heal" means that the doctor has the power or ability to activate a process of change through which someone goes from being ill to being healthy, or that the patient has the disposition to undergo the same process. This sort of possibility cannot be attributed to substances that are not susceptible to change with respect to their nature or state. To this invariant and fixed part of reality, Aristotle attributes the modality of necessity, which he characterizes in terms of the immutability and impossibility of being otherwise (e.g., *Metaphysics* V 5, 1015a33–1015b15).[2]

Now, as is very often the case with Aristotle, things are more complicated than I have suggested in this sketchy presentation. For one thing, it is debatable whether Aristotle took every possibility to be reducible to a correspondent potentiality. In fact, there is at least one passage in which Aristotle affirms that some meanings of the term "*dynatón*" are not related to the notion of potency, thereby suggesting that the extension of "possible" is wider than that of "potential" (*Metaphysics* V 12, 1019b27–35). Moreover, scholars have often highlighted that the potency-based account of possibility is just one of several modal paradigms that are mixed together in Aristotle's texts, and that such paradigms are not always consistent with one another.[3] Nevertheless, at least in *De Interpretatione*—the main Aristotelian source on modalities that medieval logicians up to the early twelfth century had to hand—Aristotle seems quite consistent in analyzing every possibility in terms of a corresponding potentiality. In Chapters 12 and 13 of this work, Aristotle only takes into account examples of possibility-propositions whose meaning is spelled out in terms of a substance's capacities or dispositions, such as "it is possible for fire to warm," "it is possible for someone to walk," or "it is possible for this cloth to be cut in two pieces." Here, the philosopher claims that the truth conditions of propositions of this sort depend on the fact that the corresponding potencies may come to actuality in present or future situations (e.g., *De interpretatione* XIII, 23a6–15). The connection between possibility and potentiality is further strengthened by the association established in *De*

---

[1] See, for example, Knuuttila (1993, 19–30 and 46 ff.), Knuuttila (2017, Sections 1–2). In this article I will use the terms "powers," "capacities," "tendencies" as being roughly synonymous, and I will not enter into the (in other respects, important) discussions concerning the differences between them.

[2] See Knuuttila (1993, 8–9).

[3] See, in particular, Knuuttila (1993, 11–12).

interpretatione (e.g., XIII, 23a21–23) between the notions of necessity and actuality: in this way, the pair of concepts *actual/potential* mirrors that of *necessary/possible*.

The potency-based account of possibility was further consolidated in Boethius' commentaries on *De Interpretatione*. Here, Boethius treats possibility as a genus within which we may distinguish different aspects or species. In his *divisio*, "possible" is first said of what is actually the case (*quod iam est*), as when we say that it is possible for someone to write because he/she is currently writing: Boethius calls this "*possibile actu*" or "*verum possibile*." Second, possible is what, though not actually being the case, might nonetheless be (*quod cum non sit esse potest*): this is referred to as "*possibile extra-actum*" or "*forsitaneum possibile*".[4] Within the first species of possibility, Boethius further distinguishes possibilities that are always in act (which are, in fact, necessities) and possibilities that are presently in act, but which have existed in potency before being realized and which may return to being *extra-actum* in the future, for example, the possibility for a person to walk insofar as he/she is now walking, even though there are past or future times in which he/she is seated.[5] This second aspect of *possibile actu*, Boethius remarks, is specific to the mutable and corruptible part of the world, in which potency and actuality alternate. The second species of the possible—which includes things that are potentially, though not actually, the case—is itself divided into two categories. On the one hand, there are possibilities that will actualize in the future, and, on the other hand, there are those that will never be actualized but will remain perpetually *in potestate*.[6] Except for the part of the possible coinciding with the necessary, all possibilities considered by Boethius in this *divisio* are interpreted as being grounded in a certain potentiality or power (*potentia, potestas*) possessed by a substance. The relation between possibility and potentiality is implicitly assumed throughout the entire commentaries on *De interpretatione* 12–13, where Boethius often goes from talking about the *possibilitates* pertaining to a certain individual to the *potestates* or *potentiae* that this individual has to bring something about. On a few occasions, Boethius makes the connection between possibility and potentiality explicit, for instance, where he affirms that the signification of the term "possible" is derived from that of potency (*possibile a potestate traductum est*).[7] Similarly, in some passages he draws a parallel between the multiple meanings of the term *potestas* and the semantic ambivalence of the modal term *possibile*.[8] Boethius also thinks that when we affirm that something non-actual is possible, the affirmation is true if the predicated possibility corresponds to a potency or power that exists in a latent, "inactive" state (*potestate tantum*) in the subject. According to his view, certain capacities remain

---

[4] See Boethius (1880, 411 ff. and 454 ff.).

[5] See e.g. Boethius (1877, 203).

[6] Of this last category, there are some possibilities that remain eternally unrealized because of some contingent development of events, such as that of the cloth that can be cut in half but which will wear out first (the example is from Aristotle's *De interpretatione* IX 19a12–14), and there are some that are in principle unrealizable, for instance, the existence of an infinite number (Boethius 1877, 207; 1880, 463).

[7] See Boethius (1880, 453).

[8] See Boethius (1877, 201–202).

in their bearer even when they are not manifest, and, because of this we can say, for instance, that it is not impossible for a man who is not walking to walk, insofar as he possesses the *potestas* to do so.[9]

A special case of *extra-actum* possibilities are those that may be antecedently attributed to a bearer before their manifestation. Indeed, Boethius claims that every property or action manifested in a (contingent) subject is preceded by the corresponding potentiality to have that property or perform that action. For instance, he affirms that before I started to write, a *potentia scribendi* existed in me, which eventually came to actualization, thus allowing me to write.[10] At times, Boethius also speaks as if antecedent possibilities of this sort could be truly predicated of subjects that do not yet exist. He says that Socrates "can be" (*potest esse*) in those times that precede his existence, just as all other mortal beings are first potentially and then actually alive.[11] He also considers the case of an, as yet, non-existing house, saying that the potency of it being built exists before the house's construction, and it is by virtue of this potency that we could have truly said that "It is possible for the house to be," when in fact there was no such house.[12]

Borrowing the terminology used in contemporary debates on the metaphysics of modality, we could say that Aristotle's and Boethius' account of possibility is *realist*, in the sense that "it recognizes [possibility] as a real and mind-independent feature of reality [...] Because possibility is grounded in potentiality, and potentialities are [...] real and mind-independent properties of real, mind-independent objects."[13] In recent years, the interpretation of possibility in terms of potentiality has come back into fashion in contemporary metaphysics of modality, where it is usually defended by claiming that it is more intuitive than other accounts of possibility (e.g., possible-worlds semantics) and "firmly rooted in everyday life," since potentialities are "ubiquitous in our ordinary thought about, and dealing with, the world.".[14] A further advantage of this interpretation is that grounding possibilities in potentialities, which are distinctive features of individuals, would allow for a

---

[9] See, for example, Boethius (1880, 203): "nam et quae actu quidem non est, esse tamen poterat, ut homo cum non ambulat, ambulandi tamen retinet potestatem, non est eum impossibile ambulare."

[10] See Boethius (1880, 413).

[11] See Boethius (1880, 411): "Possibilis duae sunt partes: unum quod cum non sit esse potest, alterum quod ideo praedicatur esse possibile, quia iam est quidem. Prior pars corruptibilis et permutabilis propria est. In mortalibus enim *Socrates potest esse cum non fuit, sicut ipsi quoque mortales, qui sunt id quod antea non fuerunt*. Potest enim homo cum non loquitur loqui et cum non ambulat ambulare" (my emphasis).

[12] See Boethius (1877, 206): "quae actu sunt cum potestate, id est quae et actum habent et aliquando habuerunt potestatem, ut fabricata iam domus aliquando potuit fabricari et prius habuit potestatem secundum tempus, postea vero actum." One may interpret this passage as if the relevant potentiality were attributed to some substance other than the house, for example, the builder of the house or the materials with which the house will be fabricated. However, Boethius does not explicitly appeal here to these explanations.

[13] See Vetter (2018, 291).

[14] See Vetter (2018, 292).

finer-grained treatment of modality than the one permitted by the more classical possible-worlds framework.[15]

As Knuuttila has pointed out, many philosophers throughout the Middle Ages interpreted modalities by applying the potency-based account of possibility inspired by the works of Aristotle and Boethius.[16] As I mentioned above, however, this understanding of possibility was received with some discomfort by early medieval logicians, who were particularly interested in the nature of *extra-actum* possibilities predicated of future individuals or non-existent substances. Their reflection on this sort of modality uncovered some gaps in the traditional reading of possibilities in terms of potency. For instance, they thought that such a framework could not satisfactorily explain the meaning of propositions such as "it is possible for my future son to exist," in which a possibility is attributed to an object that—not yet being in existence—can bear no capacity or disposition. As I propose in Sect. 2, concerns of this sort were already advanced in the works of Anselm of Canterbury, who asked whether we may antecedently attribute potencies to future objects, as when we say that a house can exist before its fabrication or that the world can exist before creation. In Sects. 3 and 4 I show that Anselm's examples, together with many other cases of the same sort, return in the texts of Abelard and his contemporaries in the early twelfth century. I then present some of the doubts that logicians of the time raised concerning the possibilities of future individuals and non-existent things, and their attempts at a new interpretation of possibility that could account for such cases.

2 Anselm on the Predication of Antecedent Possibilities

One early discussion on the relation between potency and possibility—and on how this relation should work when possibilities are ascribed to non-things—is the one found in the *Philosophical Fragments*, an incomplete late work by Anselm of Canterbury. In the opening lines of this work, which has the form of a dialogue between a student and a master, the former poses a question concerning the notions of potency and impotency (*potestas et impotentia*) and possibility and impossibility (*possibilitas et impossibilitas*). One problem emerges from the fact that, on some occasions, we predicate possibilities of things that cannot bear any *potestas*. For instance, we say that a house that does not yet exist "can be" (*potest esse*), thus attributing possession of a certain potency (*potestas*) to it. Indeed, as Anselm remarks, every ascription of possibility involves the ascription of a correspondent potency, for no one doubts that whenever we say that something "can," it does so by virtue of a potency (*Nullus enim negat omne, quod potest, potestate posse*).[17] The case under consideration is

---

[15] See Borghini (2016, 159).

[16] See Knuuttila (2017, Section 2).

[17] See Anselm of Canterbury (1969, 341, 1–12): "DISCIPULUS. Plura sunt, de quibus tuam diu desidero responsionem. Ex quibus sunt postestas et impotentia, possibilitas et impossibilitas, necessitas atque libertas. Quas idcirco simul quaerendo connumero, quia earum mihi mixta videtur

the same as that used by Boethius in his first commentary on *De interpretatione* to exemplify possibilities *extra-actum*. Anselm's student—who explicitly declares that no *"potest"* phrase could be true without an ascription of *potestas* being implicitly involved—seems to adopt the Boethian understanding of possibility as potency described in Sect. 1.

The student then proceeds by construing an argument *ad absurdum*, stating that since a non-existing thing cannot bear any potency, it possesses neither the potency to exist nor the potency not to exist. From this, it follows that (i) *it is not possible for it to exist* and that (ii) *it is not possible for it not to exist*. But from (i), it follows that *it is necessary for it not to exist*, and from (ii) that *it is necessary for it to exist*, by virtue of the inferential rules holding among modal propositions. We may then conclude that for the same (non-existent) thing, it is necessary both to be and not to be, which is absurd.[18]

Because the *Fragments* are interrupted before the master in the dialogue advances his solution to the student's puzzle, one can only try to reconstruct Anselm's way out of the paradox. As Serene notes, one way to solve cases such as the one raised by the student would be to "transfer" the ability that cannot be ascribed to the non-existent thing (in this case, the future house) to some other subject. In this way, one could paraphrase the modal proposition about the house's possibility as being in fact about some other subject's capacities, for example, the builder of the house. This, however, does not seem to be Anselm's strategy here, as he relies instead on distinguishing between proper and improper ascriptions of the term "possible." Anselm does not reject the traditional view, according to which the term "possible" properly expresses a potency or capacity existing in a subject; rather, he suggests that in certain cases the modal term is used improperly. This happens when the thing to which the capacity is attributed could be considered a subject only in an improper sense, for example, when it is a non-existing being. Rather than understanding the details of Anselm's solution, what is important here is to acknowledge that the argument raised by Anselm in the *Philosophical Fragments* unveils a fragility in the traditional, potency-based interpretation of possibilities, as it suggests that the interrelation between possibility and potency needs to be further explained in cases in which non-existing beings are used as the subjects of modal claims.[19]

---

cognitio. In quibus quid me moveat, ex parte aperiam, ut cum de his mihi satisfeceris, ad alia, ad quae intendo, facilius progrediar. Dicimus namque potestatem esse aliquando, in quo nulla est potestas. Nullus enim negat omne, quod potest, potestate posse. Cum ergo asserimus, quod non est, posse esse, dicimus potestatem esse in eo, quod non est; quod intelligere nequeo, velut cum dicimus domum posse esse, quae nondum est. In eo namque, quod non est, nulla potestas est."

[18] See Anselm of Canterbury (1969, 341, 12–39); for a detailed analysis of this argument, see Serene (1981) and Knuuttila (2004).

[19] For this reconstruction of Anselm, see Serene (1981, 120–121. Cf. in particular p. 121): "The paradoxical status of the future house is symptomatic of two gaps in the Aristotelian-Boethian view of modalities: the lack of a systematic explanation of the relationship between capacity and possibility, and the lack of an adequate treatment of antecedent ascription of capacity or possibility to particular subjects."

Similar concerns about the ascription of possibilities to things that are yet to exist are raised by Anselm in Chapter 12 of *De casu diaboli*. Here, while seeking an answer to the problem of whether an angel has the ability to will even though he has never exercised it, another discussion between a master and a student begins about possibilities that are predicated antecedently to their actualization (*de potestate quae praecedit rem*). According to the student, everything that is, at a certain time, actual was possible before being actualized. Indeed, as he points out, if it were not possible for something to be, this would never have actualized (*Si enim non potuisset, numquam esset*). The master replies that if something did not exist at an earlier time, it did not possess either the potency to exist or any other potency (*potestas*), for something that does not exist is nothing, and therefore it would seem that no possibility could be predicated of it (*quod nihil est omnino nihil habet, et ideo nullam habet potestatem, et sine potestate omnino nihil potest*). This discussion is then applied to a specific case: Was it possible for the world to exist before its creation? On the one hand, one would intuitively say that it was, since there could be no actuality without a former possibility of something being the case. On the other hand, because the world was entirely nothing before its creation, it had no capability, and therefore no "*potest*" phrase about it could have been true.

The teacher solves the puzzle by saying that it was both possible and impossible that the world existed before creation: impossible because no potency could be predicated of a non-existing subject; and possible because another agent (God) had the ability to cause the world to exist before he actually created it.[20] Therefore, a modal proposition such as "it is possible for an S to be P" is said to be true in two senses: when "S" has the potency (*potestas*) to be P; or when some other thing has the potency to actualize what the proposition "S is P" says.[21] The teacher remarks, however, that the latter is an improper use of the term "possible," because possibility is properly used to ascribe a certain ability to the proposition's subject, that is, to the thing denoted by "S." If we speak accurately, then, no possibility can be predicated of subjects that do not yet exist.[22]

Both examples of antecedent predications of possibility employed by Anselm in the *De Casu Diaboli* and the *Philosophical Fragments*—that is, the possibility of a house being fabricated before its actual existence and the possibility of the world existing before creation—return in logical sources from the early twelfth century, together with new cases involving the possibilities of non-things, such as chimaeras,

[20] Anselm of Canterbury (1946–1961, vol. I, 253): "Et possibile et impossibile erat antequam esset. Ei quidem in cuius potestate non erat ut esset, erat impossibile; sed deo in cuius potestate erat ut fieret, erat possibile. Quia ergo deus prius potuit facere mundum quam fieret, ideo est mundus, non quia ipse mundus potuit prius esse." This reference to God's ability to create the world as existing "before" (*prius*) creation should perhaps be interpreted as a natural priority rather than a temporal one. The latter interpretation would in fact commit Anselm to the assumption that there existed time before creation, which he does not explicitly state here.

[21] Anselm of Canterbury (1946–1961, vol. I, 253–254): "Ita ergo quidquid non est, antequam sit sua potestate non potest esse; sed si potest alia res facere ut sit, hoc modo aliena potestate potest esse."

[22] For this analysis of Anselm's argument, see Serene (1981, 126), Knuuttila (2004, 119).

goat-stags or future sons. Similarly to Anselm, early twelfth-century authors asked what is signified by the term "possible" in such cases, where there is no proper subject in which to ground the relevant potentialities. Differently from what Anselm suggests in *De casu diaboli*, though, early twelfth-century logicians did not resort to the idea that these sorts of possibility could be ontologically grounded in God and his power. As far as I know, there is only one text in which this strategy is mentioned, namely, the *Dialectica* of Garlandus, a logical treatise whose dating is still uncertain but which was probably written at the turn of the twelfth century.[23] While distinguishing between absolute and determinate modal propositions, Garlandus considers the claim "it is possible for birds to fly" (*possibile est avem volare*), which was employed in Boethius' *De hypotheticis syllogismis* as an example of *absolute* modal proposition, namely, as a proposition in which the modal term is not qualified by any temporal determination and which, as such, must be applied to all times ("*omni tempore*").[24] For the proposition to be true, Garlandus notes that it must be possible for birds to fly even in those times in which they do not exist. Garlandus maintains that this proposition is indeed omnitemporally true, because even when there are no birds it is still *possible for God* to create them and make them fly.[25] Not dissimilarly from what Anselm proposes in *De casu diaboli*, then, Garlandus suggests that the possibilities of non-existent things may be ontologically grounded in the power that God has to bring them about. To my knowledge, however, this is the only case in which such an explanation is put forward in early twelfth-century logical sources. Rather, the common strategy that we find in this period to deal with modalities of non-things and *extra-actum* possibilities was to substitute the traditional view of possibilities as potencies or powers embedded in things with a new interpretation of the term "possible," as I show in the next section.

3 Early Twelfth-Century Logicians on the Signification of Modal Terms

Many sources from the first decades of the twelfth century raise questions concerning the proper interpretation of the modal terms "possible," "impossible," and "necessary." A doubt about the signification of modes that often returns in logical texts of this time is whether or not the term "*possibile*" has denotation, that is, whether there

[23] The authorship of this logical textbook is still debatable. Iwakuma (1992, 47–54) argued that the author should be identified with Gerlandus of Besançon, who died after 1148, and not with Garlandus Compotista, who was believed by de Rijk to be the author. Because of the uncertainty concerning the authorship, the dating of the *Dialectica* also remains an open question. Marenbon suggests that the text could have been written any time between the 1080s (or even earlier) and the 1120s (see Marenbon 2011, 194–196 on this).

[24] See Boethius (1969, 238).

[25] See Garlandus Compotista (1959, 84): "Item possibile est quod absolute omni tempore contingere potest, ut 'possibile est avem volare': licet enim avis omni tempore non sit, potest tamen contingere ut fiat a Deo et ut volet."

is some existing *res* to which it refers. Specifically, Abelard's contemporaries were interested in whether this term denotes a form or property existing in things, so that when we say "it is possible for Socrates to be a man," what we mean is that a *possibilitas* of some sort inheres in Socrates. A discussion of this topic may be found in Abelard's *Dialectica* (*circa* 1110–1115), as well as in an anonymous commentary on *De Interpretatione* labeled *H9*,[26] and in a brief treatise on modalities, also anonymous but containing references to masters of the time, such as William of Champeaux and Joscelin of Soissons, labeled *M3*.[27] These last two texts are preserved in the same manuscript and are probably to be dated around the same time in which Abelard wrote the *Dialectica*, with *H9* supposedly being the oldest among the three texts, and perhaps a common source for both Abelard and the author of *M3*. Despite maintaining very different opinions with respect to the logic of modal propositions, the authors of these three texts answer in the negative as to whether the term "possible" signifies a property inhering in things, and they employ similar arguments to justify their position. This may lead us to suppose that this was a shared opinion in their time, and indeed there is no evidence of people defending the opposite view.

In the *Dialectica* Abelard reports that it is the opinion of "some" that by predicating terms such as "possible" and "necessary," one attributes a certain property (*aliqua proprietas*) to a substance. This cannot be the case, Abelard continues, for if modal terms signified an intrinsic feature of things, then every modal proposition about non-existent objects should turn out to be false, since non-things can bear no property. However, evidently there are many such propositions that are true, such as "it is possible for a future son to exist" (*filium futurum possibile est esse*)[28] or "it is necessary for a chimaera not to be human" (*necesse est chimaeram non esse hominem*). Therefore, he concludes, nothing is attributed to non-things by means of modal terms:

> We shall now investigate whether any property is predicated by means of nominal modes,[29] as some people want. They say that by the name "possible" a possibility is predicated, and a necessity by the name "necessary", so that when we say "It is possible (or necessary) for Socrates to be" we attribute a certain possibility or a certain necessity to him. But this is false.

---

[26] *H9*: Orléans, Bibl. Municipale, 266, pp. 5a–43a; Assisi, Bibl. Conv. Franc., 573, fols. 48rb–67vb. A catalogue of twelfth-century logical texts, including some unpublished sources, to which I will refer in this article, may be found in Marenbon (1993) (republished and updated in Marenbon 2000a).

[27] *M3:* Orléans, Bibl. Municipale, 266, pp. 252b–257b.

[28] In the glosses on *De interpretatione* contained in the *Logica Ingredientibus*, Abelard offers a different reading of such modal propositions, stating that all claims about possibility and necessity (with the exception of those that are impersonal in both grammatical construction and meaning) have an implicit existential import. Appealing to the same example used in the *Dialectica*, he affirms that the proposition "it is possible for my future son to exist" is false if there is no actual object to which the subject term refers (Abelard 2010, 417). On this, see Binini (2018).

[29] In logical sources of this time, nominal modes (*casuales modi*) are opposed to adverbial ones. Propositions containing adverbial and nominal modes are considered by some authors to have a different nature and different semantics. Other authors, such as Abelard, argue instead that every nominal proposition, despite a few exceptions, may be rephrased as having a corresponding adverbial form. On the relation between these two categories of modal, see Binini (forthcoming).

> There are many affirmations of this sort that are true even though they are about non-existent things, which—being non-existent—admit no property of accidents. Indeed, what does not exist cannot bear anything existent. Of this sort of modals [that is, nominal modal claims], the following are true: "It is possible for a future son to exist", "It is possible for a chimaera not to exist" or "It is necessary for a chimaera not to be human"; nonetheless, nothing is taken to be attributed to non-existent things by means of these propositions.[30]

Moreover, Abelard continues, paradoxical consequences would follow from interpreting modes as attributing properties to substances, for instance that "if S will exist, then S presently exists." Suppose that a certain subject S will exist in some future time. We may then infer that it is possible for S to exist (Abelard does not justify this inference here, but he uses it elsewhere by arguing that "whatever is future is possible").[31] But if we take the term "possible" to refer to a certain property of *possibilitas* existing in the subject, this implies that "S has the *possibilitas* to exist" (*possibilitatem existendi habet*) and therefore that it exists now, which would contradict the premise. Similarly, we may argue that "if S does not exist, it exists," for if the subject does not exist, then it is possible for it not to exist (it would seem, by virtue of the principle—commonly accepted in early medieval logic—that whatever is actual is possible), and therefore S has the possibility not to exist, which implies that S exists.[32] Note that even though Abelard uses these arguments to deny that the term "possible" refers to a form of "*possibilitas,*" what they in fact show is more general, namely, that modal terms do not denote *any property* or intrinsic feature of the things that are denoted by the proposition's subject.

On at least two other occasions, Abelard denies that the predication of a possibility amounts to the ascription of a property to a thing. In one passage of the *Dialectica* he writes that when we say that someone *is able* (*potens*) to do or be something, we understand no form as being posited by this term, but we only intend that being in a certain way is not repugnant to the nature of the subject.[33] The same idea is repeated once more in the same work, where Abelard says that, when affirming that "it is possible for Socrates to be a man," we do not attribute any property to anyone

[30] Abelard (1970, 204): "Nunc autem utrum aliqua proprietas per modalia nomina, ut quidam volunt, praedic[ar]etur, persequamur. Aiunt enim per '*possibile*' possibilitatem praedicari, per '*necesse*' necessitatem, ut, cum dicimus: '*possibile est Socratem esse vel necesse*', possibilitatem aut necessitatem ei attribuimus. Sed falso est. Multae verae sunt affirmationes huiusmodi etiam de non existentibus rebus, quae, cum non sint, nullorum accidentium proprietates recipiunt. Quod enim non est, id quod est sustentare non potest. Sunt itaque huiusmodi verae: '*filium futurum possibile est esse*', '*cbimaeram possibile est non esse*', vel '*necesse est non esse hominem*'; nihil tamen attribui per ista his quae non sunt, intelligitur."

[31] See Abelard (1970, 196): "*Quod futurum est, possibile est.*"

[32] Abelard (1970, 204): "Alioquin haberemus quod, si erit, tunc est, vel, si non est, est. Quod sic ostenditur: 'si erit, possibile est esse'; unde 'et possibilitatem existendi habet', unde 'et est'; qua re 'si erit, et est'. Sic quoque: 'si non est, est', ostenditur: 'si non est, possibile est non esse'; unde 'et possibilitatem non-existendi habet'; unde 'est'; 'si non est, est'."

[33] Abelard (1970, 98): "Sic quoque et potentiae non esse album, cum sit actus non esse album, ipsi tamen universaliter subdi non potest, ut videlicet dicamus omne quod non est album potentiam illam habere, sed fortasse ita: '*potens non esse album,*' ut nullam formam in nomine '*potentis*' intelligamus, sed id tantum quod naturae non repugnet; in qua quidem significatione nomine '*possibilis*' in modalibus propositionibus utimur."

(*non aliquam alicui attribuimus proprietatem*); we simply say that the content of the proposition "Socrates is a man" is one of the things that nature allows (*unum de his quae natura patitur esse*).[34]

The idea that the predication of modal terms does not amount to a predication of forms or properties of things is also proposed in commentary *H9* (whose arguments Abelard might in fact be rehearsing), in which the author concludes that the modal terms "possible," "impossible," and "necessary" "do not posit the existence of anything in the substances that modal propositions are about" (*nichil ponunt in rebus de quibus agitur in propositionibus illis*). The author of *H9* proposes an argument that starts by considering a case in which no thing exists. He then infers the proposition "it is not possible for Socrates to be a stone," the truth of which seems to be implicitly derived by the truth of "Socrates does not exist" and the consequent "Socrates has no *possibilitas.*" The author continues stating that "it is impossible for Socrates to be a stone," by virtue of the laws of equipollence among modes, replacing "not possible" with "impossible." Finally, he says that if we intend this affirmation as positing the existence of an *impossibile* in the subject, then we must admit that the subject actually exists, which contradicts the premise. For this reason – the author concludes – we should say that, when "possible", "impossible" and "necessary" are predicated in modal claims, they signify possibility, impossibility and necessity, but "they do not posit the existence of anything in the substances that modal propositions are about (*nihil ponunt circa res de quibus agitur in propositionibus modalibus*)."[35]

In the treatise *M3*, where the author advances a theory on the signification of modal propositions that is in many respects opposite to that of Abelard,[36] we again find the question about whether the existence of any property is predicated by means of modal terms (*utrum aliqua proprietas per modalia nomina ponatur*). And, again,

---

[34] Abelard (1970, 205): "Similiter et quando dicimus: '*possibile est Socratem esse hominem,*' non aliquam alicui attribuimus proprietatem, sed id dicimus quod id quod dicit haec propositio: '*Socrates est homo,*' est unum de his quae natura patitur esse." The use of the term "*natura*" in this and similar contexts is still not yet entirely understood. On some occasions, Abelard and other authors of his time use "nature" to talk about the nature of individual substances (e.g., *natura Socratis*). Elsewhere, they talk about the nature of species of genera (e.g., the nature of human beings), or even about nature in a more general sense, such as "*Natura rerum.*" In the passage in question, it seems to me that Abelard is using "*natura*" in this latter and wider sense, but this is open to speculation. On the notion of nature in Abelard, see, for example, King (2004) and Binini (2021).

[35] See *H9*: Orléans, Bibl. Municipale, 266, p. 37a–b: "Notandum etiam quod iste voces '*possibile*', '*necessarium*' et alii modi qui predicantur, nichil ponunt in rebus de quibus agitur in propositionibus illis. Si enim ponerent, sequeretur: '*si nichil est, aliquid est*' hoc modo. Verum est enim '*si non est possibile Socratem esse lapidem, tunc impossibile est Socratem esse lapidem*'. Et si quia non est possibile Socratem esse lapidem, impossibile est Socratem esse lapidem, et quia non est possibile Socratem esse lapidem, Socrates habet impossibile, et ita Socrates est. Et si quia non est possibile Socratem esse lapidem, Socrates est, et quia nichil est, Socrates est – ab antecedenti, quia si nichil est, Socrates non est; si Socrates non est, non habet possibile, et ita non est possibile eum esse lapidem. Quare si nichil est, aliquid est. Quare dicendum est – quando '*possibile*' et '*impossibile*' et '*necesse*' in modalibus praedicantur – quod significant possibilitatem et impossibilitatem et necessitatem, sed nihil ponunt circa res de quibus agitur in propositionibus modalibus."

[36] For the analysis of the theory of modals included in *M3* and a comparison with Abelard, see Binini (forthcoming).

we find the idea that this view leads to paradoxical consequences, such as that "if something does not exist, it exists" (*si non est, est*) or "if something will exist, it exists now" (*si erit, est*).[37] Having offered this and other arguments, the author of *M3* insists that clearly many inconveniences follow if we admit that the existence of something is posited by modal words (*videamus utrum aliqua proprietas per modalia nomina ponatur. Si enim per ea aliquid ponitur, multa sequentur inconvenientia*).

In another passage of *M3*, the author considers modal propositions about non-things and points out that they reveal important aspects concerning the signification of modal terms. He says that there are some who interpret modal claims such as "for every human it is possible to be an animal," as if the modal term posited the existence of something (a *possibilitas*) possessed by the subject. This interpretation, however, fails to account for the many cases in which modal claims are true despite their dealing with *non-existent things*, such as "it is possible for a chimaera not to be a goat-stag" or even "it is possible for the world to be created" (*possibile mundum fieri*), if uttered before the creation of the world. The author takes both propositions to be evidently true, and yet he wonders which thing would be the "possessor" of the possibility, if nothing exists:

> There are some who expound [nominal propositions such as] "for every man to be an animal is possible" in this way: "things have the possibility that every man is an animal". But this cannot be right. Indeed, it is true that "for chimaeras not to be goat-stags is possible". [To which] we will say: How would things have the possibility that chimaeras were not goat-stags? After all, there are no things having that possibility, because neither chimaeras nor any other thing exist. And yet the proposition is true. In the same way, if before the creation of the world one said: "It is possible for the world to be made", this proposition would be true, but which things would have the possibility of the world being made, if nothing whatsoever existed? Thanks to this and many other examples, it can be shown that their exposition [of modals] is incorrect.[38]

The author of *M3* seems to suggest not only that modal terms fail to denote the *possibilitates* existing in the modal proposition's subject, but also that we cannot "relocate" these possibilities as existing in any other subject. In a case in which the world did not yet exist, there would be no *res* bearing the relevant potency, but it would still be true to predicate the possibility of its existence. The example concerning the antecedent possibility of the world to exist is the same as the one

---

[37] See *M3*, p. 254b: "Investigato sensu modalium, videamus utrum aliqua proprietas per modalia nomina ponatur. Si enim per ea aliquid ponitur, multa sequentur inconvenientia. [...] Item "*si erit, et est*", sic: Si Socrates erit, possibile est esse Socratem; et sic Socrates habet possibilitatem existendi; et ita est. Item si non est, non possibile est esse, quia si est possibile esse, et est. Si Socratem esse est possibile, Socrates habet possibilitatem existendi; et ita possibilitas est in Socrate; et ita est."

[38] See *M3*, p. 254b: "Sunt qui exponant ita '*Omnem hominem esse animal est possibile*': res habent possibilitatem quod omnis homo sit animal. Sed hoc nihil est. Vera est enim '*chimaeram non esse hircocervum est possibile*'. Dicemus: quomodo [*corrected from: dicemus modo quod*] res habent possibilitatem quod chimaera non sit hircocervus, quippe nullae res habent illam possibilitatem, quia neque chimaera neque alia, tamen vera est illa propositio. Item antequam mundus fieret, si diceretur '*possibile mundum fieri*', vera esset talis propositio; sed cum nulla res esset, quae res habebant possibilitatem ut mundus fieret? His et multis aliis exemplis nulla esse ostenditur illa expositio." I thank Professor Wciórka for suggesting me this reading of the text.

brought up by Anselm in *De casu diaboli*, as seen in the previous section. However, differently from what Anselm says there, the idea that God's power may grant the truth of propositions such as "it is possible for the world to exist" is not mentioned here; nor is it generally put forward in other early twelfth-century logical sources. On the contrary, their authors seem to lean towards the idea that some possibilities could be admitted without being analyzed in terms of properties or intrinsic features that substances possess, thereby abandoning the traditional potency account of possibility.

Were early twelfth-century authors aware that, by rejecting the traditional potency account, they were also discarding part of Aristotle's legacy on modalities? At least Abelard seems to realize this, as in the *Logica Ingredientibus* he counters Aristotle's view of possibility with his own, saying that Aristotle interpreted the term "*possibile*" as referring to some *possibilitas* or *potestas*, that is, to some form or property existing in a substance, and therefore he made "possible" "a name of things" (*nomen rerum*), that is, a name denoting some real component of reality. Abelard claims to have abandoned this idea, and he thinks that no form or property is understood by the terms "possible" or "necessary":

> Note that, from the words of Aristotle, when he speaks of "possibilities" (*potestates*) it seems that in the name "possible" he understands a certain form, that is, a certain potency or possibility, which seems to make [the name "possible"] a name of things (*nomen rerum*)—a position that we have rejected above. We, on the contrary, do not understand any form when speaking of "possibility" or "necessity", but we expound [these terms] according to the meaning of modals.[39]

In the *Logica* Abelard further stresses this "*de-reification*" of the notion of possibility. Appealing to his twofold theory of signification, he argues that modal terms have no denotation (*nominatio*), for they do not refer to substances or to any form possessed by a subject, and that they do not have signification (*significatio*) either, for no image is caused in the hearer's mind when they are uttered outside of a context.[40] After rehearsing what he had already said in the *Dialectica*—namely, that we cannot take modal terms "as forms inhering in things" (*quasi formas aliquas in rebus*), otherwise, modal propositions about non-existent beings could not be true—Abelard goes on to say that "possible" and "necessary" only signify when they are considered in the linguistic context in which they are embedded, as they express a *way of conceiving* the things that are conjoined to them:

> Since [the modal nouns] "possible" and "necessary" are not derivative (*sumpta*) expressions, and they neither contain any thing by denoting it nor determine any form, it should be asked what is it that they signify. Indeed, when it is said "It is possible for what is not to be", or "It is necessary for God to exist" or again "It is necessary for chimaeras not to exist", we do not intend this in the sense that certain forms exist in such things. We say that in propositions of

39 Abelard (2010, 472): "Nota etiam quod ex verbis Aristotelis, cum ait '*potestates*', videtur ipse in hoc nomine '*possibile*' (quod etiam nomen rerum facere videtur) potestatem sive possibilitatem, quandam formam, intelligere, cum ipsum in modalibus propositionibus ponit; quod supra negavimus. Nos tamen, cum dicit '*potestatem*' vel '*necessitatem*,' nullas intelligimus formas sed iuxta sensum modalium omnia exponimus."

40 On the notion of *consignificatio* in Abelard and in the grammatical tradition of the late eleventh-century *Glosulae* on Priscian, see for example, Rosier-Catach (2003).

this sort [the terms] "necessary" and "possible" co-signify rather than having a signification of their own, because nothing is understood in them unless they are applied to the phrase (*oratio*) that is the subject. And therefore, these terms *express a way of conceiving* the things that the subject phrase is about, just as an interposed verb or the conjunction "if" (which expresses a necessity of conjunction) would do. And, just as in the case of these last expressions no image is created in the understanding, but by means of the verb or the conjunction the mind captures a certain way of conceiving those things that are adjoined to them, the same happens for the terms "possible" and "necessary". And here with "necessary" the meaning is what is inevitable, and with "possible" what is not incompatible with nature.[41]

4 A New Understanding of Possibility

At the end of the passage from the *Logica Ingredientibus* quoted above, Abelard mentions his idea, already encountered in other passages, that the modal term "possible" should be understood in terms of "what is not incompatible with nature" (*non repugnans naturae*). This characterization of possibility often returns in both the *Dialectica*[42] and the *Logica Ingredientibus*,[43] and it is at the basis of Abelard's theory of modalities.[44] Indeed, as Martin recently pointed out, the same definition is used in other logical sources from the early twelfth century and seems to be a standard characterization in this period.[45] In Abelard's *Dialectica*, the definition of possibility as non-repugnancy with nature is presented in connection with his explanation of unrealized possibilities: in a well-known passage, Abelard states that the truth value of propositions about possibility does not depend on the actual happening (or non-happening) of things, for there are certain things that are possible even though they are never actualized, for instance, that Socrates is a bishop. What is needed for a proposition such as "it is possible for Socrates to be a bishop" to be true is the absence of an incompatibility relation (*non repugnantia*) between what is expressed by the predicate and the nature of the thing denoted by the subject, namely, Socrates.[46]

[41] See Abelard (2010, 407–408): "At vero cum '*possibile*' vel '*necessarium*' sumpta non sint nec res aliquas nominando contineant nec formas determinent, quid significent quaerendum est; non enim, cum dicitur: '*Id quod non est possibile est esse*' vel: '*Deum necesse est esse*' vel: '*Chimaeram necesse est non esse*' quasi formas aliquas in rebus accipimus. Dicimus itaque necessarium sive possibile in huiusmodi enuntiationibus magis consignificare quam per se significationem habere; nil quippe in eis est intelligendum nisi subiectae orationi applicentur, et tunc modum concipiendi faciunt circa res subiectae orationis sicut facit verbum interpositum vel coniunctio *si*, quae ad necessitatem copulat; ac, sicut in istis nulla imagine nititur intellectus sed quendam concipiendi modum anima capit per verbum vel per coniunctionem circa res earum vocum quibus adiunguntur, ita per *possibile* et *necessarium*. Et est hoc loco necessarium pro inevitabili, possibile quasi non repugnans naturae."

[42] See, for example, Abelard (1970, 98; 176; 196–198; 200–204; 385).

[43] See, for example, Abelard (2010, 266; 408; 414–415).

[44] On Abelard's paradigm of possibility as non-repugnance with nature, see Knuuttila (1993), Martin (2001, 2004), Thom (2003) and Marenbon (2000b).

[45] See Martin (2016, 121).

[46] See Abelard (1970, 193–194): "'*Possibile*' quidem et '*contingens*' idem prorsus sonant. Nam '*contingens*' hoc loco <non> quod actu contingit accipimus, sed quod contingere potest, si etiam

Similarly to Abelard, Garlandus advances the same understanding of possibility in connection to the notion of *extra-actum* possibilities: while commenting on Aristotle's idea at the end of *De Interpretatione* 13 concerning the existence of some "pure potencies," Garlandus states that these are said to be "possible," though never actualized, insofar as they are not incompatible with nature (*nec natura repugnat nec tamen umquam erit*).[47] Early twelfth-century logicians were particularly attracted to possibilities that remain perpetually *extra-actum*, and indeed we may suppose that they may have favored the explanation of possibility in terms of non-repugnancy with nature in order to account for unactualized (and unactualizable) possibilities.

At least two other mentions of this same definition of possibility can be found in sources of the time. Though brief, these mentions are interesting because they suggest that this particular interpretation of possibility might have been elaborated in the early twelfth century in response to the ontological concerns related to "problematic" possibilities, such as those that are predicated of non-existent or future things, which could not be accounted for in terms of properties or potentialities inhering in substances. The first occurrence is in the commentary *H9*, where the anonymous author, while presenting Boethius' *divisio* of the many species of "possible," argues that some *possibilia* may come to actualization after having existed in potency at an earlier time. Borrowing the example that Boethius uses in his first commentary on *De interpretatione*, the author of commentary *H9* claims that a house that now exists in act was already existing *potestate*, as an *extra-actum* potency, before it was fabricated. Perhaps willing to further explain how possibility can be predicated of something that does not yet exist, the author remarks that things of this sort are said to be possible before their existence (*prius potuerunt existere quam fuerunt*), in the sense that their existence is not incompatible with nature (*ita quod natura non repugnat*).[48] Although this passage is so brief that any conclusion drawn from it should be regarded as speculative, it is interesting that the analysis of possibility as

numquam contingat, dummodo natura rei non repugnaret ad hoc ut contingat, sed patiatur contingere; ut, cum dicimus: '*Socratem possibile est esse episcopum*', etsi numquam sit, tamen verum est, cum natura ipsius episcopo non repugnet; quod ex aliis eiusdem speciei individuis perpendimus, quae proprietatem episcopi iam actu participare videmus. Quicquid enim actu contingit in uno, idem in omnibus eiusdem speciei individuis contingere posse arbitramur, quippe eiusdem sunt omnino naturae."

[47] See Garlandus Compotista (1959, 83–84): "Potentia vero extra actum quam effectus non consequitur, est illa cui nec natura repugnat nec tamen umquam erit, ut cum dico: '*possibile est Iarlandum fieri episcopum*', numquam tamen episcopus erit." Notice that, differently from Abelard in the *Dialectica*, Garlandus speaks here not of the nature of a thing but of nature in general.

[48] See *H9*, p. 39b: "Possibilia alia sunt in actu, alia numquam in actu. Subdividit ea etiam que sunt in actu, sic: quod alia sunt in actu sine precedente potestate, ut divine substantie, alia vero sunt in actu cum precedente potestate, idest prius habuerunt potestatem quam actum, ut fabricata domus. [...] Que, scilicet ea que sunt in actu, priora sunt et digniora scilicet potestatibus natura, idest per naturam ipsius actus. Actus namque natura et dignitate precedunt solas potestates, sed vera sunt posteriora in tempore ipsis potestatibus. Potestas namque, ut dictum est, eos actus secundum tempus precedit. Vel sic. Que priora sunt natura, idest naturaliter, prius potuerunt existere quam fuerunt et *ita quod natura non repugnat*; tempore vero, idest secundum tempus existendi actu, sunt posteriora se ipsis quantum ad hoc quod natura prius potuerunt existere. Alia vero numquam sunt, sed potestate sola, ut quod rusticus fiat episcopus vel rex." (my emphasis).

"non-incompatibility with nature" comes up in connection with the same example that puzzled Anselm's student in the *Philosophical Fragments.* As we saw in Sect. 2, the student wondered how one could account for possibilities that are antecedently ascribed to a particular subject, given that no capacity or power could be provided to ground them. While treating the same example, the author of *H9* seems to suggest that possibilities of this sort need not be ontologically grounded in some modal property embedded in things, for the term "possible" merely expresses the absence of an incompatibility between a certain predication and "nature" taken in a general sense.

A passage of the treatise *M3* also suggests that this new account of possibility was developed to avoid the embarrassment of providing an ontological foundation for possibilities based on modal properties, therefore allowing for predication of modalities to non-things. As I mentioned above, the author of this treatise rejects the idea that modal terms signify properties existing in substances, because he thinks that propositions such as "it is possible for Socrates to be an animal" are true, even though Socrates does not actually exist and therefore cannot bear any property. Once he has presented the many difficulties that would follow from interpreting possibilities as forms existing in things, the author mentions a strategy designed by a certain "Master W." (probably, William of Champeaux) to expound the signification of modes without an undesired ontological commitment to special kinds of property or to these properties' bearers. According to Master W's interpretation, propositions such as "it is possible for Socrates to be an animal" should be expounded "in a negative sense" (*in negativo sensu*), that is, to mean that: "It is not repugnant to the nature of the thing that Socrates is an animal" (*Socratem esse animal est possibile, id est non repugnat natura rei Socratem esse animal*).[49] Again, saying that something is possible amounts to saying that no relation of incompatibility exists between a certain predication and the nature of things, and this could be the case even if the thing in question does not exist.

5 Conclusion

The interest in the modalities of non-existent things or future things pushed logicians from the early twelfth century to reconsider the semantics of modal propositions and the nature of possibility. As seen in Sect. 2, concerns about the signification of the term "possible" were already present in the works of Anselm of Canterbury, who wondered how the construal of possibilities in terms of potencies, inherited by Aristotle and Boethius, could be compatible with attributing possibilities to not yet

[49] See *M3*, p. 255a: "Investigato sensu modalium, videamus utrum aliqua proprietas per modalia nomina ponatur. Si enim per ea aliquid ponitur, multa sequentur inconvenientia. [...] Item si non est, non possibile est esse, quia si est possibile esse, et est. Si Socratem esse est possibile, Socrates habet possibilitatem existendi; et ita possibilitas est in Socrate; et ita est. Quare si possibile est esse, et est. Quare '*si non est, non est possibile esse*' haec et plura alia inconvenientia, si per modales voces aliquid ponatur, sequi manifestum est. Unde m. W. exponebat eas in negativo sensu, ut istam: '*Socratem esse animal est possibile*', id est non repugnat natura rei Socratem esse animal."

existing things, for example when affirming the possibility for a future house to be fabricated or the possibility for the world to exist before creation. Anselm did not abandon the idea that the term "possible" properly denotes a potentiality or a power embedded in a subject, but he unveils an unsolved issue in the traditional explanation of possibilities as potencies, suggesting that this explanation could hardly account for cases in which possibilities are truly predicated of non-existent objects.

As I proposed in Sect. 3, in early twelfth-century logical sources examples concerning the modalities of non-things and future beings are multiplied, highlighting their authors' interest in the ontology of modalities. Logicians of this time wanted to speak about things having certain possibilities without "anchoring" them in the individuals and without committing themselves to the existence of either modal properties or the bearers of these properties, and in order to do so they unanimously ruled out the idea that modal nouns denote properties existing in substances. To defend this view, authors of the time have recourse to arguments having a similar structure: they admit that certain modal propositions are true even though they predicate the possibilities of a non-existent subject, such as "it is possible for my future son to exist" or "it is not possible for (a non-existing) Socrates to be a stone," and they show that if the term "possible" is taken to refer to a form inhering in the subject, paradoxical consequences will follow. They therefore deny that the predication of a possibility amounts to the ascription of a property to a thing, and they say that, in fact, terms such as "possible" and "necessary" do not posit the existence of anything in the substances that modal propositions are about ("*nihil ponunt circa res de quibus agitur in propositionibus modalibus,*" as the author of *H9* claims). Interestingly, there is agreement on this thesis even among logicians that offer, in other respects, a very different doctrine on modal propositions. The inclination that these authors show toward a "de-reified" understanding of possibility is further stressed by Abelard in the *Logica Ingredientibus*, where he claims that the modal terms "possible" and "necessary" have entirely no denotation or signification when taken in isolation from a context, and that they simply convey to the mind a certain "way of conceiving" the things of which they are predicated.

In Sect. 4 I advanced the idea that the definition of possibility as "non-repugnancy with nature," which we often find in Abelard's texts and in other sources from the early twelfth century, might have been developed by logicians of Abelard's time as a way out of problems related to the ontology of possibilities. This new account of possibility—according to which possibilities are not grounded in the individuals as their real constituents, but analyzed as non-contradictoriness relations holding between certain predicates and the natural laws governing creatures—enabled early twelfth-century logicians to offer an analysis of the possibilities of non-things, possibilities of future states of affairs, and generally every sort of *extra-actum* possibility.

Acknowledgements This research has received funding from the European Union's Horizon 2020 research and innovation program, under the Marie Skłodowska-Curie grant agreement n° 845061. I am very grateful to Yukio Iwakuma and C.H. Kneepkens for sharing with me their transcriptions of some unedited texts that will be considered in this article, and to Wojciech Wciórka for his valuable help in the interpretation of several manuscript passages. I am also grateful to the anonymous referees for their useful suggestions. I dedicate this article to Professor Massimo Mugnai, who first introduced me to the study of Abelard and twelfth-century modal logic. For this and for his supervision during the first steps of my research, I am extremely thankful.

References

Primary Texts

Anselm of Canterbury. (1946–1961). In F. S. Schmitt (Ed.), *S. Anselmi Cantuariensis Archiepiscopi Opera Omnia*, 6 vols. Nelson.

Anselm of Canterbury. (1969). Philosophical fragments. In R. W. Southern & F. S. Schmitt (Eds.), *Memorials of St. Anselm* (pp. 334–351). Oxford University Press.

Boethius. (1877). In C. Meiser (Ed.), *Commentarii in librum Aristotelis Perihermeneias* (vol. I). Teubner.

Boethius. (1880). In C. Meiser (Ed)., *Commentarii in librum Aristotelis Perihermeneias* (vol. II). Teubner.

Boethius. (1969). In L. Orbetello (Ed.), *De hypotheticis syllogismis.* Paideia.

Garlandus Compotista. (1959). In L. M. de Rijk (Ed.), *Dialectica.* Van Gorcum.

Peter Abelard. (1970). In L. M. de Rijk (Ed.), *Dialectica* (2nd ed.). Van Gorcum.

Peter Abelard. (2010). Glossae super Peri Hermeneias. In K. Jacobi & C. Strub (Eds.), *Petri Abaelardi Glossae super Peri Hermeneias.* Brepols.

Secondary Literature

Binini, I. (2018). "My future son is possibly alive." Existential presupposition and empty terms in Abelard's modal logic. *History and Philosophy of Logic, 39*(4), 341–356.

Binini, I. (2021). Some further remarks on Abelard's notion of nature. In I. Chouinard et al. (Eds.), *Women's perspectives in ancient and medieval philosophy.* Springer.

Binini, I. (forthcoming). The "De re"-"De dicto" distinction: A twelfth-century logical discovery. *Vivarium.*

Borghini, A. (2016). *A critical introduction to the metaphysics of modality.* Bloomsbury.

Brower, J. E., & Guilfoy, K. (Eds.). (2004). *The Cambridge companion to Abelard.* Cambridge University Press.

Iwakuma, Y. (1992). "Vocales", or early nominalists. *Traditio, 47*, 33–111.

Jacobi, K. (2004). Philosophy of language. In Brower–Guilfoy 2004, 126–157.

King, P. (2004). Metaphysics. In Brower – Guilfoy 2004, 65–125.

Knuuttila, S. (1993). *Modalities in medieval philosophy.* Routledge.

Knuuttila, S. (2004). Anselm on modality. In B. Davies (Ed.), *The Cambridge companion to Anselm* (pp. 111–131). Cambridge University Press.

Knuuttila, S. (2017). Medieval theories of modality. In E. N. Zalta (Ed.), *The Stanford encyclopedia of philosophy* (Summer 2017 Edition). https://plato.stanford.edu/archives/sum2017/entries/modality-medieval/.

Kristina, E., & Michael, Q. (Eds.). (2018). *Handbook of potentiality.* Springer.

Marenbon, J. (1993). Medieval Latin commentaries and glosses on Aristotelian logical texts, before ca.1150 A.D. In C. S. F. Burnett (Ed.), *Commentaries and glosses on Aristotelian logical texts: The Syriac, Arabic and Medieval Latin traditions* (pp. 77–127). Warburg Institute, University of London (republished and updated in Marenbon 2000a, 77–140).

Marenbon, J. (Ed.). (2000a). *Aristotelian logic, Platonism, and the context of early medieval philosophy in the West*. Routledge.

Marenbon, J. (2000b). Abelard's concept of possibility. In Marenbon 2000a, 595–609.

Marenbon, J. (2011). Logic at the turn of the twelfth century: A synthesis. In I. Rosier-Catach (Ed.), *Arts du langage et théologie aux confins des XIe-XIIe siècle* (pp. 181–218). Brepols (Studia Artistarum, 26).

Marmodoro, A. (2018). Potentiality in Aristotle's metaphysics. In Engelhard–Quante 2018, 15–43.

Martin, C. J. (2001). Abaelard on modality: Some possibilities and some puzzles. In T. Buchheim, C. H. Kneepkens, & K. Lorenz (Eds.), *Potentialität, und Possibilität. Modalaussagen in der Geschichte der Metaphysik* (pp. 97–124). Frommann-Holzboog.

Martin, C. J. (2004). An amputee is bipedal! The role of categories in the development of Abelard's theory of possibility. In I. Roseier-Catach & J. Biard (Eds.), *La tradition médiévale des Catégories* (pp. 225–242). Editions de l'Institut Supérieur de Philosophie.

Martin, C. J. (2016). Modality without the prior analytics: Early twelfth century accounts of modal propositions. In M. Creswell, E. Mares, & A. Rini (Eds.), *Logical modalities from Aristotle to Carnap* (pp. 113–132). Cambridge University Press.

Rosier-Catach, I. (2003). Priscien, Boèce, les Glosulae in Priscianum, Abélard: Les enjeux des discussions autour de la notion de consignification. *Histoire Epistémologie Langage, 25*(2), 55–84.

Serene, E. F. (1981). Anselm's modal conceptions. In S. Knuuttila (Ed.), *Reforging the great chain of being: Studies of the history of modal theories* (pp. 117–162). Reidel (Synthese Historical Library 20).

Thom, P. (2003). *Medieval modal systems: Problems and concepts*. Ashgate.

Vetter, B. (2018). From potentiality to possibility. In Engelhard–Quante 2018, 279–301.

Ockham on Abstract Pseudo-Names

Claude Panaccio

In the winter of 2005 Massimo Mugnai held a seminar on William of Ockham's *Summa logicae*, with each session of it being dedicated to a particular chapter of the work. When Massimo kindly invited me to give a talk there, I chose to deal with the status of abstract terms in Part One, chapter 8 of the *Summa*. The present paper is a revised version of this presentation.[1] Although little studied, the topic of abstract terms is indeed crucial for any nominalistically inclined semantics, and especially for Ockham's. If only individual entities are accepted in the ontology, terms like 'horseness' or 'triangularity' cannot be taken to refer to common natures or general properties. How are they to be accounted for, then? It might be held that some of them should be dealt with in the same way that concrete general terms are: the concrete term 'horse', for example, simultaneously signifies all singular horses for Ockham, and the abstract term 'whiteness' similarly signifies all singular whitenesses. This is because the latter are accepted in his ontology as distinct entities of their own: they are singular qualities—or 'tropes' in the language of today's metaphysics.[2] They can therefore be distributively referred to by way of general terms just as individual substances can. This does not hold, however, for all abstract terms. While there are whiteness tropes in the world, according to Ockham, there are no horseness tropes as distinct from individual horses, and no triangularity tropes either as distinct from triangular objects. The corresponding abstract terms, then, have to be semantically accounted for in some other way and *Summa logicae* I, 8 puts forward an interesting proposal for this. I will first situate the chapter in its surroundings. Secondly, I will explain Ockham's idea that certain abstract terms are abbreviations for complex

[1] I also dealt with connotative terms in another meeting of the seminar. This is a subject I extensively discussed in other publications. See e.g. Panaccio (2000) and Panaccio (2004, c. 4–6, 63–118).

C. Panaccio (✉)
University of Quebec at Montreal, Montreal, QC, Canada
e-mail: panaccio.claude@uqam.ca

[2] Panaccio (2008) and Panaccio (2015).

F. Ademollo et al. (eds.), *Thinking and Calculating*, Logic, Epistemology, and the Unity of Science 54, https://doi.org/10.1007/978-3-030-97303-2_8

phrases with syncategorematic terms in them. Thirdly, I will show that a contextual analysis is called for in such cases. And finally, I will comment on Ockham's own application of this approach to the interpretation of Avicenna's famous statement that "horseness is nothing but horseness".[3]

1 Concrete and Abstract Terms

Chapters 2 to 13 of Part One of *Summa logicae* have to do with the main distinctions that can be drawn among simple terms. Strikingly enough, five of these twelve chapters are dedicated to concrete and abstract terms,[4] while such important distinctions as those between categorematic and syncategorematic terms or between absolute and connotative terms receive only one chapter each. One may wonder: why is this particular distinction so special that it requires a much longer development than the other ones? The answer is that from a semantic point of view it is not a single unified distinction for Ockham. Its apparent unity is but a surface phenomenon due to morphological rather than semantic features of spoken and written languages. At the very start of his development on the subject, Ockham indeed characterizes the distinction in purely morphological terms: "Abstract and concrete names", he writes, "are names which have the same stem but different endings [...] Always (or at least frequently) the abstract name has more syllables than the concrete."[5] What he is talking about here are such pairs as white/whiteness, just/justice, horse/horseness or courageous/courage. Ockham's idea is that the members of these pairs can be semantically related with one another in various ways.

In certain cases—discussed in *Summa logicae*, I, cc. 6–7—the two members of the pair are synonymous with one another: "nothing is in any way signified by one of the terms which is not in the same way signified by the other".[6] That there are two terms rather than one is simply due in such cases to a desire to "embellish style or something of that nature".[7] 'Long' and 'length' or 'extended' and 'extension' are examples of this for Ockham: insofar as he denies that the length of a piece of wood, for instance, is a thing distinct from this piece of wood itself, and more generally that extension is something distinct from the spatially extended substances or qualities, it comes to the same in his view to say that this piece of wood is one-meter long or that it is a one-meter length, and that it is extended or that it is an extension.

[3] All references to Ockham's works will be to the critical edition of the Franciscan Institute of the Saint-Bonaventure University in two series: *Opera Philosophica* in 7 volumes (henceforth *OPh*), and *Opera Theologica* in 10 volumes (henceforth *OTh*). For the *Summa logicae* I will also frequently mention Michael Loux's translation of Part One (Loux (1974)).

[4] *Summa logicae*, I, cc. 5–9, *OPh* I, 16–35.

[5] *Sum. log.*, I, c. 5, *OPh* I, 16,5–10; transl. Loux, 56.

[6] *Sum. log.*, I, c. 6, *OPh* I, 19,10–11; transl. Loux, 58.

[7] *Sum. log.*, I, c. 6, *OPh* I, 20,34–35; transl. Loux, 58.

A different kind of case—introduced in *Summa logicae*, I, c. 5—is when one of the two terms signifies or connotes something that the other neither signifies nor connotes. 'White' and 'whiteness' provide a good example. 'Whiteness' in Ockham's analysis signifies nothing but singular whiteness tropes, while 'white' on the other hand signifies the singular substances that underlie such whiteness tropes and it connotes the whiteness tropes thus possessed by these substances. 'White', then, signifies some things that 'whiteness' neither signifies nor connotes, namely the white substances.

In still another sort of case—briefly dealt with in *Summa logicae*, I, c. 9—one of the two terms refers to a collection of things while the other stands for particular members of the collection. 'People' and 'popular' are given as examples by Ockham: a people is a group of persons, some of whom might individually be said to be popular. And we might also think of 'cavalry' and 'cavalryman', insofar as a cavalry is a group of cavalrymen. This case does not reduce to the previous one for Ockham because he does not take a collection of individual things to be another thing in the world in addition to its individual members. In his view a collective term such as 'people' or 'cavalry' refers to nothing but the individuals that constitute the group, but it does so collectively rather than distributively. The difference is not an ontological one, as in the case of a white substance and its whiteness trope, but a mere semantic one: 'cavalry' and 'cavalryman' refer to the very same individuals in the world, although in different ways.

The fourth case—which actually comes third in Ockham's enumeration—is presented in *Summa logicae*, I, c. 8. It is the one I will be interested in in the rest of this paper.

2 Pseudo-Names

It sometimes happens, Ockham claims, that the concrete term and the corresponding abstract one refer directly or connotatively to the same things in the world but fail to be synonymous with one other, although neither of them is a collective term. His explanation of this phenomenon is that one of the two terms in such cases—usually the abstract one—is actually an abbreviation for a complex phrase that includes the other member of the pair plus certain syncategorematic determinations. Syncategorematic terms, such as logical connectors, quantifiers and prepositions, refer to no special objects in the world in Ockham's view[8]: 'if', 'and' 'every' and 'not', for example, are not referential expressions. Yet this is not to say that syncategorematic terms have no semantic role to play. They usually affect in important ways the truth-conditions of the sentences in which they occur and the referential functions of the categorematic terms that they accompany. The complex phrase 'every man', for instance, does not signify any extra thing in addition to those that are signified by 'some man' or by 'man' taken alone, but the truth-conditions of 'Every man runs' are not the same as

[8] *Sum. log.*, I, c. 4, *OPh* I, 15.

those of 'Some man runs', and the mode of reference of the term 'man' is not the same either in the two sentences.[9] Ockham's point in *Summa logicae*, I, c. 8 is that some abstract terms are conventionally used as abbreviations for phrases that include, when explicitly displayed, at least one categorematic term (usually the corresponding concrete term) and at least one syncategorematic term.

This is possible because, as Ockham points out: "the speakers of a language can, if they wish, use one locution instead of several".[10] A speaker, he says, can decide to use '*a*' in place of 'every man', or '*b*' in place of 'man alone'. It might happen, of course, that the abbreviation in such cases does refer to certain things that are not referred to by any particular constituent of the abbreviated sequence. Should I decide to use '*c*' in place of 'non man', for instance, '*c*' would refer to things that are not referred to by 'man' although 'man' is the only categorematic constituent of the abbreviated phrase. But this is far from being always the case. In the above examples, every real thing that is referred to by the complex sequences 'every man' and 'man alone' is also referred to by 'man' and vice versa, since 'man' is said by Ockham to signify all the men there are (*SL* I, c. 33). The terms '*a*', '*b*' and 'man', therefore, have exactly the same ontological import. Yet they are not synonymous. They are not substitutable to each other *salva veritate* even in normal transparent contexts: '[A] man runs' can be true while '*a* runs' (every man runs) and '*b* runs' (man alone runs) are not. And they are not always interpredicable either: '[A] man is *a*' is false, for example, since it is false that a man is every man.

It is important to note that the abbreviated phrase can be *any* sequence at all. In particular, it does not need to be an expression that can occur as subject or predicate in a grammatically well-formed sentence. 'Horseness', for example, is sometimes used as an abbreviation for 'a horse necessarily', Ockham thinks. This would presumably happen because this particular sequence of words, 'a horse necessarily', was expected by the abbreviators to occur frequently in their discourse. Yet even though it can correctly be part of a well-formed sentence, the expression 'a horse necessarily' is not a well-formed subject term. It occurs, for instance, in a sentence such as "A horse necessarily is an animal", but the adverb 'necessarily' in this sentence is not part of the subject term in Ockham's analysis, it qualifies the copula. Thus understood, the morphologically simple term 'horseness', then, is a pseudo-name. This is Ockham's explicit diagnosis when he deals with such nouns as 'point', 'line' and 'surface' in his treatise *De quantitate*. These, he claims, are not genuine designative terms. Since they are "equivalent in signification to something complex composed of a noun and a verb or a conjunction or an adverb, or the pronoun 'which', or to something composed of a verb and some oblique case [e. g. a genitive or an accusative etc.]",

---

[9] Ockham says that 'man' has confused and distributive supposition in "Every man runs", and determinate supposition in "Some man runs". This has direct consequences, for example, on which valid inferences can be drawn from these sentences. Thus, we can infer from the former but not from the latter that: this man runs and that man runs and that other man runs and so on, successively pointing at all the men there are. See *Sum. log.*, I, c. 70, *OPh* I, 210–212.

[10] See *Sum. log.*, I, c. 8, *OPh* I, 29–30,12–13; transl. Loux, 65.

they "cannot properly speaking be conjoined to a verb as its grammatical subject", and they "do not have precisely the force of a name".[11]

Thus conventionally introduced in the language for reasons of brevity or elegance, these pseudo-names are nevertheless given the grammatical status of nouns and this can easily become misleading when it is forgotten what they are abbreviations for. This is especially true in philosophy and theology, where such abbreviations often take the form of abstract terms such as 'horseness', 'humanity' or 'divinity': "Although abstract names are seldom or never employed in everyday speech as equivalent in signification to several expressions", Ockham writes, "they are frequently used in this way in the writings of saints and philosophers".[12] It is crucial, then, when interpreting or discussing authoritative texts, not to forget the abbreviative function of this particular kind of abstract terms, lest we are led to wrongly imagine that there exist special abstract entities corresponding to them.

3 Contextual Analysis

Ockham's approach to this kind of abstract terms interestingly anticipates a view of philosophical analysis as providing rephrasals for certain expressions so as to make it explicit what is abbreviated in them and thus to reveal their real ontological significance.[13] This reflects an important concern of his general philosophical practice. Ockham's theory of nominal definitions for connotative terms, for example, primarily aims at showing that terms like 'white', 'father' or 'cause' do not commit their users to the ontological acceptance of anything but singular substances and singular qualities.[14] Connotative terms, however, are not mere abbreviations for their definitions in Ockham's semantics and some of them correspond to simple concepts in the mind.[15] What is special about the abstract terms we are now interested in, by contrast, is that they are but conventional abbreviations for certain sequences of words, and that in many cases what they abbreviate cannot be seen as a designative expression and cannot serve as subject or predicate in well-formed sentences.

Since these abstract terms nevertheless take the grammatical form of nouns, a rather odd situation is generated in the conventional languages where they occur: contrary to the expressions that they abbreviate, they can serve as subjects or predicates in grammatical sentences. Suppose that 'horseness' is an abbreviation for 'every horse', as the one-letter term '*a*' was supposed to be in one of Ockham's examples. A sentence such as "Horseness is an animality" will then mean the same as: "Every horse is an animal". Although 'every horse' is not the subject of the latter

[11] *De quant.*, q. 1, *OTh* X, 23–24,404–439; my transl.

[12] *Sum. log.*, I, c. 8, *OPh* I, 31, 51–54; transl. Loux, 66.

[13] See e. g. Beaney (2007).

[14] See Panaccio (2004, 85–102).

[15] See Panaccio (2000) and Panaccio (2004, 63–83).

sentence,[16] everything turns out to be all right in this particular case because 'horseness' and 'every horse' can occupy the same place in the corresponding spoken or written sentences (leaving aside for a moment the transformation of the abstract predicate 'animality' into the concrete term 'animal' in this rephrasal – I will come back shortly to this sort of complementary modification). But since 'horseness' is itself to be counted as a noun, complex expressions such as 'some horseness' or 'pure horseness' consequently turn out to be grammatically acceptable. The problem is that replacing the abbreviation by its explicit counterpart in such phrases yields incongruous sentences such as "Some every horse is an animal" or "Pure every horse is an animal". Because of this, Ockham says, "sentences in which such terms occur should not be taken literally, they are figurative ways of speaking".[17]

The relevant stylistic figure, according to the *De quantitate*, is what grammarians call 'hypallage', which they describe as "a transposition of [grammatical] cases and construction, or of the whole sentence sometimes".[18] The term 'hypallage', which Ockham probably inherited from a successful manual of grammar of the thirteenth century, the *Doctrinale* of Alexander of Villedieu, is still in use today, pretty much in the same sense, to designate a way of speaking in which a certain predicate is associated with a term that it does not fit with semantically, so as to indirectly suggest that this same predicate, or a related one, is to be associated with something else, normally well-identified in the context. The sentence "Mary had a sad day yesterday", for example, is not normally used to say that the day itself was sad. A day, after all, is not the sort of thing that emotions can literally be attributed to. To correctly understand the statement, we have to mentally reorganize it into something like 'Mary was sad yesterday', where the term 'sad' is predicated of Mary in connection with a particular period of time. Something like this is usually what happens, according to Ockham, in philosophical or theological sentences with abstract pseudo-names in them. He gives a number of examples in Part Three of *Summa logicae* when dealing with the 'second mode of amphiboly', and he is explicit there that they can all be reduced to cases of hypallage.[19] Suppose that 'horseness' is an abbreviation for 'each horse' (this is not one of Ockham's own examples, but it will be useful in the present context). A sentence such as "The capacity to neigh is exclusive to horseness" could not be made explicit, then, by simply replacing the abstract term 'horseness' in it with what it is an abbreviation for. That would yield: "The capacity to neigh is exclusive to each horse". But this is literally false: each horse is not the only animal that is capable of neighing since all other horses are as well. In order to get the right meaning, the sentence has to be reorganized so that the capacity to neigh is exclusively associated not with each horse, but with horses by contrast with non horses. The original sentence in

[16] In Ockham's analysis, the subject of "Every horse is an animal" is 'horse' rather than 'every horse' since sentences such as "Every horse is an animal", "Some horse is black" and "No horse is a goat" must be taken to have the same subject term in Aristotelian logic. The quantifier in such examples determines the subject-term, but it is not part of it.

[17] See *De quant.*, q. 1, *OTh* X, 23, 408–410; my transl.

[18] See *De quant.*, q. 1, *OTh* X, 23, 411–413; my transl.

[19] See *Sum. log.*, III–4, c. 6, *OPh* I, 776–777, 154–165.

this case turns out to be what medieval logicians called an 'exclusive proposition' such as "Only man is capable of laughing", which, according to Ockham, is to be analyzed into the conjunction of an affirmative and a negative sentence.[20] The correct analysis of "The capacity to neigh is exclusive to horseness" would thus require a reorganization of the sentence into the following conjunction: "Horses are capable of neighing and non-horses are not", where the phrase 'capable of neighing', that was absent from the original sentence but is closely related with 'the capacity to neigh' that did occur in it, is affirmatively predicated of horses and negatively predicated of non-horses.

There is no general recipe, in Ockham's view, for transforming a figurative sentence with an abstract pseudo-name in it into its literal counterpart. As he says:

> Not only can one word be sometimes equivalent in signification to several words, but when it is added to another expression, the resulting whole is equivalent to yet another complex expression, in which sometimes the part that is added will have a different case, mood, or tense, and sometimes it will simply be eliminated by being analyzed away and what it conveys being explicated.[21]

One of Ockham's own examples is the sentence "The generation of a form occurs in an instant", which, he says, should be understood as equivalent to "No part of a form comes to be before another but all come to be together".[22] Not only is the pseudo-name 'generation' replaced by the verbal expression 'comes to be' (or 'is generated') in this analysis, but the term 'instant' is eliminated as well, a reference to the parts of the form is squeezed in, and the original elementary sentence is transformed into a conjunction. How to reach the right result is obviously a case-by-case affair and depends on the particularities of the context.

Moreover, since the pseudo-name is put in place of a complex phrase that is not by itself a well-formed designative expression, it can have semantic interest only if it is taken in the context of an even larger expression, usually a sentence. In Ockham's semantics a genuine general name such as 'horse' or 'horseman' is meaningful even when taken alone: it signifies all horses or all horsemen, exactly as it does in the context of a sentence where it is used literally.[23] If 'horseness', by contrast, is an abbreviation for '[a] horse necessarily' (as it sometimes is according to Ockham), it can only occur in the context of a sentence in which it makes sense to combine the categorematic term 'horse' with a verb modally determined by the adverb 'necessarily' (or something closely related to it). The analysis of abstract pseudo-names, consequently, always requires a rephrasal of whole sentences, and this is to be done on a case-by-case basis under the guidance of a tentative hypothesis

[20] See *Sum. log.*, II, c. 17, *OPh* I, 296–307.

[21] *Sum. log.*, I, c. 8, *OPh* I, 33, 106–110; my transl. I rephrase Loux's translation of this sentence by following the text more closely than he does (Loux 1974, 68).

[22] *Sum. log.*, I, c. 8, *OPh* I, 33, 114–116; transl. Loux, 68.

[23] See e.g. *Sum. log.*, I, c. 33, *OPh* I, 95. I have argued elsewhere for an atomistic understanding of Ockham's semantics (Panaccio 1984). The signification of a univocal term in his approach stays the same whether the term is taken alone or in a sentence. What varies is the *suppositio*, i. e. the particular referential function of the term in the context of a sentence (Panaccio 1983, 1999).

about what the pseudo-name in question might be an abbreviation for in this particular context.

4 Avicenna on Horseness

Philosophical and theological authorities often have to be glossed along these lines according to Ockham. The main benefit of being able to correctly identify these abstract pseudo-names and what they are abbreviations for is indeed, in his view, that it allows for interpretations of authoritative texts that are more faithful and less extravagant than if these texts were taken literally. In *Summa logicae*, I, c. 8, Ockham develops an instructive example of this in relation to a famous passage by Avicenna. Closely looking at how he deals with the example and unfolding some of the implicit features of his treatment of it, will help to make it clear how in general this interpretative process is supposed to work. "Horseness is nothing but horseness", Avicenna's passage goes, "for by itself it is neither one nor many; nor does it exist in sensible things, nor in the soul".[24] What is at stake here is especially important to Ockham, since Avicenna can easily be interpreted as attributing a special ontological status to common natures such as horseness. Duns Scotus for one had used the passage to support his point that common natures can of themselves indifferently exist within the individuals or within the mind, but with a less than numerical kind of unity in both cases: "although it [horseness] is never really apart from some of these", Scotus writes after having quoted Avicenna's lines, "of itself, nevertheless it is not any of them, but rather is naturally prior to them all".[25] Many recent students of medieval philosophy have also given great importance to Avicenna's statement, often seen as the *locus classicus* of the theory of the 'indifference of essence'. Alain de Libera, for one, proposes a fascinating discussion of it in *L'art des généralités*.[26]

Now, Avicenna is a highly respected authority for Ockham and should be taken seriously. Yet the suggestion that common natures have some existence in themselves must be unambiguously rejected, he thinks. Ockham, then, puts forward a deflationist interpretation of Avicenna. The passage in question, according to him, does not endow common natures with any special ontological status. What Avicenna meant is simply that "'horse' is not *defined* as being either one thing or many, nor as existing in the soul or in the things outside", so that "none of these notions is contained in the definition of 'horse'".[27] "Thus", Ockham adds, "in that passage he [Avicenna] means to use the name 'horseness' in such a way that it is equivalent

[24] See Avicenna, *Metaphysics* V, as quoted by Ockham in *Sum. log.*, I, c. 8, *OPh* I, 31, 55–57; transl. Loux, 66, slightly amended.

[25] See John Duns Scotus, *Ordinatio* II, d. 3, p. 1, q. 1, ed. Vaticana VII, 403,6–8, n. 32; transl. Tweedale, 177.

[26] See de Libera (1999, 499–607).

[27] *Sum. log.*, I, c. 8, *OPh* I, 31, 57–60; transl. Loux, 66, with my italics.

in signification to several expressions".[28] Ockham does not explicitly specify which complex expression exactly would 'horseness' be an abbreviation of in Avicenna as he interprets him. Presumably it would be something like 'horse by definition'. The sentence "Horseness is neither one nor many", for example, would thus amount to "Horse by definition is neither one nor many", or even more explicitly to: "The definition of 'horse' does not include the notion of unity nor that of multiplicity". Ockham's strategy in this case is to transform what looks like a first-order statement about a particular sort of entity into a metalinguistic assertion about the definition of a certain term. Avicenna should not be taken to mean "that horseness is some entity which is neither one thing nor many and neither outside the soul nor in the soul"; for this, Ockham pertinently remarks, "is both impossible and absurd".[29]

A difficulty arises at this point, though, that it will be instructive to discuss in some detail although Ockham does not do so in *Summa logicae*, I, c. 8. The problem is the following. Is not it part of the very meaning of the term 'horse' that each of its referents is a unity by itself? And since 'horse', according to Ockham, signifies the horses themselves, isn't it part of its meaning too that it signifies extramental entities? What is the point, then, of saying that the definition of 'horse' neither includes the notion of unity nor that of extramentality? In order to reach a satisfactory understanding of what is going on in Ockham's reconstruction of Avicenna, we have to be clear about what sort of definition Ockham himself has in mind here, even though he is not explicit about it. In a later chapter of the *Summa*, Ockham distinguishes between nominal definitions (*definitio exprimens quid nominis*) and real or essential definitions (*definitio exprimens quid rei*).[30] In the case we are now interested in, we must realize that the definition he is talking about in *Summa logicae*, I, c. 8 is the *real definition*, although he does not say so in so many words. This is patent, because 'horse' for him is an absolute term, and absolute terms, Ockham insists, only have real definitions.[31] Absolute terms in Ockham are pretty much what we call today natural kind terms such as 'human being', 'cat', 'flower' and 'animal'. Ockham contrasts them with 'connotative' terms such as 'white', 'father' and 'horseman', and one of the main differences he mentions between them is that all connotative terms have a nominal definition, while absolute terms do not.

A nominal definition, for him, explicates the linguistic meaning of a term by making it clear in appropriate ways what it refers to and what it merely connotes. The nominal definition of 'white', for example, would be 'a substance that has a whiteness', thus making it clear that 'white' refers to certain substances while connoting their whitenesses; and the nominal definition of 'father' would be 'a male animal having engendered a child', making it clear that 'father' refers to certain male animals while connoting their children.[32] But such semantic analyses, obviously, are

[28] *Sum. log.*, I, c. 8, *OPh* I, 31, 60–62; transl. Loux, 66.

[29] *Sum. log.*, I, c. 8, *OPh* I, 31, 63–65; transl. Loux, 66.

[30] See *Sum. log.*, I, c. 26.

[31] See *Sum. log.*, I, c. 10, *OPh* I, 35–36.

[32] Panaccio (2004, 89–93).

only possible for connotative terms. Absolute terms do not connote anything. Whatever thing they signify is signified by them in only one way. There is no semantic distinction within them between various modes of signification. 'Horse', for example, signifies all horses, and nothing but horses, and it signifies each one of them in exactly the same way. Put in modern language, this comes down to saying that the meaning of 'horse' reduces to its extension. 'Horse', consequently, cannot have a nominal definition.

Absolute terms, on the other hand, always have a real definition, Ockham says. And he has very precise ideas about what real definitions should be in Aristotelian science. Real definitions, according to him, should reveal the internal structure of the individuals that are signified by the defined absolute terms.[33] When completely laid out, a real definition should first indicate the most general genus which the individual referents of the defined term belong to. In the case of 'horse', that would be the genus 'substance'. Next, the definition should successively identify the essential parts of the individuals in question. In Ockham's hylomorphic conception, the essential parts of a thing are its matter (if we are talking of material things) and a number of substantial forms.[34] For the term 'horse', the complete real definition would thus have to specify that the substances it refers to are:

- material;
- vegetative: since they are living beings, horses must have a vegetative substantial form as one of their essential components;
- sensible: since they have a capacity for perception, horses must have a specific sensitive substantial form as well.

The complete real definition of the absolute term 'horse', consequently, will be something like: a substance which is material, vegetative and equinely sensitive.

This definition does not specify that each such individual substance is one, or that it exists outside the mind. For Ockham, this is all true, of course, but it must not be mentioned in the complete real definition of 'horse', since neither unicity nor extramentality are distinct essential parts of horses. Avicenna being a good Aristotelian philosopher, this must be what he meant, according to Ockham, by saying that horseness by itself is "neither one nor many, nor does it exist in sensible things, nor in the soul".

Ockham's interpretative strategy in this case can be seen to have three main elements to it:

First, it implicitly rests on what we call today the Principle of Charity. This principle says that, when plausibly possible, we should favor an interpretation that

[33] See *Sum. log.*, III–3, c. 24, *OPh* I, 683–688.

[34] Contrary to Thomas Aquinas, for example, Ockham subscribed to the doctrine of the plurality of substantial forms: certain material creatures, he thought, have several such forms. Human beings, in particular, are endowed with a vegetative substantial form, a sensitive substantial form, and an intellective substantial form (Adams 1987, 647–667; Maurer 1999, 451–460).

maximizes the truth of what the speaker says.[35] Such a charitable approach is especially important for Ockham when dealing with authoritative texts. Their authors can be mistaken of course, there is no denying it, but we should not attribute bad philosophical errors to them if we can plausibly interpret them otherwise. In the case under consideration, the thesis that horseness is an entity that is neither one nor many, and neither outside nor inside the mind, is "both impossible and absurd", Ockham notes.[36] Implicitly using a Principle of Charity, he wants us to conclude that an alternate interpretation of Avicenna is therefore required.

Secondly, the abstract term 'horseness' in Avicenna's text is seen by Ockham as an abbreviation for a complex expression that is not by itself a designative expression, as previously theorized in *Summa logicae*, I, c. 8, and the whole sentence must be rephrased accordingly. 'Horseness' indeed cannot, in this particular context, be taken to belong to any of the other three categories of abstract terms. It is certainly not synonymous with the concrete term 'horse': Avicenna cannot sensibly mean that horses are neither one nor many, and neither outside nor inside the mind. It cannot charitably be taken to designate real tropes either, as 'whiteness' might be, since there is no such things in the world as a horseness trope. And it clearly cannot be a collective term. The only charitable possibility left for Ockham is that 'horseness' in Avicenna's passage is an instance of these pseudo-names that are frequently used by theologians and philosophers as abbreviations for complex expressions with syncategorematic terms in them.

Determining what exactly 'horseness' abbreviates in a particular context is of course the most delicate part of the reconstruction. The reader should try to make sense of the text in the best possible way, but there is no mechanical method to achieve this result. In the case of Avicenna's passage, Ockham transposes it into a metalinguistic assertion. This is the third main element of his interpretation. It is not, of course, the approach to favor in all cases of abstract pseudo-names. 'Generation', for example, is an abstract pseudo-name for Ockham, but what it is an abbreviation for has no metalinguistic component to it.[37] Nevertheless, it is quite typical of Ockham that he often finds a metalinguistic aspect in sentences that superficially look like first-order statements. 'Animal is a genus' is a case in point. The term 'genus', Ockham repeatedly insists, is a 'term of second intention': it designates generic terms such as 'animal' or 'flower'.[38] For the sentence 'Animal is a genus' to be true, consequently, the subject-term 'animal' must be understood in it as being in material supposition, and the sentence must be taken to mean that the *term* 'animal' is a genus, – a generic term, in other words. This is the kind of metalinguistic construal that Wilfrid Sellars saw as a "major breakthrough" on Ockham's part.[39] 'Genus', admittedly, is not a pseudo-name for Ockham, but the metalinguistic transposition is also a commendable strategy, he suggests, when dealing with certain pseudo-names.

---

[35] Davidson (1984, 136–137) and Feldman (1998).

[36] *Sum. log.*, I, c. 8, *OPh* I, 31, 63–65; transl. Loux, 66. See above, note 29.

[37] *Sum. log.*, I, c. 8, *OPh* I, 33.

[38] *Sum. log.*, I, c. 12, *OPh* I, 43–44; I, c. 20, *OPh* I, 67–69.

[39] Sellars (1970, 62).

Implicitly using the Principle of Charity again, Ockham presumes that Avicenna had what he, Ockham, takes to be a correct understanding of the structure of real definitions. Avicenna's famous statement, then, is taken to assert something not about some mysterious common nature out there, but about the real definition of the term 'horse'. Ockham thus anticipates Sellars's own approach to such abstract terms as 'triangularity', which is normally used, Sellars contends, to convey something about the *term* 'triangular'.[40]

As the example nicely illustrates, *Summa logicae*, I, c. 8 thus proposes, at least in its outlines, a way of keeping the ontology parsimonious when faced with the occurrence of abstract pseudo-names in sentences that we would like to endorse as true for some reason or other. The recommendation is the following: look for a complex expression that the abstract term in question might plausibly be taken to be an abbreviation for. As Ockham stresses, this might not be a well-formed designative expression. It will often be an artificially assembled sequence of words, with syncategorematic terms in it. And in many such cases, it will not be sufficient for a good analysis to merely replace the abstract term with the longer unabbreviated expression while leaving the rest of the sentence intact. Certain modifications elsewhere in the sentence might be needed, and it might even be necessary to restructure the whole of it. As a rule, Ockham suggests, the discourses that make use of abstract pseudo-names should be considered as non-literal, and the use of hypallage, in particular, should be suspected. This is good advice, it seems to me.

References

Primary Texts

John Duns Scotus. 1973. *Ordinatio*, II, dd. 1–3, ed. C. Balić et al., in Ioannis Duns Scoti *Opera omnia*, t. VII. Civitas Vaticana: Typis Polyglottis Vaticanis.
William of Ockham. 1974–1988. *Opera Philosophica* [= *OPh*], I–VII, ed. Ph. Boehner et al. St. Bonaventure, NY: The Franciscan Institute.
William of Ockham. 1967–1986. *Opera Theologica* [= *OTh*], I–X, ed. G. Gál et al. St. Bonaventure, NY: The Franciscan Institute.

Secondary Literature

Adams, M. M. (1987). *William Ockham*. University of Notre Dame Press.
Beaney, M. (2007). The analytic turn in early twentieth-century philosophy. In M. Beaney (Ed.), *The analytic turn: Analysis in early analytic philosophy and phenomenology* (pp. 1–30). Routledge.
Davidson, D. (1984). *Inquiries into truth and interpretation*. Clarendon Press.

[40] Sellars (1963, 650 ff).

Feldman, R. (1998). Charity, principle of. In P. D. Klein & R. Foley (Eds.), *Routledge Encyclopedia of Philosophy*, Routledge. https://www.rep.routledge.com/articles/thematic/charity-principle-of/v-1. Accessed 31 January 2021.

de Libera, A. (1999). *L'art des généralités*. Aubier.

Loux, M. (1974). *Ockham's theory of terms. Part I of the Summa logicae*. University of Notre Dame Press.

Maurer, A. (1999). *The philosophy of William of Ockham in the light of its principles*. Pontifical Institute of Mediaeval Studies.

Panaccio, C. (1983). Guillaume d'Occam: Signification et supposition. In L. Brind'Amour & E. Vance (Eds.), *Archéologie du signe* (pp. 265–286). Toronto: Pontifical Institute of Mediaeval Studies.

Panaccio, C. (1984). Propositionalism and atomism in Ockham's semantics. *Franciscan Studies, 44*, 61–70.

Panaccio, C. (1999). Semantics and mental language. In P. V. Spade (Ed.), *The Cambridge companion to Ockham* (pp. 53–75). Cambridge University Press.

Panaccio, C. (2000). Guillaume d'Ockham, les connotatifs et le langage mental. *Documenti e Studi Sulla Tradizione Filosofica Medievale, 11*, 297–316.

Panaccio, C. (2004). *Ockham on Concepts*. Ashgate.

Panaccio, C. (2008). L'ontologie d'Ockham et la théorie des tropes. In C. Erisman & A. Schniewind (Eds.), *Compléments de substance. Études sur les propriétés accidentelles offertes à Alain de Libera* (pp. 167–181). Vrin.

Panaccio, C. (2015). Ockham's ontology. In G. Guigon & G. Rodriguez (Eds.), *Nominalism about properties* (pp. 63–78). Routledge.

Sellars, W. (1963). Abstract entities. *Review of Metaphysics 16*, 627–671. Repr. in Sellars, W. 1977. *Philosophical perspectives: Metaphysics and ontology*, 49–89. Ridgeview Publ. Co.

Sellars, W. (1970). Toward a theory of the categories. In Lawrence Foster & J. W. Swanson (Eds.), *Experience and theory*, (pp. 55–78). Duckworth.

Tweedale, M. M. (1999). *Scotus vs. Ockham. A medieval dispute over universals*, 2 vols. The Edwin Mellen Press.

Ockham and Chatton on the Origin of Logical Concepts

Fabrizio Amerini

In Ockham's vocabulary, syncategoremata designate logical operators: connectives (*et*, *vel*, *sed*, etc.), quantifiers (*omnis*, *quidam*, *aliquis*, *nullus*, etc.), exceptive, exclusive, reduplicative particles (e.g. *preter*, *solum*, *in quantum*, etc.), the copula (*est*), adverbs and all the modes, grammatical as well as logical (e.g. *necessario*, *possibile*, *per se*, *formaliter*, etc.), which affect the semantics of terms and, in consequence, the truth-value of propositions.[1] They express concepts that are called 'logical' or 'syncategorematic'. Many scholars have argued that Ockham's change of position on the nature of concepts entailed a change of position on the nature of syncategorematic concepts as well.[2] If, in his first theory of concepts (the so-called *fictum*-theory), Ockham describes syncategorematic concepts as concepts that signify conventionally since they refer to the syncategoremata of spoken or written language from which they have been abstracted, in his second theory (the so-called *actus*-theory), he opts for describing them as natural signs of the mind that signify naturally.

It is not certain, however, that during his career Ockham modified his teaching on syncategorematic concepts. The impression is that in his later works Ockham did not abandon his early account of the origin and nature of such concepts.[3] Ockham deals with syncategorematic concepts for the first time in *Ordinatio*, I, d. 2, a. 8, a text where

[1] In this paper, I shall not consider the case of the copula, which is the most controversial because of the existential import it can have. Ockham gives arguments for the syncategorematic value of the copula in categorical propositions in *Quaestiones in librum secundum Sententiarum*, II, q. 1, *OTh* IV, 17,16 ff. On the logical value of the copula in Ockham, see Panaccio (2004, 151 ff.).

[2] On this, see Amerini (2017) and the bibliography quoted therein. On Ockham's theory of concepts and its evolution, see Panaccio (2004). On the chronology of Ockham's life and works, see Spade (1999) and Courtenay (1999).

[3] Amerini (2013, 2017).

F. Amerini (✉)
Department of Humanities, Social Sciences, and Cultural Industries, University of Parma, Parma, Italy
e-mail: fabrizio.amerini@unipr.it

F. Ademollo et al. (eds.), *Thinking and Calculating*, Logic, Epistemology, and the Unity of Science 54, https://doi.org/10.1007/978-3-030-97303-2_9

he still advocates the first theory of concepts. In a text dating to a later period of his career, i.e. *Quodlibet* IV, q. 35, when Ockham has definitively moved to the second theory of concepts, he again addresses the question of the origin of syncategorematic concepts. My interpretation is that in that quodlibetal question Ockham does not depart from the explanation of the *Ordinatio*. I shall reconsider Ockham's position in the first part of this paper, while offering an overview of syncategoremata in Ockham's texts.[4] I shall argue that his early account of the origin and nature of syncategorematic concepts may be reconciled with his second theory of concepts. On the other hand, someone who was not of this opinion was Walter Chatton, a confrère of Ockham at London and Oxford. Chatton vigorously reacted to Ockham's early explanation of syncategorematic concepts.[5] According to Chatton, such concepts do not signify conventionally, but naturally. Chatton considers the *actus*-theory of concepts to be incompatible with the *Ordinatio* explanation of the origin and nature of syncategorematic concepts. I shall reconstruct Chatton's criticism in the second part of this paper.

1 Origin and Nature of the Logical Concepts in Ockham

If scholars disagree on Ockham's evolution with respect to the origin of syncategorematic concepts, they instead agree that Ockham never changed his mind about the nature of syncategoremata. Ockham follows tradition and accounts for syncategoremata as a special category of terms of spoken or written language. They are special particles of conventional languages whose function is to modify the signification of the so-called categorematic terms, namely of those terms that have, as Ockham says, 'a definite and certain signification'.[6]

Ockham makes two fundamental claims about syncategoremata.

The first is that syncategorematic terms do not signify anything on their own; nonetheless, when they are added to the categorematic terms, they are said to be taken significatively (*significative*). They are like zero with respect to numbers: it does not signify anything on its own, by when added to a number (e.g. '1'), it is taken significatively in that it modifies the signification of that number (e.g. it transforms '1' into '10').[7] For example, the universal quantifier 'every' (*omnis*) works in this way.

---

[4] In the first part, I draw upon and occasionally clarify or expand what I said in Amerini (2017).

[5] For an introduction to Chatton's life and works, see Keele and Pelletier (2018).

[6] *Summa logicae*, I, c. 4, *OPh* I, 15,6–8.

[7] Ockham makes this first claim in several works. See e.g. *Ordinatio*, I, d. 2, q. 8, *OTh* II, 289,16 – 290,3; d. 25, q. un., *OTh* IV, 138,14–22; *Qu. Sent.*, q. 1, *OTh* V, 20,3 – 23,5, esp. 22,17–20; *Sum. log.*, I, c. 1, *OPh* I, 9,59–65; c. 4, *15–16*,9–31; II, c. 4, 259,29–36; III–4, c. 10, 798,193–204; *Quaestiones in libros Physicorum*, q. 56, *OPh* VI, 548,57–66; *Quodlibeta*, I, q. 19, *OTh* IX, 193,19–22. The source of this claim is Boethius's second commentary on the *De interpretatione*, on which see Ockham, *Expositio in librum Perihermenias*, I, c. 1, § 1, *OPh* II, 378–9,38–60; § 5, 381–2,10–20. On the connection between the syncategorematic terms and zero, see Mugnai (2004).

It does not signify anything on its own (anything such as a mysterious every-ness, *omneitas*, as some realists before Ockham called it),[8] but when it is added to 'man', it universally quantifies the term and makes it distribute over each singular man. Owing to the way it modifies the function of the signification of terms, 'every' may be said to co-signify. To be precise, syncategoremata do not modify the absolute signification of the terms to which they apply, but only what Ockham calls the supposition of terms. This is the semantic function that terms have when they occur in propositions, namely their reference. For example, the term 'man' does not change its signification when it is taken absolutely or with a universal quantifier; it continues to signify every man, existent in the present, past or future, and also possible men. But when the syncategorema 'every' is added to 'man' in a proposition, it modifies the supposition of 'man' by modifying the reference of the term and thereby changing the truth-value of the proposition. With respect to the actual world, for example, while "some man is white" is true, "every man is white" is false.[9]

Ockham also makes a second claim about syncategoremata: such terms, when they are taken significatively, cannot occur as extremes of a proposition nor, consequently, can they be endowed with supposition. To be precise, Ockham distinguishes three kinds of supposition. Terms in a proposition can be taken significatively or non-significatively. When they are taken in the first way, they refer to what they signify, and Ockham calls this 'personal supposition'. When terms are instead taken in the second way, they do not refer to what they signify, and in this case, they can refer either to a concept ('simple supposition'), or to a term of spoken or written language ('material supposition'). In the light of this distinction, Ockham says that syncategoremata cannot be endowed with the first kind of supposition, i.e. personal supposition. Since syncategoremata do not signify anything, it follows that, even if they are taken significatively, they cannot stand for what they (do not) signify. Nonetheless, they can occur in a proposition as extremes if they are taken in material supposition. Cases of syncategoremata having material supposition occur in instances such as in the proposition "the term 'every' is a noun" or "the term 'if' is a conjunction".[10]

---

[8] Ockham complains such linguistic uses; they express trivial reifications, which result from ignoring the different signification of categorematic and syncategorematic terms. See e.g. *Ord.*, I, d. 31, q. un., *OTh* IV, 405,9 – 406,7, esp. 406,2–7: "Et similiter signa syncategorematica importabunt alias res, sicut 'omnis' unam 'omnitatem' et 'aliquis' unam 'aliquitatem', quae omnia videntur absurda et procedunt ex ignorantia differentiae inter categorema et syncategorema et ex ignorantia differentiae inter nomina et verba et alias partes orationis indeclinabiles in significando." See also *Expositio in libros Physicorum*, III, c. 4, § 6, *OPh* IV, 473–4,78–97; § 10, 495,167–95, esp. 172–80.

[9] See e.g. *Ord.*, I, d. 25, q. un., *OTh* IV, 138,9 – 139,14, esp. 138,14–22: "sic dicendo 'aliquis homo', ponitur illud quod [ponitur] quando dicitur 'homo', sed sibi additur unum syncategorema quod non habet aliquod per se significatum, sed consignificat cum alio adiunctum. Et ita 'aliquis homo' non plus significat singulare vagum quam 'ornnis homo' significat singulare vagum. Nec distinguuntur quantum ad significata; sed propter diversa syncategoremata addita, quae sunt signum universale et signum particulare, potest aliquid verificari de uno et non de alio."

[10] For this second claim, see e.g. *Qu. Sent.*, q. 1, *OTh* V, 22,10–21; *Sum. log.*, I, c. 2, *OPh* I, *9–10*,15–25; c. 69, 208,3–14; also *Quod.*, IV, q. 35. On the different kinds of supposition, see Ockham, *Sum. log.*, I, 64, *OPh* I, 195–6. Ockham's claim in the *Summa logicae* that 'every' is a noun can at first

In his works, Ockham gives various examples to clarify the above claims. I shall consider some of them in the next section.

The Logical Function of Syncategoremata

On many occasions, Ockham says that syncategorematic terms can modify categoremata explicitly or implicitly. The term 'every' (*omnis*), for example, explicitly modifies the supposition of 'man' when is added to it in a proposition like "every man is an animal". But in other cases, this happens implicitly. For example, Ockham notes that the abstract term 'humanity' (*humanitas*) is semantically synonymous to the concrete term 'man' (*homo*), if they are considered with respect to the things they signify. Contrary to Aquinas, who held that the two terms signify two different things, namely, on the one hand, 'man' signifies a compound of matter and form, and, on the other, 'humanity' signifies the formal principles in virtue of which a man is a man, Ockham holds that the two terms satisfy all the conditions for being synonymous: they signify the same thing and in the same way. Nevertheless, they are often taken as non-synonymous, especially in theology, because according to the customary usage of theologians, 'humanity' is supposed to include some syncategorema that makes it semantically non-synonymous to 'man'. In fact, Ockham notes that theologians commonly take 'humanity' signify what a man *necessario* or *per se* is; for this reason, theologians assume that the proposition "man is humanity" is false. This may be accepted, Ockham says, but it must be clear that this is due to pragmatic and not to semantic reasons. The non-synonymy of abstract and concrete terms is matter of linguistic (especially, theological) agreement, but semantically speaking, such terms must be accounted for as synonymous.

Ockham not only considers categoremata that include syncategoremata, but also syncategoremata that include categoremata. The term 'everything' (*quidlibet*) is an example of this sort. It includes by equivalence (*per aequivalentiam*) the thing it distributes over: it is semantically synonymous to something like 'every being'. Accordingly, a proposition such as "everything is a being" is true because it is equivalent to the proposition "every being is a being", which is true.[11]

---

surprise. There can be two ways for making sense of it. The first is to take it mean that a quantifier grammatically is nothing but a part of a noun: 'every' is a noun to the extent to which it is part of the noun to which it applies. The second is to presuppose a bland distinction between categorematic and syncategorematic nouns: 'every' is a noun of the syncategorematic kind. Ockham does not clarify the point. Whatever interpretation one follows, however, it is clear that for Ockham 'every' may be endowed with material supposition. For further discussion of this point, see Crimi (2014).

[11] See *Sum. log.*, I, c. 7, *OPh* I, 23–25, 21–67; 27,115–128; c. 8, 29,8 ff.; c. 45, 142,72–8; III–1, c. 4, 375–6,309–31; c. 5, 381,120–30 and 391 ff. (for the discussion of invalid syllogisms due to syncategoremata-variations); c. 28, 432,33–42; III–3, c. 7, 613,84–5; *Brevis summa libri Physicorum*, III, c. 1, *OPh* VI, 42–3,81–112; *Ord.*, I, d. 2, q. 7, *OTh* II, 260,18 – 261,6; d. 5, q. 1, *OTh* III, 39,20 – 44,7; d. 7, q. 2, 145,20 – 146,7; *Quod.*, V, q. 9, *OTh* IX, 515,56–60. For more on the special case of the synonymy between concrete and abstract categoremata, see also Claude Panaccio's chapter in the present volume.

Being careful about syncategoremata is important, because the incorrect use of them can lead to wrong conclusions, especially in theology. A follower of Duns Scotus, for example, could defend the idiosyncratic doctrine of the formal distinction by arguing as follows: if a property P is truly (i.e., absolutely) predicated of a thing *a* but not of a thing *b*, then *a* and *b* can be said to be really distinct. In Trinitarian theology, one might say that the divine wisdom is truly the divine wisdom, but that the divine goodness is truly not the divine wisdom, and from that conclude that the divine wisdom and the divine goodness are really distinct. But if P is predicated of a thing *a* but not of a thing *b*under a syncategorematic condition as it may be an adverbial mode, one might no longer draw the false conclusion that they are really distinct. The argument Ockham considers is the following: from saying that the divine wisdom is *formally* the divine wisdom but the divine goodness is not*formally* the divine wisdom, one may reach the conclusion that the divine wisdom and the divine goodness are *formally* distinct. Ockham finds this argument formally valid but captious because such adverbial modes do not infer any real modification on the things signified by the subject-term and the predicate-term of the above propositions; being nothing but syncategoremata, such adverbial modes simply modify the supposition of the terms to which they apply and consequently the truth-value of the propositions in which they occur. As a result, neither the formal nor the real distinction of things can be inferred from the use of adverbs such as *really* and *formally* in propositions.[12]

In his works, Ockham mostly discusses two cases: categoremata that include syncategoremata and syncategoremata that include categoremata. But there is also a third case he mentions, namely that of terms that can be understood both categorematically and syncategorematically. This double understanding can concern syncategorematic terms as well as categorematic terms. In the case of syncategoremata, distributive and exclusive syncategoremata especially cause problems.

A first case examined by Ockham is that of the distributive syncategorema ‘whole’ (*totus*), as in the phrase ‘the whole thing’ (*tota res*). Ockham says that this case is frequently discussed in logic. If ‘whole’ is taken categorematically, it signifies all the parts of a thing (*quaelibet pars*), that is, the thing in its entirety, but if is taken syncategorematically, it simply distributes over the single things of a collection, namely over every thing. According to the meaning we attach to the syncategorema, a proposition including that phrase can be true or false. For example, the proposition “the whole thing is white” can be true if ‘whole’ is taken in the first sense, while it is false if it is taken in the second sense.[13]

More intricate is Ockham’s discussion of the case of the exclusive syncategorema ‘only’ (*solus*). In theology, when one for example says that “only the Father is God”, the term ‘only’ can be understood in either way, i.e. categorematically (as meaning ‘alone’) or syncategorematically (as meaning ‘exclusively’). When it is understood categorematically, the proposition is clearly false, because the Father is God not only

[12] See *Ord.*, I, d. 2, q. 1, *OTh* II, 16, 15 – 17, 7; also q. 6, 175, 11 – 176, 10; q. 11, 375, 3–13.

[13] See *Sum. log.*, I, c. 8, *OPh* I, 32–33,93–122; II, c. 6, 267–8,4–12 and 20 ff.; *Ord.*, I, d. 8, q. 3, *OTh* III, 215,1–14; q. 4, 224,17 – 225,2; d. 30, q. 1, *OTh* IV, 302,9–18.

when Father is taken alone. When the term is instead understood syncategorematically, we need to distinguish, because it can be understood exclusively or precisely. In the first sense, we mean that the predicate-God is truly predicated of the subject-Father and that it is truly negated of any other thing of which the subject-Father is truly negated. In this case, the proposition can be explained by two propositions, one affirmative and one negative: (i) the Father is God and (ii) any other thing that is not the Father is not God. In the second sense, we need to distinguish again, because the predicate-God can be predicated of the subject-Father understood either precisely, namely as a separate subject, or exclusively, namely as excluding any other thing. Applying these distinctions to the case at stake, Ockham concludes that "only the Father is God" is false if the term 'only' is taken exclusively, because the inference "only the Father is God, then every God is the Father" is false, since the proposition "every divine person is the Father" is false. But it may be true if it is taken precisely in the first sense, because the inference "the Father is God, therefore precisely the Father is God" is valid; it is false, however, if taken in the second sense, because it is false that exclusively the Father is God given that the Son and the Holy Spirit are also God.[14]

No systematic treatment of the logic of syncategoremata is present in Ockham's works, but his scattered discussions of the above cases are nonetheless important. They show that, for Ockham, syncategoremata as well as categoremata can be equivocal, and the different ways that categorematic and syncategorematic terms signify can lead to the fallacy of the figure of speech.[15] Thus, syllogisms and inferences involving syncategoremata need to be disambiguated. All that shows that, for Ockham, it is very important to specify the logical role of syncategoremata both to *interpret* correctly propositions, especially in theology, and to *decide* correctly about the truth-value of propositions and the validity of inferences.

The Origin and Nature of Logical Concepts

If there are syncategoremata in spoken and written language, and if every significant term of conventional language is subordinated to a concept, it follows that there must be syncategorematic concepts. But what it is the origin and nature of these concepts? As mentioned earlier, the interpreters of Ockham gave different answers. Nevertheless, it is certain that Ockham does not doubt that syncategorematic concepts exist. In the *Ordinatio*, he explicitly mentions concepts of this kind. He describes them as those concepts that correspond to spoken or written syncategoremata, and explains that such concepts express special operations of the mind. They have no counterpart in the extramental world; nonetheless, they are required to express features that do not depend on our mind. Every human being, for example, is able to laugh, and this

[14] See *Sum. log.*, II, c. 7, 296,6–16, and 297,17 ff.; *Ord.*, I, d. 21, q. un., *OTh* IV, 40,14 – 43,20. See also *Sum. log.*, I, c. 16, *OPh* I, 56,66 ff.

[15] See e.g. *Sum. log.*, III–4, c. 10, *OPh* I, 813,638–44, and 817,763 ff.

fact does not depend on our mental act of quantifying universally over human beings; but we cannot accomplish such a mental act of universal quantification except by way of the syncategorematic concept 'every' (*omnis*).[16]

In the *Ordinatio*, Ockham identifies the syncategorematic concept 'every' with the mind's act of quantifying universally over human beings. But Ockham's statement that "every human being is able to laugh even without this concept" could induce one to think that to some extent the syncategorematic concept 'every' is caused by the external things themselves, namely by the human beings' common property of being able to laugh. All human beings are able to laugh and we express this universal feature of them through the universally quantified proposition "every human being is able to laugh". Despite the appearances, I think one should resist this realist interpretation. In other texts, Ockham unequivocally holds that syncategorematic concepts are *sui generis* concepts precisely because they do not have any foundation in the extramental world and do not signify anything outside the mind. As noted, Ockham emphasizes the absurdities that would follow from thinking that, like categoremata, syncategoremata signify something in the extramental world.[17] Our universally quantified propositions certainly have a foundation in the extramental world, otherwise they would be false, but the act of universally quantifying does not capture any real feature of extramental things. All men are able to laugh, but there is nothing like the human beings' common property of being able to laugh, for Ockham. The act of universally quantifying is only an act of the mind.

Ockham also admits the existence of syncategorematic concepts in his later works. In the Prologue to the *Expositio in librum Perihermenias*, Ockham repeats that every significant term of a spoken or written proposition, whether categorematic or syncategorematic, is subordinated to a corresponding distinct concept in the mind. Ockham even adds that mental syncategoremata fulfil by nature the same function that the syncategoremata of spoken or written language fulfil by convention.[18] Moreover, in the *Summa logicae*, Ockham reminds the reader that the distinction between categorematic and syncategorematic terms holds both for spoken words and for concepts.[19]

All these texts reveal that Ockham does not change his conviction about the existence of concepts of the syncategorematic kind. But what is their origin and nature?

---

[16] See *Ord.*, I, d. 30, q. 1, *OTh* IV, 317,22–26: "Sicut iste conceptus vel intentio 'omnis' est tantum quoddam syncategorema in anima, et tamen sine isto conceptu omnis homo est risibilis. Quod tamen omnis homo, sine omni conceptu, sit risibilis non possumus exprimere nisi per conceptum syncategorematicum."

[17] See above, note n. 8.

[18] See *Exp. Per.*, prol., *OPh* II, 356–7,129–67, esp. 357,153–7: "Et tunc cuilibet voci significativae, sive sit categorema sive syncategorema, correspondet una intellectio vel potest correspondere, quae eundem modum significandi respectu eiusdem habeat naturaliter qualem habet dictio prolata ex institutione". As noted by Panaccio (2003, 157, note 177; 2004, 151, note 24), Ockham anticipated this view in *Ord.*, I, d. 2, q. 8, *OTh* II, 289,12 – 290,11.

[19] See *Sum. log.*, I, cc. 1 and 4, *OPh* I, 9,59–65 and 15–6,4–31; also II, c. 4, 259,29–35, and III–4, cc. 2 and 10, 753,57–61, and 798,193–204. See also *Quod.*, II, q. 19, *OTh* IX, 193,14–22.

Ordinatio I, d. 2, q. 8

The text where Ockham first and most clearly explains the origin of syncategorematic concepts is *Ordinatio* I, d. 2, q. 8.[20] This short text, dating from the period of his first theory of concepts, is well-known to the interpreters of Ockham.[21] Ockham introduces the issue as a doubt, viz. the fourth of a series of doubts concerning the *fictum* status of concepts. The doubt arises because on the one hand, we cannot derive all concepts from the extramental world. This is the case, precisely, of syncategorematic, connotative, and negative concepts. If these concepts also had an extramental foundation, we could not distinguish them from their opposites, categorematic, absolute and positive concepts, respectively. But on the other hand, their mental existence is required, because to every proposition of spoken language corresponds a distinct proposition in mental language. Thus, we are required to identify a different source for these concepts.[22] In his response, Ockham acknowledges that syncategorematic concepts cannot be abstracted from external reality, for none of them, by their very nature, can signify an extramental thing and stand for it in a proposition. Accordingly, the advocates of the *fictum*-theory should conclude that no concept is syncategorematic except by mere institution (*nisi tantum ex institutione*).[23]

What does the clause "except by mere institution" mean? As I suggested elsewhere,[24] we could understand this exception in the sense that we can form syncategorematic concepts by abstracting them from spoken (or written) syncategoremata, which alone exist by mere institution. But if one wanted to avoid anticipating at this point what Ockham will say only later in his response, we could take the clause "except by mere institution" not as a clause concerning concepts, but as an exception introducing the spoken vs. mental syncategoremata divide. The sense could be that syncategoremata only belong to spoken language and for this reason they are by mere institution; no natural concept is by mere institution; therefore, no natural concept is of the syncategorematic kind.

Regardless of which interpretation is the right one, each of them fits well with what Ockham says in the body of the response. Indeed, a few lines later Ockham observes that we can impose or abstract syncategorematic concepts from spoken syncategoremata and this happens "actually or always or commonly".[25] The procedure envisaged by Ockham is not difficult to follow. It has been reconstructed in detail by Claude Panaccio. Ockham explains that from spoken words that signify syncategorematically (for example, from the spoken word 'every'), the intellect abstracts

[20] See *Ord.*, I, d. 2, q. 8, *OTh* II, 282,13 – 286,22.

[21] For a close examination of this text, see Panaccio (2003, 2004, 146 ff.).

[22] See *Ord.*, I, d. 2, q. 8, *OTh* II, 282,13–21. I shall leave aside here the problem of the origin of connotative and negative concepts. On such concepts, see Panaccio (2004, 63 ff.).

[23] See *Ord.*, I, d. 2, q. 8, *OTh* II, 285,11–6, esp. 14–6: "Et ideo dicerent quod nullus conceptus syncategorematicus nec connotativus nec negativus, – nisi tantum ex institutione."

[24] Amerini (2017).

[25] See *Ord.*, I, d. 2, q. 8, *OTh* II, 285,20–1: "Possunt autem tales conceptus imponi vel conceptus abstrahi a vocibus, et ita fit de facto vel semper vel communiter."

common concepts that are predicable of such words and imposes these concepts to signify the same things as what these external spoken words signify.[26]

This text offers a reasonable explanation of the genesis of syncategorematic concepts. Ockham, however, does not clarify the details of the derivation process and this leaves room for conjecture. First and foremost, Ockham does not explain why the genesis of syncategorematic concepts happens "actually or always or commonly" in this way, nor what this means. Is Ockham here suggesting that syncategorematic concepts can be also abstracted from another source or in a different way, although they "actually or always or commonly" are abstracted from spoken syncategoremata? What other cases could Ockham have in mind?

Ockham's text is also unclear on what exactly the concepts that we abstract from the spoken syncategoremata amount to. On the one hand, the fact that Ockham says that our intellect "imposes these concepts to signify the same things as what these external spoken words signify" leads us to think that, for example, from the spoken syncategorema 'every', we derive the syncategorematic concept of *every*. This concept may be said to signify the same things signified by the spoken syncategorema 'every'. What does it mean? Since syncategoremata do not signify anything on their own, 'to signify the same things' must be understood in this case as 'to signify in the same way', i.e. syncategorematically. The concept of *every* works exactly in this way. It is truly syncategorematic precisely because it functions syncategorematically, i.e. in the same way as the spoken syncategorema 'every' from which it has been abstracted. On this interpretation, the mental language to which Ockham refers should be understood as a mere internalization of the spoken language.[27] But, on the other hand, the fact that Ockham says that our intellect "abstracts common concepts that are predicable of such words" leads one to think that, from the spoken syncategorema 'every', we instead abstract the syncategorematic concept of *noun*. In fact, only this concept can be properly predicated of 'every', for example, as seen above, when we take 'every' in material supposition.[28] We may say "'every' is a noun", meaning in the case of concepts that the concept of *every* is a noun-kind concept, while we might not say meaningfully "'every' is every", namely that the concept of *every* is an every-kind concept. In this case, the concept of *noun*, insofar as it is predicable of 'every', would not be syncategorematic, for it would not function syncategorematically. It describes what 'every' is, but it does not work in the same way as the spoken syncategorema 'every'. It looks like a categorematic concept rather than a syncategorematic one. We will return to this apparent tension at the end of this paper.

---

[26] See *Ord.*, I, d. 2, q. 8, *OTh* II, 286,5–8: "Tunc ab istis vocibus sic significantibus abstrahit intellectus conceptus communes praedicabiles de eis, et imponit istos conceptus ad significandum illa eadem quae significant ipsae voces extra."

[27] This is what Ockham elsewhere (e.g. *Ordinatio*, I, d. 27) calls improper mental language. This kind of mental language follows the spoken language, it is just the internalization of this latter, while the proper mental language precedes any spoken language. It is the language of thought of the first man. On such a distinction, see Robert (2009).

[28] The claim that 'every' is a noun recurs in *Sum. log.*, I, c. 2, *OPh* I, 9–10,15–25. On this, see above, note n. 10.

Summa logicae, I, c. 12

Ockham deals with syncategoremata again in chapters 11 and 12 of Part I of the *Summa logicae*. Ockham devotes chapter 11 to illustrating a distinction that only pertains to the terms of the spoken language, namely the distinction between nouns of first and second imposition.[29] Ockham classifies syncategoremata among nouns of first imposition broadly understood, namely among nouns that are not nouns of second imposition. While these latter are nouns that are imposed to signify signs instituted by convention and limited to when such signs are considered in this way (this is the case of the noun 'noun'), nouns of first imposition broadly understood are defined as those names that are not imposed to signify signs instituted by convention.[30]

Yet, Ockham puts syncategoremata outside the distinction between nouns of first and second intention. This distinction only concerns nouns of first imposition strictly understood, which are categoremata.[31] In particular, nouns of second intention are said to be nouns that are precisely (*praecise*) imposed to signify the intentions of the soul, or precisely the intentions of the soul that are natural signs and other signs that are instituted by convention (*et alia signa ad placitum instituta vel consequentia talia signa*).[32] For syncategoremata, this classification has a precise meaning. First: syncategoremata *are nouns of first imposition*, that is, nouns introduced in a language in the first instance to signify things that are not signs instituted by convention. Second: they nevertheless *are not nouns of first intention*, for they do not signify any extramental thing that is not a sign instituted by convention. Third: they *are not even nouns of second intention*, for they do not signify any intention of the soul or any sign instituted by convention.[33]

If in chapter 11 Ockham examines syncategoremata from the point of view of nouns belonging to spoken language, in chapter 12 he considers the background of the distinction between nouns of first and second intention, which subdivides the nouns of first imposition strictly understood. Here, Ockham no longer focuses on what a noun has been imposed to signify, but on what allows a noun to signify in the first place. Ockham assumes that a spoken word can signify only if it is subordinated to an intention of the soul, since intentions are what allow spoken words to be imposed and then to signify: the spoken words signify secondarily (and by institution) what the intentions of the soul signify primarily (and naturally).[34]

In this chapter of the *Summa*, Ockham describes an intention of the soul as something existing subjectively in the mind, as a sign that is able to signify naturally something else for which it can also stand when it occurs in a mental proposition.[35] Such a mental sign is twofold: a first intention is a mental sign that signifies

[29] See *Sum. log.*, I, c. 11, *OPh* I, 38,4–6.

[30] See *Sum. log.*, I, c. 11, *OPh* I, 39,9–11 and 39–40,36–45.

[31] See *Sum. log.*, I, c. 11, *OPh* I, 40,46–8.

[32] See *Sum. log.*, I, c. 11, *OPh* I, 40,48–63.

[33] See *Sum. log.*, I, c. 11, *OPh* I, 40,65–71.

[34] See *Sum. log.*, I, c. 1, *OPh* I, 7–8,26–42.

[35] See *Sum. log.*, I, c. 12, *OPh* I, 41,8–9 and 43,40–3.

something that is not a mental sign, while a second intention is a mental sign that signifies another mental sign.[36] *Man* is an example of a first intention, while *species* is an example of a second intention. Ockham further subdivides the first intentions: strictly speaking, only mental categoremata are first intentions because they alone signify something for which they can also stand in a mental proposition. But broadly speaking, insofar as first intentions are signs existing in the mind that do not signify precisely (*praecise*) another intention or sign, even mental syncategoremata can be called first intentions.[37] The *Summa logicae* is one of the few texts where Ockham explicitly speaks of 'mental syncategoremata'. Although in the *Summa* Ockham does not give any explicit answer to the question of the origin of syncategoremata, he seems to think that, *qua* intentions, the mental syncategoremata are natural signs of the mind.

Quodlibet IV, q. 35

Ockham also illustrates the distinction between first and second intentions in *Quodlibet* IV, q. 35, a disputation that probably took place during Advent in 1323 in London. In that question, Ockham definitively moves from the first to the second theory of concepts, accepting the reduction of concepts to acts of cognition.[38] This text is likely contemporary to the first part of the *Summa logicae*. In the first article of the question, Ockham proposes the same definition and articulation of first intentions as that given in the *Summa logicae*. Ockham says that, broadly speaking, a first intention is "an intentional sign existing in the soul that does not precisely (*praecise*) signify other intentions or concepts in the soul or other signs." This broader sense includes categoremata as well as syncategoremata. Strictly speaking, instead, a first intention is a mental noun precisely (*praecise*) disposed by nature to occur as an extreme of a mental proposition and to stand for a thing that is not a sign, and in this stricter sense only categoremata can be called first intentions.[39] Again, Ockham explains that syncategorematic concepts are first intentions not because they can stand for some extramental thing, as categoremata do, but because, when they are united to other terms, they modify the supposition of those terms. In themselves, syncategoremata do not signify anything, neither an extramental thing, nor an intention of the soul.[40]

---

[36] See *Sum. log.*, I, c. 12, *OPh* I, 43,44–6 and 59–60.

[37] See *Sum. log.*, I, c. 12, *OPh* I, 43,49–58.

[38] See *Quod.*, IV, q. 35, *OTh* IX, 474,115–20; for a tentative dating of *Quodlibeta* see *OTh* IX, pp. 36*–38*.

[39] Cf. *Quod.*, IV, q. 35, *OTh* IX, 469–70,15–40.

[40] Cf. *Quod.*, IV, q. 35, *OTh* IX, 469–70,15–37, esp. 470,30–7: "conceptus syncategorematici [...] licet non supponant per se accepti pro rebus, tamen coniuncti cum aliis faciunt eos supponere pro rebus diversimode. Sicut 'omnis' facit 'hominem' supponere et distribui pro omnibus hominibus in ista propositione 'omnis homo currit', et tamen hoc signum 'omnis' per se nihil significat, quia nec rem extra nec intentionem animae."

Ockham notes that second intentions can also be understood broadly or strictly. Understood in the stricter sense, a second intention is a concept that precisely (*praecise*) signifies the first intentions that signify naturally.[41] Understood in the broader sense, a second intention is a concept that signifies not only the first intentions that are natural signs of the extramental things (that is, first intentions strictly understood, i.e., the mental categoremata), but also "the mental signs that signify by convention" (*signa mentalia ad placitum significantia*), which are the mental syncategoremata. Ockham notes that in this case "maybe" (*forte*) only something vocal corresponds to a second intention.[42]

This latter text is difficult to interpret and I do not want to linger on it here; I have analysed it in detail elsewhere.[43] Here it suffices to underscore that Ockham explicitly speaks of the conventional signification of mental syncategoremata. Beatrice Beretta argued that the above formulation shows the survival of Ockham's first account of mental syncategoremata in his second theory of concepts.[44] Claude Panaccio rejected this interpretation. He suggested understanding Ockham's reference to the conventional signification of mental syncategoremata as an early account that Ockham failed to adjust in the final version of the *Quodlibeta*. The fact that this formulation does not recur in the otherwise parallel passages of the *Summa logicae* where Ockham defines the wide sense of 'second intention' shows that this is the right interpretation.[45] Elsewhere I proposed an interpretation of the passage that reconciles Panaccio's and Beretta's readings.[46] My point was that Ockham never abandons the explanation of the origin of syncategoremata he gives in the *Ordinatio*. Possible oscillations in his works rather seem due to the fact that different understandings of syncategorematic concepts are in play, and Ockham does not distinguish them clearly. Distinguishing them, however, is important: it permits explaining why Ockham likely could not consider the *Ordinatio* account in conflict with his second theory of concepts.

Ockham's claim that mental syncategoremata are "mental signs that signify by convention" becomes understandable if we hypothesize that Ockham had distinguished, albeit implicitly, three ways in which a syncategorema can be involved in a mental act. A linguistic syncategorema is the sign of a mind's operation. Well, that

---

[41] Cf. *Quod.*, IV, q. 35, *OTh* IX, 471,50–2; also 470,41–2.

[42] Cf. *Quod.*, IV, q. 35, *OTh* IX, 471,44–9: "Similiter large accipiendo, dicitur intentio secunda conceptus animae qui significat non solum intentiones animae quae sunt signa naturalia rerum, cuiusmodi sunt intentiones primae stricte acceptae, sed etiam potest signa mentalia ad placitum significantia significare, puta syncategoremata mentalia. Et isto modo forte non habemus nisi vocale correspondens intentioni secundae."

[43] See Amerini (2017).

[44] See Beretta (1999, 172 ff.).

[45] See Panaccio (2003, 157, note 17); also Panaccio (2004, 145–73, esp. 151 and 160, note 23). In fact, Ockham could say the same thing in the *Summa logicae*, I, c. 11, if we would read the sentence "other signs that are instituted by convention" (for the reference, see above, note 32) as "other <mental> signs that are instituted by convention" (*et alia signa <mentalia> ad placitum instituta vel consequentia talia signa*). On this, see Amerini (2017).

[46] Amerini (2017).

operation can be considered in three ways: (i) as in use, (ii) as cognized or (iii) as object of linguistic reference.

i. When we mentally combine two or more natural concepts through negation or quantification, thinking for example that this is not that, that a man is not white or that every man is able to laugh, we are naturally accomplishing a certain mental operation and therefore *using* some syncategoremata (the negation and the universal quantifier in our examples). Considering things from this angle, it becomes reasonable to say that mental syncategoremata are signs of the mind that are naturally significant, although they do not signify naturally anything outside the mind. Understood in this way, mental syncategoremata indicate mental operations in use, which we may assess first, as natural and innate to us, and second, as preceding any spoken syncategoremata.
ii. Since we cannot abstract the concepts of these operations directly from the extramental world, we have to suppose that "actually or always or commonly" we *form* the concepts of the mental syncategoremata as understood in sense (i) by abstracting them from the spoken language we instituted to express outwardly such natural mental operations. In our example, by reflecting upon the spoken signs 'not' or 'every', through which we expressed outwardly our act of negating or quantifying universally, we can form syncategorematic concepts. Ockham does not clarify what kind of concepts are those derived from the linguistic syncategoremata. They can be considered as linguistic concepts and probably they are the concepts of *not* and *every*, which we use when we think in a language. If so, they just result from the internalization of a given spoken language. They perform the same function in mental propositions as the spoken syncategoremata perform in spoken propositions. But we may even suppose that they are not different from the concepts of syncategoremata we naturally have, viz. those that allow the externalization of our thought in a given spoken language. The mental syncategoremata so understood are nothing but the linguistic formulation in the mind of the naturally logical concepts we ordinarily use in our thinking. Probably, Ockham thinks that the formation of the mental syncategoremata happens "actually or always or commonly" in this way because we usually begin to think through the mediation of a language, which we learn in the linguistic community in which we live. But it is not impossible, at least in principle, for someone to form the concepts of the logical operations she performs in thinking without the mediation of a language. Ockham, however, does not say anything about this. We can only conjecture its possibility and how it could come about.
iii. In any case, it is only after we have formed such concepts that we can *refer to* them through some second intention. For example, we can reflect on the syncategorematic concepts of *not* and *every*, and form the second-intention concepts of *negation* or *quantification* or even, through further abstraction, form the general second-intention concept of *syncategorema*. The second-intention concepts of syncategoremata (if any) would be meta-linguistic and descriptive properly speaking, while the first intention concepts of syncategoremata would

be linguistic in two senses. First, in function, because when mental syncategoremata are taken according to sense (i), they naturally perform the same logical functions in the mental language that the spoken syncategoremata, subordinating to them, conventionally perform in the spoken language. Second, in content, because when they are taken according to sense (ii), they are obtained by abstraction from the spoken syncategoremata and consequently provided with conceptual content. Only when the mental syncategoremata are taken in sense (iii) can we properly predicate them of linguistic syncategoremata.

Accomplishing a natural mental operation, forming the concept of that operation and referring to that concept are, thus, three distinct ways in which a syncategorema can be involved in a mental act. If this distinction really gives the background to Ockham's claim that mental syncategoremata are "mental signs that signify by convention", then *Quodlibet* IV, q. 35, tells us that Ockham has not abandoned the early account of the origin of mental syncategoremata. The quodlibetal question would suggest that we naturally accomplish syncategorematic operations (i) but that we can form syncategorematic concepts (ii) only in the way illustrated in the *Ordinatio*, i.e. by abstracting them from a spoken language. This process may be obvious in the case of grammatical syncategoremata (since they are syncategoremata that express grammatical features of spoken language such as 'singularly', 'verbally', and the like), but nothing prevents us from extending it to the case of logical syncategoremata too.

My proposal about ways to distinguish various understandings of syncategorematic concepts could have, however, one point of difficulty. Let us return to the apparent tension I underscored above. It is not sure that Ockham would accept the distinction between (ii) and (iii). This distinction presupposes that syncategorematic concepts work in a syncategorematic way when they are understood in sense (ii), but categorematically when they are understood in sense (iii). Only as taken according to sense (iii) can the mental syncategoremata be *predicated* of spoken syncategoremata. But only as taken according to sense (ii) can mental syncategoremata be *abstracted* from spoken syncategoremata and be imposed "to signify the same things as these external spoken words signify". Our distinction seems to presuppose the presence of two different mental acts, one of abstraction and one of predication, which concern two different concepts: in the case of 'every', the concept of *every* and the concept of *noun*, respectively. That Ockham has differentiated the act of abstraction from that of predication is doubtful, though. The text from *Ordinatio* I, d. 2, q. 8, seems to imply that the concepts that are abstracted from spoken syncategoremata are the same as those that are predicated of them. If so, we should revise our interpretation and give up on sense (ii). We could suppose that Ockham is there searching only for a *categorematic* concept of spoken syncategoremata, namely for a logical concept of a logical operation. If this were the case, we should maintain only sense (iii). But on the other hand, giving up on sense (ii) could conflict with Ockham's intention: in his response, in fact, he claims he means to explain the origin of *syncategorematic* concepts, that is, of concepts that function syncategorematically, and it is only when syncategorematic concepts are understood in sense (ii) that they can function in this way.

I am in doubt about which sense, (ii) or (iii), is the one Ockham is searching for in *Ordinatio* I, d. 2, q. 8. Probably both, although they merge with each other. What I may say is that neither sense could conflict with my interpretation that, in his later works, Ockham did not depart from his early account of the origin of syncategoremata. This account may be reconciled with Ockham's second theory of concepts. However, this is not how one of his main opponents, the Franciscan Walter Chatton, assessed Ockham's views of the origin and nature of logical concepts. For Chatton, Ockham's genuine position is that formulated in the *Ordinatio*, but this position conflicts with the reduction of concepts to acts of cognition. In his Commentary on the *Sentences*, Chatton precisely rejects the position that mental syncategoremata signify by convention.

2 Walter Chatton on the Origin of Logical Concepts

The reconstruction of Ockham's position I proposed above is conjectural in many respects. Apart from *Ordinatio* I, d. 8, q. 2, there is no other text where Ockham directly tackles the question of the origin of syncategorematic concepts. As we have seen, Ockham again mentions the conventional signification of syncategorematic concepts in *Quodlibet* IV, q. 35. If one takes the text of the quodlibetal question be correct, one may conclude that Ockham has not changed his initial position on the linguistic origin of syncategorematic concepts. It is precisely this account of syncategoremata that Walter Chatton criticizes.

In his *Lectura*, I, d. 3, q. 1, a. 2 and a. 3, Chatton accurately reconstructs Ockham's *Ordinatio* position. He notes that, according to the *actus*-theory of concepts, there is a sense according to which syncategorematic concepts can be described as natural mental signs that naturally co-signify. Chatton notes that Ockham did not subscribe to that position in the *Ordinatio*.[47] In fact, when, in the third article, Chatton defends the reduction of concepts to acts of cognition, he points out that an opponent could reject this reduction by noting that, if such a reduction were true, syncategorematic concepts would signify by nature; but this is false, because they signify by convention.[48] As he makes clear in his answer to this article, this is the position Ockham holds in the *Ordinatio*.[49]

Chatton however thinks that the two positions cannot be reconciled. He gives two very short arguments against the position that syncategorematic concepts signify by

[47] Cf. *Lect.*, I, d. 3, q. 1, a. 2, 31,14–23.

[48] Cf. *Lect.*, I, d. 3, q. 1, a. 3, 54,24–5: "Nono, tunc conceptus syncategorematici significarent naturaliter, quod est falsum, quia solum significant ex institutione."

[49] Cf. *Lect.*, I, d. 3, q. 1, a. 3, 68,27–32: "Ad nonum dubium, de conceptibus syncategorematicis, dicunt aliqui, distinctionis 2, quaestione 8, in solutionibus argumentorum, quod tales conceptus syncategorematici, cuiusmodi sunt 'si', 'per se', 'in quantum', et huiusmodi, solum significant ad placitum ex voluntaria institutione, sicut modi grammaticales, ut homo est singularis numeri."

convention. First, even if no spoken language were invented, a mental proposition as "a human being is per se a human being" would be true. Second, mental propositions do not signify by convention, so not even the logical mode 'per se' signifies by convention, since it is a property of propositions.[50]

Chatton's arguments appear scarcely compelling because they fail to distinguish the different understandings of syncategorematic concepts I illustrated above. Consider the first argument. It presupposes that our mind can elaborate propositional thoughts that have a foundation in the extramental world. Things are made in such a way that it is always true that a human being is per se a human being. The fact that a human being is per se a human being does not depend on our act of predicating 'per se' of the proposition "a human being is a human being". Speaking in this way, Chatton seems to believe that our thought is naturally linguistic, that we do not need an internalized spoken or written language for thinking linguistically. But if the threefold distinction I proposed above really gave the background to Ockham's claim that mental syncategoremata are "mental signs that signify by convention", Ockham could simply agree with Chatton. The syncategoremata exist in our mind before a spoken language has been invented to express them outwardly when they are understood in sense (i). This explanation of the nature of syncategorematic operations is fully compatible with the *fictum*-theory of concepts as well as with the *actus*-theory.

The second argument does not add anything new, but simply corroborates the first. It recalls that the syncategorema 'per se' is a property of propositions. Clearly, this argument could not hold for other syncategoremata, like connectives or quantifiers, which are not properties of propositions. Chatton does not dwell on this point; he limits himself to a generalization with the purpose of making his position clear. He argues that if one were not disposed to understand the syncategorematic concepts as natural signs, one should conclude that, as every syncategorematic concept signifies by convention, any other concept would do.[51] This may likely be concluded because, when we think, we always experience that we think in a language. Chatton points out that it is especially the missed distinction between grammatical and logical syncategoremata which led Ockham to the wrong position that the syncategorematic concepts signify by convention. As already noted, although this may be true in the case of the syncategoremata of the grammatical kind, it is not in that of the syncategoremata of the logical kind.

But for Chatton, this cannot be the case. The syncategorematic concepts of the grammatical type signify by convention, because obviously we cannot have them before a spoken language has been invented. But the syncategorematic concepts of the logical type (as are 'per se', 'primarily', and the like) signify by nature—as do

[50] Cf. *Lect.*, I, d. 3, q. 1, a. 3, 68,33 – 69,3: "Contra: circumscripta omni institutione voluntaria, natura rei est talis qua haec propositio in mente est vera 'homo est per se homo perseitate primi modi dicendi per se'. – Secundo: propositio mentis non significat ad placitum, igitur nec 'per se vera' ad placitum, quia perseitas est propria passio propositionis."

[51] Cf. *Lect.*, I, d. 3, q. 1, a. 3, 69,4–7: "Ideo videtur dicendum quod sicut significant vel consignificant, ita naturaliter significant tales conceptus mentis, quia eadem ratione poneretur et sustineretur de quolibet conceptu, quod significaret ad placitum."

the propositions to which they apply.[52] With respect to the three understandings of syncategoremata I proposed above, although Chatton is not as explicit as one could expect him to be, nonetheless by dismissing that syncategorematic concepts signify by convention, he could suggest that we can reach a concept of the syncategorematic kind by reflecting directly on the syncategorematic operations we naturally perform.

Chatton's second criticism too sounds ungenerous. In fact, the parallel between grammatical and logical modes does not seem to be Ockham's reason for attributing conventional signification to syncategorematic concepts. Ockham distinguished syncategoremata strictly understood from the syncategorematic modes—both grammatical and logical—that can affect a proposition. The reason why the advocates of the *fictum*-theory of concepts could hold that syncategorematic concepts signify by convention is that these concepts, having no distinct foundation in the outer world, cannot be abstracted from the extramental things. Given this motivation, the advocates of the *actus*-theory of concepts could agree on the conventional origin of the concepts of syncategoremata. Unfortunately, Chatton does not clarify the details of the process of syncategorematic concepts formation, neither here nor elsewhere. Not having distinguished the different understandings of syncategorematic concepts seems to have been the main reason of the incomprehension between Chatton and Ockham.

3 Conclusion

In his reply to Ockham, Chatton does not focus on the extramental foundation of the syncategorematic concepts. Nor does he explain why these concepts can be said to signify by nature. A thing can one nonetheless learn from their debate: the question of the origin and nature of the syncategorematic concepts could be distinguished from the question of the origin and nature of the concepts of syncategoremata. Syncategorematic concepts indicate some logical operations that our mind naturally accomplishes with natural concepts, and Chatton and Ockham seem to agree on this point. They instead seem to disagree on the origin and nature of the concepts of syncategoremata: for Ockham, they are concepts derived from the linguistic operators by which we expressed outwardly our mental logical operations, for Chatton, probably, they are directly derived from the mental logical operations themselves. Ockham and Chatton could agree on the understanding (i) of the syncategorematic concepts, but disagree on the understandings (ii) and/or (iii).

Unfortunately, neither Ockham nor Chatton clearly distinguishes between, on the one hand, the linguistic nature of thought before a spoken or written language has been instituted to signify the operations of thought and, on the other, the linguistic articulation of thought after a spoken or written language has been introduced, learned, and internalized. But that distinction, to which Ockham refers on other occasions, is significant. It is a presupposition that human beings perform logical operations

[52] Cf. *Lect.*, I, d. 8, q. 1, a. 3, ad dubium 2, 27,13–20.

by nature; if it were not so, they would be simply unable to symbolize such operations through special particles of spoken or written languages. At the same time, a spoken or written language is of utility for gaining full-fledged concepts of such logical operations. This is more evident in the case of grammatical concepts, but it may also obtain for logical concepts. In fact, when we are thinking of a quantifier such as 'every', we are not thinking of a thing, but of a linguistic item that performs a certain logical function (distributive, in this case). It is one thing to use a syncategorematic concept, i.e. to perform a syncategorematic operation, and it is another thing to mention it, i.e. to have a concept of that operation. As we have seen, Chatton objects that if the syncategorematic concepts were derived from the spoken language, nothing would prevent us from saying that the categorematic concepts are also obtained in this way. But from Ockham's perspective, this cannot be inferred: categorematic concepts can be abstracted from the extramental reality, but there is nothing in the outer world that permits us to abstract syncategorematic concepts. This is the reason why Ockham assumes that the syncategorematic concepts are abstracted from the spoken language.

The three understandings of syncategoremata I introduced may help to settle the different formulations one finds in Ockham texts as well as Chatton's criticism of Ockham. While Chatton especially stresses sense (i), Ockham in his first theory of concepts emphasizes sense (ii) and/or (iii), but in his second theory, after Chatton's criticism, sense (i) probably becomes central for him too.

References

Primary Texts

Walter Chatton. (2008). *Lectura super Sententias. Liber I, distinctiones 3–7* (J. C. Wey & G. E. Etzkorn, Eds.). PIMS.

William of Ockham. 1974–1984. *Opera Philosophica* [= *OPh*], I–VI (Ph. Boehner et al., Eds.). The Franciscan Institute.

William of Ockham. 1967–1980. *Opera Theologica* [= *OTh*], I–IX (G. Gál et al., Eds.). The Franciscan Institute.

Secondary Literature

Amerini, F. (2013). Thomas Aquinas on mental language. *Medioevo, 38*, 73–106.

Amerini, F. (2017). Ockham on mental syncategoremata. In J. Pelletier & M. Roques (Eds.), *The language of thought in late medieval philosophy* (pp. 149–168). Springer.

Beretta, B. (1999). *Ad aliquid. La relation chez Guillaume d'Occam*. Éditions Universitaires.

Courtenay, W. J. (1999). The academic and intellectual worlds of Ockham. In P. V. Spade (Ed.), *The Cambridge companion to Ockham* (pp. 17–30). Cambridge University Press.

Crimi, M. (2014). Significative supposition and Ockham's rule. *Vivarium, 52*, 72–101.

Keele, R., & Pelletier, J. (2018). Walter Chatton. In *The Stanford Encyclopedia of Philosophy (Fall 2018 Edition)* (E. N. Zalta, Ed.). https://plato.stanford.edu/archives/fall2018/entries/walter-chatton/. Accessed 25 March 2022.

Mugnai, M. (2004). Termini "sincategorematici" e "cifra" in un passo della "Summa logicae" di Ockham. *Rivista di storia della filosofia, 59*(2), 515–517.

Panaccio, C. (2003). Guillaume d'Ockham et les syncatégorèmes mentaux: la première théorie. *Historie Épistémologie Langage, 25*(2), 145–160.

Panaccio, C. (2004). *Ockham on concepts.* Ashgate.

Robert, A. (2009). Les deux langages de la pensée. A propos de quelques réflexions médiévales. In J. Biard (Ed.), *Le Langage mental du Moyen Âge à l'âge classique* (pp. 145–184). Vrin.

Spade, P. V. (1999). Introduction. In P. V. Spade (Ed.), *The Cambridge companion to Ockham* (pp. 1–16). Cambridge University Press.

William of Heytesbury and Peter of Mantua on Demonstrative Pronouns in Epistemic Contexts

Riccardo Strobino and Simo Knuuttila

In "On Knowing and Doubting", the second chapter of his *Rules for Solving Sophismata* (c. 1335), William Heytesbury argues that nothing that is in doubt for a person is known by that person and *vice versa.*[1] Heytesbury aims to refute seven counter-arguments to this thesis by using the rules of obligations logic and semantic points pertaining to epistemic terms, signification, and demonstrative pronouns. He states in a general introductory reply (13ra–va) that the basic tool for dealing with these arguments is the distinction between the compounded and divided senses of epistemic propositions, but he often leaves the details of the application of this distinction unexplained.[2] In the first part of this chapter, the principles of Heytesbury's epistemic logic and their applications are analyzed (Sects. 1 and 2) as well as his considerations about the signification of demonstrative pronouns in epistemic contexts (Sect. 3). In the second part, Peter of Mantua's discussion of Heytesbury's arguments in his *Logica*

[1] *Regulae solvendi sophismata* (Venice: Bonetus Locatellus, 1494); the second chapter *De scire et dubitare* (*SD*) is translated in Heytesbury (1988).

[2] For medieval epistemic logic, see Boh (1993); for Heytesbury, pp. 67–76, and Peter of Mantua, pp. 101–115. Heytesbury was well informed about the logical theories in obligations logic of his time as well as about modal logic and the English discussions of epistemic logic and the theories of epistemic attitudes. Obligations logic dealt with rules for formal disputations in which various statements were put forward by an opponent and evaluated by an answerer who was obliged to accept an initial statement and then to evaluate whether the opponent's new statements could be consistently accepted. For obligations logic, see Yrjönsuuri (2001); for modal logic, see Knuuttila (2008).

R. Strobino (✉)
Department of Classical Studies, Tufts University, Medford, MA, USA
e-mail: riccardo.strobino@tufts.edu

S. Knuuttila
University of Helsinki, Helsinki, Finland
e-mail: Simo.knuuttila@helsinki.fi

F. Ademollo et al. (eds.), *Thinking and Calculating*, Logic, Epistemology, and the Unity of Science 54, https://doi.org/10.1007/978-3-030-97303-2_10

is presented with a study of his epistemic principles and a detailed analysis of his arguments that sometimes differ from those in Heytesbury (Sects. 4–7).[3]

1 Epistemic Principles in Heytesbury's First Argument (13vaẓ–14vp, Trans. 446–455)

In the first critical argument, which receives the longest treatment in Heytesbury's treatise, it is assumed that you know that A is one or the other of these two propositions: "God exists" and "A man is a donkey" (*SD* 12va). You obviously know which of these is necessary and which is impossible while it is hidden from you which of them is A. The case can be illustrated as follows. The propositions are written on separate cards which you see in front of you; on the backside of one card the name of the proposition (the letter A) is written, but you do not see it. Everything that holds of both propositions holds of A, for example, that it is not contingent and that it is not in doubt to you. Your ignorance of which is A is irrelevant in this respect. Keeping the identificational doubt separate from what you know about A is the basis of refuting the arguments against the original thesis—lots of things can be known about an unknown object. None of the criticized arguments show that the same is known to be true or known to be false and to be in doubt, doubt pertaining only to the truth of the deictic proposition "This is A". One does not know this proposition to be true or false, but it does not follow that A one does not know to be true or to be false because this is true in the compounded sense and false in the divided sense. Heytesbury gives some examples of a similar distinction: "A I know to be true" (in the divided sense) does not imply "I know that A is true" (in the compounded sense). (See *SD* 13rb–13va, trans. 444, cf. 14ra, trans. 450.)

The division between modal propositions in the compounded and divided sense was systematically applied in early fourteenth-century modal logic: in the compounded sense the modal notion modalized the assertoric content of a modal proposition and in the divided sense it modalized the copula—how the subject was what it was said to be. Similarly, a compounded knowledge proposition meant that the truth of the proposition was known and the divided proposition that about the subject it was known what was predicated of it, the subject being outside the scope of knowledge. Compounded knowledge propositions were typically read internally, from the point of view of the knowing subject, and divided knowledge propositions were read externally.[4]

---

[3] Sections 1–3 are written by Knuuttila and Sects. 4–7 by Strobino.

[4] Knuuttila (2008, 533–536, 551–559). According to Heytesbury, epistemic modals in the compounded sense did not imply those in the divided sense or *vice versa*. The only exception to this rule is provided by singular epistemic predications with respect to the demonstrative pronoun "this" used in a definite way (*SD*, 13rb–va, 14va, trans. 444–446, 454). See also his *The Compounded and Divided Sense*, translated in *The Cambridge Translations of Medieval Philosophical Texts*, 426–432. As in fourteenth-century modal logic in general, the compounded and divided senses were taken

In addressing the first counter-argument, Heytesbury puts forward various epistemic principles which were of systematic significance in his approach. The logical properties of epistemic notions are largely analogous to those of alethic modal notions; in fact, Heytesbury assumed like William Ockham and some others that the logic of the notions of knowledge and belief was basically the same as that of alethic modalities.[5] If Kp stands for "a knows that p", Bp for "a believes that p" and D for "a is in doubt whether p", it holds that

$$Kp \rightarrow p, \tag{1}$$

as with the notion of necessity,

$$Kp \rightarrow Bp, \tag{2}$$

as with the notions of necessity and the broad notion of possibility (not necessarily not), and

$$-Kp \ \& \ -K-p \leftrightarrow Dp \tag{3}$$

for propositions not subjectively certain, as with the notion of contingency: not necessarily p and not necessarily −p.[6]

Treating knowledge and belief as compatible attitudes was preferred by fourteenth-century authors who discussed epistemic notions as analogous to modal notions. They regarded logical possibility as the basic modal notion and similarly belief in the sense of assent as a basic epistemic notion, separating it from weak belief and doubt which included uncertainty that made them incompatible with knowing.[7]

In criticizing the first argument, Heytesbury first explains that belief may be hesitant or unhesitant, with subjective certainty or uncertainty. These are introspective qualities of an epistemic attitude recognized when one considers one's attitudes. Belief with subjective doubt is not a central notion for Heytesbury because obligations logic treats it simply as a lack of knowledge. However, unhesitant and certain beliefs (B*p) are of some interest because they accompany knowledge in Heytesbury. The introspective awareness of −Kp & −K−p is *prima facie* sufficient for answering with doubt to a number of propositions in obligations disputations, but doubt is not compatible with subjective certainty:

to be equivalent in this case. Ockham (1974), II.10 (276–279); III-1, 32 (448); III-3, 10 (632–634) and Buridan (1976) II.7, 16 (75–76).

[5] See Knuuttila (2015).

[6] *SD* 13vb, trans. 446–447.

[7] William Ockham treated the notion of belief in a broad sense as the basic judicative act directed to the content of an apprehensive act; see OPh 4 (1985), 5.29–6.50; see also Holcot (1518), I.1.6. The same view was formerly put forward by Robert Grosseteste who called this the general notion of opinion, as distinct from belief as an assent with fear that the opposite might be true. See Boh (1993, 26–28).

$$Dp \rightarrow -B^*p. \tag{4}$$

The notion of subjective certainty is included in Heytesbury's broad notion of knowledge. When it is directed towards how things are, the attitude can be called knowledge. This is said to pertain to various contingent states of affairs and accidental sensible characteristics[8]:

$$B^*p \;\&\; p \rightarrow Kp. \tag{5}$$

After having argued that certain belief about something excludes doubting it, Heytesbury goes on to refute the view that one may be in doubt about whether one knows. In dealing with iterated attitudes of this sort, he operates with two notions: perception (*perceptio*) and consideration (*considerati*o). Perception or apprehension (*apprehensio*) is a direct awareness about sensory things or about one's intentional attitudes, particularly epistemic ones, when they are attended to by considering them. Consideration is an introspective scanning activity and introspective perception is an actual or virtual awareness that accompanies all epistemic acts and attitudes. The intellectual faculty is wholly transparent in this sense.[9]

In this part, Heytesbury discusses an example formulated in obligations logic terminology with the conclusion that one should doubt that one knows that the king is in London. After a detailed description of the argument, Heytesbury states that it leads to impossibility. He first elaborates the assumption of knowing as follows: You know that p; therefore you perceive that p. And it also follows: you perceive that p and you are considering whether you perceive that you perceive p; therefore you perceive that you perceive. You do not perceive that p if you, on the same basis of evidence, do not perceive that you perceive that p. For that reason, one should accept the following: You perceive that you perceive that p; therefore you perceive that you know that p. Perceiving is here treated as a form of knowledge.[10] As for the doubt in the argument, Heytesbury argues that when you consider whether you perceive that p and you do not perceive that you perceive that p, you do not perceive that p. And when you consider whether you know that p and you do not perceive that you know that p, you do not know that p. Therefore, if you doubt whether you know that p and consider this, you do not perceive that you know that p and hence know that you do not know that p. But as far as you know that p, you do not know that you do not know that p. It follows that doubting that one knows is impossible: it contains the contradictory opposition between K−Kp and −K−Kp. The former derives from one's alleged awareness of not knowing and the latter from knowing that p.

[8] *SD* 13vb, trans. 447. For Heytesbury's notion of knowledge and its similarity to that of E. Gettier, see Hilpinen (2017, 135–151).

[9] *SD* 13va–13vb, trans. 446–448. Heytesbury's notion of perception shows some similarities to what was called intuitive cognition by Scotus and Ockham; see Cross (2014) and Panaccio (2004).

[10] This is Heytesbury's version of the epistemic KK-thesis; for the background, see Martin (2007, 93–108).

The following forms about introspective perception (P) are used here[11]:

$$Kp \rightarrow Pp \tag{6}$$

$$Pp \rightarrow PPp \tag{7}$$

$$-PPp \rightarrow Pp \tag{8}$$

$$PPp \rightarrow PKp \tag{9}$$

$$-PKp \rightarrow -Kp \tag{10}$$

$$-Kp \rightarrow K-Kp. \tag{11}$$

In (10) and (11), −PKp implies K−Kp, which is based on Heytesbury's view that the introspective consideration is complete: if one does not know about one's knowing or perceiving that p when considering one's attitudes, one does not know or perceive that p and knows that he does not. The logical behavior of perception is here the same as that of knowledge and both notions are closed under known consequence[12]:

$$K(p \rightarrow q) \,\&\, Kp \rightarrow Kq. \tag{12}$$

2 Heytesbury's Solutions of Further Counter-Arguments

The second argument (*SD* 14vb−15ra; trans. 455–458) deals with an example, a little different from the first, that the answerer knows that A is the true one of two contingent contradictory propositions (p and −p); these are in doubt to him and he does not know which one is A. The opponent tries to argue that the same (A) is known to be true and it is also in doubt. Heytesbury explains that a person's knowing

[11] SD13vb−14ra, trans. 447–448. The questions of iterated epistemic notions were dealt with by previous English authors before Heytesbury in the controversy about whether knowing that one knows that p is an act separate from knowing that p, as William Ockham thought, or whether it is included as a non-intentional awareness in first-order acts, as was argued by Walter Chatton. See Yrjönsuuri (2007), Brower-Toland (2012), Schierbaum (2016), and Schierbaum (2016). Heytesbury's view seems to be roughly the same as that of Ockham.

[12] *SD* 13vb, trans. 448. The closure principle was used in propositional modal logic for all modalities in medieval times. For its application to knowledge in known consequences in the fourteenth century, see, for example, Pseudo-Scotus, *In librum primum Priorum Analyticorum Aristotelis quaestiones*, q. 36, in Scotus (1891).

that the proposition "A is true" is true does not imply that A is known to be true by him because neither p nor -p is known to be true by him. In discussing the argument: I know that the proposition "This is A" is true; therefore I know that this is A, Heytesbury explains that this follows only if one knows that "This is A" signifies precisely that this is A when A is indicated, but since the answerer does not know how it signifies, the conclusion should be doubted. The demonstrative proposition is said to signify non-precisely to the answerer when he does not know which of the alternatives it picks. Heytesbury thinks that a proposition with a demonstrative pronoun has a different signification depending on the background information of the answerer; for example, "This is A" can be treated as a referentially indefinite utterance type or a definite utterance token. The demonstrative pronoun is taken to indicate a particular object that is unknown in the former case and known in the latter.[13]

In the third argument (*SD* 15ra–b, trans. 459–460), it is posited that you know that this is this (indicating Socrates) and that the proposition "This is this" precisely signifies that this is this and is true about the indicated thing. The proposition "This is Socrates" is said to be in doubt to you because you do not know Socrates when he is indicated and the proposition is not included in the precise signification of "This is this". However, it is allegedly known to you that this is Socrates because "this is this" also signifies that this is Socrates. According to Heytesbury, no proposition signifies precisely that this is the case and not in another way; therefore, "this is this" principally signifies that this is this and by implication the things implied by the primary signification. Therefore the first part of the argument, maintaining that one does not know that this is Socrates because of what is signified, is based on a mistaken view of signification, although it might be true. The second part claiming that one knows that this is Socrates mistakenly assumes that one knows what is signified by implication when "This is this" is proposed and Socrates is indicated. The main point in this reply is that one's knowledge of the signification of a demonstrative pronoun depends on one's knowledge of the thing indicated. The signification itself virtually includes whatever follows from the concept of the thing—the knowledge of these things depends on the background information of the answerer. This is an assumption of semantic closure with respect to signification.

In the fourth argument (*SD* 15rb–va, trans. 460–462), it is assumed that you know what is indicated by "this" in "This is either Socrates or Plato", but you are in doubt about who is who. Based on this disjunction, you allegedly know that this is Socrates when he is indicated but you are also in doubt about this. As for knowing the object, Heytesbury says that the interchangeability of propositions is said to be formal when the propositions follow each other through a necessary middle like in "This is" and "This is Socrates" through the middle "This is this". This notwithstanding, the propositions do not primarily and principally signify the same and, therefore, are not interchangeable in an epistemic scope. This is a further example of the limitations of the knowledge by deictic indication.[14]

[13] For utterance-types and tokens, see Nuchelmans (1973).

[14] According Ockham, formal consequences hold in virtue of an intrinsic middle; see Martin (2012).

The fifth argument (*SD* 15va–16ra, trans. 462–466) assumes that you know what is indicated by the subject of the proposition "This is a man", that the proposition signifies precisely as its terms suggest, and something you know to be a man and nothing do you doubt to be a man. On this basis it is argued that you know the proposition "This is a man" or you know that it is false. Heytesbury answers that if the answerer considers these alternatives, which is not said in the hypothesis, the conclusion follows on the condition that he does not unhesitantly believe the false proposition "This is a man", which is then neither known to be true nor known to be false and not doubted. If he unhesitantly believes a true proposition, the conclusion follows.

Heytesbury also denies the argument for your knowing the proposition "This is a man" to be true or false on the premises that this you know to be a man or this you know not to be a man and that you know what is indicated by "this" and that the proposition "This is a man" primarily and principally signifies in keeping with the combination of it terms. He explains this criticism by stating that knowing what "this" indicates in "This is a man" does not make one know that the proposition signifies that this is a man nor to know that the proposition is true or false. On the basis of what he later says about the definiteness of signification, he seems to mean that if knowing what is indicated is merely an awareness of an ostensive object that is indefinite in this sense, it does not make one to know the signification.

Heytesbury then distinguishes between doubting a proposition and being in doubt about how it signifies. He states that when one is in doubt regarding the proposition "I know that this is a man", he is not necessarily in doubt whether he knows that this is a man, but he is in doubt about the way the proposition signifies. One may be in doubt regarding a proposition with a demonstrative noun when it does not signify precisely a predication about a definite thing even when one knows that what it says is true about that thing (15vb, trans. 464).

After these remarks, Heytesbury repeats what he said in discussing the first counter-argument: it is impossible to be in doubt whether one knows something. He states more generally that one should never say "I doubt it" regarding first-person compounded propositions expressing one's intentional attitudes such as to know, to doubt, to believe, to perceive, to want, to want not, or to desire, except that one may sometimes be in doubt about epistemic propositions in a compounded sense when these include a demonstrative noun and one does not know how that term signifies, for example, I may be in doubt about the proposition "I know that this is a man". This does not apply to propositions with demonstrative pronouns that supposit for an altogether ununderstood term; these predications are known to be false (ibid.).

The sixth argument is a variant of the first (*SD* 16ra, trans. 466–468). It is posited that A, B, and C are three propositions of which A and B are known to you and C is in doubt for you and it is hidden from you which is A, B or C. It is argued that all these propositions are doubted when they are proposed, but A and B are known to you; therefore the same thing is known by you and in doubt for you. Heytesbury argues that this is wrong because you should answer by doubting any of these propositions when it is proposed, but you are not in doubt about which of them is in doubt for you although you are in doubt which of them is C which is said to be in doubt to you.

In the seventh criticized argument (*SD* 16ra–b, trans. 468–469), it is argued that you know that this is Socrates and you are in doubt whether this is Socrates (the same thing having been indicated). This is derived from the hypothesis that you saw Socrates yesterday and you still know that the man you saw yesterday is Socrates; now you see Socrates and it is hidden from you that he is Socrates. Therefore, you know that this is Socrates (indicating the one you saw yesterday). And (indicating the same one) you are in doubt whether this is Socrates. According to Heytesbury, it does not follow from the hypothesis that I know that this is Socrates. Rather, I am in doubt whether this is Socrates when he is pointed. However, I grant that this is Socrates, not because I know that it is Socrates, but because it follows from the hypothesis. In this way, one may know that this Socrates and be in doubt whether he is Socrates.

3 The Signification of Demonstrative Pronouns in Epistemic Contexts in Heytesbury

Most of the questions Heytesbury addresses pertain to knowledge about singular things and propositions about them in concrete contexts. Because of this perspective, demonstrative pronouns play a central role in his treatise. Even though Heytesbury does not deal with them in a separate chapter, as he does with the compounded and divided senses, he considers the signification of the pronoun "this" in most of his critical remarks on interpreting epistemic propositions and also sketches some principles about this matter. On the basis of the above summaries of the arguments in Heytesbury's treatise, one can distinguish between three uses and analyses of "this" in epistemic theory: generic, particular, and identificatory. These ideas are also found in Paul of Venice's treatise *On Knowing and Doubting*, chapter 22 of the first part of his *Logica magna*. He follows the main lines of Heytesbury's treatise, sometimes trying to make it more understandable to students, sometimes leaving out what he regards as too demanding for them or without an explanation, and sometimes introducing new examples of his own.[15]

According to the comment on argument 2, it is assumed that A is the name of the true one of a pair of two contradictory contingent propositions that are in doubt to you. In a disputation on these guidelines, you know that the proposition "This is A" is true, even though you do not know which of the propositions is A. It does not follow that you know that "This is A" is true when A is indicated in a particular case; you know the proposition "This is A" as a referentially indefinite proposition type that signifies that this or this is A. "This" is here used *in a generic sense* and not in an actual deictic sense in a token sentence with respect to what is indicated (arguments

[15] Paul of Venice (1981). Paul of Venice interpreted Heytesbury's theory of virtual reflexive knowledge in terms of less elaborated psychological acts. See op. cit., 131; cf. *Logica magna* (Venice, 1499), 80v–82r.

2, 5). Paul of Venice calls this a distinction between using "this" in a universal and confused manner and a particular and distinct manner (52–54, 116).

When "this" is used as a subject or predicate in a non-generic occurrent sentence, it has the function of *indicating a particular ostensive object* and it signifies that specific object in a way determined by the context. When it is posited that you know that this is this (Socrates being indicated) and that the proposition "This is this" primarily signifies that this is this, it does not follow that you know that this is Socrates, because you may not know or recognize him and you may not be informed about other non-deictic identity predications with respect to the subject, although they are virtually included in the secondary signification by implication (*SD* 15rb, trans. 459). There was a well-known sophism "You know this to be everything that this is", also alluded to by Heytesbury, which included an expansion of a demonstrative pronoun used alone.[16] "This is Socrates" is not interchangeable with "This is this" in an epistemic scope because these propositions do not primarily and principally signify the same (arguments 3 and 4). There is a longer discussion of this case in Paul of Venice (144–148).

In the fifth argument, it is assumed that you know what is indicated by the subject of the proposition "This is a man" and that the proposition signifies as its terms suggest. This is not regarded as a sufficient basis for knowing that the proposition is true or false, because one may know what is indicated and be in doubt about what this is. When a demonstrative pronoun is used alone and not attached to a classificatory noun, it may signify *definitely or indefinitely*, depending on what is known about what is indicated. Heytesbury assumes that knowing an indefinite deictic object is an awareness of it without further specification and knowing it definitely includes an identificatory interpretation of the object. The known signification of a demonstrative pronoun is dependent on the intention of the speaker or listener; therefore it is always possible to doubt propositions with a demonstrative pronoun as a subject in obligational disputations (cf. *SD* 15vb, trans. 464; Paul of Venice (114–116)).

When Heytesbury discusses the obligations logic case about Socrates whom you saw yesterday but do not recognize now, he says that on the basis of the hypothesis one should grant: "This is Socrates" when he is indicated. One does not grant it because one directly knows that this is Socrates but because it follows from the background information that it is Socrates, not that one recognizes or identifies this as Socrates. When Heytesbury says that the answerer does not know that this is Socrates, he means *recognition by acquaintance* that could be called perspectival identification. To grant the proposition "This is Socrates" in the example case is required by the non-perspectival facts from which it follows. This acceptance is no less veridical than direct knowledge, but it is based on non-perceptual identification (DS 16ra–b, trans. 468; cf. Paul of Venice, 142–148). The example shows some similarity to contemporary discussion of perspectival and public identification.[17]

---

[16] See Kilvington (1990); cf. Heytesbury, *SD* 15rb, trans. 459.

[17] See Hintikka and Symons (2003).

4 Peter of Mantua on Knowing and Doubting

Peter of Mantua's (d. 1399) views on epistemic logic are primarily discussed in his treatise on knowing and doubting (*De scire et dubitare*), a chapter of his *Logica*, which dates from the early 1390s. The treatise is divided into the following sections[18]: (1) preliminary assumptions (K1ra–K2rb); (2) rules (K2rb–K3rb); (3) objections to the assumptions and to the rules (K3rb–K3vb); (4) replies to the objections (K3vb–K4rb); (5) first doubt (*dubium*), articulated through thirteen arguments (K4rb–L4ra), followed by (6) a set of general qualifications, remarks, and distinctions (L4ra–L4+1ra) which are instrumental for (7) the solution to the thirteen arguments (L4+1ra–L4+3ra); (8) second doubt, articulated through four arguments (L4+3rb–L4+3va), followed again by (9) a set of general remarks which serve as further preliminary qualifications (L4+3va) for (10) the solution to the four arguments (L4+3va–L4+4ra).

The first *dubium* specifically deals with the question of whether something may be known and be in doubt for one and the same person in the same respects. The thirteen arguments purport to show, in different ways, that this may be the case, which Peter categorically sets out to deny. Even on a superficial reading, it becomes immediately clear that Peter is responding in the first *dubium* to the same set of problems discussed by William Heytesbury in his *Regulae*. First, clear correspondences may be established between the hypotheses of the seven arguments in Heytesbury's treatise and the cases discussed by Peter, who incorporates them in various forms into his list, as well as adding a number of additional arguments not explicitly to be found in Heytesbury. Secondly, their method is also, for all intents and purposes, practically identical, as the arguments are invariably developed and understood in the context of obligational settings and governed by principles of obligations logic. Yet, while many of the basic epistemic principles discussed by Heytesbury are also adopted by Peter (notwithstanding some differences), the relation between their solutions is more complex. On the one hand, the overarching goal is the same, namely to deny in all cases that one and the same thing may be known and in doubt for the same person in the same respects. And their general strategy, too, is similar to the extent that it involves for both authors either (1) a rejection of the initial hypothesis (*casus*) or, (2) when the hypothesis is admitted, (2.1) a rejection either of the arguments in support of knowledge or (2.2) a rejection of the arguments in support of doubt

[18] Peter of Mantua's *Logica* is still inedited. The manuscripts that contain the treatise *De scire et dubitare* (henceforth, DSD) are: (1) [O]xford, Bodleian Library, *Canon. misc.* 219, ff. 85va–95va; (2) [B]erlin, Staatsbibliothek Preussischer Kulturbesitz, *Hamilton* 525, ff. 88vb–93va; (3) [M]antova, Biblioteca Comunale, Ms. 76 (A III 12), ff. 72ra–78vb; (4) Venezia, Archivio dei [P]adri Redentoristi di Santa Maria della Fava, Ms. 457, ff. 57ra–63va; (5) [V]enezia, Biblioteca Nazionale Marciana, L.VI.128 (2559), ff. 62rb–68vb; (6) Città del Vaticano, Biblioteca Apostolica Vaticana, Vat. [L]at. 2135, ff. 55ra–59vb. The early printed editions are: (7) [E1]: [Johannes Herbort, Padua 1477], sig. K1ra–L4+4rb; (8) [E2]: [Antonius Carcanus for] Hieronymus de Durantibus, Papie 1483, sig. K2va–L3ra; (9) [E3]: Bonetus Locatellus, Venetiis 1492, sig. F4+1rb–G2ra; (10) [E4]: Simon Bevilacqua, Venetiis 1492, sig. I3vb–K3va. In this paper, all references to sections, arguments or direct quotations are from E1 (with occasional emendations).

or (2.3) a specification of *different* respects in which one and the same thing may be known and in doubt. However, when it comes to the tools introduced to achieve their goal and the details of specific replies to the seven arguments, some interesting differences seem to emerge.

5 Epistemic Principles in Peter of Mantua's *De scire et dubitare*

In Peter's *De scire et dubitare*, the main sources for general theoretical principles of epistemic logic are the preliminary assumptions (sections 1–3 of the treatise) and the preambles to the solutions of the two *dubia*, especially the first (sections 6 and 9). Other useful observations of theoretical significance are scattered here and there in the formulation of the arguments and their individual solutions. Moreover, while the basic epistemic principles concern, unsurprisingly, the relation between notions such as knowledge, belief, and doubt, qualified in various ways, Peter is also keen, especially at the beginning of the treatise, to lay out certain important presuppositions about cognition, signification, and intellection. Among the significant principles derived from Heytesbury are the theses that knowledge is veridical and that knowledge implies belief. The notion of subjectively certain belief as incompatible with doubt and as part of the broad notion of knowledge are also ideas taken from Heytenbury, as is the standard closure principle for knowledge (see forms 1–5, 12 above).

The core issues in Peter's basic theory of cognition and signification discussed in the preliminary assumptions concern, by contrast, (1) the relations between concepts, terms and propositions, (2) the notions of signification and understanding, along with the identification of the proper object of epistemic attitudes, which are directed to things signified by propositions and terms, not to propositions and terms as such, and (3) the distinction between an expression and the corresponding mental concept, either common or discrete, to which the expression is subordinated. Further assumptions include the following contentions: (4) for an expression to signify something is for it to represent that thing to the cognitive faculty; (5) true and simple intellection is a kind of cognition; (6) every complex notion (to be taken here in the sense of proposition) presupposes incomplex notions (i.e., concepts corresponding to its terms); and knowing (*cognoscere*) X implies grasping (*apprehendere*) X by means of a concept to which the term "X" is subordinated; (7) knowing X incomplexly (where X is a thing) implies that X is considered and that one has a concept that signifies it incomplexly; while knowing X complexly (where X is a fact) implies that X is considered and that one has a complex notion of it (i.e., a mental proposition); (8) no written or spoken term is understood unless a concept or an act signifying it is generated in the intellect; (9) one ought to respond only to propositions that are understood; and (10) verbs expressing mental acts, including epistemic and volitional verbs (*verba pertinentia ad actum mentis*) bring about propositions in the compounded sense and in the divided sense, depending on the scope of the verb, which yields in turn different kinds of supposition.

In the seventh assumption Peter introduces an important principle concerning the identification of the proper object of our epistemic attitudes, which he frequently relies on in the treatment of sophisms, namely the view that *things* (and only things) are *known* through concepts and propositions, and that concepts and especially *propositions* are not the direct object of our knowledge (or doubt), but rather only a *vehicle* for them. Even though we frequently (and inappropriately) refer to propositions as being know or in doubt, Peter contends that what we really mean is, properly speaking, that *through* those *propositions* we *consider* and *complexly* know or are in doubt about the *things* signified by those propositions. And if a proposition is said to be known or in doubt, it is only known or in doubt as a signified proposition, and not as a sign (*signum*) or as signifying proposition (*significans*). In several cases below, Peter rejects the initial argument for both knowing and doubting one and the same "thing" because the thing in question is inappropriately identified with a proposition, rather than with what the proposition signifies. This will frequently be the strategy used to dismiss what he regards to be a superficial and fundamentally incorrect formulation of the problem.

But there is also another way in which propositions turn out to be central in the analysis of epistemic attitudes, when they are taken (correctly, in this case) as that *through* which we know or are in doubt about things. In the analysis of the arguments, Peter often emphasizes that what appears to be a single proposition through which we supposedly seem to both know and be in doubt about one and the same thing can—and indeed must—in fact be resolved into a pair of propositions; and that it is *through one* of them that we *know* something, and *through the other* that we are in *doubt* about something (else). This process of disambiguation, which is the standard strategy of solution for most arguments below, may involve showing that something is known with respect to a common or general concept and in doubt with respect to a discrete or singular concept; or that it may be known or in doubt according to what turn out to be disparate terms; or that it may be known if a term is taken merely substantially and in doubt if the same term is taken accidentally (as interchangeable with a certain name or description); or, again, that it may be known in virtue of a peculiar kind of intellective ostension and in doubt with respect to perceptual identification.

6 Analysis of Peter of Mantua's *first dubium*

In this section, I will focus exclusively on the seven arguments in Peter of Mantua's *De scire et dubitare* that have a direct relevance for, or seem to reveal the influence of, Heytesbury's discussion, and examine his approach with a view to identifying what is peculiar to Peter's philosophical commitments and conceptual vocabulary.

6a. A is a Singular Name of One of the Following: "God Exists" "A Chimera Exists" (PM9, WH1)

The counterpart of Heytesbury's first argument (WH1) is Peter of Mantua's ninth (PM9), which presents a shorter and much more condensed—if not convoluted—version of the former.[19] The hypothesis is that you know that A is a singular name of one of these two propositions, namely "God exists" or "A chimera exists", the first of which is known to be necessary and the other impossible. Also, by hypothesis, a proposition is said to be in doubt as a proposition (*dubia ut propositio est*) when it is understood as a proposition and is neither believed to be true nor believed to be false. The argument purportedly aims to show that one of these propositions is known by you and that both are in doubt for you, which would entail that something known by you is in doubt for you.

That one of them is known by you is shown as follows: "God exists" is known by you; and "God exists" is one of them; therefore one of them is known by you. By contrast, if one of them is in doubt for you (according to the characterization in the hypothesis), for the same reason the other must be too; therefore both are in doubt for you. This follows from the assumption that A is one of these propositions and that A is in doubt for you, where A may refer to either proposition, as it is hidden to you which proposition is A. From these two claims, it follows that one of them is both known by you and in doubt for you. Peter considers a first potential counter: the assumption that there exists no other proposition might be incompatible with the hypothesis. If we drop it and carry on with the disputation, assuming that the proposition "A is one of these two" is put forward, one ought to concede it because if follows from the hypothesis. If A itself is put forward next, Peter considers various reasons why A should be in doubt for you (because you do not know which proposition it is) and shows how, even under the revised hypothesis, it should not be in doubt for you. First, whatever A is, it is not something in doubt for you, as you either know it to be necessary or know it to be impossible. Secondly, if A is doubted, suppose that "God exists" and "A chimera exists" are put forward in turn. According to the rules of obligations logic presupposed by Peter here, the first must be conceded and the second denied. But then, since A is one of them, A is either conceded or denied, which is incompatible with A being in doubt. Another reason adduced in support of this view introduces an important concept, which plays a significant role for Peter's general approach to problems of identification involving singular terms and demonstratives in epistemic contexts. He considers the conditional claim that if A is in doubt for you as a sign (*prout est signum*), then you understand A as a term. Since the consequent is false, then A cannot be in doubt for you as a sign. The reason why the consequent is false is that by means of the utterance A, no discrete concept signifying something which A signifies *ad placitum* is generated in the mind, such that A is subordinated to it. In other words, since by hypothesis you do not know

[19] Peter of Mantua, DSD, sig. L2 rb–L3ra (argument), L4+1vb–L4+2rb (solution).

what A refers to, you cannot understand it as a term. But this is in turn required for you to understand A and be in doubt about it as a sign.

Peter's solution to this argument entertains first the possibility of not admitting the hypothesis on account of its alleged impossibility. But even if we ignored the fact that no proposition can be in doubt (under the hypothesis in question), there would still be a deeper problem, according to Peter. It is the very characterization of a *proposition* as being in doubt, i.e., as being the genuine object of an epistemic attitude, rather than merely a vehicle (for beliefs that are about what is signified by it), that is fundamentally flawed. Just as no proposition is *known as a proposition*, in the same way no proposition is *doubted* as a proposition. While it is only *through* propositions that we form epistemic attitudes in virtue of which we are said to know or be in doubt about things, the objects of our knowledge and doubt are not propositions *as such,* but that which they signify.

Furthermore, the argument turns on the question of whether "A is a singular name of one of these two" is a proposition or not. Peter argues that the hypothesis ultimately ought not to be admitted because the above expression can only be understood in virtue of a previous imposition which determines completely what A signifies (the singular term A in this case behaves just like a generic demonstrative whose deictic content has not been specified). In keeping with one of the central contentions stated in the preamble of the treatise, Peter notes in his solution to this argument that one ought not to respond to anything unless it is understood. In this case A is not understood because we lack, by hypothesis, the required identificatory knowledge, as it is hidden from us which proposition is A, and hence it is hidden from us what A signifies.

While A is either necessary or impossible and hence ought to be conceded or denied, you do not know which proposition it is and therefore lack the necessary information to respond correctly. If A stands for "one of these two", its signification is indeterminate. As noted above, in this argument the use of the demonstrative is generic in the sense that the expression "You know one of these two" is a referentially indefinite sentence type.

At the same time, by knowing that A stands for one of these two (and nothing else) you know something about it. For, since these two are necessarily true or false propositions, and A is one of them, A cannot be in doubt to you, whatever it is. A is being used differently at different stages of the argument. When you concede "God exists" or deny "A chimera exists", i.e., the two propositions that A might refer to, and then respond to A by doubting it, the sense in which you are in doubt about A concerns its identification with one proposition or the other, not their respective content. You are not in doubt about the content and the resulting evaluation of "God exists" or "A chimera exists" but rather about which one is A, when A is put forward without further specification of what is being thereby indicated (in other words, you are in doubt about A *through* the proposition "This is A", where the denotation of "this" is unknown to you because you do not know which one is being indicated). The rules of an obligational disputation would force you to concede A if you knew that A is (or signifies) "God exists," and to deny A, if you knew that A is (or signifies) "A chimera exists". In neither case should you be in doubt about A. But here you are epistemically in no position to know whether you should concede or deny A (in other words, you are in no position to know A to be true or A to be false) because you do not know what A is. Thus, your being in doubt about A merely reflects your inability to idenfity which proposition is A.

By contrast, if A is taken in the sense of "one of these two", then you know that neither of these two ought to be doubted; so there is a way to conceptualize A according to which part of the information in the hypothesis may be accounted for, thereby giving you a basis to make a stronger claim, namely that no matter what A is, since it can only be either a necessary or an impossible proposition, you ought not to be in doubt about it. (This is based on the external information captured by the hypothesis and understood mentally when A is taken in the sense of "one of a pair including a necessarily true and a necessarily false proposition").

6b. A is the True One of the Two Contradictories "A King is Seated," "No King is Seated" (PM4, WH2)

The second of Heytesbury's arguments corresponds to Peter of Mantua's fourth (with further significant parallels in the eleventh).[20] In PM4, it is assumed that the proposition "This is true" is known as a sign (i.e. that the proposition insofar as it signifies, rather than what it signifies, is the object of the relevant epistemic attitude), that this mental proposition refers to the true one of two contradictory propostions, say "A king is seated" and "No king is seated", that A is a singular name for it, that you sufficiently consider "This is true" in your mind and that you know that "this" indicates the true one of those two propositions, even though you do not know which one of them is the true proposition. The claim is that at the same time you doubt this to be true and you know this to be true (indicating in the latter case the true one of them); therefore something is both known by you and in doubt for you.

Peter considers two arguments in support of the view that you know that this is true. First, one could argue that "You know that 'This is true' is true, therefore you know that this is true" is a valid consequence whose antecedent follows from the hypothesis (since "this" picks out the true one of *p* and *not-p*, and hence, whatever that is, "This is true" is true by hypothesis). On this reading, the argument presupposes that you know the *proposition* "This is true." Second, if A is the name of the true one of them, the consequence "A is true, therefore this is true" is also a good consequence, which you know to be good. But since you know its antecedent and understand its consequent, this (along with two further conditions that may be disregarded here) entails that you know the consequent as well. But if you know the consequent, namely "This is true", it follows in turn that the following consequence is good:

[20] Peter of Mantua, DSD, sig. L1ra–rb (fourth argument), L4+1rb–L4+1va (solution), L3rb (eleventh argument), L4+2va (solution).

(C1) "'This is true' (*hoc est verum*) is known by you, and you know that 'This is true' adequately signifies that this is true, therefore you know this to be true (*hoc esse verum*)."

The argument for doubt, by contrast, runs as follows. The proposition "This is true" is in doubt for you no matter which one of the two contradictories "A king is seated" and "No king is seated" is indicated. The following consequence will then be a good consequence, in analogy with C1:

(C2) "'This is true' is in doubt for you in both cases; and you know that 'This is true' adequately signifies this to be true; therefore, you doubt this to be true."

To solve the problem, Peter first contends that the hypothesis ought not to be admitted under the condition that the proposition be known *ut signum*, because this is incompatible with the general framework of his analysis (propositions are not the objects of epistemic attitudes, so any argument built on the assumption that *p* is known or in doubt, where *p* is a *proposition* rather than what the proposition signifies, is unsound). At the same time, Peter concedes (here and elsewhere) that one could turn a blind eye on this (improper) manner of speaking and tolerate it for the sake of argument as a widely accepted *modus loquendi*. Thus, if the clause is removed, the solution consists in conceding that you know this to be true in one sense and doubt this to be true in another sense. In particular, Peter formulates here a strategy that we encounter frequently in the analysis of the *dubium* and whose elements are codified in the preamble to the solutions: something may be known to the intellect and in doubt for the senses. The distinction encapsulates two different types of cognition. For the purposes of this argument, your conceptualization of A as "the true one of these two" is what matters to the intellect, as a privileged way in which you understand A. But at the same time, you doubt this to be true perceptually (*ad oculum* or *ad sensum*) on account of your identificatory ignorance, and this is not incompatible with the rest of the hypothesis.[21] Thus, the solution involves a distinction between knowing intellectually (through one proposition, namely "(the true) one of these two is true") and being in doubt with regard to perceptual identification (through another proposition, namely "this is true", where it is hidden from you what "this" refers to). Again, this hypothesis involves the use of a demonstrative pronoun in a generic sense, whereby doubt is the result of not knowing what the term actually refers to because "This is true" is a referentially indefinite sentence type whose signification is specified in different ways depending on the context, that is to say depending the actual denotation of "this" (which is what you are in doubt about). The problem emerges even more clearly at another turn in the argument. The two consequences:

> (C1) "'This is true' (*hoc est verum*) is known by you and you know it to signify adequately that this is true, therefore you know this to be true (*hoc esse verum*)"
> (C2) "'This is true' is doubted by you and you know it to signify adequately that this is true, therefore you doubt this to be true"

---

[21] On these two different types of ostension, especially in the context of grammar, see Sirridge and Fredborg (2013).

must both be denied even though their consequents must be conceded with regard to the different deictic content associated with "this" in the two possible circumstances of evaluation (the true contradictory in one case, the false contradictory in the other). This is because "this" in "This is true" is fundamentally understood in two different ways:

i. in one case as "what is being indicated now" (of which you cannot say whether it is true or not because you do not know whether "A king is seated" is true or "No king is seated" is true, and these are the two things that "this" can refer to)
ii. in the other case as "the true one of these two contradictories" (of which you trivially know, by hypothesis, that it is true)

When "this" is used in sense (i), its content varies depending on what is being indicated, and hence you are in doubt about the resulting signification of the proposition "This is true." When "this" is used in sense (ii), by contrast, its signification is univocally determined in virtue of the initial stipulation (and in a sense, it is no longer dependent on contextual deictic information) through an external assumption which guarantees that "this" picks out the true proposition of a pair of contradictories, whence the evaluation of "This is true" does not fluctuate but remains always, and trivially, the same. In this case the indexical uncertainty is neutralized through a description that provides a way for the intellect to conceptualize a proposition and legitimately claim to know that what it signifies is the case, keeping the issue of the identification on a separate level. Interestingly, Peter does not make use of the distinction between compounded and divided sense in this context, but introduces the distinction between the level of intellection and that of perceptual cognition to interpret two claims that are both in the compounded sense and resolve their apparent conflict. It is no longer a problem for you to know A and to be in doubt about A at the same time, if knowledge is, in one case, a function of the semantic content associated with A (when it is conceptualized as "the true one of these two contradictories,") while doubt depends, in the other case, on the veil of ignorance that characterizes the deictic process under the hypothesis.

6c. You Know This to Be Socrates and Doubt This to Be Socrates (PM3, WH3)

The third argument in Peter of Mantua bears a remote similarity to the third argument in Heytesbury.[22] While the difficulty still involves for both a conflict between conceptualization and identification, here the two approaches are slightly different, even in the very formulation of the hypothesis. We assume that Socrates is in front of you and that you have every possible concept of him, except that you do not know him to be called Socrates. Furthermore, you consider sufficiently whether this is

[22] Peter of Mantua, DSD, sig. K4vb–L1ra (third argument), L4+1rb (solution).

Socrates. You know that Socrates is some man but you do not know which man is Socrates.

That you should be in doubt about whether this is Socrates is said to follow directly from the hypothesis and the assumption that you neither firmly believe this to be Socrates nor firmly believe this not to be Socrates.

By contrast, the argument for your knowing that this is Socrates is rather elaborate, but its key assumption is fairly simply stated in terms of a distinction between taking a term merely substantially and taking a term accidentally, which also provides the basis for Peter's solution.[23] If we assumed in addition to the hypothesis that you know this to be called Socrates, it would follow that you would know this to be Socrates. However, Peter shows by *reductio* that it is not in virtue of this additional assumption (in conjunction with the hypothesis) that you know or begin to know that this is Socrates; hence you know (or knew all along) this to be Socrates even without knowing that this is called Socrates. The required knowledge that justifies the compounded sense claim about your knowing this to be Socrates (namely, a piece of identificatory knowledge) is somehow implicit in the hypothesis and turns on the idea that you have a privileged initial way to conceptualize Socrates (irrespective of how he is called) which fixes him for your intellect in a stable way.

Here is the reasoning: if you knew this to be Socrates beacause you knew this to be called Socrates, "This is Socrates" would signify the same as "This is called Socrates." On this reading, however, "Socrates" would then no longer be a substantial term; rather, it would become an accidental term equivalent to "called Socrates". If "Socrates" and "called Socrates" were genuinely interchangeable terms (as they are assumed to be in arguing that it is because of your knowing that this is called Socrates, which is the supplemental assumption, that you know this to be Socrates, a position Peter rejects here), the same would apply to any common term (for example "donkey" and "called donkey"). By the same token, "A man is a stone" and "A man is a goat" would not be impossible propositions because they could be taken to signify that what is called a man (e.g. a stone, if we adopted a new imposition) is a stone. Peter's general conclusion is that obliterating the distinction between substantial and accidental terms would undermine the notion of a necessarily true proposition.

Peter's solution to the argument contends that (i) the hypothesis should be admitted, we ought to (ii) concede that you know this to be Socrates and that "This is Socrates" is true in your mind taking the term "Socrates" merely substantially (*mere substantialiter*) (which involves some direct way for you to pick out the object without knowing its name), (iii) and deny that you doubt this to be Socrates, if "Socrates" is taken in the same sense. If "Socrates" is taken accidentally, however, (in which case it would signify the same as the complex expression "a man called Socrates"), then one should grant that you doubt whether this is Socrates, because you do not know whether he is called Socrates or not (the hypothesis is that you

---

[23] The distinction is a distant echo of *El. Soph.* 24, 179a26–b33, where Aristotle discusses the famous case of Coriscus and the man approaching in the context of his analysis of the fallacy of accident.

know everything about him except that he is called Socrates). Doubting the proposition "This is Socrates" when the term is taken accidentally (i.e. when the proposition is equivalent to "This is a man called Socrates") is not inconsistent with knowing Socrates and knowing that this is Socrates (*cognoscis Sor et scis hoc esse Sor*), where the term is taken merely substantially.

A term taken merely substantially in the intellect does not connote any extrinsic naming or nomination nor an extrinsic likeness (*non connotat vocationem seu nominationem aliquam extrinsecam neque similitudinem aliquam extrinsecam*). In this sense, it is possible to know (*cognoscere*) Socrates without knowing that the same individual is called Socrates or without understanding the term "Socrates." Any term in the category of substance is said to behave in this manner (I take this to express some kind of direct way to refer to an object or concept for substantial terms).

A term, say the utterance "Socrates," is taken accidentally if it is subordinated to another concept which is connotative of a name and of an extrinsic likeness (*conceptus est connotativus nominis et similitudinis extrinsecae*), in such a way that the intellect only apprehends by means of this term what it apprehends by means of the complex "called Socrates" or by means of the complex "something of such a quantity and quality." In this sense, the term no longer falls in the category of substance but becomes a genuinely *accidental* term (and is equivalent to descriptions or characterizations such as "called Socrates" or "thing of a certain quantity and quality" (*aliquid simile tantae quantitatis et talis qualitatis*). The accidental sense is characterized by Peter as an ordinary way of speaking (*modus vulgaris*); it pertains to a lower faculty (sense perception), which is modified by sensible objects and receptive of their images. The intellect understands the substance of a thing beyond its extrinsic accidents, and since we do not understand as easily or as quickly as we perceive, we do not have as quickly in our intellects *determinate substantial concepts* (common or singular) in the way we have *vague concepts* (common or singular).

Last, this argument introduces a different use of the demonstrative pronoun "this." In this case, you are no longer uncertain about the identification of Socrates: there is only one object that is deictically identified, and that object is Socrates. The problem is rather that, at least in one sense (accidentally, that is to say in terms of what Peter characterizes as an accidental property such as having a certain name), you cannot conceptualize what "this" univocally refers to in an adequate way and hence are correctly said to be in doubt about it. But if we prescind from the accidental characterization of "this" as something that is "called Socrates" (a piece of knowledge that is not available to you, by hypothesis), there is still a sense in which you know this to be Socrates, as "Socrates" taken merely substantially is just a proxy for "the individual that is identical with Socrates," regardless of the way it is named or described.

6d. You Know This to Be Socrates and Doubt This to Be Socrates (WH4, PM2, Same Sentence as 6c)

The second argument in Peter of Mantua corresponds to Heytesbury's fourth and is about the same sentence "This is Socrates" as in 6c.[24] The assumptions are that Plato and Socrates are in front of you, you know one of them to be Socrates, you know one of them to be Plato, but you do not know which one is Socrates and which one is Plato. The claim is that you know this to be Socrates and are in doubt about whether this is Socrates.

The argument in support of the claim that you know this to be Socrates rests on the assumption that every complex (propositional) knowledge or notion (*notitia complexa*) presupposes incomplex knowledge or cognition (*notitia incomplexa*) of the terms occurring in the proposition. And that in turn requires (stable) concepts in the mind to which the latter are subordinated. Thus, you can only know that one of them is Socrates if you have the mental propositional or complex concept "one of them is Socrates". The latter in turn cannot be in your mind unless the term "Socrates" is also in your mind, but this is all that having a simple concept or notion of Socrates means. However, it also follows from the hypothesis that whoever is indicated, you doubt whether he is Socrates (since you do not know which one is Socrates and which one is Plato); therefore, you do not know this to be Socrates.

This is one of the few cases in which Peter acknowledges that the hypothesis should not be admitted because it leads to an impossibility, namely that at the same time you know (*cognoscis*) and do not know (*non cognoscas*) Socrates. Peter accepts the argument for knowing that this is Socrates, and then argues that in order for you to be in doubt about whether this is Socrates, you must have a simple notion of Socrates and hence know Socrates.

The solution is to reject the hypothesis because it effectively produces a contradiction, in conjunction with a principle that Peter accepts, namely that complex, propositional knowledge (here in the compounded sense) presupposes knowledge of the relevant term(s). If a proper name falls within the scope of an epistemic verb (*scio* or *dubito*) in a true proposition, this presupposes that knowledge of the signification of the proper name in question is available; otherwise, the proposition as a whole would not be understood.

6e. This Is a Man (PM12, WH5)

The next two arguments in Heytesbury (WH5 and WH6) are both covered by Peter of Mantua's twelfth (PM12).[25] The assumption is that you know what is indicated by the subject of the proposition "This is a man" (i.e. you know the referent of "this"),

[24] Peter of Mantua, DSD, sig. K4va–vb (second argument), L4+1ra–rb (solution).

[25] Peter of Mantua, DSD, sig. L3rb–vb (twelfth argument), L4+2va–vb (solution).

that you sufficiently consider whether what is indicated by that subject is a man or not a man, that you are not deceived, that something you believe to be a man and something (else) you believe not to be a man, and finally that you know that "This is a man" signifies precisely that this is a man. If we assume next that "This is in doubt for you 'This is a man'" is proposed to you in the disputation, and it is true and not incompatible with the hypothesis, the rules of obligations tell us that it ought to be granted (in other words, one ought to grant "This is in doubt for you 'This is a man'". At the same time, however, "This is a man" is either known to be true or known to be false, and hence something known (either way) is granted to be in doubt for you.

Just as in the third argument, the use of the demonstrative pronoun "this" here is not generic but deictically determined. You know *what* individual thing is being indicated. But you do not know whether it is a man or not. The problem is that the hypothesis assumes that *if* this is a *man*, you *know* this to be a man, and that *if* this is *not a man*, you *know* it not to be a man. And this is in conflict with the fact that you do not know (and therefore are in doubt about) whether the thing being indicated is a man or not a man. As a result, Peter denies that you are under an obligation to respond either to "This is a man" or to "You know this to be a man," before it is known what thing is being indicated by the term "this," which seems to suggest that even if you know which object is being indicated, the signification is still incomplete, if it is not known to you whether it has or not a relevant property.

The solution, however, seems to presuppose that the demonstrative is being used in the generic sense, and offers a general principle to address any case involving a demonstative pronoun whose signification has not been specified completely. The principle in question is that no proposition is understood as a proposition, unless each and every one of its terms is understood as a term, and for a term to be understood as a term, a concept must be generated in the intellect. But "this", on the hypothesis under consideration, does not seem to have a definite signification.[26]

6f. A Is "God Exists," B is "A King is Seated," C is "A Chimera Exists" (WH6, PM12 Solution)

The sixth argument in Heytesbury, as noted above, is a variant of the first. Interestingly, in Peter of Mantua, a similar case is discussed in the context of the solution to a different sophism (PM12, which corresponds to WH5).[27] The solution is said to work in response to a common argument (*et per hec dicta et prius potest patere solutio ad argumentum quod facere solent*) showing that the same thing is both known and in doubt. The hypothesis involves three propositions:

---

[26] Peter here seems to assume that it is not sufficient for you to know what is being indicated by "this" but that you also need to know that it is qualified in a certain way, and might be elaborating on Heytesbury's point that your knowing what the subject indicates does not help to know how the proposition signifies.

[27] Peter of Mantua, DSD, sig. L4+2vb.

A:	"God exists"	known (to be necessary)
B:	"A king is seated"	in doubt
C:	"A chimera exists"	known (to be impossible)

If the three propositions are arranged in such a way (*moveantur et taliter disponantur*) that you do not know which one is A, B, or C, one can argue that A is at the same time known by you and in doubt for you. That A is known by you follows from the fact that A is "God exists", which you know to be necessary. That A is in doubt for you follows from the assumption that you doubt whether this is in doubt for you, where "this" actually refers to A but you do not know that A is being indicated (for instance, you do not know whether what is being indicated is B, in which case you would know that you are in doubt about it). Peter's reply is that A, insofar as it signifies something (*ut significans est*), is not known. To the extent that A is not taken with regard to what it signifies (that God exists) but rather as sign or as a name of a proposition, it is not the object of an epistemic attitude. And if the focus shifts instead onto the things that are signified, the difficulty is then solved by appealing to the distinction introduced above between knowing something intellectually and being in doubt about it with regard to perceptual identification. The argument would then simply conclude that you know this to be known to you, indicating it to the intellect, and that you doubt this to be known to you perceptually (because you cannot identify it). The case is another example of Peter's use of a referentially indefinite sentence type "A is known by you," whose signification depends on deictic information you do not have, which explains why it is, in one sense, true to say that you are in doubt about it (for you do not know which proposition is being indicated). In another sense, however, it is known to you that A is known to you because you know A to be the same as (or stand for) "God exists," which Peter associates with the notion of intellectual ostension (the act that encapsulates the initial assumption of the hypothesis, which stipulates that A is the name of "God exists").

6g. "This is Socrates" is Known by You and in Doubt for You (PM6, WH7)

The last argument in Heytesbury bears a remote resemblance with Peter of Mantua's sixth.[28] In Peter the hypothesis is more complex. It is assumed that Socrates is in front of you and that you know this to be Socrates, when he is indicated by means of the mental proposition "This is Socrates." The mental proposition is identified as usual by a singular name A and is assumed to be known to you during the next hour. At some point in the next hour, Socrates moves away (which Peter contends must be possible, unless we are prepared to deny that any proposition of the form "This

[28] Peter of Mantua, DSD, sig. L1va–L2ra (sixth argument), L4+1va (solution).

was here" can be known, when the subject refers to an object that is not present). It is also assumed that after half an hour, say at time *t,* you encounter Socrates and that he looks different, so much so that, at *t*, you (falsely) believe him to be Plato.

It is argued that you do not know this to be Socrates at *t* (and hence are in doubt about it). This is because, on the one hand, the proposition "This is Socrates" is not known to you, because at *t* Socrates you believe (albeit falsely) to be Plato and hence it is not the case that you believe this to be Socrates. Moreover, if "This is Socrates" is known to you, and you know that it signifies adequately that this is Socrates, then you know that this is Socrates. But the consequent is false, for you do not believe him to be Socrates at *t* (rather, you then believe him to be Plato).

By contrast, since "This is Socrates" is A, and the hypothesis stipulates that A is known throughout the next hour, "This is Socrates" is known to you even at *t*. The assumption that you continue to know A as you did at the beginning of the hour is justified by the fact that you firmly believe it in exactly the same way, your relation to A has not changed in any way, and A is true just as it was before (because in fact, at *t*, this happens to be Socrates once again, even though you falsely believe him to be Plato).

The first stage of the solution turns (as in Peter's fourth argument) on exposing the incorrect identification of the proper object of our epistemic attitudes. We should not admit that the proposition "This is Socrates" is known or in doubt *as a proposition,* because nothing is known or in doubt in that sense. Once this conceptual inaccuracy is eliminated, we can concede with the rest of the hypothesis that you know this to be Socrates, if he is being indicated *to the intellect* (on account of the familiar distinction between two types of deictic acts, one intellectual and the other perceptual) and that you doubt this to be Socrates, with respect to sense perception. This is because, perceptually, you are in doubt about what "this" actually refers to, whereas in the former case "this" seems to be treated as a proxy for Socrates, whom you then of course trivially know to be Socrates, in the same way you know that A is true, when A is understood as "the true one of two contradictory propositions" (even if you do not know which one is A).

7 Conclusion

It seems that Peter of Mantua's approach is ultimately motivated by many of the same concerns that characterize Heytesbury's discussion. At the same time, however, their solutions tend to differ in terms of conceptual vocabulary. In sum, Peter's approach involves

i. a pervasive—if not ubiquitous—adoption of obligational hypotheses, settings and rules;
ii. the distinction between

i. cases that ought not to be admitted (e.g., PM2, PM9), because they inevitably lead to contradiction, and none of the escape routes listed below is readily available) and
ii. cases that ought to be admitted; if a case is admitted, the task is to show either
    i. that of two inconsistent claims (one and the same thing is both known and in doubt) one is true and the other is false;
    ii. or that they are both true but in different senses (and hence not inconsistent); i.e., that it is not really one and the same thing that is known and in doubt, as the hypothesis implicitly presupposes a plurality of senses and distinctions that need to be made explicit. This in turn involves a series of possibilities:
        i. a distinction between types of cognition: intellectual and sensitive, general or singular, common or discrete (e.g., PM4, PM6, PM11, PM12)
        ii. a distinction between taking a term merely substantially and taking it accidentally, involving connotative (accidental) determinations that may include names or descriptions (e.g., PM3)
        iii. a distinction between what is a real object of epistemic attitudes (proposition as *significatum*, or that which a proposition signifies) and what is merely an apparent object of epistemic attitudes (proposition as a sign, or *significans*) (e.g., PM4, PM6, PM12)
        iv. the contention that in most cases we may know and be in doubt about the same thing according to disparate terms, or different concepts, or through different propositions.

We conclude by raising a few questions.[29] Can a closer correspondence be identified between the three types of uses of demonstrative pronouns in Heytesbury and the different conceptual tools introduced by Peter of Mantua, most notably the distinction between intellectual and perceptual identification? What is the exact nature of this distinction and how does it relate to the distinction between taking a term substantially or accidentally? What is the precise role of the distinction between a proposition as such and a proposition as a vehicle of meaning in the economy of the treatise? It often seems to be advocated just to dispel potential ambiguity and shift the focus onto the correct level of analysis, to which the distinction between intellectual and

[29] Some such questions should be explored in connection with the second doubt in Peter of Mantua's *De scire et dubitare*, which focuses on the notion of appearance and asks whether a man may appear (to be) a donkey. In this context, Peter introduces a distinction that is especially relevant for the problem of intellectual and perceptual identification. He contends that every appearance may be twofold, one relating to the intellect, the other to the senses, just as every internal cognition (*noticia interior*), whether complex or incomplex, may be either intellectual or perceptual. He then goes on to illustrate the issue by means of the following example. The sky appears to be blue to the senses, but the intellect knows it not to have any color, which is in turn inferred from the principles of physics (something is colored if and only if it is a composite, substantially or qualitatively). Thus, the senses and the intellect may have different *notitiae*, perceptual and intellectual (where intellectual is spelled out inferentially here, rather than in terms of identification).

perceptual knowledge and doubt is then properly applied, but is Peter responding to a specific set of concerns in one of his sources (possibly not Heytesbury, in this case) or is he introducing this strategy on his own? What are, more generally, the implications of Peter's theory of imposition and naming with regard to modality?

The analysis of William Heytesbury's and Peter of Mantua's epistemic logic reveals, perhaps unsurprisingly, yet another dimension of the influence of English logic in Italy in the fourteenth century.

References

Boh, I. (1993). *Epistemic logic in the later middle ages*. Routledge.

Brower-Toland, S. (2012). Medieval approaches to consciousness: Ockham and Chatton. *Philosophers' Imprint, 12*, 1–29.

Buridan, J. (1976). *Tractatus de consequentiis* (H. Hubien, Ed.). Publications Universitaires and Vander-Oyez.

Cross, R. (2014). *Scotus's theory of cognition*. Oxford: Oxford University Press.

Heytesbury, W. (1988). *The verbs "know" and "doubt."* In N. Kretzmann & E. Stump (Eds.), *The Cambridge translations of medieval philosophical texts. Volume one: Logic and the philosophy of language* (pp. 435–479). Cambridge University Press.

Hilpinen, R. (2017). Sed ubi Socrates currit? On the Gettier problem before Gettier. In R. Borges, C. de Almeida, & P. R. Klein (Eds.), *Explaining knowledge: New essays on the Gettier problem* (pp. 135–151). Oxford University Press.

Hintikka, J., & Symons, J. (2003). Systems of visual identification in neuroscience: Lessons from epistemic logic. *Philosophy of Science, 70*, 89–104.

Holcot, R. (1518). *In quatuor libros Sentiarum quaestiones*. Lyon

Kilvington, R. (1990). *Sophismata* 45. In *The Sophismata of Richard Kilvington* (Trans. with introduction and commentary by N. Kretzmann and B. Kretzmann). Cambridge University Press.

Knuuttila, S. (2008). Medieval modal theories and modal logic. In D. M. Gabbay & J. Woods (Eds.), *Handbook of the history of modal logic 2: Mediaeval and renaissance logic* (pp. 505–578). Elsevier.

Knuuttila, S. (2015). Epistemic logic in Paul of Venice's commentary on the beginning of *Posterior Analytics*. In. J. Biard (Ed.), *Raison et démonstration. Les commentaires médiévaux sur les Seconds Analytiques* (pp. 185–198). Brepols.

Martin, C. (2007). Self-knowledge and cognitive ascent: Thomas Aquinas and Peter Olivi on the *KK-Thesis*. In H. Lagerlund (Ed.), *Forming the mind* (pp. 93–108). Springer.

Martin, C. (2012). Logical Consequence. In J. Marenbon (Ed.), *The Oxford handbook of medieval philosophy* (p. 306). Oxford University Press.

Nuchelmans, G. (1973). *Theories of the proposition: Ancient and medieval conceptions of the bearers of truth and falsity*. North-Holland.

Ockham, W. (1974). *Summa logicae* (P. Boehner, G. Gàl, & S. Brown, Eds.). St Bona-venture University.

OPh 4. (1985). *Expositio in libros Physicorum Aristotelis* (V. Richter & G. Leibold, Eds.). St. Bonaventure University.

Paul of Venice. (1981). *Logica magna I, tractatus de scire et dubitare*, edited with an English translation *On knowing and Being Uncertain* by P. Clarke. The British Academy/Oxford University Press.

Panaccio, C. (2004). *Ockham on concepts*. Ashgate.

Schierbaum, S. (2016). Chatton's critique of Ockham's conception of intuitive cognition. In C. Rode (Ed.), *The opponents of Ockham* (pp. 15–46). Brill.

Schierbaum, S. (2016). Subjective experience and self-knowledge: Chatton's approach and its problems. In J. Kaukua & T. Ekenberg (Eds.), *Subjectivity and selfhood in medieval and early modern philosophy* (pp. 143–156). Springer.

Scotus, D. (1891). *Opera omnia* (Vol. II, p. 175). Vivès.

Sirridge, M., & Fredborg, M. (2013). Demonstratio ad oculum and demonstratio ad intellectum: Pronouns in Ps.-Jordan and Robert Kilwardby. In *Logic and language in the middle ages* (Festschrift Ebbesen) (pp. 199–220). Brill.

Yrjönsuuri, M. (Ed.). (2001). *Medieval formal logic: Obligations, insolubles and consequences.* Kluwer.

Yrjönsuuri, M. (2007). The structure of self-consciousness: A fourteenth-century debate. In S. Heinämaa, V. Lähteenmäki, & P. Remes (Eds.), *Consciousness: From perception to reflection in the history of philosophy* (pp. 141–152). Springer.

Poncius contra (Dicta Mastrii contra (Dicta Poncii))

Fabrizio Mondadori

[***Note of the Editors****: The present paper reproduces the paper that Fabrizio Mondadori read at the conference in honour of Massimo Mugnai, "Logic, Ethics and Modalities", hosted by Mario Piazza at the Scuola Normale Superiore (Pisa) on May 21, 2019, and that he then submitted, at the end of 2019, for publication in the present volume. The paper focuses on John Duns Scotus's notion of the possible and the interpretations proposed by two of the most prominent seventeenth-century Scotist philosophers: the Italian Franciscan Bartolomeo Mastri (1602–1673) and the Irish Franciscan John Punch (1603-1661). In his "Disputationes theologicae" (1655) Mastri rejected some interpretations of Scotus proposed by Punch in his "Cursus philosophiae" (1643). The paper discusses Punch's reactions to Mastri's criticism.*

Sadly, the unexpected death of Fabrizio Mondadori prevented him from revising the paper. We decided to publish it as he sent it to us. We limited ourselves to minor editing interventions and to making explicit the references to the editions of Scotus's works contained in the paper (we list them in footnote 1, in italics and between square brackets).]

1 Interpreting an Interpretation

Poncius contends that

(**a**) "[…] *esse diminutum non producitur per actum divini intellectus.* (*PMSCI*, p. 903b: the passage continues as follows: "Haec […] videri posset esse contra Scotum […], sed non est tamen, ut postea videbitur").

There are actually two contentions here. The first concerns the (notion of) "esse diminutum": the second—"Haec […] videri posset […]"—concerns (what Poncius takes to be) Scotus' conception of the possibility of the possible. What kind of claim should we take "esse diminitum non producitur per actum divini intellectus" to be? An ontological claim (in which case "esse diminutum" comes to be an ontological status, i.e., the ontological status of a *possibile*)? Or a modal claim (in which case

F. Mondadori (✉)
University of Wisconsin-Milwaukee, Wisconsin, WI, USA

F. Ademollo et al. (eds.), *Thinking and Calculating*, Logic, Epistemology, and the Unity of Science 54, https://doi.org/10.1007/978-3-030-97303-2_11

"esse diminutum" is rather a modal status, i.e., the possibility—not the reality—of a *possibile*)? Definitely the latter: there is (virtually) conclusive textual evidence for taking it to be a modal claim. First: Poncius says that *possibilia* have an "esse aliquod possibile a seipsis, independenter ab actu intellectus divini" (*PMSCI*, p. 904a).

This is part of an argument in favour of (**a**) above: now talk of "esse possibile" is plainly talk of "possibilitas" (that is, of the being possible—a modal status—of a *possibile*). Second: there is textual evidence for this too. In the course of the same argument, Poncius maintains that "[...] ergo praesupponitur possibilitas aliqua, saltem remota [...] in homine, ante intellectionem divinam": "ante" should be taken to mean "independently of" (this is the very view (**a**) above puts forth). Third, Poncius considers the following objection to (**a**) above: "Obiicies primo Scotum [...] dicentem quod esse possibile creaturarum competit ipsis per intellectum divinum; ergo conclusio [see (**a**) above] est contra ipsum". His reply to the objection is this: "Respondeo distinguendo antecedens, esse possibile proxime, concedo; remote, nego" (*PMSCI*, p. 904a). (**a**) comes accordingly to mean that, according to Scotus (according to Poncius), the possibility of the possible "remote" (see footnote 2) is formally independent of the divine intellect. Needless to say, objection and reply only make sense provided talk of "esse diminutum" is really talk of a modal status.

Now, contrary to what Mastrius, Crescentius Krisper (1679–1749), and Antonius Ruerk (fl. first half of eighteenth century) implicitly assume, Poncius gives "a se" (in "possibile a se") a quite different sense than the sense Scotus gives it. So, if we are correctly to evaluate his interpretation of Scotus, we must draw a distinction between, on the one hand, interpreting Poncius' own views on the possible (which crucially rely on his use of the qualification "a se"), and, on the other, interpreting his interpretation of Scotus' views (which also crucially relies on the qualification "a se"). Preparatory to passing judgment on Poncius' interpretation of Scotus, I proceed, first, to discuss Scotus' conception of possibility, and, second, to interpret Poncius' own conception of possibility.

2 "Possibilis ex Se Formaliter" and "Possibilis Principiative per Intellectum Divinum"

In *Ordinatio* I, d. 43, q. u., ed. Vaticana, VI, n. 7, Scotus contends that the possible is possible "ex se formaliter", and that it is (also) possible "principiative per intellectum divinum". This distinction involves two sub-distinctions: between "ex se" and "per intellectum divinum", on the one hand; and between "formaliter" and "principiative", on the other. The first sub-distinction, which pits "ex se" against "per intellectum divinum", suggests that "ex se" should be understood to mean "in virtue, and only in virtue, of itself": that which is possible "ex se formaliter" cannot (also) be possible "(ex se formaliter) ab alio/per aliud".

Now the claim that the possible is possible "principiative per intellectum divinum" may be interpreted in at least three ways. First way:

(A) the divine intellect is the principle—a "principium effectivum", an efficient cause of sorts—of the formal status of quiddities, hence of the mutual compossibility of quiddities, and, hence, of the possibility of the possible.

(A) is a dubious claim. An efficient cause is typically the cause of the ontological status of quiddities, or of the ontological status of *possibilia*: it could hardly be the cause of the formal status of the former, or of the modal status of the latter.

Second way:

(B) the divine intellect is the "principium effectivum" (an efficient cause of sorts) of the ontological status of *possibilia*, i.e., of the ontological (not the modal!) status of that which is possible "ex se formaliter possibilis".

In his commentary on Scotus, *Ordinatio*, I, d. 43, q. u., Poncius claims that

Resolutio Doctoris [...] consistit in hoc, quod possibilitas rerum non habeatur ab omnipotentia per se primo, etiam principiative, sed ab ipsa natura rerum formaliter, et ab intellectu principiative, quod intelligo iuxta dicta in Metaph. de possibilitate proxima, non de remota quae praecedit intellectionem, ut terminatur ad possibile [...]. (*Commentarii theologici* ... I, Parisiis 1661, p. 1192, n. 3)

This yields the third interpretation of the claim I have been discussing:

(C) the divine intellect is the "principium effectivum"—the efficient cause—of the possibility of the "possibile proximum". (This clearly, if implicitly, suggests that the divine intellect is not the "principium formale" of the possibility of the "possibile remotum".)[1]

Three points are worthy of notice here. First, Poncius ascribes (C) to Scotus. Second, we shall see later, he takes Scotus to have rejected (A). Third, if we reject (A), we have no choice but to hold that the formal status of quiddities, and the possibility of the possible are independent—principiatively—of the divine intellect (that is: the divine intellect is not their "principium effectivum"). Finally, a powerful argument in favor of the view that the possibility of the (formally) possible is (formally) independent of the divine intellect is to be found (we shall soon see) in Scotus'

[1] I remark in passing that the distinction between "possibile proximum" and "possibile remotum" is equivalent to the distinction Scotus draws between "possibilitas obiectiva" and "possibilitas logica". The title of this paper is a variation on the title of a paper by Heinrich Schepers, "Holkot contra dicta Crathorn", *Philosophisches Jahrbuch* 79 (1972): 320–354. The following abbreviations are used in the text: *Rep.* IA for *Reportatio* IA, MS Biblioteca Apostolica Vaticana, Borgh. Lat. 325; *PMSCI* for Johannes Poncius, *Philosophiae ad mentem Scoti cursus integer*, Lugduni 1672; *DLM* for Bartholomaeus Mastrius, *Disputationes in XII Aristotelis libros Metaphysicorum*, disp. 8, q. 1, Venetiis 1647; *DT* I for Bartholomaeus Mastrius, *Disputationes theologicae* I, Venetiis 1719; *TSS* for Crescentius Krisper, *Theologia scholae scotisticae* I, Augsburg-Innsbruck 1748; *CTS* for Antonius Ruerk, *Cursus theologiae scholasticae* ... I, Valladolid 1746. *[For the quotations of Scotus have been used the following editions: John Duns Scotus. 1950–2013. Ordinatio, I-IV. In: "Opera omnia", vol. I-XIV, ed. C. Balić et al. Vatican City: Typis Polyglottis Vaticanis (=ed. Vaticana); John Duns Scotus. 1960–2004. "Lectura", I-III. In: "Opera omnia", vol. XVI-XXI, ed. C. Balić et al. Vatican City: Typis Polyglottis Vaticanis (=ed. Vaticana); John Duns Scotus. 1639. "Reportata Parisiensia". In: "Opera omnia", XI, ed. L. Wadding. Lyon (=ed. Wadding); John Duns Scotus. 1639. "Quaestiones quodlibetales". In: "Opera omnia", XII, ed. L. Wadding. Lyon (=ed. Wadding); John Duns Scotus. 1997. "Quaestiones super libros Metaphysicorum Aristotelis, Libri VI-IX", ed. G. Etzkorn et al. St. Bonaventure, N.Y.: The Franciscan Institute (=ed. Etzkorn)].*

contention that, if "ex" and "ab" express the same kind of cause, then nothing can be necessary (or possible) both "ex se (formaliter)" and "ab alio (formaliter)"—from which it follows that, if the possible qualifies as such "ex se formaliter", then it cannot also qualify as such "formaliter ab alio", hence, in particular, it cannot so qualify "formaliter per intellectum divinum" (see below, footnote 3).

With minor and obvious modifications, the same contention holds for the impossibility of the impossible *simpliciter*, and for the (in)compossibility of *(in)compossibilia*. For, since possibility is a matter of compossibility, if the possible is possible "ex se formaliter", *compossibilia* must also be compossible "ex se formaliter", and the same is true of the "constituents" of the possible, i.e., quiddities, be they quiddities that are "simpliciter simplices", or quiddities that are "simplices", but not "simpliciter simplices". The textual evidence for the claim just made is quite convincing: according to Scotus, "[...] aliquid est compossibile vel incompossibile alicui, quia ipsum est tale in se" (*Ordinatio*, I, d. 13, q. u., ed. Vaticana, V, n. 18; see also *Rep.* IA, fol. 42v: "[...] nihil repugnat alicui nisi quia ipsum est ipsum").

I take "quia ipsum est tale in se" to mean "because its quiddity has the formal status it has 'ex se formaliter'", and take "quia ipsum est ipsum to mean "because its quiddity has the formal status it has 'ex se formaliter'". Quiddities, accordingly, are the *relata* of the relation of compossibility. (Note that "quia ipsum est ipsum", "quia ipsum est tale in se", and "quia hoc est hoc" are all equivalent.) That line of reasoning applies to *differentiae* (each *differentia* is the *differentia* it is "ex se formaliter"), and it applies, generally, to all cases in which Scotus draws a distinction between something's being *F* (e.g., *ratum*) "ex se formaliter" and its being *F* "causaliter ab alio".

In *Lectura* I, d. 2 pars 2, q. 3, ed. Vaticana, XVI, n. 148, Scotus reports an argument whose conclusion is that "nihil potest esse necesse esse ex se et ab alio". His objection to that conclusion is this:

> [...] quando arguitur quod 'nihil est necessarium ex se et ab alio', dicendum quod aliquid potest dici necessarium ex se dupliciter: uno modo formaliter [...]; alio modo dicitur aliquid ex se esse necessarium effective [...]. Si tunc accipitur in eodem genere causae esse ex se et alio, sic impossibile est aliquid esse necessarium ex et ex alio; sed tamen in alio genere causae non est inconveniens quod aliquid sit ex se formaliter necessarium, et tamen necessarium ex alio effective (*Lectura* I, d. 2 pars 2, q. 4, ed, Vaticana, XVI, n. 186; see also *Rep.* IA, fol. 14r).).

Why is it "impossible" that something *s* should be possible both "formaliter ex se" and "formaliter ab alio"? The following answer naturally suggests itself: if we take "ex" and "ab" to express the same kind of cause, a contradiction follows. For, were we so to take them, *s* would be both necessary "formaliter ex se" (resp. necessary "formaliter ab alio") *and* not (necessary "formaliter ex se") (resp. not (necessary "formaliter ab alio")): that is: if *s* is necessary "formaliter ab alio" (resp. necessary "formaliter ex se"), it cannot also be necessary "formaliter ex se" (resp. necessary "formaliter ab alio").[2] Needless to say, the argument only works if "ex" and "ab" are

[2] With obvious and minor modifications, the same argument applies to the claim that nothing can be possible both "formaliter ex se" and "formaliter ab alio": it applies, in general, to all claims

taken to express the same kind of cause: if they do, "formaliter ex se" and "formaliter ab alio" are clearly incompatible. Not so if they are taken to express different kinds of causes: "formaliter ex se" and "effective ab alio" are compatible.

There is some (fairly) convincing, albeit indirect, textual evidence in favor of the answer I have given in the previous paragraph: consider Scotus' claim that

> [...] idem est necessarium a se formaliter et non necessarium a se, sed ab alio vel originaliter; et ista non contradicunt. (*Rep.* IA, fol. 15r)

There is a lot more to this passage than meets the eye. First, to claim that "ista non contradicunt" is to claim that "necessarium a se formaliter" and "necessarium ab alio originaliter" can both be true of one and the same thing. Second, "ista non contradicunt" implicitly involves a contrast: "these" do not contradict one another, but others do ("ista non contradicunt" is actually shorthand for "ista non contradicunt, sed alia ab istis contradicunt)". For instance: "necessary 'ex se formaliter'" and "necessary "formaliter ab alio'" do contradict one another (we have just seen why). Third, the reason why "these" are not mutually contradictory is that different kinds of causes are at play: a formal cause in "a se formaliter" and an efficient cause in "ab alio originaliter". Hence, fourth, from the fact that "ista non contradicunt [sed alia ab istis contradicunt]" it follows that a contradiction would indeed come about if the same kind of cause were at play in "necessarium ex se" and "necessarium ab alio".[3]

Scotus maintains, we have seen, that the possible is possible "ex se formaliter" and (also) possible "principiative per intellectum divinum": there is no contradiction here, since "ex" and "per" (i.e., "ab") express different kinds of causes. He also holds, we have just seen, that nothing can be necessary (or possible) both "ex se formaliter" and "formaliter ab alio". Now let the "aliud" be the divine intellect: if the possible is possible "ex se formaliter" (which it in fact is), then it cannot (also) be possible "formaliter ab intellectu divino". The formal grounds of the possibility of the possible "ex se formaliter" are not, therefore, to be found in the divine intellect: the latter may well endow the possible with an ontological status. But, when it comes to the modal status of the possible, it plays no modal role whatsoever. The possibility of the possible "ex se formaliter" is (modally as well as formally) independent of

in which the qualification "formaliter ex se" is implicitly or explicitly employed by Scotus. In all those claims, "formaliter ex se" is explicitly or implicitly taken by Scotus to be incompatible with "formaliter ab alio".

[3] Cf. the following passage: "Nec hic est contradictio quod aliquid quasi orginaliter vel causaliter habeat ab alio hoc quod sibi convenit formaliter" (*Ordinatio*, I, d. 26, q. u., ed. Vaticana, VI, n. 94, *textus interpolatus*). A contradiction would of course come about if "quasi originaliter vel causaliter", in the passage just cited, were replaced by "formaliter".

the divine intellect.[4] This view is centrally bound up with (it is, in fact, one and the same view as) the view Scotus puts forth in the following passage:

> S) "[...] [1] quare homini non repugnat [*scil.* esse aliquid] et chimaerae repugnat, est, [2] quia hoc est hoc et illud illud, [3] et hoc quocumque intellectu concipiente, quia [...] [4] quidquid repugnat alicui formaliter ex se, repugnat ei, et quod non repugnat formaliter ex se, non repugnat". (*Ordinatio*, I, d. 36, ed. Vaticana, VI, q. u., n. 60)

Now: "quia" in [2] is clearly meant to provide an answer to the question [1] raises. It means, "because this has the formal status it has 'ex se formaliter', and that has the formal status it has 'ex se formaliter'" (a chimera has of course no quiddity, nor does it have a formal status: but it has "parts", each "part" has a quiddity as well as a formal status, and the relevant quiddities are mutually incompossible). Scotus regards "hoc est hoc" ("ipsum est ipsum", "ipsum est tale in se") as a formal principle: in *Quodlibet*, q. 5, ed. Wadding, XII, n. 25, he remarks, for instance, that "Etiam in genere principii formalis est status, hoc enim accipit infinitatem, illud non, quia hoc est hoc et illud est illud [...]: repugnantia enim formalis, et non-repugnantia, primo reducitur formaliter ad rationem eius, cui dicitur esse repugnantia". Two points here.

First, "quia", in "quia hoc est hoc", clearly suggests that "hoc est hoc" is indeed a "principium formale": my reason for adding the qualification "ex se formaliter" to "hoc est hoc" in [2] above is, first, that "hoc est hoc" is intended to explain why it is

---

[4] Consider, for instance, Scotus' claim that "Licet aliquid secundum quidlibet sui sit causatum a causa extrinseca, tamen potest praedicationem alicuius immediate recipere formaliter, ac si non esset causatum [...]. Exemplum: formaliter creatura est ens, licet totaliter a Deo" (*Quaestiones super Metaphysicam*, VII, q. 1, ed. Etzkorn, n. 27: there is an important *adnotatio posterius interpolata* here: "[...] quando dicitur quod 'accidens non est ens nisi quia entis', dicendum quod haec coniunctio 'quia' non mediat inter praedicatum 'ens' et subiectum quod est accidens, quasi reddens causam formalem entitatis ipsius accidentis, [...]. Et ideo non sequitur quod essentia accidentis sit includens in suo conceptu quiditativo ipsam inhaerentiam. Verbi gratia, si dicam sic 'creatura est quia Deus, ergo formale esse creaturae est dependentia ad Deum', non sequitur; [...]"). A number of points are worthy of notice here. First, "licet" marks the (implicit) presence of a contrast. Second, the contrast is a contrast, not so much between "totaliter a Deo" and "formaliter", as, more precisely, between "totaliter a Deo" and "ex se formaliter" (a similar contrast crops up in Scotus' contention that "[...] formaliter ratum seipso est ratum formaliter, [...], et causaliter est tale a Deo; [...]", *Reportata Parisiensia*, II, d. 1, q. 2, ed. Wadding, XII, n. 16: I note in passing that "seipso est ratum formaliter" is just another way of saying, "est ratum formaliter ex se"; and that "tale" is a cross-refererence to "ratum"—*not* to "formaliter ratum"). The point of "[...] formaliter creatura est ens [...]", just cited, is then this: a(ny) given creature *c*, in spite of its being "totaliter a Deo", is "ex se formaliter" "an ens" (recall that "ens", in its most general sense, means "cui non repugnat esse", *Ordinatio*, IV, d. 8, q. 1, ed. Vaticana, XII, n. 2). That is: the possibility of *c* is formally independent of God, although God is the (efficient) cause of the ontological status of *c*. That is: whether or not *c* is formally an "ens" is formally independent of whether or not it is "totaliter a Deo" (cf., "ac si non esset causatum"). Third, "Licet totaliter a Deo" is shorthand for "Licet secundum quidlibet sui sit causatum a Deo". Fourth, to assert that "creatura est quia Deus, ergo esse formale creaturae [...]" (cited above) is a *non-sequitur* is to make an independence claim: since "quia Deus" is shorthand for "quia Deus est causa efficiens" (of the "esse creaturae": not of its "esse formale"), and "creatura est" is shorthand for "creatura est dependens ad Deum", a creature will be formally independent of God with respect to its "esse formale", although it will depend on him with respect to its "esse" (i.e., its ontological status, which is an "esse simpliciter" or, as the case may be, an "esse secundum quid").

not repugnant to a man to exist, and, second, that, in [4], Scotus claims that "quod non repugnat formaliter ex se, non repugnat": in both cases, non-repugnancy is possibility, possibility is compossibility, and compossibles are mutually compossible because each of them has the formal status it has "ex se formaliter".

There is textual evidence for the claim last made: witness the following two passages:

> [I] [...] aliquid est compossibile vel incompossibile alicui, quia ipsum est tale in se (*Ordinatio*, I, d. 13, q. u., ed. Vaticana, V, n. 18).
>
> [II] [...] quaelibet forma seipsa est talis forma, nec est alia ratio intrinseca quare est talis forma (*Ordinatio*, I, d. 26, q. u., ed. Vaticana, VI, n. 88).

In [I], Scotus implicitly draws a distinction between "tale in se" and "tale ab alio". On his view, to assert that something—in this case, a quiddity—is "tale in se" is to deny that it is "tale ab alio", i.e., more precisely, that it is "tale ab alio effective". But to deny that it is "tale ab alio effective" is in fact to assert that it is "tale ex se formaliter". Further, to contend that a quiddity is "tale" is to contend that "it has such-and-such formal status": which means that a(ny) given quiddity has the formal status it has "ex se formaliter". In [II], "seipsa" is a formal variation on "ex se formaliter": for a "form"—or, for a quiddity—to be "seipsa talis forma" is for it to have the formal status it has "ex se formaliter". The form, or the quiddity, will also have a "causa extrinseca": but its "causa extrinseca" does not make it the kind of form/quiddity it is; it just bestows an ontological—not a formal!—status on it).

Second, "quia hoc est hoc" is plausibly interpreted to mean, "because this has the formal status it has 'ex se formaliter'": the "ratio" Scotus alludes to in "reducitur formaliter ad rationem eius" is accordingly a formal status. All of this is couched in purely formal terms: so are [1], [2] and [4] above. And so is [3]: at least if we regard it as an independence claim—which is what it turns out to be, if we take it to mean, "and this—viz., that the quiddity of e.g. man has the formal status it has 'formaliter ex se'—is the case/ holds for any conceiving intellect". But, we have seen, if it has the formal status it has "formaliter ex se", it cannot also have it "formaliter ab alio". There is, therefore, no variation in the formal status of quiddities from conceiving intellect to conceiving intellect: nor is there variation in the modal status of *possibilia* from conceiving intellect to conceiving intellect.

Virtually the same view can be found in the following passage:

> [...] propositio non dicitur per se nota quia ab aliquo intellectu per se cognoscitur (tunc enim si nullus intellectus actu cognosceret, nulla propositio esset per se nota), sed dicitur per se nota quia quantum est de natura terminorum nata est habere evidentem veritatem in terminis etiam in quocumque intellectu concipiente terminos. (*Ordinatio*, I, d. 2, pars 1, q. 1–2, ed. Vaticana, II, n. 22)

It may be wondered what is the point of "etiam" in "etiam in quocumque intellectu concipiente terminos". Answer: it is meant to be contrasted with "quantum est de natura terminorum", as follows. A "propositio" qualifies as "per se nota" because (and only because) "quantum est de natura terminorum nata est habere [...]": and this is the case (recall "et hoc quocumque intellectu concipiente" in [3] above) "etiam in quocumque intellectu concipiente". So: the fact that a "propositio per se nota" is "nota [...] etiam in quocumque intellectu concipiente" has no bearing on the question of what—if anything—makes it so that it is "per se nota". Its being "per se nota" (like a possible's being possible, or a quiddity's having the formal status it has) is, accordingly, naturally prior to, and formally independent of, its being known (or conceived of) by an(y) intellect: I suggest that "etiam"—like "et hoc" in [3] above—is meant by Scotus to make precisely this point. I remark, finally, that "etiam" implicitly involves an independence claim: whether or not a "propositio" is "per se nota" is (formally) independent of whether or not it is known by an(y) intellect.

I turn now to a discussion of [4] above. First: what is the point of it? Note that the scope of "quia" in [4] is "et hoc quocumque intellectu concipiente": the point of [4], accordingly, is to explain why [3] is true. The crucial role in the explanation is played by the qualification "ex se": if non-repugnancy is non-repugnancy "formaliter ex se", it cannot (also) be non-repugnancy "formaliter ab alio" (and hence not "formaliter ab intellectu divino"). Now non-repugnancy is compossibility: and, if the former qualifies as such "formaliter ex se", so must the latter. But the *relata* of the relation of compossibility are quiddities: hence, if compossibility is compossibility "formaliter ex se", each of its *relata*—i.e., quiddities—has the formal status it has "formaliter ex se", and therefore independently—formally speaking—of (in particular) the divine intellect. In other words: it has the formal status it has "quocumque intellectu concipiente".

So much for the claim that the possible is possible "ex se formaliter". What of the claim that the possible is (also) possible "principiative per intellectum divinum"? In view of the fact, discussed earlier, that—if "ex" and "ab"/ "per" express the same kind of cause—nothing can be possible both formally "ex se" and formally "ab/per alio", to claim that the possible is possible "per intellectum divinum" is to claim that the divine intellect is the "principium effectivum"—the efficient cause—of the ontological status of *possibilia*, and hence that the possible "principiative" is possible "effective ab alio".

I must now discuss a passage which appears to provide not only very convincing evidence in favor of the view that the divine intellect produces (formally) the formal status of quiddities, but, also, very convincing counterevidence to the claim, made earlier, that that which is possible "ex se formaliter", as well as the formal status of quiddities, are (formally) independent of the divine intellect: the passage is this:

> [a] [...] cum ipsius impossibilis sit aliqua ratio prima, [...] oportet primam rationem impossibilitatis [...] inquirere. [b] Ad hoc dicendum est quod impossibilitas in impossibili habet reduci ad intellectum divinum, non quod in Deo sit prima impossibilitas, [...], sed ut in ipso respectu partium repugnantium impossibilis invenitur prima ratio principiationis. Nam partes ipsius incompossibilis simul sunt incompossibiles et in se formaliter repugnantes, ut album et nigrum. [c] Primum esse possibile quod habent, ab intellectu divino habent principiative et per consequens ab intellectu divino principiative habent suam incompossibilitatem, sicut et suas rationes formales. [d] Sed ex se formaliter sunt talia, circumscripto quocumque alio quod est extra illa, et ideo impossibilitas huius 'album est nigrum', non reducitur ad Deum ut ad causam privativam [...], sed [e] reducitur ad intellectum divinum ut ad causam positivam a quo sunt primo principiative partes formales in esse possibili ipsius impossibilis et per consequens incompossibilitas totius. (*Rep.* IA, fol. 91v)

I make the following remarks here. First, reference to a "ratio prima" in [a] is reference to a "prima ratio extrinseca" (as contrasted with a "prima ratio intrinseca/formalis": see *Lectura*, I, d. 43, q. u., ed. Vaticana, XVII, n. 16, and *Ordinatio*, I, d. 43, q. u., ed. Vaticana, VI, n. 6: "[...] reducendo quasi ad primum extrinsecum principium, intellectus divinus erit illud a quod est prima ratio possibilitatis in lapide"). The terms "extrinseca" and "effective ab alio" designate but two sides of the same coin, and so are "intrinseca" and "ex se formaliter". Second, "reduci" in [b] is shorthand for "reduci principiative": not for "reduci formaliter" (as "Sed" in [d] shows: it marks a contrast—an opposition—between "ex se formaliter sunt talia" and "ab alio principiative sunt talia"). The "reduction" Scotus alludes to in [b] is accordingly a principiative—not a formal—reduction: thus note that in [c] "habent principiative"—not: "habent formaliter"—is employed by Scotus. The two kinds of reduction just alluded to go hand-in-hand with two kinds of origin, i.e., a principiative origin, and a formal origin: nothing reduces, formally, to the divine intellect, nor is the divine intellect the formal origin of anything. It is, however, the principiative origin of something: it cannot be the origin of the formal status of quiddities (the latter's origin is a purely formal matter). As [c] clearly suggests, it is the origin of the ontological status of quiddities: Scotus characterizes it (see [c]) as "Primum esse possibile"—the ontological, not the formal, status of a *possibile.* More precisely: "a quod sunt primo principiative partes formales in esse possibili", should be read, "a quo partes formales sunt in esse possibili primo principiative". The point of the qualification "in esse possibili" is that that which is "primo ab intellectu divino" are the "partes formales" in "esse possibili": for such "parts" to be "in esse possibili" is for them to possess an ontological status.

A further distinction should be drawn here: between "formaliter talia" (see [d]) and "principiative talia". The distinction goes hand-in-hand with Scotus' claim, in [d], that "Sed ex se formaliter talia sunt, circumscripto quocumque alio quod est extra illa". The point of "Sed" here, I have said earlier, is to draw a sharp wedge between the purely formal aspect, and the principiative aspect, of quiddities. A quiddity is formally what it is in virtue of its formal status (that is: in virtue of itself): it has the formal status it has "ex se formaliter"; it is principiatively what it is—i.e., it has the ontological status it has—"per intellectum divinum" ("ab alio").

It will be objected that Scotus' claim that "[...] sicut Deus suo intellectu producit possibile in esse possibili, ita producit duo entia formaliter (utrumque in esse possibile), et illa 'producta' se ipsis formaliter sunt incompossibilia" (*Ordinatio*, I, d. 43. q. u., ed. Vaticana, VI, n. 16) clearly suggests that the "esse possibile" of *possibilia* is formally produced by the divine intellect. I doubt it. The production at play here is not ("formaliter" notwithstanding) a formal production: the "esse possibile" Scotus alludes to is the "esse"—the ontological status—of the formal status of an "ens". So: God not only produces a *possibile* in "esse possibili" (i.e., he endows that *possibile* with an ontological status); he also produces the "esse possibile" of the formal status of that *possibile.* Why "formaliter" (in "duo entia formaliter")? Because the divine production of the possible in "esse possibili" is (also) a production (in "esse possibili") of something that is possible "ex se formaliter": "formaliter" does not qualify "producit", but "entia". There is a reason for this. On Scotus' view, that production is governed by the Principle of Contradiction (henceforth: PC): the divine intellect produces all and only that which does not flout PC, hence, in particular, that which is possible "ex se formaliter"—so that "duo entia formaliter (utrumque in esse possibili)" comes to mean, "two *possibilia*, each of which is possible 'ex se formaliter'". The rest of the passage appears to confirm this: "se ipsis formaliter sunt incompossibilia" is intended to be contrasted with "ab alio/ab intellectu divino formaliter sunt incompossibilia". Since the two claims cannot both be true, and since, according to Scotus, the first of them is true, the second must be false.

A similar interpretation also applies to Scotus' claim, in *Reportata Parisiensia*, I, d. 43, q.u., ed. Wadding, XI, n. 14, that the divine intellect "est tota ratio [scil tota ratio extrinseca] esse formalis partium": the "esse formalis partium" is both an ontological and a formal status. The divine intellect, by producing those possible "parts", bestows on them an ontological status: but, since they are possible "ex se formaliter", those parts also possess a formal status. The latter is not, however, of God's own making: the "parts" have an "esse formalis" in and of themselves, independently of the divine intellect. To claim that God's intellect is the "ratio" of that "esse" is just to claim that it is the "ratio" of the ontological status of "parts", each of which possesses "ex se", not "ab alio", the formal status it has. (This amounts to rejecting (A) on p. 3 above.)

I conclude that, in the three passages I have been talking about, there is no (implicit or explicit) suggestion that the formal status of quiddities, and (hence) the possibility of that which is possible "ex se formaliter", are grounded in the divine intellect. The distinction between "ex se formaliter" and "principiative ab alio", and Scotus' contention that nothing can be possible both "ex se formaliter" and "formaliter ab alio" suggest, in fact, that the very opposite view is, in all likelihood, the view we should ascribe to Scotus: the formal origin of the formal status of quiddities, and of the possibility of the possible, are not to be found in the divine intellect. As we shall presently see, some of the counterpossibles Scotus makes use of provide a strong argument in favor of the claim last made.

3 Counterpossibles and Adversative Conjunctions

Counterpossibles are typically even-if counterfactuals with an impossible antecedent: this characterization applies to (most or all of) the counterpossibles Scotus employs. A question naturally arises here: what are counterpossibles intended (by Scotus) to show? Especially when modal notion are at play, some, possibly even all, of the counterpossibles he employs involve—at least implicitly—a question (e.g., "Utrum prima ratio impossibilitatis sit [...]"), to which they are meant to give the answer, which typically is– or strictly implies—an independence claim. Consider, for instance, the following counterpossibles:

α) "[...] etsi Deus non esset, contradictoria contradicerent." (*Reportata Parisiensia*, I, d. 43, q. u., ed. Wadding, XI, n. 9: "contradicerent" is of course shorthand for "contradicerent ex se formaliter").

β) "[...] si per impossibile starent differentiae per se sine efficientibus, formaliter distinguerentur." (*Rep.* IA, fol. 43r).

γ) "[...] illa entitas qua constituitur aliquid in esse specifico, ex se formaliter est indivisibilis in plures species species, etiam dato per impossibilile quod esset incausata, licet modo istam indivisionem habeat causaliter unde est causaliter." (*Ordinatio*, I, d. 26, q. u., ed. Vaticana, VI, n. 94, *textus interpolatus*).

I note, first, that α)-γ) all involve adversative conjunctions. Implicitly present in the consequent of α), is "adhuc". Implicitly present in β) are "etsi" ("si" is shorthand for "etsi" here) and "adhuc": thus, "[...] etsi per impossibile [...], adhuc [...]". Finally, in γ), "adhuc" is implicitly present in "ex se formaliter [...]" (so that the latter reads, "adhuc ex se formaliter esset [...]"), and "etsi" is quasi explicitly present in "etiam dato" (given, at any rate, that "etiam dato" means, "even if we were to [...]", and that the Latin for "even if" is "etiamsi"). I note, next, that adversative conjunctions play, explicitly or implicitly, a crucial role in counterpossibles (or in the corresponding even-if counterfactuals): but exactly what is an adversative conjunction? In *Totius latinitatis lexicon*, one of the senses "nihilominus"—an adversative conjunction—is given is this: "Nihilo minus [...] significat id, quod dictum est [...], non impedire quin alterum fiat". This is also true of "adhuc", and of all adversative conjunctions. So: buried in "adhuc" and "nihilominus", and in adversative conjunctions in general, is what might be called an adversative thought—a thought that is as it were "adverse" to what is conventionally implied, but not outright asserted, by the antecedent.

For example: "adhuc" in α) above clearly involves the idea of something's being the case that might not be thought to be the case, given what is implied by the antecedent. The latter implicitly involves a view Scotus means to reject in α): the view that the divine intellect is the formal origin of the contradictoriness of mutual contradictories. The claim that contradictories would still be mutually contradictory even if God did not exist (and, plainly, no God, no divine intellect), is, therefore, "adverse" to, or a rejection of, that view. The contradictoriness of mutual contradictories should be regarded, then, as formally independent of the divine intellect. What of β)? A similar interpretation applies to it: whether or not *differentiae* are distinguished "formaliter" (i.e., "ex se formaliter") is formally independent of whether or not they have an efficient cause. The same view Scotus puts forth in γ): whether or not the "entitas" *e* "qua constituitur aliquid [...]" is indivisible "ex se formaliter" in several species does not formally depend on whether or not it has a cause (i.e., an efficient cause). Cause or no, *e* would still be indivisible in several species "ex se formaliter". I remark that "licet" (a concessive conjunction) is intended to draw a contrast of sorts between "ex se formaliter" and "ab alio causaliter": the "indivisio" *e* has (its being undivided "causaliter ab alio" is a consequence of its being indivisible "ex se formaliter"), it has from its (efficient) cause, whereas it has its indivisibility "ex se formaliter". The concede, then, that *e* is undivided "causaliter ab alio" is in no wise to concede that its being indivisible "ex se formaliiter" is not (formally) independent of whether or not it has a cause.[5]

---

[5] In this connexion, the following passage should also be considered: "[...] nec etiam clarum est quod idem est principalius respectu esse, et cujuscumque perfectionis consequentis esse, siquidem

This is just half of the story (as regards counterpossibles). The other half is this. First, α)–γ) all implicitly involve a question. Second, they all explicitly involve a *positio impossibilis*: how is the latter to be understood? Third, there is an implicit view α)–γ)—more precisely, their consequents—are meant to reject. Fourth, there is a point to the question. I shall just deal with α) here. The question it implicitly raises is,

> δ) whether the contradictoriness of contradictories is formally grounded in the divine intellect, or in the formal status of contradictories themselves.

(With minor and obvious modifications, the same questions crop up in the case of the (im)possibility of the (im)possible "ex se formaliter", in that of the (in)compossibility of (in)compossibilia "ex se formaliter", and in that of the formal status of quiddities.) As concerns the second point: the impossible *positio* that God does not exist must be relativized to a context that is relevant to the *positio*. In the case of e.g. contradictories, the relevant context is the question in δ) above. In order to answer it, I conceptually bracket—or suspend judgment about—the existence of God: such a bracketing is a purely—for-the-sake-of-argument bracketing. When I bracket the existence of God, I am of course not asserting that God does not exist: I am just asking whether or not, God's existence having been bracketed, contradictories would still qualify as contradictory. There is a point to my question: I ask it with a view to bringing out the "primum distinctivum" of contradictoriness—is it to be found in the divine intellect, or in the formal status of quiddities? (See *Ordinatio*, I, d. 11, q. 2, ed. Vaticana, V, n. 60: "Movetur enim quaestio [...]".) It turns out that it is to be found in the latter: Scotus makes, albeit only implicitly, precisely this claim in *Ordinatio*, I, d. 36, q. u., ed. Vaticana, VI, n. 60 (see p. 7 above).[6]

Now, as concerns α) above, it will be objected that, if God did not exist, there would be no contradictories. The objection is worthless, since it fails to distinguish between, on the one hand, a formally correct account of what the contradictoriness of contradictories consists in, and, on the other, a non-vacuous ascription of contradictoriness to mutually contradictory e.g. quiddities. I can give a formally correct

---

aliquid potest esse causa alicuius in essendo, et tamen illud aliud per nullam aliam causam recipit; imo si esset incausatum, reciperet: sicut Deus est causa trianguli in essendo, tamen si triangulus esset incausatus, seipso haberet tres, etc." (*Ordinatio*, IV, d. 49, q. 2, ed. Vaticana, XIV, n. 14). Hiquaeus' commentary on the passage just cited is extremely interesting: "[...] non sequitur ex eo quod unum sit causa alicuius in essendo, sit etiam causa, ex qua tertium convenit causato, v.g. [...] causa albedinis non est ratio, per quam albedini conveniat esse disgregativum visus; neque causa trianguli est causa, per quam formaliter conveniat triangulo habere tres, quia etiamsi esset improductum, haberet tres angulos. [...] quodlibet est primum, et incausabile in suo ordine, licet in diverso ordine sit causatum, forma causatur ab efficiente [...] tamen non ideo [forma] causat in suo genere, quia sic [scil ab efficiente] causata; sed causat primo, et ex propria ratione formali, et sic est prima, quia [est prima] etiamsi non sic [scil ab efficiente] causaretur." (Antonius Hiquaeus (1581–1641), *Quaestiones in librum IV Sententiarum*, ed. L. Vivès, XXI, Paris 1892, p. 25).

[6] The strategy of bracketing is employed by Scotus in (among others) the following passage: "[...] si circumscribatur ab homine animalitas—quod tamen includit incompossibilia—et quaeratur an hoc circumscripto possit homo distingui ab asino, videtur quod determinate posse responderi quod sic, quia non per animalitatem conveniebat homini sic distingui, sed per rationalitatem" (*Ordinatio*, I, d. 11, q. 2, ed. Vaticana, V, n. 30).

account of what it is for mutually contradictory quiddities to be mutually contradictory without having to assume that quiddities have an ontological status, like this: for two mutually contradictory quiddities to be mutually contradictory is for them, jointly taken, to involve a flouting of PC. This is a strictly formal claim: its formal correctness depends in no way on whether or not quiddities have an ontological status. An ascription of contradictoriness to two mutually contradictory quiddities is, however, quite a different story: such an ascription is non-vacuous iff quiddities have an ontological status. The non-vacuousness of an ascription does not, plainly, guarantee its formal correctness: for it may well be a non-vacuous misascription.

If I may summarize. I ascribe to Scotus (and so does Poncius) the following views:

I) The formal origin of that which is (im)possible "ex se formaliter", or of the (in)compossibility of that which is (in)compossible "ex se formaliter", is not to be found in the divine intellect, but in the formal status of quiddities.

II) The divine intellect is not the efficient cause of the possibility of the possible, nor is it the efficient cause of the compossibility of compossible quiddities, nor, finally, is it the efficient cause of the formal status of quiddities.

III) The divine intellect formally produces neither quiddities and their formal status, nor the (im)possibility of the (im)possible "ex se formaliter", nor the (in)compossibility of mutually (in)compossible quiddities.

IV) Each quiddity is the (kind of) quiddity it is in virtue, and only in virtue, of its formal status, and its formal status is the formal status it is "ex se formaliter".

V) The divine intellect is the efficient cause of the ontological status of *possibilia* as well as of quiddities.

I point out in passing that III) and V), taken together, yield C) on p. 3 above.

I proceed now to discuss Poncius' own conception of the possible and his interpretation of Scotus' conception of it.

4 "Esse diminutum"

What is the "esse diminutum" Poncius talks about when he says that "[...] esse, quod habent creaturae ab aeterno, est esse quoddam diminutum" (*PMSCI*, p. 903b)? Is it an ontological status (i.e., the "esse diminutum" of a *possibile*)? A modal status (i.e., possibility)? A formal status (i.e., the formal status of a nature)? In the course of explaining what kind of "esse diminutum" (possible) creatures have "ab aeterno", Poncius remarks, first, that "Ideo Petrus non repugnat, et chymera repugnat, quia Petrus habet naturam distinctae rationis a natura chymerae; ergo non repugnantia Petri praesupponit naturam ipsius" (*PMSCI*, p. 903a: the view that "non repugnantia Petri praesupponit naturam ipsius" is a restatement of Scotus' view that "[...] nihil repugnat alicui nisi quia ipsum est ipsum", *Rep.* IA, fol. 42v); second, that "[...] non possunt creaturae cognosci ut distinctae rationis, per hoc quod cognoscatur non repugnantia eorum; ergo debent habere aliquod aliud esse praeter non repugnantiam" (*PMSCI*, p. 903a-b). The two passages just cited clearly suggest that the "esse diminutum" Poncius alludes cannot be an ontological status: witness, also, his contention that "[...] non potest esse dubium, quin quaelibet creatura fuerit ab aeterno possibilis" (*PMSCI*, p. 902b: i.e., "[...] non potest esse dubium, quin quaelibet creatura habuerit ab aeterno esse possibile").

In view of the passages just cited, I suggest that the "esse diminutum" Poncius talks about is both a modal, and a formal, status: not an ontological status. True, Poncius often makes reference to "creatures" (and hence to possible creatures: *possibilia*): it might accordingly be thought that the "esse" *possibilia* have "ab aeterno" is an ontological status. It is not: explicit reference to *possibilia* is in fact implicit reference either to their modal status (i.e., their being possible: possibility), or to the formal status of natures. We must distinguish here between *possibilia* and their ontological status, on the one hand, and the *possibility* of *possibilia*, and the formal status of natures, on the other. The distinction I have just drawn involves three different kinds of (in)dependence: ontological (in)dependence; modal (in)dependence; and formal (in)dependence. For example: to say, of the possibility of a *possibile*, that it is independent of the divine intellect is not at all to say- nor is it to imply, nor is it to suggest- that that *possibile* is itself independent of the divine intellect. This is because, in the case of possibility, the independence at play is strictly formal, whereas in the case of *possibilia* it is purely ontological, and neither implies the other. Similarly, the claim that *possibilia* do not acquire their "esse [...] possibile per operationem intellectus divini" (*PMSCI*, p. 904a) is really a claim about the possibility of *possibilia*: it means, first, that the possibility of *possibilia* is formally independent of the divine intellect (Scotus' view, it will be remembered); and, second, that the formal status of natures is also so independent.

The "esse quoddam diminutum" Poncius talks about, then, is a modal (as well as a formal) status. Of that "esse diminutum", Poncius says that "non producitur per actum divini intellectus" (*PMSCI*, p. 903b). But "producitur" is ambiguous. We may be dealing with a formal production: or we may be dealing with an ontological production. Most of the arguments Poncius puts forth in favor of the view that the "esse diminutum" "non producitur per actum divini intellectus" involve—implicitly at least—the possibility of *possibilia*, and the formal status of natures, not *possibilia* themselves. This suggests that a formal, not an ontological, production is at play here. More precisely, it suggests that that which is not (formally) produced "through an act of the divine intellect" is the possibility of *possibilia* and the formal status of natures: the possible is possible "ex se formaliter", and each nature has the formal status it has "ex se formaliter".

I should now like briefly to discuss some of the claims Poncius makes regarding (the notion of) formal, and (that of) principiative, independence. I will then attempt to show that Poncius' contention that "[...] esse diminutum non producitur per actum divini intellectus" and Scotus' contention that "[...] hoc est hoc [...], et hoc quocumque intellectu concipiente" are but two sides of one and the same coin.

5 "A se"

Consider the following passages:

> [1] "[...] illud esse diminutum [scil. esse possibile] creaturarum e[s]t independens a Deo, quantum ad hoc quod [...] compet[i]t ipsis *a seipsis* formaliter, sine cooperatione ullius causae efficientis (*PMCSI*, p. 905a—italics mine).
>
> [2] [...] esse possibile convenit ipsis remote et fundamentaliter *a seipsis* sine sine eo quod habeant illud esse a divino intellectu aut voluntate (*PMCSI*, p. 905a-b—italics mine).

In [1]-[2], Poncius puts forth a view to the effect that the possible qualifies as possible "a se" (i.e., "a se formaliter"), not "ab alio principiative". Its not qualifying as possible "ab alio principiative" implies in no way, of course, that it qualifies as possible "a se principiative". Mastrius raises the following objection to Poncius' view: the first part of it is as follows:

> [...] Doctor [...] ait [...] per particulam *ex se* explicatur habitudo causae formalis, et per particulam *a se*, vel *ab alio*, denotatur habitudo cause efficientis; quare cum Poncius [...] asserat creaturas ab aeterno habere esse possibile non tantum ex seipsis, ut loquitur Scotus, sed etiam a seipsis, innuit consequenter habere illud ex seipsis non tantum in genere causae formalis, sed etiam in genere causae efficientis. (*DT* I, pp. 120b-121a)

Mastrius' objection relies on the mistaken assumption that Poncius gives "a se" the same sense Mastrius gives it. He does not: witness [1]-[2] above, where "sine cooperatione ullius causae efficientis" in [1], and "sine eo quod habeant illud esse [...]" in [2], are clearly epexegetic. They are meant by Poncius to explain how "a seipsis" is to be interpreted: for the possible to be possible "a seipso formaliter" is for it to be possible ("formaliter") "sine cooperatione ullius causae efficientis" (see also above, footnote 8). So: the divine intellect is not the efficient cause—the "principium effectivum"—of the possibility of the possible. With obvious and minor modifications, the same analysis applies to [2] above.[7]

I turn now to a discussion of the second part of Mastrius' objection:

---

[7] In the course of answering the question whether or not 'unbegotten' ('ingenitum') is a constitutive principle of the divine essence, Scotus considers a view to the effect that "[...] *ingenitum* non importat negationem simpliciter, sed aliquid pertinens ad dignitatem, quia hoc quod est habere, esse a se" (*Quodlibet*, q. 4,, ed. Wadding, XII, n. 8). He objects: "[...] cum dicitur a se, aut intelligitur quod haec praepositio *a* importat circumstantiam causae, vel principii positivi, et statim patet contradictio, [...]; aut intelligitur negative tantum, quia non habet aliquid pro principio, vel causa" (*Quodlibet*, q. 4, ed. Wadding, XII, n. 8). Now "a", in the context of the claim that a *possibile p* is possible "a se", has the second of the two senses Scotus describes. It means, "non habet aliquid pro principio, vel causa": which means, in turn, that *p* is possible "sine cooperatione ullius causae efficientis"; which means, finally, that its possibility is independent of the divine intellect. Cf. also, "[...] prima persona dicitur 'a se' non positive, quasi principiative habeat esse a se (quia nihil principiat se ipsum [...]), sed dicitur esse ex se formaliter et quasi negative ex alio, quia non principiatur ex alio" (*Rep.* IA, fol. 72v); "[...] prima persona dicitur esse a se, non quia positive habeat esse a se principiative [...] sed quia negative principiatur, id est, non est ab alio, ut sic accipiatur negative esse a se, quia non ab alio, non dico formaliter, sed principiative, quia principiative non habet esse ab alio, formaliter vero a se habet esse" (*Reportata Parisiensia*, I, d. 28, q. 2, ed. Wadding, XI, n. 10).

> [...] secundum [scil the view that 'res habent possibilitatem a seipsis'] est prorsus falsum, quia significat, quod illam intrinsecam possibilitatem habeant a seipsis, et non communicatam ab altero, sed independenter a quocumque extrinseco. (*DT* I, p. 120b)

But this is a questionable claim.[8] The view that the possible is possible "independenter a quocumque extrinseco" was in fact Scotus' own view: witness for instance Scotus' contention, in *Rep.* IA, fol. 91v, that "[album et nigrum: in fact: their quiddities] [...] ex se formaliter talia [scil. incompossibilia] sunt, circumscripto quocumque alio quod est extra illa" (formal independence). This obviously applies to any pair of mutually (in)compossible quiddities: to assert that two mutually (in)compossible quiddities qualify as such "independenter a quocumque extrinseco" (formal independence) is to assert that they qualify as such "circumscripto quocumque alio quod est extra illa"—and conversely ("extrinsecum" and "extra" are centrally bound up with one another).[9] It will be objected that Scotus also claims, in *Rep.* IA, fol. 91v, that "[...] impossibilitas huius 'album est nigrum' [...] reducitur ad intellectum divinum, a quo sunt principiative partes formales in esse possibili ipsius impossibilis". This proves nothing: the reduction at play here is a principiative, not a formal, reduction. And so is the production "in esse possibili" of the "partes formales": what is being produced in "esse possibili" is not the formal, but the ontological, status of those "parts"—which is hardly an objection to the claim, made a few lines back, that any two mutually incompossible quiddities are mutually incompossible "circumscripto quocumque alio quod est extra illa" (the divine intellect is "extrinsic" to the formal status of quiddities: so is its principiative activity).

---

[8] So is Mastrius' claim that (according to Scotus) "[...] si nullus daretur intellectus potens judicare aliqua extrema esse inter se compossibilia vel incompossibilia, nulla daretur possibilitas, vel impossibilitas logica, quia haec attenditur in ordine ad intellectum sic vel sic judicantem, nulla etiam daretur, si non darentur termini invicem ex suis rationibus formalibus convenientes, vel repugnantes [...]" (*DLM*, p. 72, n. 38). Mastrius relies here on Scotus' contention that "[...] potentia logica [...] est modus compositionis factus ab intellectu, causatus ex habitudine terminorum, scilicet quia non repugnant" (*Quaestiones super libros Metaphysicorum*, IX, q. *1–2*, ed. Etzkorn, n. 18). The claim that a "modus compositionis" is "factus ab intellectu" and that it is (also) "causatus ex habitudine terminorum" does not quite mean what Mastrius takes it to mean. Reference to an intellect is intended by Scotus to guarantee that ascriptions of (im)possibility are non-vacuous, i.e., that they are ascriptions of (im)possibility to something: it does not at all mean that, without an intellect, the possible would not be possible (it would still be possible, but it would be devoid of an ontological status). Talk of a "modus compositionis" *qua* "causatus", in turn, is intended by Scotus to guarantee that ascriptions of (im)possibility are formally correct. The distinction between "factus" and "causatus" is centrally bound up with the distinction between a possible's being possible "principiative per intellectum divinum" and the possible's being possible "ex se formaliter".

[9] Krisper' objection to Poncius fares no better than Mastrius': "Esse possibile creaturarum nequit esse omnino independens a Deo, ita ut nec fiat per intellectum divinum effective, sed ut competat ipsis a seipsis formaliter sine cooperatione ullius causae efficientis" (*TSS*, p. 149). Nor does Ruerk's: "Possibilitas logica creaturarum extrinsece et principiative, seu in genere causae efficientis, desumitur unice et totaliter ab intellectum divinum. Est contra Herreram et Poncium" (*CTS*, p. 384). Two minor points here. First, if, in Krisper's passage just cited, we replace "nequit esse" with "est", the resulting view is as clear (and correct) an interpretation of Scotus' conception of the possible as one could possibly ask for. Second, Ruerk's contention is not just "contra Herreram et Poncium": it is "contra Scotum" as well.

According to Poncius, then, the possible is possible both "ex se formaliter" *and* "a se formaliter". To hold that it is possible "ex se formaliter" is specifically to hold that

(a) the divine intellect is not the "principium formale" of the possibility of the possible (but the possible has a "principium formale" of its possibility: viz., the formal status of the relevant quiddities).

To hold, on the other hand, that it is possible "a se formaliter" is specifically to hold that

(b) the divine intellect is not the "principium effectivum" of the possibility of the possible.

(In fact, there is no "principium effectivum" of the possibility of the possible: the negative claim that there is no such "principium" follows from the affirmative claim that the possible is possible "independenter a quocumque extrinseco" and, hence, from the—related—affirmative claim that the possible is possible "ex se formaliter". There is a point to the negative claim: it is intended to underline the fact that the divine intellect has no role whatever to play when it comes to the question of the grounds of the possibility of the possible.)

I turn now to an analysis of Poncius' contention that

DP1) "Illud esse diminutum non producitur per actum divini intellectus" (*PMSCI*, p. 903b).

One of the arguments he gives in favour of DP1) is that

DP2) "[...] nisi creaturae haberent esse aliquod possibile a seipsis, independenter ab actu intellectus divini, non esset ratio quare homo potius esset possibilis quam chymera" (*PMSCI*, p. 904a: the argument is restated as follows in *PMSCI*, p. 907a: "[...] quaero a quo [lapis] habeat principiative quod dicat non repugnantiam; si dicas quod ab intellectu: cur similiter chymera non haberet non repugnantiam? Si dicas quod ex ipsamet natura sua intrinseca et non ex intellectu: ergo habetur intentum".)

A few points are worthy of notice here. First, "independenter" is ambiguous: the independence at play here may be principiative independence (to say that the possibility of the possible is principiatively independent of the divine intellect is to say that the latter is not the efficient cause of the former); or it may be modal/formal independence. Second, the "esse" Poncius alludes to is an "esse possibile". Third, "creaturae" is shorthand for "creaturae possibiles" (in a word: *possibilia*). Fourth, "[...] haberent esse aliquod" must be taken to mean, "[...] essent possibiles [...]". Fifth, "nisi creaturae haberent aliquod esse possibile, independenter ab intellectu divino" is really a claim, not about *possibilia*, but about the (formal grounds of) the *possibility* of *possibilia*. Sixth, to contend that a *possibile* is possible "a seipso" is not at all to imply that its ontological status—an "esse diminutum"—is ontologically independent of the divine intellect. The independence at stake here is modal *and* principiative, not ontological: to assert that the possibility of a *possibile* is modally (*and* principiatively) independent of the divine intellect is to contend that the latter is neither its "principium effectivum" nor its "principium formale". Seventh, "independenter ab actu intellectus divini" is epexegetic: to say that a *possibile* is possible "a seipso" is just to say that it does not have an efficient cause of its possibility (note that, from the fact that a *possibile* is possible "a seipso", it follows in no way that it is, therefore, possible "a se effective").

The argument now. A number of points should be made here. First, two competing views are implicitly at stake in DP2): the view, on the one hand, that

> (**A**) the (im)possibility of (*im*)*possibilia* entirely depends—formally or, as the case may be, principiatively—on the divine intellect;

the view, on the other hand, that

> (**B**) the (im)possibility of (*im*)*possibilia* is formally grounded in the intrinsic nature of (*im*)*possibilia* themselves.

Next, Poncius (implicitly) assumes that the "ratio quare homo potius esset possibilis quam chymera" is to be found either in the divine intellect, or in the (nature of the) possible itself. Next, Poncius (implicitly) contends that, if (**A**) is true, then "[…] non esset ratio quare homo […]". This can be generalized: if (**A**) is true, there is no reason why the divine intellect should make the possible possible rather than impossible, and the impossible, impossible rather than possible. We cannot, of course, appeal to something *other* than the divine intellect with a view to providing an explanation of why the divine intellect makes the possible possible rather than impossible: for, according to (**A**), the (im)possibility of the (im)possible entirely depends on God's intellect.

But (Poncius implicitly assumes) the claim that there is no reason why a man, and not a chymera, is possible is absurd (or, at any rate, wholly implausible), since there *is* a reason why this is so—and the reason is that a man is possible in virtue of its intrinsic nature, i.e., "ex se formaliter". The "ratio", then,"quare homo potius esset possibilis quam chymera" is to be found, not in the divine intellect but, in the intrinsic nature of the possible itself: this is precisely the point of "nisi creaturae haberent esse aliquod possibile a seipsis, […], non esset ratio quare […]"—"nisi" signals that we have a sort of invited inference here (*possibilia* have an "esse aliquod possibile a seipsis": we are invited to infer that this is the reason why they qualify as possible rather than impossible). Now the view that *possibilia* have an "esse aliquod possibile a seipsis" is an immediate consequence of the view that their possibility is not produced (principiatively) by the divine intellect: which is the very view DP1) puts forth. There is nothing circular here. The truth of DP1) is arrived at by showing that, and why, what looks like a plausible alternative to (**B**) (viz. (**A**) above) is in fact not plausible at all: it has an absurd consequence, and should accordingly be rejected. Poncius' (implicit) conclusion is that the divine intellect is not the "principium effectivum" of the possibility of *possibilia*.

It is not its "principium formale" either: according to Poncius, "[…] praesupponitur possibilitas aliqua, saltem remota […], in homine ante intellectionem divina" (*PMSCI*, p. 904a). This is of course true of a(ny) given *possibile*: not that it is ontologically independent of the divine intellect; rather, that its modal status is (formally) independent of it (I take "ante" to mean "independently of": the independence at play here is purely formal). There is textual evidence for this: witness the following passage:

DP3) "[...] si non esset Deus, non esset creatura possibilis [scil. proxime], et tamen Deus non dat creaturae esse possibile [scil. remote], saltem fundamentaliter, sed habet illud esse a seipsa formaliter [...]" (*PMSCI*, p. 904b).

Two remarks here. First, the qualification "formaliter" is meant to suggest that the possibility of a *possibile* is formally independent of the divine intellect. Second, why "a seipsa" (rather than "ex seipsa")? Because "a seipsa" rules out the divine intellect as the "principium effectivum" of the possibility of the possible: "formaliter" is actually shorthand for "ex seipsa formaliter", which rules out God's intellect as the "principium formale" of the possibility of the possible. So: a *possibile* is possible both "a seipso formaliter" *and* "ex seipso formaliter".

I have said earlier, it will be remembered, that, in DP1), Poncius puts forth essentially the same view Scotus puts forth in S) (see p. 7 above). The claim that the possibility of the possible is not produced by the divine intellect, and the claim that "[...] quare homini non repugnat [...] est, quia hoc est hoc [...], et hoc quocumque intellectu concipiente" are both independence claims (and so is "[...] praesupponitur possibilitas aliqua [...] ante intellectionem divinam", cited above): it follows from all of them that the possibility of the possible is (formally) independent of the divine intellect. Further, "[...] et hoc quocumque intellectu concipiente" is plausibly taken to make the same point Poncius makes when he contends that

[...] nihil absurdius videri posset quam [creaturae] ideo non repugnant quia intelliguntur. [...] Et sane ex terminis apparet ridiculum quod homo habeat etiam principiative esse possibile remotum [...] ex eo quod intelligatur a Deo: quid enim facit ad hoc intellectus? (*PSMCI*, p. 907a: "ad hoc" is obviously shorthand for "ad hoc quod habeat esse possibile remotum")

The point is this: the formal origin of the possibility of the possible cannot be traced back to the divine intellect, since *possibilia* are possible *independently* of it. A *possibile* is not possible because God's intellect conceives of it as possible. It is the other way around: God's intellect conceives of it as possible because it is ("already") possible "ex se formaliter". This being so, it is clear that the question "quid enim facit ad hoc intellectus?" (which concerns the modal, not the ontological, status of *possibilia*) can only have one answer, viz., nothing whatsoever[10]: how can the divine intellect formally ground the possibility of that which qualifies as possible

[10] Mastrius disagrees: "Respondeo [what follows is Mastrius' reply to the question, 'Quid enim facit ad hoc intellectus?'], eam possibilitatem vel non repugnantiam [creatura] habere partim ex seipsa et partim ab intellectu divino, ex seipsa quidem in genere causae formalis, ab intellectu vero divino principiative, ac in genere causae efficientis, ne ipsi concedatur aliquod esse reale, etiam diminutum, a Deo prorsus independens, ac impartecipatum, neque hoc ex terminis apparet ridiculum, sed omnino necessarium, ac evidenter deductum ex essentialissima dependentia creaturarum a Deo in quocumque esse reali quantumvis minimo; et cum potentia logica sit quidam modus compositionis factae ab intellectu inter terminos invicem non repugnantes ex suis rationibus formalibus, non est impertinens intellectus ad hanc potentiam constituendam in rebus" (*DT* I, p. 123, n. 63). But this is hardly a reply to Poncius' question, for the following reason: "non est impertinens intellectus [...]" is ambiguous. It may be shorthand for "impertinens principiative", or for "impertinens formaliter". Let us suppose it is the former: then to claim that the divine intellect is not "impertinens principiative"

independently of it? This also explains why the claim is "absurd" that the possible is possible because it is though of ('intelligitur") by the divine intellect: it is "absurd" because it reverses the correct order of explanatory dependencies (the possibility of the possible, not its being thought of by the divine intellect, is explanatorily "first"). Scotus puts forth a very similar view: to hold that "[...] quare homini non repugnat [...] est, quia hoc est hoc [...], et hoc quocumque intellectu concipiente" is to hold—or, at least, it is strongly to suggest—that the possibility of the possible, and quiddities and their formal status are (formally) independent of the divine intellect.

6 Independence

I conclude. As I have said earlier, a distinction must be drawn between the ontological, the principiative, and the formal independence (of the possibility of the possible with respect to God's intellect), and between two quite different senses of "principiative": the claim that the possible is (also) possible "principiative per intellectum divinum" is (I have remarked on pp. 3–4 above) ambiguous. It may be taken to mean that.

(a) God's intellect is the efficient cause of the ontological status of *possibilia*;

or it may be taken to mean that

(b) God's intellect is the efficient cause of the possibility of *possibilia*.

I have argued above that Scotus accepts (a), and rejects (b). So, we have seen, does Poncius. Further, both Scotus and Poncius hold (we have also seen) that the possibility of the possible is formally independent of the divine intellect. The question now arises: is the ontological status—i.e., the "esse diminutum"—of *possibilia* and possible essences ontologically independent of God? Poncius appears to maintain that it is: if this is his view (it is by no means clear that it is), this is where he and Scotus part ways. For, according to Scotus, *possibilia* and possible natures acquire the ontological status they have—if they have one—from the divine intellect: there is no question, therefore, of their being ontologically independent of God. They are formally independent of the divine intellect as concerns their modal status, but ontologically dependent on it as concerns their ontological status.

What kind of view should we ascribe to Poncius? We may not ascribe to him Scotus' view (not, anyway, as regards the ontological dependence of *possibilia*): the textual evidence against such an ascription is conclusive (see *PMSCI*, p. 904a). Should we ascribe to him the view that the ontological status of *possibilia* and

---

"ad hanc potentiam constituendam [...]" is just to claim that God's intellect is the efficient cause of the ontological status of the possibility of the possible (on Mastrius' view, "principiative" and "effective" are equivalent). This is true, but devoid of any modal interest. Let us suppose, then, that "impertinens" is shorthand for "impertinens formaliter": in which case the resulting conception of the possibility of the possible is virtually self-refuting, since (as Mastrius acknowledges) it is not the divine intellect, but the intelligible content of the possible that provides the formal foundation of the latter's possibility. Poncius' question still remains to be answered.

possible essences is ontologically independent of God? Unclear: there is little or no textual evidence in favor of such an ascription. Let us look at the relevant passages: I consider, first, Poncius' claim (already discussed above) that "Illud esse diminutum non producitur per actum divini intellectus" (*PMSCI*, p. 903b). As I have said earlier, talk of "esse diminutum" is talk of a formal/ modal status, not of an ontological status. If, however, we were to take it to be talk *also* of an ontological status, and hence take Poncius to be saying that the ontological status of *possibilia* and possible essences is not "produced" by the divine intellect, we would have no choice but to ascribe to him the view that the ontological status of *possibilia* and possible essences is ontologically independent of God: can we so take it?

I doubt it. In all the arguments Poncius puts forth in favor of that claim, the crucial role is played by the possibility of *possibilia*, not by *possibilia*: which strongly suggests that the "esse diminutum" he has in mind is a formal/modal, not an ontological, status[11] (if it is taken to be an ontological status, it is difficult to see how a battery of arguments which are meant to show that the possibility of *possibilia* is formally independent of the divine intellect could also show that the ontological status of essences is ontologically independent of God: formal independence is no guarantee of ontological independence). I turn now to a discussion of the following passages:

> A) "Respondeo distinguendo maiorem [scil 'De creaturae ratione est quod sit ab alio']: quod sit ab alio quantum ad hoc, quod non possit habere esse reale simpliciter, nisi ab alio, concedo maiorem; quod non possit habere illud esse diminutum et possibile nisi ab alio, nego maiorem." (*PMSCI*, p. 904a—I take "et" to be epexegetic here)
>
> B) "Nulla est ratio cur ad esse creaturae requiratur maior dependentia, quam talis ratione cuius non posset habere esse reale simpliciter absque Deo: talis autem dependentia esset in creatura possibili, quamvis a Deo non acciperet esse suum possibile per operationem intellectus divini." (*PMSCI*, p. 904a; see also *PMSCI*, p. 905a: "[...] licet totum esse creaturae sit participatum, id debet intelligi quantum ad hoc [...] quod est talis naturae, ut non possit habere realem existentiam absque dependentia a Deo tanquam a causa efficienti.")
>
> Γ) "[...] essentia est independens quoad esse possibile, dependens vero quoad esse reale simpliciter." (*PMSCI*, p. 905b)
>
> Δ) "[...] creatura [possibilis] [...] potest habere independentiam quoad esse essentiae, seu possibile, quamvis haberet dependentiam quoad esse reale existentiae" (*PMSCI*, p. 906b; cf. also, "ego vero non pono nisi esse diminutum, seu possibile", *PMSCI*, p. 906a)

In A)–Δ), Poncius puts forth the view that the dependence of *possibilia* on God is an ontological dependence of sorts: *possibilia* ontologically depend on God "quoad esse reale simpliciter", or "quoad esse reale existentiae". Their dependence on God comes of this, that—without God—they could never acquire an "esse simpliciter": which is indeed a kind of ontological dependence. It is, in fact, the only kind of ontological dependence Poncius explicitly ascribes to *possibilia* (see e.g. B above). Does this imply that *possibilia*, insofar as their ontological status is concerned, are

---

[11] Note that Poncius' characterization of the "esse, quod habent creaturae ab aeterno [...]" (*PMSCI*, p. 903a) could hardly be regarded as the characterization of an ontological status: and that, in most of the passages in which he ostensibly talks about *possibilia*, Poncius is really talking about the possibility of *possibilia*.

therefore independent of God? It does not: it just leaves the possibility open that their ontological status should be so independent. But this proves little or nothing: the question whether or not the ontological status of *possibilia* is, according to Poncius, independent of God remains open.

I proceed now to discuss Γ) and Δ) above. I remark, first, that, whereas "[essentia est] dependens [...] quoad esse reale simpliciter" is an ontological, not a formal claim, "essentia est independens quoad esse possibile" is a formal, not an ontological, claim: the latter claim means no more than that the possibility of essences is formally independent of God's intellect. I remark, second, that talk of "esse essentiae, seu possibile", just like talk of "esse diminutum, seu possibile" (see Δ) above), is talk of a formal/ modal status, not of an ontological status: what Poncius means to say here is that the possibility of *possibilia* is formally independent of the divine intellect. He puts forth the same view in A), where the distinction between ontological and formal (in)dependence plays a crucial role[12]: his "distingu[end]o maiorem" is a "distinguo" between ontological and formal (in)dependence (when he concedes the major, he has in mind ontological dependence; when he denies it, formal independence). Now, in particular, to deny that a *possibile* has "ab alio" the "esse diminutum et possibile" it has, is to assert that possession, by it, of an "esse diminutum et possibile"—in short: a modal status—does not formally depend on the "aliud" (i.e., the divine intellect). Nor, we have seen, does it principiatively depend on it.

The textual evidence in favor of the claim that claim Poncius reifies *possibilia* (possible essences) is, accordingly, virtually non-existent. The textual evidence in favor of the claim that, on his view, the possibility of *possibilia* is absolutely independent—formally speaking—of the divine intellect is, on the other hand, conclusive: but absolute independence is not—nor does it imply (although it may suggest)—reification.

[12] See also the following passages, where a similar distinction is drawn by Poncius: "[...] essentia et existentia dependent efficienter a causa efficiente, licet formaliter intrinsece non dependeant" (*PMSCI*, p. 901b); "[...] dicuntur res quoad essentiam non dependere a causa efficiente" (*PMSCI*, p. 901b). Considered together, these two passages suggest that the ontological status of essences (ontologically) depends on an efficient cause, whereas their formal status is formally independent of it (and of the divine intellect as well).

Leibniz

Possibility vs Iterativity: Leibniz and Aristotle on the Infinite

Monica Ugaglia

1 Introduction

I will show how Leibniz's notion of syncategorematic infinite can be traced back to Aristotle's notion of potential infinite, which is conceived and described by Aristotle in an operational way.

At the beginning of *Physics* III (III.1 200b17-18) while he is introducing the subject, Aristotle says that the infinite manifests itself primarily in the infinite divisibility of the continuum. Take a segment, or a material line, and divide it into two parts; then divide every part into two parts, and so on, iterating the action of dividing. Since the procedure of division is potentially infinite—every action of dividing can be repeated, over and over again, without limit—the segment is said to be potentially infinite too. In Sect. 2 I will focus on this constructive meaning of the adjective "potential". Hereinafter, I will refer to this kind of potentiality, which is distinctive of Aristotle's infinite, as *iterativity* (or A-*potentiality*): my purpose in doing so will be both to emphasize the dependence of Aristotle's notion of infinite on that of iterative procedure and to distinguish it from the different kind of potentiality that Leibniz will attach to the notion of infinite.

Indeed, although he recognizes the iterative nature of Aristotle's infinite, and starts from the same procedure of division of the continuum, Leibniz does not think that iterativity is the ultimate source of the potentiality of the infinite. To Leibniz, a segment is potentially infinite primarily because there are infinitely many ways (or possibilities: I call this L-*potentiality*) of dividing it—according to the infinitely many ways of 'locating' the cuts—and only secondarily because the division is potentially infinite (iterative, A-*potentially*) [Sect. 3].

It is for this reason that in physics, where the location of the cuts is fixed, Leibniz can conceive the seemingly contradictory notion of what has been called a "syncategorematic actual" infinite [Sect. 4].

M. Ugaglia (✉)
Independent Scholar, Montecassiano, Italy

F. Ademollo et al. (eds.), *Thinking and Calculating*, Logic, Epistemology, and the Unity of Science 54, https://doi.org/10.1007/978-3-030-97303-2_12

2 Potentiality as Iterativity: Aristotle's Infinite

As is well known, Aristotle denies any form of infinite in act; the cosmos, that is to say the sum of all existing things, is finite in magnitude and in multiplicity: it is a sphere of fixed finite diameter which contains an actually finite number of objects. So, the infinite cannot exist as a thing, or as the attribute of a thing[1]; instead, it can exist "in virtue of" things:

> in general, the infinite is in virtue of another and another thing being taken, over and over again (*aei*); and what is taken is finite, over and over again (*aei*); but it is a different thing, over and over again (*aei*) (*Ph.* III 6, 206a27-29)

Clearly this is not a conventional Aristotelian definition, but an operational characterization. It does not describe an object but an action—'taking' something—and the iteration of the action, signaled by the adverb *aei*. For each thing we take (*first step*), there is another thing to take (*second step*), and another to take beyond that (*third step*), and another beyond, and so on, over and over again (*aei, next steps*).

The main example of potential infinite, to which Aristotle refers any other manifestation of the infinite, is the continuum, characterized as "infinitely divisible" or, equivalently, as "divisible into what is always (*aei*) divisible".[2] Take for instance a segment, or a physical line: it is divisible (*first step*), and what one obtains from this division is still divisible (*second step*), and what one obtains from this division is still divisible (*third step*), and so on, over and over again (*aei, next steps*). More generally, any mathematical proof by infinite iteration—reciprocal subtractions, exhaustion, asymptotical properties...—can be used as an example of potential infinity as meant by Aristotle: not surprisingly, he formulates his definition, and his examples, using the same algorithmic scheme and terminology.[3]

Leaving aside more complicated mathematical constructions, all the crucial features of Aristotle's potential infinite clearly emerge in the procedure of division

---

[1] Aristotle shows that the term "infinite" cannot be employed as a subject—"the infinite" *per se* does not exist (*Ph.* III 4, 202b36–203a16; cf. III 5, 204a8–34; III 4, 203a6–16 and III 1 200b27), nor as the attribute of a subject—"an infinite thing" does not exist either (III 4, 203a16-b3; cf. III 5, 204b4–206a7).

[2] Aristotle defines continuity as a relational notion: two parts of a whole are said to be continuous with one another when they are in contact and their ends therefore become one (*Ph.* V 3, 227a11–12; see. *Ph.* V 4, 228a29–30; *Cat.* 6, 4b20–5a14). Accordingly, Aristotle calls continuous a whole composed of parts continuous with one another, that is to say parts that have an end in common (*Ph.* VI 1, 231b5–6; *GC* I 6, 323a3–12; *Metaph.* V 13 1020a7–8). But in order to have an end in common, parts must be congeners (*Ph.* IV 11 220a20–21; *Ph.* V 4, 228a31-b2) so that they must lose their individuality, and form a homeomeric whole, free of inner limits (*Ph.* 5 IV, 212b4–6; *Metaph.* VII 13 1039a3–7 see V 26 1023b33–34; VII 16, 1040b5–8). This leads to Aristotle's peculiar characterization of the continuum as "infinitely divisible" (*Ph.* III 1, 200b18; *Ph.* I 2, 185b10–11; VI 6, 237a33; VI 8, 239a22) or "divisible into what is always (*aei*) divisible" (*Ph.* VI 2, 232b24–25; see IV 12, 220a30; VI 1, 231b15–16; VI 6, 237b21; VIII 5, 257a33–34; *Cael.* I 1, 268a6–7). Indeed, if there are no individual parts, and no actual limits, then there are no actual parts (*GC* I 2, 316a15–16) and the continuum can be divided everywhere.

[3] I have compared mathematical and Aristotelian texts in Ugaglia (2009), showing in particular that the adverb *aei* must be taken in its iterative atemporal meaning, typical of mathematics.

of the continuum. The first among such features is the fact that the attribute "infinite" applies primarily to processes, and only secondarily to objects: the parts into which a segment can be divided are called infinite *only* because the procedure of dividing the segment is infinite. This makes clear that for Aristotle the infinite parts are not the result of the process of dividing: they exist, as potentially infinite, only insofar the iterative process which produces them exists. If one stops the procedure, of course one obtains a partial result—a finite set of finite parts actually divided—but this result has nothing to do with the infinite, regardless of the step at which the iteration has been stopped.

The same holds for numbers, which Aristotle constructs as the "names" for the action of dividing (*Ph.* III.7, 207b1-15). Take again the process of dividing the segment, but now imagine attaching a label to each division, that is, counting the steps. Since each step in the iterative procedure of division implies that it will be followed by another step, whatever the number assigned to an arbitrarily chosen step, it will be followed by another number, and so on, without limit. In this sense, and *only* in this sense, we can speak of an infinite number: the number is infinite not because there is in act an infinite set of things to be numbered (the segment *is not* made up of points that can be counted), but because the continuum is infinitely divisible. As a consequence, such an infinite number is not something in itself, separated from the process that defines it, but exists only in the process, which is to say that it only exists within the possibility for the process to go on[4]:

> But in the direction of more it is always possible to conceive of a larger number, since halvings of a magnitude are infinite. So it is potentially infinite, but not actually infinite; but *the thing taken always exceeds any definite number*. (*Ph.* III 7, 207b10–13)

In more general terms, Aristotle's potential infinite is not separable from the process it qualifies. Indeed, its potentiality ultimately consists, and manifests itself, in the possibility, at any step, of moving on to the next step: divide, and divide, and divide... Since this holds in the continuum, as well as in any other genuine iterative process, we will call this kind of infinite *iterative*, and *iterativity* (or A-*potentiality*) its form of potentiality.[5]

Alternatively, we might call Aristotle's infinite *syncategorematic*, which is in fact the term employed by certain medieval thinkers.[6] In this case it is important to point

[4] Of course, if we decide to stop the process at a given point, we get an actual number, independent of the process and separable from it, but once again this has nothing to do with the infinite. On the dependence of number on the division of the continuum see Wieland (1970, §18).

[5] For a more detailed discussion of Aristotle's iterative notion of infinity see Ugaglia (2018).

[6] The distinction between categorematic and syncategorematic infinite, dating back to the Middle Ages, has a semantic origin: *categoremata* were called the parts of speech which possess a definite meaning of their own (typically nouns, adjectives and verbal forms); *syncategoremata*, or *cosignificantia*, such parts as depend on *categoremata* in order to acquire a definite meaning (e.g. conjunctions and prepositions). A wide debate originated around the notion of *syncategoremata*; the adjective "syncategorematic" was extended to cover also *categoremata*, when employed in some particular sense. In particular, the distinction between categorematic and syncategorematic use of a term became one of the standard ways of resolving *sophismata*. The sentence *infinita sunt finita* was a paradigmatic example and the way of resolving it by recourse to the categorematic/syncategorematic

out that syncategorematicity and potentiality are synonymous *only if* potentiality is meant in the peculiar sense of A-potentiality.

Indeed, as Aristotle himself rightly observes at the beginning of his analysis, this kind of potentiality does not coincide with the notion of potentiality he usually deploys, and opposes to actuality, when he speaks of objects.[7] When the notions of actuality and potentiality are applied to an object, indeed, they qualify two different but not incompatible ways of being. Being potentially presupposes and implies being actually: i.e. a piece of marble is potentially a statue because it could be, and in the right situation will be, actually a statue—and the act is logically and ontologically anterior to the power, i.e. the image of the statue is in the mind of the sculptor before he sculpts the piece of marble.[8] On the contrary, no actuality can be related to potentiality in Aristotle's sense of *iterativity*. For, as I pointed out above, no final result can be obtained by an infinite iterative procedure—i.e. the procedure of dividing the continuum is potentially infinite because it can never reach any final actual result whatsoever—nor can such a result be thought of without contradiction, either before or after the procedure.[9]

Moreover, it is clear that the potentiality of Aristotle's infinite, once it is properly understood in terms of mathematical iterative procedures, has nothing to do with time.[10]

---

opposition made the distinction between a categorematic and a syncategorematic use of the word infinite a very common one. The infinite must be intended as a categorematic attribute, when it refers to a whole, or a totality, and hence to the number which counts its elements, but as a syncategorematic attribute when it does not refer to anything but is employed distributively: to say that the parts of a physical body are syncategorematically infinite simply means that, for every finite number *n*, these things are more than *n*. Given the traditional Aristotelian distinction between actual and potential infinite, meant as philosophical notions, it is not surprising that the logico-grammatical level and the philosophical one were somehow conflated in such a way that people could speak of a categorematic and a syncategorematic infinite *tout court*. This conflation was all the easier since logicians themselves, introducing the grammatical distinction, referred to Aristotle's philosophical discussion of the infinite. For an explicit connection between the infinite used for "a syncategorematic expression, that in itself indicates distribution" (*dictio sincathegorematica importans in se distributionem*), and Aristotle's potential infinite, see Peter of Spain, *Summulae Logicales* XII.36–38 (Peter of Spain (2014), 504–509). The denial of any connection, ascribed to Peter in recent literature, without explicit reference—see for instance Uckelman (2015, 2368), quoting Moore (1990, 51)—is probably based on Duhem's reading of *De exponibilibus*, a treatise which De Rijk excluded from his critical edition of Peter's *Summulae* as inauthentic. See Duhem (1985, 50; 516). In general, on the distinction, and on the identification syncategorematic/potential and categorematic/actual, see for example Kretzmann (1982). On the meaning of syncategorematicity in Leibniz see Antognazza (2015).

[7] "We must not take potentiality here in the same way as that in which, if it is possible for this to be a statue, it actually will be a statue, and suppose that there is an infinite which will be in actuality" (*Ph.* III 6, 206a18-21).

[8] *Metaph.* IX 8, *passim*; see also *Ph.* II 1, 193b7-8.

[9] In my notation, the notion of A-*actuality* is meaningless, by definition.

[10] Sometimes, when contrasted with Leibniz's infinite, Aristotle's potential infinite is erroneously described in term of unending temporal processes. See for instance Arthur (2019) p.107: "So long as there is a law according to which further elements are generated (as in the case of the Euclidean generation of primes), an infinity of elements can be understood without any commitment to either an

Iterativity and Possibility

As we have seen, for Aristotle the potentiality of the infinite coincides with the possibility, at any step of a procedure, of moving on to the next step. I have called this kind of potentiality *iterativity*, or A-*potentiality*, in order to suggest that this is not the only form in which potentiality crops up during the division of the continuum. Indeed, the more familiar notion of potentiality—namely, the potentiality which preludes to actuality—is involved in the procedure as well.

Take again a segment: not only can it be infinitely divided, but the procedure of division can be carried out in infinitely many ways, according to the potentially infinite ways we have of "locating" the successive cuts. Begin to divide the segment. "Where?", we could ask. It does not matter, because a continuum is divisible everywhere. So, let's start by dividing it in the middle. Now divide it again. Where? Again, it does not matter, because a continuum is divisible everywhere. So, let's divide it in the middle, again. And so on. This way of always cutting the segment into two equal parts—the so-called dichotomic division, or simply dichotomy—represents for Aristotle the paradigmatic way of dividing the continuum. Of course, however, it is not the only *possible* one. Instead of dividing the segment in the middle (1/2), and then in the middle of the middle ($1/4 = 1/2^2$), and then in the middle of the middle of the middle ($1/8 = 1/2^3$), and so on ($1/2^4, \ldots 1/2^n, 1/2^{n+1}, \ldots$), we could opt for dividing it at one third of its length (1/3) and then at one third of one third ($1/9 = 1/3^2$), and so on ($1/3^3, \ldots 1/3^n, 1/3^{n+1}, \ldots$), or in infinitely many other ways.[11]

So there is another sense, a non-iterative one, in which we could say that the divisions (or rather the laws of division) of the segment are potentially infinite. Before we start dividing it is possible to choose among an infinite number of possible cuts, but only one of them—the one which is actually chosen—will be actualized.

This second kind of potentiality—which I will call instead *possibility*, or L-*potentiality*—has nothing to do with Aristotle's *iterative* infinite. Moreover, its passage to actuality—that is to say the decision of making actual one of the potentially infinite laws of division—has nothing to do with any sort of actuality of the infinite. To come back to our previous example, the piece of marble is potentially a Zeus, a Hermes, a horse or anything else, but once the sculptor has made his choice, that potentiality has vanished: the marble will actually be a Hermes, and nothing else.

---

infinite collection (Weierstraß, Cantor) or an unending temporal process (Aristotle, Kant, Brouwer, Weyl)." It is surprising to find Aristotle's own definition of infinite in terms of iteration—namely, a law according to which further elements are generated—contrasted with Aristotle's supposed idea of infinite as unending temporal process. I have discussed the timelessness of Aristotle's infinite in Ugaglia (2009, 196) and (2012, 27–28).

[11] In modern terms, any convergent series can be chosen. Aristotle explicitly acknowledges the possibility of choosing different laws of division of a finite magnitude—e.g. a segment—provided that they follow a geometrical sequence. "For if, in a finite magnitude, one takes a definite amount and goes on taking in the same proportion [...] one will not traverse the finite magnitude" (*Ph.* III 6, 206b7–9). Note that the condition that one will not traverse the finite magnitude is our condition of convergence.

Of course, the second kind of potentiality implies some actuality—hereinafter L-*actuality*—but not an actual infinite. For when one fixes (= L-*actualizes*) a specific law of division, one does not actualize the "totality" of the cuts described by the law—which remain infinite in a purely potential (= *iterative*, A-*potential*) way—but only the rule for making them.

In other words—and this is a crucial point—for Aristotle no cuts are present, in any form, *before* they have been made. When he says, for example, that in Zeno's path there are potentially infinite halves, he simply means that the action of dividing the path in half can be performed endlessly, not that there is "something" already present, namely potential divisions, just waiting for someone to actualize them.[12]

Getting back to our terminology, the potentiality of the parts of the continuum is a kind of A-*potentiality*, to which no actuality corresponds. A continuum, be it mathematical or physical, is a whole, which pre-exists (physically, logically and ontologically) its parts: it can be divided into potentially infinite parts, but there is no way of constructing it starting from infinite parts, since to do so the parts would *already* be actual.

3 Potentiality *as* Possibility: Leibniz's Mathematical Infinite

Whether Leibniz directly knew Aristotle's texts on the infinite, or he had an indirect knowledge of them through ancient commentators and late scholasticism,[13] he acknowledges the importance of *iterative* infinite,[14] an infinite which coincides with the possibility of further progress (*ulteriori progressus*) in doing something, and which, adopting the scholastic terminology, he prefers to call *syncategorematic*:

> *There is a syncategorematic infinite* or passive power for having parts, namely, the possibility of further progress (*ulteriori progressus*) in dividing, multiplying, subtracting, or adding. [*Leibniz to Des Bosses* 1706. G II 314-315].[15]

This is particularly evident in Leibniz's treatment of infinite number, and infinite lines:

> I concede the infinite plurality of terms but this plurality itself does not constitute a number or a single whole. It means nothing, in fact, but that there are more terms than can be designated

[12] *Ph.* VIII 8, 263a4-b9. On Leibniz about Achilles' paradox see the letter to Foucher of January 1692 [A II 2, 491–2].

[13] A direct knowledge, even if partial, of Aristotle's works is suggested by Leibniz's correspondence with Thomasius [A II 1^2, 16–44], where the claim that scholastics have distorted Aristotle's own theories is discussed extensively: "tenebras Aristotelis a scholastico fumo esse, Aristotelem ipsum Galilaeo, Bacono, Gassendo, Hobbesio, Cartesio, Digbaeo mire conformari" [A II, 1^2, 18].

[14] "Aristotle a fort bien expliqué le plein et la division du continu contre les atomistes" [*Remarques sur la doctrine Cartesienne* A VI 4, 2047].

[15] "Datur infinitum syncategorematicum seu potentia passiva partes habens, possibilitas scilicet ulterioris in dividendo, multiplicando, subtrahendo, addendo progressus".

by a number. Just so there is a plurality or a complex of all numbers, but this plurality is not a number or a single whole [*Leibniz to Joh. Bernoulli*, February 21 1699, GM III, 575. Transl. Loemker][16]

It is perfectly correct to say that there is an infinity of things, i.e. that there are always more than one can specify. But it is easy to demonstrate that there is no infinite number, nor any infinite line or other infinite quantity, if these are taken to be genuine wholes. The Scholastics were taking that view, or should have been doing so, when they allowed a 'syncategorematic' infinite, as they called it, but not a 'categorematic' one. [*Nouveaux essais sur l'entendement humain*, II. XVII. A VI 6 157. Transl. Remnant-Bennet].[17]

And, accurately speaking, in place of "infinite number," we should say that more things are present than can be expressed by any number; or, in place of "infinite straight line," that a line is extended beyond any specifiable magnitude, so that there always remains a longer and longer line [*Leibniz to Des Bosses*, March 11 1706, Transl. Look-Rutherford].[18]

However, Leibniz's attention is not focused (only) on *syncategorematicity*: he acknowledges the fact that the infinite division of a segment is an iterative process—for every assigned number (*n^{th} step*) it is possible to take another one (*$(n + 1)^{th}$ step*)—but this is not the primary sense in which he understands potentiality. For Leibniz, a segment is potentially infinite—sometimes he employs the term *indefinite*—primarily because of the infinite range of possible (= L-*potential*) procedures of division it admits of, and only secondarily because each of these procedures is an infinite iteration (= A-*potentiality*).

In Leibniz's terms, as long as no law of division is fixed, the parts into which a segment can be divided are only possible (Latin *possibiles*), that is to say indefinite[19]:

As for the first point, it follows from the very fact that a mathematical body cannot be analyzed into primary constituents that it is also not real but something mental and designates nothing but the possibility of parts, not something actual. A mathematical line, namely, is in this respect like arithmetical unity; in both cases the parts are only possible and completely indefinite. A line is no more an aggregate of the lines into which it can be cut than unity is the aggregate of the fractions into which it can be split up. [*Leibniz to De Volder* June 30 1704; G II, 268. Transl. Loemker][20]

---

[16] "Concedo multitudinem infinitam, sed haec multitudo non facit numerum seu unum totum; nec aliud significat, quam plures esse terminos, quam numero designari possint, prorsus quemadmodum datur multitudo seu complexus omnium numerorum; sed haec multitudo non est numerus, nec unum totum".

[17] "A proprement parler, il est vrai qu'il y a une infinite des choses, c'est à dire qu'il y en a tousjours plus qu'on n'en puisse assigner. Mais il n'y a point de nombre infini ny de ligne ou autre quantité infinie, si on les prends comme des veritables Touts, comme il est aisé de demonstrer. Les écoles ont voulu ou du dire cela, en admettant un infini syncategorematique, comme elles parlent, et non pas l'infini categorematique".

[18] "Accurateque loquendo loco numeri infiniti dicendum est plura adesse, quam numero ullo exprimi possint; aut loco lineae Rectae infinitae, productam esse rectam ultra quamvis magnitudinem, quae assignari potest, ita, ut semper major et major recta adsit. De essentia numeri, lineae et cujuscunque Totius est, esse terminatum.".

[19] On Leibniz's choice to call the potential mathematical infinite "indefinite" see for instance the letter to Thomasius, 20 April 1669 [A II 1^2, 26–27; cf. GP IV, 328; 394].

[20] "Quod primum attinet, eo ipso quod corpus mathematicum non potest resolvi in prima constitutiva, id utique non esse reale colligatur, sed mentale quiddam nec aliud designans quam possibilitatem partium, non aliquid actuale. Nempe linea mathematica se habet ut unitas arithmetica, et

> But a mathematical continuum consists in pure possibility, like numbers [*Leibniz to Des Bosses* 1713. Transl. Look-Rutherford][21]

But if one eventually fixes (= L-*actualizes*) a specific law of division, this kind of indefiniteness collapses, and the potentiality of the parts vanishes:

> Let us suppose (*ponamus*) that in a line are 1/2, 1/4, 1/8, 1/16, 1/32 etc. actually given, and that all the terms of this series actually exist [*Leibniz to Joh. Bernoulli*, 1698, GM III, 536][22]

The crucial point here is that for Leibniz, in contrast with Aristotle, by actualizing the law of the convergent series—i.e. by univocally fixing "where" the potential cuts have to be made—one actualizes *all* the terms of the series—i.e. the cuts themselves.[23] The reasons why this can happen are not obvious, but as long as we are concerned with mathematics, it never occurs that a law of division is mandatory[24]: one can choose to divide a segment following a law, but any other law could have been selected instead, and a different choice can be made at each time, so that for Leibniz, the infinite is potential, i.e. syncategorematic, as it was for Aristotle.[25]

More precisely, for Leibniz the mathematical infinite is *both* syncategorematic *and* potential, where "syncategorematic" refers to A-potentiality, i.e. *iterativity*, and "potential" refers to L-potentiality, i.e. *possibility*. The complete phrase is required in order to avoid confusion with Leibniz's physical infinite, which, as we are going to see, is at the same time syncategorematic (= A-*potential*) *and* actual (= L-*actual*).

---

utrobique partes non sunt nisi possibiles et prorsus indefinitae; et non magis linea est aggregatum linearum in quas secari potest, quam unitas est aggregatum fractionum in quas potest discerpi".

[21] "Continuum vero mathematicum consistit in pura possibilitate ut numeri".

[22] "Ponamus in linea actu dari 1/2, 1/4, 1/8, 1/16, 1/32 etc. omnesque seriei hujus terminos actu existere".

[23] Similarly, Leibniz mentions the possibility of taking into account a whole infinite series: "Est enim illa serierum infinitarum Methodus tam generalis ut ejus ope omnis quantitatis incognitae valor exprimi possit, analytice, pure, rationaliter per formulam tamen infinitam [...] Valor iste est exacte verus si totam seriem infinitam consideres, et eatenus mentem illustrat; parte autem sumta ad usum idem mirifice aptus est" [*Leibniz to Molanus* 1677. A II 1^2, 481].

[24] In fact, the mathematical example in the correspondence with Bernoulli (see note 22) refers to a physical situation. The point at issue is the existence of infinitesimals in nature: if physical matter is infinitely divided—Bernoulli says—then there must be infinitesimal portions of matter, a conclusion Leibniz does not accept. In order to convince Leibniz, on August 16 Bernoulli proposes an example: "Let us suppose that a given magnitude is divided in parts that follows the geometric progression 1/2, 1/4, 1/8, 1/16 etc." In his response, Leibniz accepts Bernoulli's example as significant: "Uteris exemplo sane ad rem accomodato. Ponamus in linea actu dari 1/2, 1/4, 1/8, 1/16, 1/32 etc. ...".

[25] Leibniz's "constructivist strand" has been pointed out in the past. Surprisingly, it has been read in the light of recent constructivist positions, like Brouwer's (see for instance Levey 1999), instead of the more straightforward, and earlier, Aristotelian position. In particular, I would like to stress the crucial role that converging series have in Aristotle's discussion (see note 11 above). Moreover, while the intuitionist's potential infinite is constructive in a deep, Kantian temporal sense, Aristotle's potential infinite is explicitly and purposely independent of time, like Leibniz's infinite.

4 Potentiality *Without* Possibility: Leibniz's Physical Infinite

As we saw in the previous sections, both for Aristotle and for Leibniz a mathematical object, namely a continuous object, can be divided everywhere, and the division is an iterative infinite process. According to Aristotle this is true not only in mathematics but also in physics, because the same rules hold for dividing mathematical and physical objects, all of which are continuous.

For a short period, in which Leibniz tries to reconcile Aristotle and the moderns on the composition of physical bodies, he shares a similar view, and does not introduce any distinction between mathematical and physical extension. In particular, he maintains that both are infinitely potentially divisible.[26]

In the end, however, Leibniz's mathematical studies, and at the same time his return to a conception of physics governed by deeply un-Aristotelian metaphysical assumptions, drove him to conceive of infinite divisibility in two completely different and mutually incompatible ways in physics[27] and in mathematics.[28] In physics, in particular, Leibniz creates the notion of what has been called a "syncategorematic actual" infinite, which implies a complete separation between physical bodies and geometric solids. According to Leibniz's mature thought, indeed, while a geometrical solid can be divided everywhere, a physical body cannot: it can still be divided ad infinitum, but one has to follow a well-defined procedure of dissecting. In physics, indeed, every object is characterized, and constrained, by a particular law of division, which is a feature of the object in itself[29]: no matter if, or when, or who will

---

[26] In this period Leibniz thinks that Aristotle's treatment of physical objects can be partially saved but also that it can be freed from any metaphysical commitment. In particular, he maintains that the basic notions of matter, form and change can be further analyzed in terms of extension, figure and local motion, i.e. in terms of purely mechanico-geometrical concepts. This early stage in Leibniz's thoughts is well represented by the letters he wrote to Thomasius in 1668–9.

[27] The dynamics of the evolution of Leibniz's thought concerning the composition of physical bodies is a broad issue and the subject of a vast literature: see for instance Garber (2009) and the bibliography cited.

[28] Thereby I am referring to the introduction of infinitesimals, and "infinita terminata", which have nothing to do with physics. On this point I endorse the non-Archimedean interpretation advanced by Sherry-Katz (2012), Katz-Sherry (2012), (2013) and developed in Katz et al. (2021). For a completely different, syncategorematic reading of the infinitesimals see for instance Levey (1998), Arthur (2013) and Rabouin-Arthur (2020). A more "neutral" interpretation was advanced in Bos (1974) and Knobloch (1999) and a survey of different approaches can be found in the collection of essays in Goldenbaum, Jesseph (2008).

[29] I understand the laws of division of physical bodies—namely the rules for constructing the i-th term of a convergent series $\sum_i a_i$ of cuts—as the physical counterpart of the laws of order of substances. The way in which the law of the series of cuts contains the position of *all* the cuts must "express" the way in which the law of order contains *all* the states of the substance:

"L'essence des substances consiste dans la force primitive d'agir, ou dans la loy de la suite des changemens, comme la nature de la series dans les nombres [*Excerpta ex notis meis marginalibus ad Fucherii responsionem primam in Malebranchium critica*, 1676, A VI 3, 326]; "chacune de ces substances contient dans sa nature *legem continuationis seriei suarum operationum*, et tout ce qui luy est arrivé et arrivera [*Leibniz to Arnauld*, 1690, A II 2, 312; cf. GP IV 518]. In particular, I think divide it, the cuts that can be made are fixed, and although the operation of cutting is only syncategorematically (= A-*potentially*) infinite, the divisions are actual (= L-*actual*). This is the only way in which we can consistently speak of a "syncategorematic actual" infinite: namely, of something that is *both* potential *and* actual. The different letter prefixes show that it is potential and actual in different respects, and no contradiction arises.

This is all well and good—except that the absence of contradiction is a necessary condition, not a sufficient one. In order truly to understand Leibniz's physical infinite we should grasp on what grounds Leibniz can say that not only matter can be infinitely divided, but it actually is. In other words, not only is the law of division actual, but also the cuts themselves, which must *already be* in the matter:

> For the continuum is not merely divisible to infinity, but every part of matter is actually divided into other parts [*Leibniz to Arnauld*, November 28-December 8, 1686. A II 2, 122. Transl. Ariew-Garber][30]

> The difference between the way in which a line is made of points and the way in which matter is made of substances, which are in it, is the following: the number of points is undetermined, but the number of substances, even if it is infinite, is nevertheless true and determined, for it comes from an actual division of matter, not from a merely possible one. Moreover, matter is not divided in any possible way, but only with respect to some established proportion, as in a mechanism, a fish pond, or a flock. A line is not an aggregate of points, albeit a body is an aggregate of substances. [*Communicata ex disputationibus cum Fardella*, A VI 4, 1673-4].[31]

> In truth, matter is not a continuous thing but a discrete one, actually divided at infinity [*Leibniz to De Volder*, October 11, 1705; GP II, 278].[32]

> To pass now from the ideas of geometry to the realities of physics, I hold that matter is actually fragmented into parts smaller than any given, or that there is no part of matter that is not actually subdivided into others exercising different motions. This is demanded by the nature of matter and motion and by the structure of the universe, for physical, mathematical, and metaphysical reasons. [*Leibniz to Des Bosses*, March 11, 1706, Transl. Look-Rutherford pp. 33-35].[33]

that in Leibniz's later metaphysics the law of division of a particular physical body must be traced back to the law of order of its dominant monad. The idea that the law of order of simple substances could be traced back to a mathematical rule has been advanced in Loeb (1981, 273 and 317 ff.), although the author does not relate this rule to the law of the convergent series of cuts. On the relation between laws of the series and substances, both from a physical and from a metaphysical perspective, see Whipple (2010).

[30] "Car le continu n'est pas seulement divisible à l'infini, mais toute partie de la matiere est actuellement divisée en d'autres parties".

[31] "Hoc interest inter modum quo Linea constituitur punctis, et quo Materia constituitur ex substantiis quae in ea sunt, quod punctorum numerus non est determinatus, at substantiarum numerus etsi infinitus sit tamen est certus ac determinatus, nascitur enim ex actuali divisione materiae non ex possibili tantum. Neque enim materia divisa est omnibus modis possibilibus, sed certis quibusdam proportionibus servatis, ut Machina, piscina, grex. Linea non est aggregatum punctorum cum tamen corpus sit aggregatum substantiarum".

[32] "Revera materia non continuum sed discretum est actu in infinitum divisum".

[33] "Caeterum ut ab ideis Geometriae, ad realia Physicae transeam; statuo materiam actu fractam esse in partes quavis data minores, seu nullam esse partem, quae non actu in alias sit subdivisa diversos motus exercentes. Id postulat natura materiae et motus, et tota rerum compages, per physicas, mathematicas et metaphysicas rationes".

> And certainly every extended thing is divisible, in such a way that parts can be assigned in it, but in matter they are actually assigned, while in extension they are only potential, as in number. [*Ad Schedam Hamaxariam*].[34]

But what does it mean that the divisions, and hence the parts, already are in the matter?

In order to give a definitive answer, we have to turn to metaphysics. Indeed, while Leibniz initially aims at reducing physics to mechanics, and hence to mathematics, in his mature philosophy the relationship between physics, mathematics and metaphysics are completely redefined. Mathematics ceases to be the foundation of physics and becomes at most its description—and a necessarily approximate description at that. On the contrary, metaphysics becomes for Leibniz the ultimate real foundation of physics. In short, if we want to understand the constitution of bodies, we must stop thinking in terms of suitably divided extensions, namely mathematical continua, and we rather have to start thinking in terms of suitably aggregated individual substances, namely discrete metaphysical entities.

According to Leibniz's mature view, indeed, only metaphysics is concerned with the state of the universe, a state that God clearly "perceives", while it is unattainable for his creatures. Physics, which on the contrary is directly accessible to us, is a purely phenomenal "expression" of metaphysics.[35] A similar relation holds between physics and mathematics, which is a purely quantitative idealization, i.e. abstraction, of physics.[36]

Like any other physical notion, therefore, Leibniz's syncategorematic actual infinite can be better understood if it is compared with its metaphysical and mathematical counterparts: on the one side the actual infinite of the individual substances,[37] on the other side the potential infinite of the mathematical continua.

---

[34] "Et sane omne extensum divisibile est, ut partes in eo assignari possint, actuque sint assignatae in materia, potentiales vero sint in ipsa extensione, ut in numero". A similar statement is reiterated a few pages later: "Est enim ut saepe dixi dispositio compossibilium phaenomenorum, Geometriaque est possibilitatum ipsarum tractatio. Et ideo extensio ipsa possibilis sive continua non habet partes determinatas, non magis quam Unitas numerica; quod secus est de physicis, seu existentibus ubi divisiones sunt actu factae. Quemadmodum non datur minima fractio in numeris, ita nec minimum in Geometria, punctumque est extremum non pars. Sed in rebus ipsis dantur minima simplices nempe substantiae. (*Ad Schedam Hamaxariam*, ad. 16, in LH IV, 3, 5c, Bl. 1–2 (1703?) Wz 1123 (1692?). Transcription by Massimo Mugnai and Heinrich Schepers). I am grateful to Massimo Mugnai for sharing with me the unpublished manuscript.

[35] "Une chose exprime une autre (dans mon langage) lorsqu'il y a un rapport constant et reglé entre ce qui se peut dire de l'une et de l'autre. C'est ainsi qu'une projection de perspective exprime son geometral" [*Leibniz to Arnauld* 1687. A II 2, 231]. The notion of "expression" is widely discussed throughout the correspondence with Arnauld from September 1687 [A II 2, 230 ff.].

[36] On the three "realms" of Leibniz's philosophical system: metaphysics, physics and mathematics, and the corresponding three degree of infinity see for instance Nachtomy (2011), Vidinsky (2008).

[37] On the actual infinity of metaphysical substances and the "hypercategorematic" infinite of God see Antognazza (2015). "Datur et infinitum Hypercategorematicum seu potestativum, potentia activa, habens quasi partes, eminenter, non formaliter aut actu. Id infinitum est ipse Deus." [*Leibniz to Des Bosses* 1706. GP II, 314–315].

Leibniz's syncategorematic actual infinite, indeed, makes its appearance in the operation of dividing a physical body, whose abstraction is the mathematical continuum (whence the syncategorematicity),[38] and whose metaphysical foundation are simple substances (whence the actuality)[39]:

> But in realities in which only divisions actually made enter into consideration, the whole is only a result or coming together, like a flock of sheep. It is true that the number of simple substances which enter into a mass, however small, is infinite, since besides the soul, which brings about the real unity of the animal, the body of the sheep (for example) is actually subdivided – that is, again, an assemblage of invisible animals or plants which are in the same way composites, outside of that which also brings about their real unity. Although this goes on to infinity, it is evident that, in the end, everything reduces to these unities, the rest or the result being nothing but well-founded phenomena. [*Remarques sur les Objections de M. Foucher*, 1695, GP IV, 492. Transl. Ariew-Garber][40]

On the one hand, therefore, although the parts of a physical infinite cannot be actual in the same way as simple substances are, nevertheless they have to "express" their individuality and self-subsistence.[41] On the other hand, although the parts of a physical infinite cannot be potential in the same way as the parts of a continuum are, physical extension must be approximated by the mathematical continuum. Take a physical object: the way in which it is composed of, and divided into, infinite parts must keep track both of how bodies "result" from actual substances[42] and of how mathematical continua can be divided into potential parts.

With these references in mind, we can re-examine the initial question about the infinite parts of a physical object and the way in which they can be in act. The idea is to consider the kind of actuality which they cannot enjoy—namely, the one which belongs to substances—and replace this real (metaphysical) form of being with something that can be consistently taken as its physical phenomenal counterpart.

[38] As a matter of fact, Leibniz's mathematics is something more: in addition to ideal mathematical objects, which have a physical counterpart, he defines also fictional mathematical objects—like infinitesimals and bounded infinities—which have no counterpart but are useful mental constructions. See note 28.

[39] The precise way in which simple substances enter the composition of physical bodies is a largely debated issue. For a comprehensive study see Garber (2009).

[40] "Mais dans les realités où il n'entre que des divisions faites actuellement, le tout n'est qu'un resultat ou assemblage, comme un trouppeau de moutons; il est vray que le nombre des substances simples qui entrent dans une masse quelque petite qu'elle soit est infini puisqu'outre l'ame qui fait l'unité reelle de l'animal, le corps du mouton (par exemple) est soubsdivisé actuellement c'est à dire qu'il est encor un assemblage d'animaux ou de plantes invisibles, composés de même outre ce qui fait aussi leur unité reelle, et quoyque cela aille à l'infini, il est manifeste, qu'au bout du compte tout revient à ces unités; le reste ou les resultats, n'estant que des phenomenes bien fondés".

[41] I do not mean, of course, that the single part "expresses" the single simple substance, which is false—as Leibniz explains in his letter to De Volder of 5 December 1702 [GP II, 252]—but that the procedure of infinite analysis of matter into parts "expresses" the state of infinite synthesis of simple substances.

[42] Among the many works discussing the composition of physical bodies I suggest Nunziante (2011), where a very unusual connection is advanced between Leibniz's and Cyrano de Bergerac's works.

Leibniz's choice is to replace unity with motion[43]: if something has a proper (*conatus* to) motion, it is in act, but it is not necessarily a whole. In this way he bypasses the problem of the coexistence of infinite divisibility and actuality: physical matter being infinitely divisible, no part of it can be a unity, but they can notwithstanding be in motion.

Now, if physical magnitudes are not only divisible, but *actually* divided into infinite parts, and if a part is something characterized by its own proper motion, the continuity that would follow by the static union of two standard homogeneous parts cannot longer be established: due to motion, the "locus" where two different motions "came in contact" necessarily remains a discontinuity. As a consequence, physical magnitudes are not continuous, like mathematical objects are: using Leibniz's words, they are only *contiguous*.[44]

Here it is important to notice that contiguity is a totally new concept, which Leibniz *never* tries to define in mathematical terms. Indeed, it is *not* a mathematical property, but a physical one: while a mathematical continuous magnitude is only potentially divisible into infinite parts, of which it is not composed, a physical contiguous object is actually divided into infinite parts, from which it "results".[45]

As you would expect, the notion of "resulting" too is never defined by Leibniz in mathematical terms, and once the notion of "part" has been re-defined in purely physical terms, by reference to motion, one has to accept the fact that Leibniz's syncategorematic actual infinite is something which is ultimately foreign to mathematics and cannot be handled with mathematical tools or translated in formulas.

Now, Leibniz was perfectly aware that his treatment of the physical infinite is not—and should not be – a mathematical one, and he explicitly warns his readers against the misunderstandings arising from the temptation to make it mathematically sound:

---

[43] "Nam omne corpus etiam quantulumcumque, meo senso, dividitur in partes actu, et quidem non tantum mente assignabiles, sed et diversitate motuum reapse discretas" [*Leibniz to Joh. Bernoulli* 1697. GM III, 447]. On motion, and dynamics, in Leibniz's theory of the infinite, see Vidinsky (2008).

[44] "Caeterum ut ab ideis Geometriae, ad realia Physicae transeam; statuo materiam actu fractam esse in partes quavis data minores, seu nullam esse partem, quae non actu in alias sit subdivisa diversos motus exercentes" [*Leibniz to Des Bosses* March 11–17 1706. GP II, 305]. On the relation between partition, motion and force, from a physical and a metaphysical perspective, see De Risi (2007, 522 ff.).

[45] "At in realibus, nempe corporibus, partes non sunt indefinitae (ut in spatio, re mentali), sed actu assignatae certo modo, prout natura divisiones et subdivisiones actu secundum motuum varietates instituit, et licet eae divisiones procedant in infinitum, non ideo tamen minus omnia resultant ex ceteris primis constitutivis seu unitatibus realibus, sed numero infinitis. Accurate autem loquendo materia non componitur ex unitatibus constitutivis, sed ex iis resultat" [*Leibniz to De Volder* June 30, 1704, GP II, 268]. On the dependence of Leibniz's actually infinite division of matter on Descartes' argument from motion in a plenum see Arthur (2001, li-lii and Appendix 2c); Arthur (2014, 81).

> It is the confusion of the ideal with the actual which has muddled everything and caused the labyrinth of *the composition of the continuum* [*Remarques sur les Objections de M. Foucher*, 1695; GP IV, 491. Transl. Ariew-Garber].[46]
>
> As long as we seek actual parts in the order of possibles and indeterminate parts in aggregates of actual things, we confuse ideal things with real substances and entangle ourselves in the labyrinth of the continuum and inexplicable contradictions [*Leibniz to De Volder*, January 19 1706; GP II 282. Transl. Ariew-Garber].[47]

If we do not want to entangle ourselves in the labyrinth of the continuum, we must come to terms with the fact that Leibniz's notion of the syncategorematic actual infinite cannot be made mathematically sound: not because it is not sound, but because it is not mathematical at all.[48] And this is not a flaw. Like any other physical notion, indeed, Leibniz's actual infinite can be idealized into a sound mathematical notion—the potential infinite analyzed in Sect. 2—which describes and approximates it.

References

A = *Sämtliche Schriften und Briefe*, herausgegeben von der Berlin-Brandenburgischen Akademie der Wissenschaften und der Akademie der Wissenschaften zu Göttingen (1923), Reihe 1–8; cited by series, volume and page.

GM = Gerhardt, ed., *Leibnizens Mathematische Schriften.* Asher and Schmidt (1849–63; reprint ed. Olms 1971), 7 vols; cited by volume and page.

GP = Gerhardt, ed., *Die Philosophischen Schriften von Gottfried Wilhelm Leibniz.* Weidmann (1875–90 reprint ed. Olms 1960), 7 vols; cited by volume and page.

Antognazza, M. R. (2015). The hypercategorematic infinite. *The Leibniz Review, 25*, 5–30.

Arthur, R. T. W. (2001). *The Yale Leibniz: The Labyrinth of the continuum. Writings on the continuum problem, 1672–1686*. Yale University Press.

[46] "Et c'est la confusion de l'ideal et de l'actuel qui a tout embrouillé et fait le labyrinthe *de compositione continui*".

[47] "Nos vero idealia cum substantiis realibus confundentes, dum in possibilium ordine partes actuales, et in actualium aggregato partes indeterminatas quaerimus, in labyrinthum continui contradictionesque inexplicabiles nos ipsi induimus".

[48] On a different, attractive interpretation of the physical, non-mathematical nature of the notion see Bosinelli (2011) and Antognazza (2015). For an attempt to prove its mathematical consistency see Richard Arthur's papers, especially (2019). Leaving aside the fact that, if the notion of the physical infinite could be exactly translated in mathematical terms, the labyrinth of the continuum would not have arisen, my difficulties with this interpretation derive from the fact that no genuinely mathematical characterization is proposed of Leibniz's syncategorematic actuality which allow to distinguish it from Aristotle's syncategorematic potentiality. If fixing the iteration rule—namely, "the law according to which further elements are generated" (see note 10)—means actualizing the elements, then Aristotle's infinite is as actual as Leibniz's, and calling it potential or actual is just a matter of taste. I think, however, that this would be unfair to Leibniz. If, on the contrary, the actuality involved in Leibniz's syncategorematic actual infinite is something more, a mathematical characterization of this "something" must be given. Instead, all the characterizations that Arthur proposes bring into play physical concepts. But this fact corroborates the hypothesis that the syncategorematic actual infinite cannot be reduced to a mathematical notion.

Arthur, R. T. W. (2013). Leibniz's syncategorematic infinitesimals, smooth infinitesimal analysis, and second order differentials. *Archive for History of Exact Sciences, 67*, 553–593.
Arthur, R. T. W. (2014). *Leibniz*. Polity Press.
Arthur, R. T. W. (2015). Leibniz's actual infinite in relation to his analysis of matter. In N. Goethe, P. Beeley & D. Rabouin (Eds.), *G. W. Leibniz: Interrelations between mathematics and philosophy* (pp. 137–156). Springer.
Arthur, R. T. W. (2019). Leibniz in cantor's paradise: A dialogue on the actual infinite. In V. De Risi (Ed.), *Leibniz and the structure of sciences—Modern perspectives on the history of logic, mathematics, epistemology*. Boston Studies in the Philosophy and History of Science 337, Springer.
Bos, H. (1974). Differentials, higher-order differentials and the derivative in the Leibnizian Calculus. *Archive for History of Exact Sciences, 14*, 1–90.
Bosinelli, F. C. M. (2011). Über Leibniz' Unendlichkeitstheorie. *Studia Leibnitiana, 23*(2), 151–169.
De Risi, V. (2007). *Geometry and monadology. Leibniz's analysis situs and philosophy of space*. Birkhäuser.
Garber, D. (2009). *Leibniz: Body, substance, monad*. Oxford University Press.
Goldenbaum, U., & Jesseph, D. (2008). *Infinitesimal differences. Controversies between Leibniz and his contemporaries*. Walter de Gruyter.
Katz, M. G., & Sherry, D. (2012). Leibniz's laws of continuity and homogeneity. *Notices of the American Mathematical Society, 59*, 1550–1558.
Katz, M. G., & Sherry, D. (2013). Leibniz's infinitesimals: Their fictionality, their modern implementations, and their foes from Berkeley to Russell and beyond. *Erkenntnis, 78*, 571–625.
Katz, M., Kuhlemann, K., Sherry, D., & Ugaglia, M. (2021). Leibniz on bodies and infinities: Rerum natura and mathematical fictions. *The Review of Symbolic Logic*, 1–31. https://doi.org/10.1017/S1755020321000575
Knobloch, E. (1999). Galileo and Leibniz: Different approaches to infinity. *Archive for the History of Exact Science, 54*, 87–99.
Kretzmann, N., Kenny, A., & Pinborg, J. (1982). *The Cambridge history of later medieval philosophy: From the rediscovery of Aristotle to the disintegration of scholasticism 1100–1600*. Cambridge University Press.
Levey, S. (1998). Leibniz on mathematics and the actually infinite division of matter. *The Philosophical Review, 107*(1), 49–96.
Levey, S. (1999). Leibniz's constructivism and infinitely folded matter. In R. J. Gennaro & C. Hueneman (Eds.), *New essays on the rationalists* (pp. 134–162). Oxford University Press.
Loeb, L. E. (1981). *From Descartes to Hume*. Cornell University Press.
Look, B. C., & Rutherford, D. (2007). *The Yale Leibniz: The Leibniz-Des Bosses correspondence*. Yale University Press.
Moore, A. W. (1990). *The infinite*. Routledge.
Nachtomy, O. (2011). A tale of two thinkers, one meeting, and three degrees of infinity: Leibniz and Spinoza (1675–1678). *British Journal for the History of Philosophy, 19*(5), 935–961.
Nunziante, A. M. (2011). Continuity or discontinuity? Some remarks on Leibniz's concepts of "substantia Vivens" and "organism." In J. E. H. Smith & O. Nachtomy (Eds.), *Machines of nature and corporeal substances in Leibniz* (pp. 131–143). Springer.
Peter of Spain. (2014). *Summaries of logic. Text, translation and notes* by Brian P. Copenhaver, Calvin Normore and Terence Parson. Oxford University Press.
Duhem, P. (1985). *Theories of infinity, place, time, void, and the plurality of worlds*. Edited and Translated by Roger Ariew. The University of Chicago Press.
Rabouin, D., & Arthur, R. T. W. (2020). Leibniz's syncategorematic infinitesimals II: Their existence, their use and their role in the justification of the differential calculus. *Archive for History of Exact Sciences, 74*, 401–443.
Sherry, D., & Katz, M. G. (2012). Infinitesimals, imaginaries, ideals, and fictions. *Studia Leibnitiana, 44*, 166–192.

Uckelman, S. L. (2015). The logic of categorematic and syncategorematic infinity. *Synthese, 192*, 2361–2377.

Ugaglia, M. (2009). Boundlessness and iteration: Some observations about the meaning of ἀεί in Aristotle. *Rhizai, 6*(2), 193–213.

Ugaglia, M. (2012). *Aristotele, Fisica. Libro III.* Carocci editore.

Ugaglia, M. (2018). Existence vs conceivability in Aristotle: Are straight lines infinitely extendible? In M. Piazza & G. Pulcini (Eds.), *Truth, existence and explanation—FilMat 2016 studies in the philosophy of* mathematics (pp. 249–272). Boston Studies in the Philosophy and History of Science 334, Springer.

Vidinsky, V. (2008). Dynamical interpretation of Leibniz's continuum. *Kaygi, 10*, 51–70.

Wieland, W. (1970). *Die aristotelische Physik.* Vandenhoeck & Ruprecht.

Whipple, J. (2010). The structure of Leibnizian simple substances. *British Journal for the History of Philosophy, 18*(3), 379–410.

Pure Positivity in Leibniz

Maria Rosa Antognazza

1 Introduction

The notion of 'pure positivity' plays a pivotal role in Leibniz's version of the ontological argument. In his view, it establishes the possibility of the *Ens perfectissimum*, thereby providing the premise missing from other versions of the argument. As he puts it very briefly around 1685: 'the most perfect Being is possible, because it is nothing other than pure positivity [Ens summe perfectum est possibile, quia nihil aliud est, quam pure positivum]' (*Definitiones notionum metaphysicarum atque logicarum*; A VI 4, 626). It may be tempting to dismiss 'pure positivity' as a hastily contrived notion introduced in a desperate attempt to support the ontological argument.

In this paper, I will discuss Leibniz's conception of pure positivity, exploring its connection with his notions of perfection, pure act, being, reality, absolute, and infinite. I will come to the conclusion that 'pure positivity', far from being a feeble, ad hoc attempt to rescue the ontological argument, constitutes a fundamental feature of Leibniz's metaphysics that aligns with well-documented key commitments of his philosophical thought.

2 The Role of Pure Positivity in the Ontological Argument

One of the merits of Leibniz's discussion of the ontological argument is its emphasis on the issue of God's possibility.[1] Anselm's and Descartes' proofs, Leibniz objects,

[1] For a full discussion see Antognazza (2018), from which I am drawing.

M. R. Antognazza (✉)
King's College London, London, UK
e-mail: maria.rosa.antognazza@kcl.ac.uk

F. Ademollo et al. (eds.), *Thinking and Calculating*, Logic, Epistemology, and the Unity of Science 54, https://doi.org/10.1007/978-3-030-97303-2_13

only show that 'If God is possible, God exists' (A VI 6, 438; A VI 4, 588). The conclusion, however, is merely conditional. It only allows the claim that *if* God is possible, God exists. But is God possible? To be unconditionally conclusive, the argument must be completed as follows:

1. If God is possible, God exists
2. God is possible
3. Therefore, God exists.

Of course, Leibniz was not the first to note that God's possibility cannot be taken for granted. His point belongs to a broader family of reasons for rejecting a priori arguments due to doubts about whether we have a genuine idea of God, that is, a sufficiently clear grasp of the essence of God. Leibniz, however, is arguably the author who most explicitly identifies the issue of God's possibility not only as the pivot of a conclusive argument but also as the only genuine objection to the ontological argument. 'Those who maintain that one can never infer actual existence from mere notions, ideas, definitions, or possible essences', Leibniz claims, 'fall back in fact' on 'the only thing one could say against the existence of such a Being, namely, deny its possibility' (GP IV 406). He was well aware of the objection that one could insert 'necessary' into the definition of anything (e.g. 'necessary man') and thereby prove its existence. His chief reply, however, is that the issue revolves on showing that this alleged definition is a consistent one, namely that it picks out a possible being (see A II 1^2, 587). In other words, the familiar objections – that existence is not a predicate, or a perfection, or a property; that things cannot be defined into existence; that one cannot legitimately move from mere concept to reality, and so on – reduce for Leibniz to the question of whether a being which exists by conceptual necessity is possible.

In the first part of his proof Leibniz argues in fact that, if possible, God conceived as the *Ens perfectissimum* implies a logically necessary being. Further, he proves by a modal argument that, if possible, a logically necessary being necessarily exists. Assuming he has been successful in establishing these two points (or, more succinctly, assuming he has shown that the *Ens perfectissimum* is the *Ens necessarium*), he can then focus on the possibility of the *Ens perfectissimum* without further worries that its essence, even if it does not involve logical contradiction, may still fail to involve existence.

The really crucial challenge is therefore a successful defence of the claim that the *Ens perfectissum* is possible. As mentioned above, the pivot of Leibniz's argument is the notion of 'pure positivity'. The *Ens perfectissimum*, he argues, is constituted by the conjunction of all purely positive qualities, that is, all qualities which express without any limits whatever they express (A VI 3, 578). Their being 'purely positive' means that they lack any negation whatever. Leibniz's key point is that all purely positive properties are compatible, 'that is, they can be in the same subject' (A VI 3, 578). From a logical point of view, he reasons, that which is purely positive cannot involve any formal contradiction. There would be a contradiction if two (or more) incompatible qualities (or, broadly speaking, properties) are attributed to the same subject (say, 'being square' and 'being round'). Such incompatibility implies that one property is the negation of the other property not merely in the sense that a property A (e.g. extension) is not another property B (e.g. thought) but in the sense that A ('being square') excludes B ('being round'). This exclusion or incompatibility would be expressed by an identity: A = not-B. A = not-B, however, is directly against the hypothesis of the pure positivity of all perfections.[2] Hence the *pure positivum*, since it does not involve any negation, cannot involve any contradiction. If it does not involve any contradiction, this means that it is possible. In short, according to Leibniz, purely positive qualities must be compatible because, as purely positive, they cannot involve negation and they cannot, therefore, involve contradiction.[3]

In paragraph 45 of the *Monadology* (1714), Leibniz advances very briefly a similar thought but with a metaphysical slant.[4] Since 'nothing can prevent the possibility of that which is without any limits, without any negation, and consequently without any contradiction' (GP VI 614), the *Ens perfectissimum* must be possible. In other words, that which is without limits cannot be limited, that is, partially negated, by anything. Therefore, nothing can restrict it or hinder its possibility. This argument, presented by Leibniz at the end of his life, echoes an earlier note of March 1676 in which the compatibility of purely positive (or 'affirmative') properties is seen as directly implied by their lack of any negation, which in turn implies their being unlimited or absolute:

> For the rest, I demonstrate that all attributes are compatible in no other way than by their being absolute, pure, and unlimited. For if they were modified by limits, they would not be affirmative but negative in a certain manner. (A VI 3, 396)

This remark brings me to the core of this paper, namely, to an exploration of the place of 'pure positivity' in Leibniz's metaphysics more generally. I propose to explore this topic by identifying connections, established by Leibniz, between 'pure positivity' and a cluster of metaphysical notions that pervade his metaphysical thought. Some of these notions have already appeared in the passages quoted above, signalling that 'pure positivity', over and above its specific appearance in the ontological argument, is of much broader significance for Leibniz's philosophical system.

3 Pure Positivity, Perfection, and Pure Act

As early as 1676, Leibniz identifies 'pure positivity' as a defining feature of what it is to be a 'perfection'. A perfection, Leibniz writes in the context of his exchange of 1676 with Spinoza, is

---

[2] See esp. A VI 3, 572 (trans. by Adams 1994: 145): 'one [property] would express the exclusion of the other, and so one of them would be the negative of the other, which is contrary to the hypothesis, for we assumed that they are all affirmative.' For an insightful discussion see Adams (1994: 142–148, see esp. 145).

[3] See A VI 3, 572, 575, 577, and 578–9. Interestingly, the same type of argument is employed by Eckhard in his exchange with Leibniz (see letter of 19 April 1677; A II 1^2, 495–6).

[4] Cf. Adams (1994: 173).

> … every simple quality which is positive and absolute, or that expresses without any limits whatever it expresses. But a quality of this kind, since it is simple, is for that reason unanalysable, or indefinable, for otherwise either it will not be one simple quality but an aggregate of a plurality [of qualities], or, if it is one, it will be circumscribed by limits, and so it will be understood through negations of further progress, contrary to the hypothesis, for it is assumed to be purely positive. (A VI 3, 578)[5]

In brief, a perfection is a quality which is simple and purely positive. As regards simplicity, a simple quality is for Leibniz a quality which cannot be resolved into, or reduced to other qualities. As regards positivity, a purely positive quality involves no negation whatsoever. Since any limitation is a negation of some 'further progress' or further degree of that which is limited,[6] pure positivity involves having no limitation at all. As Robert Adams (1994: 113) insightfully stresses, 'the most durable feature of Leibniz's conception of perfections is that they involve no negation at all'. Thus, 'while there is reason to think that simplicity is in fact dispensable as a criterion of perfections for Leibniz' (Adams, 1994: 121), pure positivity remains the essential defining feature of what it is to be a perfection.

Some twenty years later, Leibniz goes as far as equating perfection and pure positivity. In a text dated by Grua around 1695–1697, he writes: 'Perfection is pure act or [seu] pure positivity. What we commonly say of act and potency is more correctly said of the positive and privative, or of the absolute and limited' (Grua 371).[7] I will return below to the link between pure positivity and Leibniz's conception of the absolute. For now, I would like to focus on the equivalence established by this text between 'perfection', 'pure act' and 'pure positivity', and on the parallel between act and potency, on the one hand, and positive and privative, on the other hand. 'Pure positivity', Leibniz suggests, corresponds to the more traditional but less accurate notion of 'pure act'. Its opposite is 'privative', corresponding in turn to the more traditional but less accurate notion of 'potency'. In brief, Leibniz proposes to recast the traditional distinction between act and potency in terms of a distinction between positive and privative. Why? I believe Leibniz regards the latter distinction as explanatorily superior because it reduces actuality and potentiality to a metaphysically more fundamental pair: being and non-being.

---

[5] See also *Ens Perfectissimum Existit* (c. November 1676); A VI 3, 575: 'Perfectiones, sive formae simplices, sive qualitates absolutae positivae'.

[6] Cf. A VI 3, 578, quoted above.

[7] See also a key text on the infinite, probably written around 1698, recently transcribed and translated from Leibniz's manuscript by Richard T. W. Arthur and Osvaldo Ottaviani (*De Scientia Infiniti*, LH 35, 7, 10, Bl. 5r-8v): 'Porro ex his rerum praedicatis quae mens nostra intelligit, alia sunt absoluta, alia limitationem involvunt. Absoluta constant realitate pure positiva, atque adeo perfectionem indicant, solaque recensentur inter attributa substantiae supremae' ('Furthermore, of those predicates of things which our mind understands, some are absolute, some involve limitation. Absolute ones consist in a purely positive reality and therefore indicate perfection, and are the only ones counted among the attributes of the supreme substance'). I am very grateful to Richard Arthur and Osvaldo Ottaviani for drawing my attention to this text, and for generously sharing their transcription and translation, from which I am quoting. As they note, this important piece has previously appeared only in an obscure nineteenth-century publication (Gerhardt 1875: 595–608).

4 Pure Positivity, Being, and Reality

In fact, in the *Generales Inquisitiones* of 1686, he bluntly states:

> Positive is the same as *Being* [Positivum idem est quod *Ens*]. *Non-Being* is what is merely privative, that is, the privative of everything [*Non Ens* est quod est mere privativum, seu omnium privativum]. (A VI 4, 740)

In the *Definitiones notionum metaphysicarum atque logicarum* of 1685, 'pure positivity' is laconically equated to 'Being' (*Ens*) ('Ens seu pure positivum' A VI 4, 626) in the context of a very interesting explanation of existence in terms of degrees of perfection. More precisely, in this text, the *reason* of existence is identified in the degree of perfection of the possible beings which are, in turn, compossible in the most perfect series:

> *Existent* cannot be defined, any more than Being or the purely positive [Ens seu pure positivum], that is, in such a way that some clearer notion might be shown to us; one should know, however, that every possible will exist if it can, but since not all possibles can exist, as some hinder others, those exist which are more perfect. (A VI 4, 626)

The conclusion to be drawn from this metaphysical picture is straight-forward: 'Therefore, it is certainly established that what is most perfect exists [Itaque quod perfectissimum est, id certo constat existere]' (A VI 4, 626).

The equivalence, for Leibniz, of positivity, being (*ens*), and reality, and their link to perfection, is apparent also in other texts of the mid-1680s, developing a line of thought already firmly in place at least since 1677, when Leibniz engaged in a sustained discussion of the ontological argument with Arnold Eckhard.[8]

In the *Notationes Generales* of 1683–1685, perfection is equated to pure reality or [seu] positivity, while imperfection is equated to limitation ('Perfectio autem est realitas pura seu quod in essentiis est positivum atque absolutum. Contra imperfectio consistit in limitatione.' A VI 4, 556).[9] In a series of definitions penned around the end of 1687, in line with the view presented the year before in the *Generales Inquisitiones* that existence is a function of the degree of perfection of essences, Leibniz gives the following definition of 'existent': 'Existent is the possible [which is] more perfect, or com-possible with more, or that which involves more reality [Existens est possibile perfectius, seu pluribus compossibile, seu quod plus involvit realitatis]' (*Definitiones. Notiones. Characteres*; A VI 4, 875).

In turn, in another series of definitions probably written between the summer of 1687 and the end of 1696, Leibniz defines degrees of perfection in terms of degrees

---

[8] Cf. A II 1^2, 488 (April 1677): 'perfection is every attribute, or every reality'; A II 1^2, 543 (Summer 1677): 'as I would prefer to define it, *perfection* is degree or quantity of reality or [seu] essence'; *Elementa verae pietatis*, c. 1677–8 (A VI 4, 1358): '*Perfection* is degree or quantity of reality. Hence *perfectissimum* is what has the highest degree of reality'. In his exchange with Leibniz on the ontological argument, Arnold Eckhard insists that *perfectissimum* can be substituted with *realissimum*, the latter being a rougher way to express what the former expresses in a more sophisticated manner (cf. A II 1^2, 527, 541: '*Perfecti* vocabulum non est inutile, quia id Latine dicit, quod *reale* barbare effert').

[9] Cf. also LH 35, 7, 10, Bl. 5r-8v (quoted above, n. 7).

of reality or [*vel*] positivity: '*The more perfect* is what has more *reality* or positive Entity [*Perfectius* est quod plus habet *realitatis* vel Entitatis positivae]' (*Definitiones: Ens, Possibile, Existens*; A VI 4, 867) adding later on to the text: '*Perfection* is the magnitude of reality or entity [*Perfectio* est magnitudo realitatis vel entitatis].' (A VI 4, 867) In the *Definitiones notionum metaphysicarum atque logicarum* of the mid-1685, he gives a logical definition of *ens* in terms of the subject of predication of positive terms, in constrast to *nihil* as subject of predication of negative terms:

> *Being* or *Something* [*Ens* vel *Aliquid*] is that to which belong positive terms, such as A, B, C, provided of course that its explanation does not resolve in the merely privative. I consider here, however, Concrete Terms. *Nothing* [*Nihil*] is that to which belong only merely negative terms, certainly if N is not A, nor B, nor C, nor D, and so on, so that if no positive term is found which is predicated of it, then N is said to be *Nothing*. Therefore that commonplace Axiom: 'Non entis nulla sunt attributa' comprises the definition of Nothing or non-Being itself. (A VI 4, 625)[10]

Likewise, in the *Generales Inquisitiones* (1686) he states: 'The privative is non-A. Non-non-A is the same as A. The positive is A … This is what is meant when it is commonly said that the 'nothing' has no properties' (A VI 4, 740).[11]

In sum, there can be little doubt that the equivalence of positivity, being (*ens*), and reality, all of which are also employed by Leibniz to define perfection, is a well-established aspect of Leibniz's metaphysics. This equivalence and its link to perfection is, in itself, far from novel. On the contrary, it places Leibniz firmly in the long and multiform tradition of Platonism. As a number of scholars, including myself, have argued in a number of places, and as Leibniz himself indicates, Leibniz is, at heart, a Platonist.[12] Of course, this is not to deny the importance of a great deal of other philosophical traditions for his thought, not least the Aristotelian tradition. But when it comes to the mould which most fundamentally shapes his metaphysics, it seems to me that it can be identified with a Christianized version of Platonism. I hope this will become even clearer in the following discussion of the relationship between pure positivity and Leibniz's conceptions of the absolute and the infinite.

---

[10] A VI 4, 625: '*Ens* vel *Aliquid* est cui competit terminus positivus, ut A, B, C, si scilicet in explicatione non sit resolvendus in mere privativum. Adhibeo autem hic Terminos Concretos. *Nihil* est cui non competit nisi terminus mere negativus, nempe si N non est A, nec est B, nec C, nec D, et ita porro, ita ut nullus reperiatur terminus positivus qui ejus sit praedicatum, tunc N dicitur esse *Nihil*. Itaque Axioma illud vulgare, Non entis nulla sunt attributa, continet ipsius Nihili seu non Entis definitionem'.

[11] It is worth quoting the whole passage: 'Privativum non-A. Non-non-A idem est quod A. Positivum est A, si scilicet non sit non-Y quodcunque, posito Y similiter non esse non-Z et ita porro. Omnis terminus intelligitur positivus, nisi admoneatur eum esse privativum. Positivum idem est quod Ens. Non Ens est quod est mere privativum, seu omnium privativum, sive non-Y, hoc est non-A, non-B, non-C, etc. Idque est quod vulgo dicunt nihili nullas esse proprietates' (*Generales Inquisitiones*; A VI 4, 740).

[12] Cf. for instance A VI 6, 48 / NE 48; A VI 6, 378 / NE 378; *Méditation sur la notion commune de la justice* (in Mollat / PW 45–64); Dutens IV 280 / PW 71–2. In Antognazza (2016a) I argue that the overall inspiration of Leibniz's thought is broadly Platonic (cf. pp. 114–115). See also Antognazza (2016b). Prominent Leibniz scholars who have stressed the Platonic (and/or Christian-Platonic; Neoplatonic) inspiration of Leibniz's philosophy include Rutherford (1995), Mercer (2001), Riley (1996), (2003), and Jolley (2005).

5 Pure Positivity and the Absolute

Pure positivity captures, for Leibniz, also what it is to be absolute. To be absolute is to be free from any condition and any limitation.[13] Only what is purely positive can be absolute since any condition is a sort of limitation, and any limitation is a negation. Therefore, an absolute quality must be a purely positive quality, as Leibniz already signals in the text of 1676 discussed above (A VI 3, 578).

The link between pure positivity and Leibniz's conception of absolute is in full view in later texts. In the passage from the *Notationes Generales* of 1683–1685 already quoted above, he writes: 'however, perfection is pure reality, that is, what in essences is positive and absolute. [Perfectio autem est realitas pura seu quod in essentiis est positivum atque absolutum.]' In the passage of 1695–1697 in which Leibniz proposes to reconceive the traditional distinction between actuality and potentiality in terms of a distinction between positive and privative, he also equates the latter to the distinction between absolute and limited: 'Perfection is pure act or pure positivity. What we commonly say of act and potency is more correctly said of the positive and privative, or of the absolute and limited. [Perfectio est purus actus seu pure positivum. Quae vulgo de actu et potentia, rectius dicemus de positivo et privativo, seu de absoluto et limitato.]' (Grua 371) In turn, in an earlier text of the mid-1680s, Leibniz conceives the absolute as what is, in its kind, purely positive, namely, that in which no degree of reality or being that can fall under a certain kind is lacking:

> ABSOLUTE is that the concept of which is unlimited or outside of which nothing can be assumed in the same kind, or the concept of which is capable of quantity, and yet does not involve limits. Hence it is possible to conceive absolute Extension, but not an absolute circle. It seems that, in this sense, absolute and maximum are the same. God is the absolute Being [Ens absolutum] nor indeed there is any reality or perfection which is not in God. To put it best, we say that the Absolute is purely positive in its kind.[14]

A text of 1676, written in the context of his encounter with Spinoza, throws some light on the notion of 'maximum' regarded by Leibniz, in the note above, as 'identical' to the sense in which he is taking 'absolute'. 'Maximum,' Leibniz writes, 'is everything of its kind [*omnia* sui generis], or that to which nothing can be added, like a line unbounded on both sides, which is obviously also infinite; for it contains every length.' (A VI 3, 282 / Arthur 115; trans. slightly modified). However, the 'absolutely infinite' is 'maximum in entity', that is, is everything (*omnia*) simpliciter (as opposed to '*omnia* sui generis'): 'Whatever contains everything [*omnia*] is maximum in entity

---

[13] On the meaning of 'absolute', cf. Adams (1994: 115), referring to texts in which Leibniz explicitly opposes 'absolute' to 'limited' or uses 'absolute' in the sense of 'unconditioned' or 'unqualified'.

[14] *Definitiones notionum ex Wilkinsio*, c. 1685–1686 (A VI 4, 36): 'ABSOLUTUM est, cujus conceptus est illimitatus seu extra quod nihil in eodem genere sumi potest, seu cujus conceptus est capax quantitatis, et nullos tamen involvit limites. Hinc concipi potest Extensio absoluta, sed non circulus absolutus. Videtur hoc sensu idem esse absolutum quod maximum. Deus est Ens absolutum neque enim ulla datur realitas sive perfectio, quae in Deo non sit. Optime dicemus Absolutum esse in suo genere pure positivum.' In this text, Leibniz is expanding a series of definitions given by John Wilkins (1668). On p. 35 Wilkins writes: 'ABSOLUTENESS, *Independent, Freehold.* DEPENDENCY, *Under*'.

… Likewise, that which contains everything is the most infinite [infinitissimum], as I am accustomed to call it, or the absolutely infinite [absolute infinitum].' (A VI 3, 282 / Arthur 115) The latter notion of *absolute infinitum* returns, in connection with *pura realitas*, in a virtually unknown text of 1698 (*De Scientia Infiniti*):

> whenever simple and pure reality is understood, by that very fact is constituted the Maximum possible in things, or [seu] absolute infinite, in which duration, diffusion, power, cognition and anything at all that is in it, lacks limits, and in turn anything that can lack limits is in it; but the rest originate from it, and it is called GOD. And so there is in GOD every nature capable of perfection, with each of these natures there perfectly or [seu] absolutely, to the extent that he is an absolutely absolute Being. For if there were some Being merely omnipresent, or merely omniscient, or endowed with some other definite nature capable of perfection, it would indeed be absolute in its own kind, but would not have an absolute reality.[15]

These texts suggest, in my view, that Leibniz ultimately explains what it is to be absolute in terms of pure positivity, that is, in terms of a complete fullness of being or reality which excludes any negation whatsoever. The question now becomes: is such pure positivity possible? That is, can a being that embraces the fullness of reality exclude any negation whatsoever?

6 Pure Positivity, Plurality, and the Infinite

Despite Leibniz's confident assertion that, from a logical point of view, that which is purely positive cannot contain any formal contradiction, one may still doubt whether a purely positive being constituted by the conjunction of all purely positive qualities is possible.[16] That is, even if one granted that all purely positive properties are compatible, one would still have to face the objection that a plurality of perfections implies some sort of negation for the very fact that perfection A, although compatible with perfection B, is *not* perfection B. In other words, as Spinoza and Hegel point out, *omnis determinatio est negatio*[17]: any determination implies some kind of negation (being *this* perfection and *not that* perfection). One may object, therefore, that the notion of a purely positive being constituted by a *plurality* of perfections is not a coherent notion since plurality implies negation.

---

[15] LH 35, 7, 10, Bl. 5r-8v (transcription and translation by Arthur and Ottaviani): 'ubi simplex ac pura realitas intelligitur, eo ipso constituitur Maximum in rebus possibile seu absolute infinitum, in quo duratio, diffusio, potentia, cognitio, et omnino quicquid inest, limite caret, et vicissim quicquid limite carere potest inest [;] caetera autem ex ipso oriuntur, idque DEUM appellamus. Itaque inest DEO omnis natura capax perfectionis, et unaquaeque harum naturarum perfecte seu absolute. Ut adeo sit Ens absolute absolutum. Nam si quod esset Ens tantum omnipraesens aut tantum omniscium, aut alia quadam certa natura perfectionis capace praeditum, in suo quidem genere absolutum foret, non tamen absoluta realitatis'.

[16] For a detailed discussion of the issues raised in this section see Antognazza (2015), from which I am drawing.

[17] Cf. Spinoza to Jarig Jelles, Epistola 50 in Spinoza (1925: Vol. 4), and Hegel's 1816 review of Jacobi's *Werke* in Hegel (2009: 9).

Furthermore, there is the problem of clarifying whether all the qualities or properties which are traditionally attributed to God can be purely positive. In this regard, a particularly difficult case is omniscience. Since omniscience implies knowledge of limited beings, and hence representation of beings which involve negation, one may object to a conception of omniscience as a purely positive attribute. If it is not purely positive and yet belongs to God, it seems that some form of negation is being introduced in what is supposed to be the purely positive nature of God.[18]

In fact, in 1677, Leibniz himself challenges Eckhard's claim that there can be a Being—the *ens perfectissimum* or *ens purum*—which does not participate in any non-Being[19]:

> The concept of Being [*ens*] which does not participate in non-Being [*non-ens*] involves two aspects, one: to be [*esse*], the other: *not to be nothing*, that is, to be all things [*nihil non esse*, seu omnia esse] ... But it seems impossible for there to be a Being which is all things; of such a Being it could be said that it is you, and that it is me as well; something which, I take, you will not admit. (A II 1^2, 500)

At least part of Leibniz's objection involves the question of how a Being that embraces all reality ('*nihil non esse*, seu omnia esse') can exclude all negation since any determinate thing implies the negation of other determinate things (in Leibniz's own example, 'me' is not 'you'). In other words, Leibniz raises the question of whether the notion of a purely positive being embracing a plurality of things is a coherent one.

Eckhard replies, basically, that embracing all reality does not mean to be all things in their determinate essences—that is, it does not mean that God is 'man, brute, horse, lion, dog, Peter, Paul, you, me and so on'.[20] It means that in God there is all the *perfection* or *reality* of all these things.[21] As the scholastics would say and as Leibniz notes *en passant* in his own reply, all things are in God not *formally* but *eminently*, that is, not in their own determinate form (as horse, lion, dog and so on) but in a superior form which implies, while at the same time surpassing, the degree of perfection or reality of these determinate essences.[22] It seems Eckhard's point

---

[18] Cf. Adams (1994: 122–3).

[19] Cf. Eckhard to Leibniz, 19 April 1677 (A II 1^2, 494–5): 'philosophers do not distinguish perfect from Being [*ens*] if not as a mere distinction of reason. Indeed Being and positive are opposed to non-Being [*non ens*]. ... where there is some positive entity, there is also perfection. ... pain is not something positive but non-Being [*non ens*] and negation, and indeed also imperfection. ... to be an entity, to be real, to be positive, and to exist do not differ among themselves ... Being [*Ens*] and perfect do not differ from one another, as it is shown above. Accordingly, *ens perfectissimum* is identical to *ens purum*, or that which in no manner is non-Being [*non ens*]'.

[20] A II 1^2, 515. Eckhard's letter of reply of May 1677 stretches over almost forty pages (A II 1^2, 505–541).

[21] For instance, in an earlier exchange, Eckhard notes that if there is something metaphysically real or positive in pain, pain will be *materially* in God, that is, its reality and positivity will be in God (A II 1^2, 488). It is implied that, on the contrary, pain *as such* (that is, *formally*) is not in God.

[22] Cf. Leibniz to Eckhard, summer 1677 (A II 1^2, 543): 'The concept which you assign to the *Ens perfectissimum* not only does not imply contradiction, but produces, or contains *eminently*, every other perfection' (my emphasis).

convinced Leibniz, who drops his objection and explicitly states some twenty years later, in a note of 1695: 'In the divine essence, things are contained eminently; in the intellect, they are contained somewhat more widely, indeed representatively, because in the divine intellect are represented also the imperfections or limitations of things. … Hence it is manifest that all things are in God' (Grua 355–356).

Later on, in a remark of 1706 meant for his friend, the Jesuit Bartholomew Des Bosses, Leibniz adds: 'there is a *hypercategorematic infinite*, or potestative infinite, and active power having, as it were, parts eminently but not formally or actually. This infinite is God himself' (LDB 52–53). In order to protect the pure positivity of God, Leibniz seems to turn to a metaphysical model inspired by the Neoplatonic 'One'. Only what is *beyond* all determinations (or, as Leibniz puts it, what is *hypercategorematic*),[23] while containing *eminently* all determinations, can be the ontological grounding of all things (*omnia*) without being tainted by the negation which comes with any determination. The Neoplatonic inspiration of Leibniz's thinking on the infinite is also strikingly apparent in *De Scientia Infiniti* (c. 1698). On the one hand, Leibniz insists

> that things exist by the participation of Being itself, that is, through the participation of the First Being, and that unities, good things, and beautiful things exist by the participation of the one itself, of the good itself and of beauty itself, that is, by benefit of absolute reality or goodness, which is in the prime substance.[24]

On the other hand, he stresses that, although all 'natures capable of the absolute … are contained [insunt] in the Substance of absolute reality, or of the utmost perfection', 'things are not parts of God':

> the absolute should not be thought to be like a whole which comprises limited things of its own kind (as certain people think the immense substance is the universe of things itself), for what is constituted by parts has a nature posterior to its parts, whereas the absolute is the origin of limited things.[25]

In turn, he explains the derivation of finite things in terms of the *addition* of limits,[26] suggesting, once again, a top-down Neoplatonic model characterized by the priority of the absolute or perfect[27]: limited things originate from what is purely positive (and, therefore, beyond all determinations), through the addition of determinations (and, therefore, negations).

Finally, Leibniz also seems to need some version of the Neoplatonic distinction between One and Intellect to counter the objection that God's knowledge of limited

---

[23] The Plotinian One is, of course, totally ineffable and beyond even the category of being.

[24] LH 35, 7, 10, Bl. 5r-8v (transcription and translation by Arthur and Ottaviani).

[25] LH 35, 7, 10, Bl. 5r-8v (transcription and translation by Arthur and Ottaviani).

[26] LH 35, 7, 10, Bl. 5r-8v (transcription and translation by Arthur and Ottaviani): 'certain circumscribing limits are added, excluding things beyond'; 'the absolute is found in every notion or nature, as long as nothing limiting is added to it'; variant: 'in addition to the nature of the absolute, limits are added, from which imperfection arises…'; variant: 'it is necessary that limits are added to the nature of the absolute in order for anything more restricted to arise'.

[27] LH 35, 7, 10, Bl. 5r-8v (transcription and translation by Arthur and Ottaviani): 'absolutum est fons realitatis et prius limitato'; 'Substantiae absolutae realitatis, seu summae perfectionis'.

things introduces negation in God. As he acknowledges in the note of 1695 quoted above, 'in the divine intellect are represented also the imperfections or limitations of things' (Grua 355). That is, the divine intellect includes also the representation of negations. Leibniz seems to think, however, that the mere mental representation of negations by the divine mind does not threaten the ontological pure positivity of the *essence* of God. Thanks to its containing all things purely eminently, the divine essence is, in itself, still beyond any determination. For Leibniz, this seems sufficient to ensure the absolute positivity of God's own nature.

7 Conclusion

The crucial feature which keeps created substances from matching the 'absolute infinity' of God is not activity, indivisibility, simplicity, or even infinity. Also created monads are active, indivisible, and simple. They enjoy also a kind of infinity since through their confused perceptions 'individual substances ... involve the infinite' (Dutens V 147).[28] What they lack, however, is pure positivity. As Leibniz writes in April 1679 in a variant of *Calculus consequentiarum*: 'in truth only the notion of God is purely positive, and involves no limitation or negation [Et vero sola notio Dei pure positiva est, nullamque limitationem seu negationem involvit].' (A VI 4, 223) Pure positivity is the metaphysical tool which allows Leibniz to escape (narrowly) Spinozism and pantheism. In so far as the divine essence cannot formally contain any limited perfection without losing its pure positivity, limited perfections require distinct substances as their bearers.[29] The absolute ontological incompatibility between God and any negation (*seu* limitation) whatever is what justifies (at least to Leibniz's satisfaction) the metaphysical need for limited substances distinct from God.[30] Pure positivity is what, ultimately, distinguishes the divine nature from the nature of created things. Far from being a rabbit magically pulled from a hat in a feeble attempt to rescue the ontological argument, pure positivity plays a key role in Leibniz's metaphysics, aligning with some of its most fundamental commitments.[31]

---

[28] Cf. also Leibniz to Pierre Bayle, c. 1702 (GP III 72).

[29] See Adams (1994: 123–134).

[30] Interestingly, in a text of c. 1683–1685, the notions of 'absolute' and 'limited' are employed to define the difference between God and creature under the category of 'substance': 'Substantiae[.] *Deus* Ens absolutum. *Creatura* Ens limitatum' (*Genera Terminorum. Substantiae*; A VI 4 567).

[31] I am very grateful to Richard Arthur, Vincenzo de Risi, and the anonymous referees consulted by the editors for their feedback.

Abbreviations

A	Leibniz, G. W. *Sämtliche Schriften und Briefe*. Ed. by the Academy of Sciences of Berlin. Series I-VIII. Darmstadt - Leipzig - Berlin, 1923 ff. Cited by series, volume, and page. The superscript '2' after the volume number indicates the second edition of the volume.
Arthur	Leibniz, G. W. *The Labyrinth of the* Continuum: *Writings on the* Continuum *Problem, 1672–1686*. Translated, edited, and with an introduction by R. T. W. Arthur. New Haven: Yale University Press, 2001.
Dutens	Leibniz, G. W. *Opera omnia, nunc primum collecta, in classes distributa, praefationibus et indicibus exornata*. Ed. by L. Dutens. 6 vols. Geneva: De Tournes, 1768. Cited by volume and page.
GP	Leibniz, G. W. *Die Philosophischen Schriften*. Ed. by C. I. Gerhardt. 7 vols. Berlin: Weidmannsche Buchhandlung, 1875–1890. Reprint, Hildesheim: Olms, 1960–1961. Cited by volume and page.
Grua	Leibniz, G. W. *Textes inédits d'après les manuscrits de la Bibliothèque Provinciale de Hanovre*. Ed. by Gaston Grua. 2 vols. Paris: PUF, 1948.
LDB	*The Leibniz-Des Bosses correspondence*. Translated, edited, and with an introduction by Brandon C. Look and Donald Rutherford. New Haven - London: Yale University Press, 2007.
Mollat	*Rechsphilosophisches aus Leibnizens Ungedruckten Schriften*. Ed. by Georg Mollat. Leipzig: Verlag Robolski, 1885.
NE	Leibniz, G. W. *New Essays on Human Understanding*. Ed. and trans. by Peter Remnant and Jonathan Bennett. Cambridge: Cambridge University Press, 1981.
PW	*The Political Writings of Leibniz*. Trans. and ed. with an introduction by Patrick Riley. Cambridge: Cambridge University Press, 1972.

References

Adams, R. M. (1994). *Leibniz: Determinist, Theist, Idealist*. Oxford University Press.

Antognazza, M. R. (2015). The hypercategorematic infinite. *The Leibniz Review, 25*, 5–30.

Antognazza, M. R. (2016a). *Leibniz: A very short introduction*. Oxford University Press.

Antognazza, M. R. (2016b). God, creatures, and neoplatonism in Leibniz. In Wenchao Li (Ed.), *Für unser Glück oder das Glück anderer. X. Internationaler Leibniz-Kongress*, (vol. 3, 351–364). Olms.

Antognazza, M. R. (2018). Leibniz. In G. Oppy (Ed.), *Ontological arguments* (pp. 75–98). Cambridge University Press.

Gerhardt, C. I. (1875). Zum zweihundertjahrigen Jubiläum der Entdeckung des Algorithmus der höheren Analysis durch Leibniz. *Monatsberichte Der Königlich Preußischen Akademie Der Wissenschaften, 1875*, 595–608.

Hegel, G.W.F. (2009). *Heidelberg Writing*. (Trans. B. Bowman & A. Speight). Cambridge University Press.

Jolley, N. (2005). *Leibniz*. Routledge.

Mercer, C. (2001). *Leibniz's metaphysics: Its origins and development*. Cambridge University Press.
Riley, P. (1996). *Leibniz's universal jurisprudence: Justice as the charity of the wise*. Harvard University Press.
Riley, P. (2003). Leibniz's Méditation sur la notion commune de la justice, 1703–2003. *The Leibniz Review, 13*, 67–78.
Rutherford, D. (1995). *Leibniz and the rational order of nature*. Cambridge University Press.
Spinoza, B. (1925). Carl Gebhardt (Ed.), *Spinoza Opera*, 4 vols. Carl Winter.
Wilkins, J. (1668). *Essay towards a Real Character and a Philosophical Language*.

Essentialism, Super-Essentialism and/or Anti-Essentialism in Leibniz

Stefano Di Bella

1 Is There Anything Essential to an Individual?

In his *Essay Concerning Human Understanding* (Book 3, Chapter 6) John Locke launches a vigorous attack on the essentialist assumptions underlying traditional (generally speaking: Aristotelian) metaphysical and epistemological claims. The most radical objection he raises concerns the very possibility of making sense of the attribution of "essential" qualities or properties to individuals:

> That *Essence*, in the ordinary use of the word, relates to *Sorts*, and that it is considered in particular beings, no farther than as they are ranked into sorts, appears from hence: that, take but away the abstract *Ideas*, by which we sort Individuals, and rank them under common Names, and then the thought of anything *essential* to any of them instantly vanishes ... 'Tis necessary for me to be as I am; God and Nature has made me so: but there is nothing I have is essential to me. An Accident or Disease, may very much alter my Colour, or Shape; a Fever or Fall, may take away my Reason, or Memory, or both; and an Apoplexy leave neither Sense, nor Understanding, nor life ... None of these are essential to the one or the other, or to any Individual whatever, till the Mind refers it to some Sort or *Species* of things; and then presently, according to the abstract *Idea* of that sort, something is found *essential*.[1]

In other words, every essential predication would be relative only to some (general) description of our own; were it not for our own classificatory activity, attributing an

[1] This research was funded by the Department of Philosophy "Piero Martinetti" of the University of Milan under the Project of Excellence 2018–2022 awarded by the Ministry of Education, University and Research (MIUR). Abbreviations: A = G.W.Leibniz, *Sämtliche Werke*, Akademie Ausgabe, 1923- (series and volumes); GP = G.W.Leibniz, *Die philosophischen Schriften*, ed. by C. I. Gerhardt, Berlin 1875–1890, vols. 1–7; L = G.W. Leibniz, *Philosophical Papers and Letters*, transl. by L. E. Loemker, Dordrecht, 1969; Lodge = G.W.Leibniz, The Leibniz-De Volder Correspondence, transl. and ed. by P. Lodge, Yale, 2013.

(Locke 1975, Bk. III, Ch. 6, § 4, p. 440).

S. Di Bella (✉)
Dipartimento Di Filosofia, Universita Degli Studi di Milano, Milano, Italy
e-mail: Stefano.dibella@unimi.it

F. Ademollo et al. (eds.), *Thinking and Calculating*, Logic, Epistemology, and the Unity of Science 54, https://doi.org/10.1007/978-3-030-97303-2_14

"essential" property to an object (and also, presumably, making a related *de re* necessary predication) would make no sense at all. Leibniz's response to this argument of Locke's is as neat as it is elliptical: "I believe that there is something essential to individuals, and more than there is thought to be"[2] In fact, his examples seem simply to re-propose the traditional understanding of the notion of "essential"[3]: namely, the idea that there are some properties of an individual which, though common to all members of its species, are nonetheless fundamentally constitutive of said individual's identity, so that this latter cannot fail, or cease, to possess these properties on pain of ceasing thereby to exist. This implies an intuitive distinction, within the set of all the properties of an object, between the subset of such essential properties and another subset consisting of accidental properties, which the object can cease to possess without thereby ceasing to exist. In short, we are faced with the core intuition of "old-fashioned Aristotelian essentialism", critiqued in some well-known texts of Quine's.[4]

So far so good; and certainly, Leibniz's remark concords with the willingness he exhibits, in this and in other chapters of the *New Essays*, to defend some kindred traditional tenets against Locke's massive attack on essentialism and the related threat of some sort of conventionalist drift. Nevertheless, it must be seen as at least a possibility that Leibniz's rather vague hint regarding the "more things" essential to the individual than there are usually thought to be might conceal a commitment on Leibniz's part to some more robust, and far less obvious, thesis.

A reader acquainted with his private writings and correspondence, indeed, would have known how Leibniz, approximately twenty years before, had endorsed a view whereby *all* properties of an individual would form part of its definition; nor had he stopped short of embracing the most controversial and counter-intuitive consequences of this view, such as the flat denial of counterfactual identity. The reason for his adopting this astonishing stance had been his endorsement of the "complete concept" view: that is to say, of the idea—well documented in the *Discourse of Metaphysics* and in the correspondence with Antoine Arnauld, but also in many other private drafts[5]—that to be an individual substance amounts to possessing a concept from which *all* predicates of the corresponding individual can in principle be deduced: on this account, then, the concept would function as a metaphysical individuator, grounding every identity statement concerning its corresponding substance.

---

[2] *New Essays on Human Understanding*, Bk. III, Ch. 6 (Leibniz 1996, 305).

[3] "... there are sorts or species such that if an individual has ever been of such sort or species it cannot (naturally, at least) stop being of it, no matter what great events may occur in the natural realm. But I agree that some sorts or species are accidental to the individuals which are of them, and an individual can stop being of such a sort" (Leibniz 1996, 305).

[4] "Essentialism is the doctrine that some of the attributes of a thing (quite independently of the way we refer to the thing) are essential to it, while others are accidental" (Quine 1953). As is well known, Quine thought that quantified modal logic was committed to this untenable doctrine.

[5] See in particular *Discourse of Metaphysics*, §§ 8 and 13, and the subsequent discussion with Arnauld on the thesis of § 13 in the letters from March to July 1686.

Exactly this view was dubbed by Fabrizio Mondadori, in two seminal papers published in the 1970s,[6] as “super-essentialism”. With this label Mondadori intended precisely to capture the thesis whereby *every* property of an individual would be essential to it, insofar as the individual in question could not exist without possessing it. On the basis of the textual evidence available Leibniz’s endorsement of this claim is undisputable, and Mondadori was entirely right to defend this reading against every attack, and to draw a distinction between, on the one hand, Leibniz’s substantive adherence to the claim in this form and, on the other hand, his rejection of the same claim in cases where *other* senses of “essential” were at issue[7] (cases which I shall consider in detail below). Nevertheless, even in the light of Mondadori’s clarifications, the real metaphysical (and, consequently, modal) import of Leibniz’s thesis remained to a large extent a puzzling one.

2 Super-Essentialism or Anti-Essentialism? A Possible Ambivalence in the “Complete Concept” Doctrine

Put quite simply, the label “super-essentialism” seems to suggest to the reader the idea of a reinforcing of the essentialist intuition. Is this, however, really and unambiguously the case? Or—surprising as this may be—could Leibniz’s seemingly outrageous thesis be understood, on the contrary, as the symptom or consequence of a radical weakening of this very essentialist intuition? It should be borne in mind that what is crucial, on the standard essentialist view—i.e. on the view of the “old-fashioned Aristotelian essentialism” to which, according to Quine, *de re* modal logic was fatally committed—is precisely the possibility of drawing a fundamental distinction between essential properties on one hand and accidental ones on the other. The critics of essentialism are eager to challenge exactly the possibility of legitimately drawing such a distinction. As a matter of fact, however, the so-called super-essentialist view also tends to abolish such a distinction. Are we sure, then, that the rationale behind this thesis is really so far removed from the concerns which typically come to expression in anti-essentialist criticism?

Benson Mates—an interpreter who shares (and even anticipated) Mondadori’s emphasis on world-bound individuals, as well as the related account of modality (with contingency explained as possible non-existence, and a counterpart-theoretical account of counterfactual talk)—provides an interesting confirmation of our suspicion here. In a pioneering paper in which he attempted to make sense of Leibniz’s

[6] (Mondadori 1973, 1975).

[7] Mondadori’s super-essentialist thesis has been critiqued, by appeal to several distinctions to be found in Leibniz’s texts, by (Hunter 1981) and later and especially in R. Sleigh’s influential commentary on the Leibniz-Arnauld correspondence (Sleigh 1990). For Mondadori’s detailed replies to his critics, see his (Mondadori 1985) and (Mondadori 1993). For the general discussion on Leibniz’s metaphysics of individual substance and counterfactual nonidentity, see also (Adams 1994); (Cover and Leary Hawthorne 1998); (Di Bella 2005).

paradoxical thesis,[8] he pointed out the Leibnizian view of continuity as applied to "forms" or concepts of (possible) beings. According to Mates's conjectural reconstruction, given that this continuity would not allow for any clear-cut boundary within the series of individual concepts—giving rise, in fact, to a situation such as the one illustrated by Chisholm's puzzle[9]—it would not make sense to allow for the slightest change within the properties of one and the same individual concept: "The concept of any individual, by a series of the same sort of relatively 'slight' changes that are not regarded as crucial for identity, could be gradually transformed into the concept of anything whatever. Once we allow that an individual could have had some attributes other than the ones he does have, there is no 'natural' place to draw the line. There is no plausible way of dividing the attributes of an individual into two non-empty classes, the essential and the accidental"[10] Now, it is easy to see how an alleged motivation of this kind for super-essentialism, far from resulting in a more robust version of "essentialism", would turn out to be fully in accord with typically Quinean concerns about essentialism as such.

3 Individual Essences?

I do not, admittedly, feel able, ultimately, to subscribe to Mates's conjectural reading of Leibniz's motivations, although I think that this reading really does capture some aspects of Leibniz's approach. Mondadori, for his part, provided a different explanation for Leibniz's theses, by emphasizing the role of the principle of reason (interpreted, in its turn, through the containment theory of truth) and that of the "complete concept" as a metaphysical individuator. We know, indeed, how Leibniz in *Discourse* § 8 explicitly speaks about the "complete concept" as an "haecceity", making play of this old Scotist tool: "God…in seeing the individual notion or 'haecceity' of Alexander, sees in it at the same time the basis and the reason for all the predicates which can truly be affirmed of it".[11] Notice how, in this context, an individual essence (provided that we can take 'haecceity' to indeed signify such an individual essence) does not seem to be equated with the set of all the properties of the individual but rather with a privileged subset, functioning as a principle of deduction from which all other properties can be, in principle, derived. On this reading, the lack of a distinction between essential and accidental predicates—far from simply being a result of the impossibility of drawing any metaphysically grounded division between the two—would allow for a different distinction between primitive and derivative properties—both being essential.

---

[8] (Mates 1972).

[9] See (Chisholm 1979).

[10] (Mates 1972, 117).

[11] L 308. For Leibniz's handling of Scotus's concept, from the sharp criticism in his early *Disputatio de principio individui* to the mature alleged re-evaluation, see (Di Bella 2014). On the whole issue of the development of Leibniz's views about individuation, see (Mugnai 2001).

If, in the passage quoted above, the terminology of "essence" does not, strictly and explicitly, appear, we find it nonetheless in § 16 of the same *Discourse*, where Leibniz—when dealing with the issue of miracles, which are also to be included in the "complete concept" of the substance which they befall—explicitly equates the individual concept with the essence: "... We can say that this extraordinary action of God upon this substance is always miraculous, though it is included in the general order of the universe insofar as that order is expressed by *the essence or individual concept* of this substance".[12] Interestingly enough, in this one particular context Leibniz contrasts, terminologically, the pair "individual concept/essence" with a notion of "nature" epistemologically defined.[13]

4 Criticizing "Old-Fashioned Essentialism": "Second Substance" Terms and Natural Kinds

In any event, we should consider how Leibniz's apparent approach to the idea of an individual essence proceeds from a background quite different from the traditional Aristotelian-Scholastic framework. In the latter, indeed, individual essence was a (controversial) extension of the standard notion of essence, which had been a notion focused on *specific* essences. Leibniz's view, on the contrary, seems to presuppose a generally weak status for these other types of essences. In order to clarify this, let me briefly consider his view of species concepts.

Leibniz was in fact far more sceptical, during his whole career, concerning the ontological and epistemological status of species concepts than may initially appear from his critical remarks on Locke's views in the *New Essays*. Aristotelian essentialism, as indicated above, was centred on *specific* essences corresponding, intuitively, to natural kinds such as 'man' or 'dog'—the so-called "second substances" of the ontology of the *Categories*. The nominalist tradition, however, had vigorously challenged the ontological import of such "second substances" while trying to preserve the epistemological value of the corresponding concepts. But once these concepts had been downgraded from the status of expressions of real essences to that of mere classificatory devices elaborated through our own abstractive procedures and based on similarities, the way was clearly paved for their relativization.

Now, the scholarship of recent decades has shown how much Leibniz's general attitude regarding ontological issues was shaped by a nominalistically-minded background.[14] Already his earlier writings—such as the *Preface* to Nizolius, with its

---

[12] L 313 (my italics).

[13] "Everything which is called natural ... depends on the general maxims which creatures can understand" (ibidem). In its metaphysical sense, however, the concept of individual nature is taken to be synonymous with individual essence. For more on this topic see below.

[14] See, on this question, the pioneering works of (Mates 1980, 1986, esp. Ch. 10) and (Mugnai 1992). See now also (Di Bella 2017).

explicit re-evaluation of nominalism—testify to his firm refusal to assign a positive ontological status to essences, be it either in a Platonic sense (as something over and above existing particulars) or in the sense of moderate realism (as something present within particulars themselves). But a decided weakening of the role of "second substances" and of the corresponding specific essences can be well documented also in the later, private writings of Leibniz's mature years. Thus, for instance, in an important draft from among the writings surrounding the *Discourse*, the *Notationes Generales*, he gives a standard nominalist account of general reference: general concepts simply stand for a plurality of individual things: "A universal substance stands for any singular substance having something common with others".[15]

Moreover, he very carefully distinguishes his own particular view of the *species infima*—which he equates with the individual concept—from the traditional one, modelled on the concepts of natural kinds; he also adopts a quite deflationary and relativizing attitude towards the latter: "When I say that the difference of individual men from one another is a difference of 'the lowest species', I do not use the term 'species' in its most usual sense—i.e. that of some type of being able to generate other beings similar to itself, such as is the case for the species of men, dogs, roses etc. (*although this notion of species is itself not sufficiently clear and one might wonder whether wolves and household canines, or giant Molossian and little Maltese dogs, do in fact belong to different species.*)"[16]

This attitude, admittedly, was to be greatly attenuated, as noted above, in the *New Essays*. Nevertheless, an attentive reading of the relevant chapters of that work would, in my opinion, show us that we are faced there more with a difference in emphasis (and in argumentative aim) than with a substantive change of mind. Thus, in the same chapter of the *New Essays* we find a distinction between a 'mathematical' and a 'physical' sense of "species" quite in accord with the deflationary remarks in the *Notationes Generales* of twenty years before.[17] We should note also that, a few pages prior to this, Leibniz is willing to subscribe to the emphasis that Locke placed on the continuity of forms: it is just that he refrains, for his own part, from drawing the relativizing and anti-essentialist "moral" which Locke was eager to draw from the same set of facts.[18]

[15] A VI 4, 554.

[16] A VI 3, 553–54 (my italics).

[17] "All this trouble arises from a certain ambiguity in the term 'species' or 'of different species' … One can understand 'species' mathematically or else physically. In mathematical strictness, the tiniest difference which stops two things from being alike in all respects makes them of 'different species'… Two physical individuals will never be perfectly of the same species in this manner, because they will never be perfectly alike; and furthermore, a single individual will move from species to species, for it is never entirely similar to itself for more than a moment. But when men settle on physical species, they do not abide by such rigorous standards … in the case of organic bodies—i.e. the species of plants and animals—we define species by generation…" *New Essays*, Bk, III, Ch. 6 (Leibniz 1996, 308–309). This sense of 'mathematical strictness', it should be noted, corresponds to what Leibniz calls "metaphysical rigour". Elsewhere, he will prefer to contrast mathematical concepts with individual essences (see below).

[18] See ibidem, § 12 (Leibniz 1996, 306–307).

5 Criticizing the New Essentialism: Varieties of Essentialism and Two Types of Leibnizian Essence

A dismissive attitude towards Aristotelian essentialism was not something peculiar to Leibniz. The claim to be able to capture the real essences of things through merely 'verbal definitions' had been sharply criticized by all Renaissance and post-Renaissance critics of Aristotelian Scholasticism. Nor did an empirically-based inductive working out of concepts appear sufficient to legitimate the hold of these concepts on real essences. Rather, the new Cartesian ideal of science had advanced a view of essences as modelled on the paradigm of mathematical knowledge, by which the properties of these essences were entailed as if by a kind of conceptual necessity.

Leibniz, however, was unsatisfied also with this type of essentialism. Thus in the same passage of the *Notationes Generales,* quoted above, in which he distances his *species infimae* from the species taken as natural kinds, he contrasts them also with universal concepts in general: "Nor do I mean a universal notion, i. e. a concept built up from a finite number of simpler concepts taken together".[19] It should be borne in mind that, for Leibniz, such concepts of finite complexity—typically, the type of abstract definitions suited for mathematical objects—provided the basis for procedures of demonstration, the latter being defined precisely by the possibility of their being achieved via a finite number of steps.

Now Leibniz, in discussing with Cartesian-minded authors like Arnauld and de Volder—whose intuitions rely chiefly on this latter sense of "essential"—is always keen to emphasize the need to sharply distinguish such abstract ("incomplete") concepts from the "complete" ones.[20]

Indeed, Leibniz goes so far as to distinguish, *within* the complete concept of an individual, a subset of properties corresponding to the abstract or "incomplete" concepts upon which our general knowledge and the traditional varieties of essentialism are grounded, these latter consisting either in the specific concepts which constitute the definition of a natural kind (for example: 'man is a rational animal') or in the concepts of mathematical objects (for example: a sphere), both being contrasted with their concrete particular instances (for example: 'this man' or 'the sphere on Archimedes's tomb'). It is for this subset alone that Leibniz sometimes wishes to reserve a "strict" sense of "essence" and "essential": namely, a sense which is clearly connected to a modal feature (necessity), this latter being in its turn reduced to a logico-epistemological one (demonstrability). Thus, in the draft *De libertate fato gratia* we read: "...In this complete notion of possible Peter that appears to God are contained, I concede, not only essential or necessary things (which, of course, flow from incomplete or specific notions and are thus demonstrated from terms, such that the contrary involves a contradiction) but also existential [*existentialia*] or, so to speak, contingent things."...".[21]

---

[19] A VI 3, 553–554.

[20] Leibniz to Arnauld, L 332. Leibniz to de Volder, GP II 277.

[21] A VI 4, 309.

As Mondadori has rightly pointed out, the (different) sense of "essential" which he (Mondadori) links to "super-essentialism" (that is to say, to the denial of counterfactual identity) can perfectly well coexist with this terminological distinction; and one can perfectly well read this latter, as Mondadori does, as the distinction between '*specifically* essential properties' and '*individually* essential' ones. In any event, it is important to note how the distinction is a relevant one both from a modal and from a metaphysical point of view; and indeed Leibniz prefers to reserve the terminology of 'essence/essential' for general concepts, and to associate this terminology always with a modal feature: namely, necessity.

The terminology of "existentialia" employed in the passage quoted above is a very interesting one. It does not imply actual, but only possible existence. It expresses, however, a concept of possibility which is oriented toward existence. Whereas the 'strict essential' subset corresponds to a merely abstract object, incapable of existing as such, a "complete concept" includes all features which make up a possible concrete object, susceptible of coming, or being brought, into existence as such.

It is important to take note of the fact that the difference between the two types of concepts does not concern just the richness in descriptive power of the concepts concerned, that is to say the number of predicates contained in the one and the other—these being finite in one and infinite in the other. The type of properties implied and the internal connections applying are also fundamentally different in the two cases.

Thus we can see that the term "possibility", as employed by Leibniz, is fundamentally equivocal between these two different senses,[22] being applicable—given Leibniz's ongoing equivalence between possibility and essence—respectively to general and to individual essences. Furthermore, the models in which these two senses are intuitively rooted are different: on the one hand a mathematical model which envisages the deducibility of a set of timeless properties from a definition; on the other hand, a concept modelled on the history of a concrete individual. In the second case, the 'properties' involved constitute a series of states, causally connected and temporally ordered (I shall return briefly, later on, to this temporal aspect).

That the dichotomy in question can be extended also to essences—despite Leibniz's preferring to reserve the use of this vocabulary of "essence" and "essential" for general possibilities alone—is documented by an interesting exchange in the correspondence with the Cartesian physicist Burcher de Volder. The latter had remarked that only existences, and not essences, require a cause. Leibniz replies that "the concept of a possible cause is required to conceive of its [sc. of the substance's] essence and the concept of an actual cause is required to conceive of its existence"[23]—where there is taken into account the causal-genetic account typical of the second type of concepts of possible things.[24]

---

[22] A very subtle clarification of the distinction can be found in a recent work by (Ottaviani 2018).

[23] Lodge 207, GP II 225.

[24] Leibniz's counter, to be true, applies the genetic account to a geometric example (the ellipse), hence to an incomplete concept; but at the end he refers it to complete concepts, the only properly suited for substances. See Ibidem.

6 The Essential and the Intrinsic: An (Attempted) Way Out of the Modal Problem?

Evident in certain passages is an intention on Leibniz's part to narrow the scope of necessity to the sphere of what is *strictly* essential, or to the *specific* essential properties. Accordingly, in his discussion with Arnauld, as is well known, he also tried to propose a concept of '*intrinsic* properties' which would comprise *all* the properties of an individual (including those which are, in Mondadori's phrase, "individually essential"), thus confirming the "complete concept" view while at the same time being devoid of necessity.

Is this move a plausible, or at least a sensible, one? I will venture only a remark on the margin of this vexed issue. If we take seriously (as we must do) the denial of counterfactual identity and then go on to interpret necessity in "possible-world" terms, then the undesired modal consequence is not to be avoided; from this point of view, every alleged distinction between "super-intrinsicalness" and super-essentialism turns out to be vacuous.[25] It seems, however, that Leibniz's way of approaching the issue is a bit different from ours. It is worth noting how his intrinsic/necessary distinction immediately precedes his considerations about individual identity. Surprising though it may be, therefore, he does not hesitate—even in the midst of his engagement in attempting to avoid the charge levelled by Arnauld against the alleged 'fatalistic' consequences of his doctrine of "complete concepts" and immediately after what he presents as his own solution to this problem—to propose and emphasize his counter-intuitive view regarding counterfactual non-identity. It would appear, then, that the modal threat, for Leibniz, is basically distinct from the further issue of making sense of identity in counterfactual situations.

It appears that the only thing that counts for Leibniz is the qualification of necessity—where only the (broadly) logical one is to be avoided. He is prepared to allow, it should be noted, that intrinsic connection too implies, as such, some kind of necessity; but it would be, on this account, at most a hypothetical, or in other terms a physical, one.

As for the (non-) identity statement, it seems to belong, in Leibniz's view, to another level of discourse: given a certain type of nomological pattern governing the story of an individual (the necessity of which, however, would not be a logical one), this pattern turns out to be the intuitive precondition for our making sense of said entity's sameness; that is to say, it turns out to be that which alone accounts for our being able meaningfully to talk of "the same individual". Moreover, we should not forget that in texts such as the first part of the Arnauld correspondence, or in other contexts relative to Leibniz's key problem of theodicy, the emphasis placed on the "compactness" of a ("complete") individual concept—from which there follows the inconceivability of counterfactual identity—serves a specific and precise function within a broader theodicean strategy. This inasmuch as one of the

[25] Robert C. Sleigh relied on this alleged distinction between "superintrinsicalness" (the term was coined by him) and "superssentialism", in (Sleigh 1990, esp. Ch. 4) in order to avoid the modal strictures of the latter view. For a criticism of this reading, see (Mondadori 1993).

consequences of Leibniz's emphasis on the *en bloc* character of each individual concept is, for example, that God cannot, properly speaking, be said to have chosen that Judas should commit his sin but rather only that "Judas-the-sinner" should come to be.[26] On the other hand, this strategy is fundamentally weakened in this very same reply of Leibniz's to Arnauld through Leibniz's inclusion of divine "possible decrees" within the concept. Interestingly enough, the possibility of this move—which, it should be noted, ought, in Leibniz's view, to have justified the contingency of intrinsic inclusion—is linked up by Leibniz precisely with his second type of concepts or essences, as contrasted to the specific ones: "The concept of a species does not include anything but eternal or necessary truths, whereas the concept of an individual involves, regarded as possible, what is factual and refers to the existence of things and to time."[27]

7 Making Sense of Intrinsicality

While trying to avoid undesired modal consequences, Leibniz was eager to maintain, for his concept of intrinsic connection, a sufficiently robust metaphysical significance. As indicated above, such connection, while devoid of (broadly) logical necessity, does nevertheless imply a *de re* necessity of the physical or nomological type. In a way, Leibniz's idea of intrinsicality tends to recover the original Aristotelian sense of "natural"—that is to say, the very core of Aristotelian essentialism—which was related in that tradition to "natural kinds". According to this view, indeed, the "essential" is neither what is implied by a (more or less) arbitrary definition through stipulation nor what conforms to the mathematical model of Cartesian essentialism. The "essential" is rather what constitutes the fundamental structure identifying a natural kind, typically including a determinate pattern of development and action. As a matter of fact, essentialism connects here with the other main philosophical pillar of the Aristotelian metaphysical and scientific framework: namely, hylomorphism, or the view whereby a type of substance is identified by the sharing of a substantial form, this form being the principle of sameness and unity underlying each individual belonging to the type in question, and determining a certain pattern of change. Now, this view was even more sharply challenged at the time of seventeenth century scientific revolution. We know, however, how, in those same texts of the 1680s to which we are referring, Leibniz had, surprisingly, attempted to effect a sort of rehabilitation of the decried notion of substantial form.

Without entering into the details of this Leibnizian attempt to rethink hylomorphism in connection with his discoveries in the field of dynamics, we may say that what is interesting here, at a general logico-ontological level, is Leibniz's eagerness to re-conceive of the ancient Aristotelian sense of a "nature" as a "principle of change".

[26] For this strategy, see, among many other passages, *Discourse of Metaphysics*, § 30; *De libertate fato gratia Dei*, A VI 4, 1603.

[27] Leibniz to Arnauld, GP II 39.

This intuition is documented in a draft from the mid-eighties: "...[According to Aristotle] a thing's nature is the principle or cause by virtue of which the thing in which such a nature inheres immediately and *per se* (i.e. not *per accidens*) both moves and rests. These statements would seem to be absurd and meaningless, were they not clarified through my discoveries: I have proven, indeed, that each being *per se*, or each being that is truly one, has in itself some principle from which there ensues all that does indeed ensue to it naturally or *per se*, and that this is something analogous to the soul, and is also that which Aristotle meant when he spoke of a being's 'nature'... It is indeed the case, then, that it is out of the individual nature of each individual thing that all that which ensues to this latter ensues; that is to say: nothing happens *per accidens* to the individual. Something happens *per accidens* to the species, however: so that, for example, to 'man' happens *per accidens* to be musical, while it does not happen to Peter."[28] It should be noted that Leibniz here attempts to make sense, through this concept of a thing's "nature", of the other Aristotelian concept of *per se* or *kath'hauto* predication, strictly bound up with that of essential predication.

As is clear from the last lines of the passage quoted, a major difference between Leibniz's conception and the original Aristotelian model consists, once again, in the fact that "essence", also in this sense of "principle of change", is not grasped at the level of the species, but rather at that of the individual.

8 Change, Time and the "Essence"/"Nature" Pair

Conceiving of an individual entity's "nature" as a principle of change implied a quite different approach to the dimension of temporality from that commonly associated with the idea of essence and, in particular, with certain standard varieties of essentialism. Thus for instance de Volder, being an adherent to the Cartesian model of essentialism, was committed to the view expressed in his correspondence with Leibniz: "... I have always been convinced that whatever follows from the nature of a thing is always in the thing in an invariant way and cannot be removed from it, certainly as long as the nature of the thing remains the same, since there is a necessary connection between it and the very nature of the thing. Therefore, a change which occurs while the nature of a thing remains the same must necessarily be due to an external cause."[29] Leibniz, by contrast, is prepared to accept the inclusion of temporal predicates within a nature or within its corresponding concept. This retort to de Volder, it should be noted, is one made within the context of a vigorous defence of change (and thus of action) as an *essential* feature of (individual) substances. It is just that, for Leibniz, "... a distinction must be made between *properties*, which are permanent, and *modifications*, which are transitory. Whatever follows from the nature of a thing can follow either permanently or for a time..."[30] We could say, in

[28] *De natura sive analogo animae*, A VI 4, 1505.

[29] Lodge 273–275 (GP II 256).

[30] Lodge 279 (GP II 258).

other words, that also temporal predicates are included in the "complete concept" in a tenseless way, whereas the corresponding features or events happen to the individual in the course of time.

As noted above, patterns of temporal change were included within the definition of (specific) essences also on the Aristotelian essentialist model for "natural kinds"; and in the corresponding individual substances there was contained a corresponding equipment of dispositional properties, capable of becoming active and unfolding at the due time. But here too, the model is individualized in Leibniz. Thus, the dispositional facts embodied in a Leibnizian individual are not only general developmental patterns typical of the individual in question's species, but they correspond to particular temporalized properties and events, capable of actualizing themselves at a precise date.[31]

Finally, it is worth remarking that Leibniz sometimes reserves the denomination "essential" for atemporal predications alone. We see this from his correspondence with another Cartesian interlocutor, François Lamy, who, in a similar vein to de Volder, had objected that "what follows from the nature or essence of a thing must endure like the thing itself". Leibniz's reply proposes a new terminological distinction between "essential" and "natural": "People commonly distinguish what is essential from what is natural... Properties are essential and eternal, whereas modifications can be natural, although they are changing"[32] Also in the de Volder correspondence, it should be noted, we find the terminological equation essential = perpetual (i.e., omnitemporal) set over against the equation accidental = changeable: "... everything *accidental, i.e. mutable*, must be a modification of something *essential, i.e. perpetual*"[33]

Needless to say, this is a different (more restricted) sense of "essential" as compared to the sense which also embraces temporal predicates. As usual, we are faced with a battery of terminological distinctions which can change according to different contexts. In any event, temporal predicates do belong to the nature of a thing on a par with all other predicates. In this sense once again, then, the sense of "essential" relevant for super-essentialism remains intact. Nevertheless, the difference in respect of the way of inclusion is clearly relevant, both from a metaphysical and a modal point of view.

To sum up, then: Leibniz turns out to be engaged in a complex attempt to think out a new model of essentialism, one which is focused on the individual as such—in connection, naturally, with Leibniz's own idea of the "complete concept". The standard varieties of essentialism, which had been oriented to general concepts—the Aristotelian one being modelled on "natural kinds" and the Cartesian one modelled on mathematical objects—are now vigorously relativized by reference to this new Leibnizian notion. While the latter of these two earlier models—the Cartesian one—had served to suggest to Leibniz a strict concept of essence grounded in the notion of necessity, the special type of essentialism which is actually pursued by him, with his

[31] For a sketch of this model, see (Broad 1949).

[32] GP II 258.

[33] Lodge 307 (GP II 270); itaics mine.

reflection on the "complete concept", is one that is more inspired by the Aristotelian idea of a being's "nature" as an intrinsic principle of change—but with this latter idea transferred now from the level of the species to the level of the *individual* as such. In this way, Leibniz attempts—more or less successfully—to think out a metaphysically relevant notion of "being *per se*" relatively independent of his own notion of necessity.

References

Adams, R. M. (1994). *Leibniz. Theist, determinist, idealist*. Oxford University Press.

Broad, C. (1949). Leibniz's predicate-in-notion principle and some of its alleged consequences. *Theoria, 15*, 54–70.

Chisholm, R. (1979). Identity through possible worlds, some questions. In M. Loux (Ed.), *The possible and the actual. Readings in the Metaphysics of Modality* (pp. 80–87). Cornell University Press.

Cover, J. O' & Leary Hawthorne, J. (1998). *Substance and individuation in Leibniz*. Cambridge University Press.

Di Bella, S. (2005). *The science of the individual: Leibniz's ontology of individual substance*. Springer.

Di Bella, S. (2014). Angels, matter and haecceity: Scholastic topoi for Leibnizian individuation. *Studia Leibnitiana, 46*(2), 127–151.

Di Bella, S. (2017). Some perspectives on Leibniz's nominalism and its sources. In S. Di Bella & T. Schmaltz (Eds.), *Universals in early modern philosohy* (pp. 198–219). Oxford University Press.

Hunter, G. (1981). Leibniz and the super-essentialist misunderstanding. *Studia Leibnitiana, 13*(1), 123–132.

Leibniz, G. W. (1996). *New essays on human understanding*, ed. P. Remnant and J. Bennett. Cambridge University Press.

Locke, J. (1975). *An essay concerning human understanding*, ed. Peter H. Niditch. Clarendon Press.

Mates, B. (1972). Individuals and modality in the philosophy of Leibniz. *Studia Leibnitana, 4*(2), 81–118.

Mates, B. (1980). Nominalism and Evander's sword. *Studia Leibnitiana Supplementa, 21*, 213–223.

Mates, B. (1986). *The philosophy of Lebniz: Metaphysics and philosophy of language*. Oxford University Press.

Mondadori, F. (1973). Reference, essentialism, modality in Leibniz's metaphysics. *Studia Leibnitiana, 5*, 75–101.

Mondadori, F. (1975). Leibniz and the doctrine of inter-world identity. *Studia Leibnitiana, 7*, 21–57.

Mondadori, F. (1985). Understanding super-essentialism. *Studia Leibntiana, 17*, 162–190.

Mondadori, F. (1993). On some disputed questions in Leibniz's metaphysics. *Studia Leibnitiana, 25*(2), 153–173.

Mugnai, M. (1992). *Leibniz's theory of relations*. Steiner.

Mugnai, M. (2001). Leibniz on individuation: From the early years to the discourse and beyond. *Studia Leibnitiana, 33*(1), 36–54.

Ottaviani, O. (2018). *Modality, ontology and phenomenology. Leibniz's multiple views of existence. A historical and analytical reconstruction*. PhD Dissertation, Scuola Normale Superiore, Pisa.

Quine, W. O. (1953). Three grades of modal involvement. *Proceedings of the XIth International Congress of Philosophy, 14*, 65–81.

Sleigh, R. (1990). *Leibniz and Arnauld: A commentary on their correspondence*. Yale University Press.

Leibniz's Metaphysics of Change: Vague States and Physical Continuity

Richard T. W. Arthur

It is hard to see consistency in the various things Leibniz says about continuity, especially in connection with time. Bertrand Russell did not hesitate to point this out: "In spite of the law of continuity", he complained, "Leibniz's philosophy may be described as a complete denial of the continuous" (Russell, 1900, 111).[1] Thus in the face of Leibniz's many proud endorsements of that law, e.g. in the *Nouveaux essais* of 1704–1705,

> Nothing takes place suddenly, and it is one of my great and best confirmed axioms that nature never makes leaps. I call this the Law of Continuity… (A VI 6, 56)[2]

[1] J. E. McGuire echoes Russell's criticism: "the perceptual continuum is not truly continuous", and "Leibniz explicitly denies that actual substances are continuous in the sense of containing potential and indeterminate parts. On both these scores it seems that his philosophy is a denial of the continuous" (McGuire, 1976, 311).

[2] In this article I use the following standard abbreviations: A for Leibniz (1923–), GM for Leibniz (1849–1863), GP for Leibniz (1875–1890), WFT for Leibniz (1998), LLC for Leibniz (2001), LDB for Leibniz (2007), and LDV for Leibniz (2013).

I presented earlier versions of this paper to the Dipartimento di Filosofia, Università degli Studi di Milano, on May10, 2018; to SPHERE at Paris 7, on November 7, 2018; at a workshop at the Centre for Theoretical Studies, Charles University, Prague staged by the Leibnizian Society of the Central Europe jointly with SELLF; at the Society for Exact Philosophy Annual Meeting at York University in May, 2019; to the Department of Philosophy at Harvard University on November 15, 2019; at Chapman University, February 10, 2020; and at the Mexico-Canada Early Modern Conference (on-line) at Western University, October 2, 2020. I am much indebted to members of all those audiences for their input and criticisms, among which those of Enrico Pasini (in both Milan and Prague), Jagdish Hattiangadi (at York), and Samuel Levey, Jeffrey McDonough and Jen Nguyen (at Harvard) were particularly helpful.

R. T. W. Arthur (✉)
Department of Philosophy, McMaster University, Hamilton, ON, Canada
e-mail: rarthur@mcmaster.ca

F. Ademollo et al. (eds.), *Thinking and Calculating*, Logic, Epistemology, and the Unity of Science 54, https://doi.org/10.1007/978-3-030-97303-2_15

Russell was able to quote, for example, Leibniz's insisting to Burchard de Volder that

> In fact, matter is not continuous but discrete, and actually divided to infinity, even though no assignable part of space is devoid of matter. Yet space, like time, is something not substantial, but ideal, and consists in possibilities, that is, in an order of co-existents that is in some way possible. And so there are no divisions in it but those that the mind makes, and the part is posterior to the whole. In real things, on the contrary, unities are prior to multiplicity, and multiplicities exist only through unities. [It is the same with changes, which are not in reality continuous.] (To De Volder, Oct 11, 1705; GP II 278/LDV 327)

It is worth dwelling on the analogy Leibniz presents here between divisions in matter and changes. According to Leibniz, matter is divided in such a way that every part of it is divided internally, as is each of these parts, and so on ad infinitum. But no piece of matter is divided in all possible ways, where the points of division would correspond with all possible points in the continuum. Rather, there is an actual sequence of internal divisions that has no end. In an important early manuscript, Leibniz gives the idealized example of dividing every part, and part of a part, into halves, versus dividing them into thirds:

> Thus if you bisect a straight line and then any part of it, you will set up different divisions than if you trisect it. ("Primary Truths", [1689], A VI 4, 1648)

So, if "it is the same with changes, which are not in fact continuous", how does it go for the temporal case? Leibniz is clear that "as a physical body is to space, so states or the series of things are to time" (To De Volder, June 30, 1704; GP II 269/LDV 305). This would then imply that, just as "in real things, that is, bodies, the parts are not indefinite … but actually assigned in a definite way" (GP II 268/LDV 303), so the same should apply to states. The duration of any real thing should be divided into states—into a certain infinite order of states of finite length, each of them further subdivided. In what follows we will see that this is indeed Leibniz's view.

As noted, however, it is difficult to see how to reconcile this position with the Law of Continuity. If every duration is divided into a certain ordering of discrete states, divided from one another by discontinuous changes, then surely we have no continuity; instead it would be necessary for each thing to be resuscitated at each new moment of its existence. Yet in his *Theodicy* Leibniz criticized the interpretations of continuous creation proposed by Erhard Weigel and Pierre Bayle for making what sounds like precisely this error (Leibniz, 1710, §384–393; GP VI 343). Similarly, he had written to De Volder in September 1699:

> Anyone who completely rejects continuity in things will have to say that motion is essentially nothing but successive leaps through intervals flowing forth not from the nature of the thing but as a result of the action of God, that is to say, reproductions in separate places, and would philosophize almost as if one were to compose matter from mere separate points. (GP II 193/LDV 127)

"This hypothesis of leaps cannot be refuted," Leibniz continues there (193/127), except by an appeal to the "principle of order", by which he means his Law of Continuity.

I contend that we may throw some light on this conundrum by considering Leibniz's early dialogue, *Pacidius Philalethi*, written on his way to Holland in late 1676.[3] In that dialogue Leibniz gives an analysis of change according to which, because successive states are mutually contradictory, "there is no moment of change common to each of two states, and thus no state of change either, but only an aggregate of two states, old and new" (A VI 3, 566/LLC 211). In support of this he gives the analogy with the infinite dividedness of body:

> (NB. Just as bodies in space form an unbroken connection, and other smaller bodies are interposed inside them in their turn, so that there is no place void of bodies; so in time, while some things last through a momentaneous leap, others meanwhile undergo more subtle changes at some intermediate time, and others between them in their turn. … At any rate, it is necessary for states to endure for some time or be void of changes. As the endpoints of bodies, or points of contact, so the changes of states. … Nor is any time or place empty. During any state whatsoever some other things are changing.) (A VI 3, 559/LLC 195–197).

This concise note to himself in the margins of a first draft contains most of the main ingredients of Leibniz's analysis of change. In this theory the only moments (strictly speaking, instants[4]) that are assigned are the endpoints—beginnings or ends—of finite temporal intervals, no more than two of which are ever next to one another. Thus concerning the spatial continuum, the interlocutors conclude that "the continuum can neither be dissolved into points nor composed of them, and that there is no fixed and determinate number (either finite or infinite) of points assignable in it" (A VI 3, 555/LLC 187). And the same applies to the temporal continuum regarding moments (i.e. instants). Moreover, because the only moments are endpoints of intervals, and no two intervals have an endpoint in common, there is no moment of change, or state of change. In sum,

> at any moment that is actually assigned we will say that a moving thing is at a new point. And although the moments and points that are assigned are indeed infinite, there are never more than two of them immediately next to each other in the same line, since indivisibles are nothing but bounds. (A VI 3, 565/LLC 209)
>
> Nor is there any moment of time that is not actually assigned, or at which change does not occur, that is, which is not the end of an old or beginning of a new state in any body. This does not mean, however, either that a body or space is divided into points or time into moments, because indivisibles are not parts but extrema of parts. And this is why, even though all things are subdivided, they are still not resolved all the way down into minima. (A VI 3, 566/LLC 209–211)

Leibniz is proud of this analysis of change, and continues to uphold it after 1676:

> Change is an aggregate of two opposite states in one stretch of time, with no moment of change existing, as I have demonstrated in a certain dialogue. ([Spring-Summer 1679] A VI 4, 307)

---

[3] See also Samuel Levey's (2002, 2005, 2010) and (2012) for further discussion of this dialogue, and Leibniz's treatment of time in it; for some rejoinders on the few points on which we disagree, see my discussion in (Arthur, 2018, ch. 5).

[4] In these passages Leibniz is mostly using 'moment' as synonymous with 'instant', the temporal counterpart to 'point', each being a mere endpoint, indivisible and lacking quantity. Sometimes, however, he will use the term 'moment' rather for an arbitrarily small duration, a usage that becomes more common later.

> Change is an aggregate of two contradictory states. These states, however, are understood to be necessarily immediately next to one another, since there is no third thing between contradictories. ([1683–1685] A VI 4, 556)

And in his second last letter to De Volder of 10/11/1705, Leibniz writes:

> Endpoints of a line and unities of matter do not coincide. Three continuous points in the same straight line cannot be conceived. But two may be conceived: the endpoint of one straight line [segment] and the endpoint of a second, out of which the same whole is constituted. In the same way in time there are two instants, the last instant of life and the first of death. (GP II 278/LDV 327)[5]

This analysis is as original as it is difficult to comprehend.[6] Several points are in order:

First, regarding *the discreteness of matter and change*: According to Leibniz the parts of matter are actually divided into determinate parts. This is what he means by calling the parts "discrete"—not that they are atomic or indivisible. For being always further divided precisely precludes their being indivisible. The same goes for states: each state is determinate, having a beginning and an end, and is subdivided into determinate or discrete substates, each of which also has its own boundaries.

Second, regarding *actually infinite division*: Taking Leibniz's idealized model of infinite division, we may wonder, how does a repeated bisection of each subinterval not dissolve the continuum into an actual infinity of points or instants? The answer depends on Leibniz's innovative construal of the actual infinite as *syncategorematic*. This is a conception according to which, however many subdivisions one may suppose there to be, there are in fact more; but there is no number of all subdivisions. As Leibniz urged Bernoulli to concede,

> Even if I concede that there is no portion of matter that is not actually cut, one does not for this reason come to uncuttable elements or minimum portions, nor indeed to the infinitely small, but only to portions perpetually smaller, and yet ordinary ones; similarly to how there arise perpetually larger ones in increasing. (19 July, 1698; GM III 524)[7]

In the same way, each state is further divided into substates in a determinate way by the changes within it, so that there are infinitely many substates, but no smallest.

Third, regarding *density*: Changes are therefore *dense* within any state, since no state is so small that it is not further divided into substates by changes within. Note, however, that instants (taken as endpoints) are not dense. For a change is defined as the aggregate of two contradictory states, one immediately next to the other, and instants as their endpoints, so that there are no further instants between the two instants bounding such contiguous states, the end of one and the beginning of the other. Even so, such pairs of contiguous instants do not exhaust the continuum, since they are always separated by the substates they bound, which have finite but arbitrarily small

---

[5] The example of the last moment of life and the first of death implicitly alludes to his discussion in the *Pacidius* of this riddle from Sextus Empiricus. See (Levey 2002) for further discussion.

[6] See Samuel Levey's important essay (2010) for a penetrating analysis.

[7] The letter is dated July 29, 1698, OS. I am giving all dates in this paper under the new system, NS, which is ten days earlier.

duration. A duration is therefore an infinite aggregate of states, each with its own distinct endpoints. Likewise, even if every state is further divided into substates, this does not suffice for true continuity. In fact, as we shall see below, well in advance of everyone else, Leibniz *explicitly distinguishes density from continuity*.

Given all this, I believe, one can discern the mereological structure of change in Leibniz, where continuous duration consists in a series of states, separated by actual changes, governed by the law of the series. But first we must confront two outstanding difficulties.

The first is that there is ambiguity if not contradiction in Leibniz's claims about states. Guided by the analogy with the actual division of bodies, we have depicted them as divided by changes occurring within them to infinity; yet we have seen him claim that "it is necessary for states to endure for some time or be void of changes". In the *Pacidius* Leibniz gestures at a solution to this discrepancy, comparing the leaps to the infinitesimals of his recently completed calculus. In a passage from the first draft that he subsequently deleted, he wrote:

> Whence you will understand that if it is a miracle for someone to be transferred from Paris to Rome in a moment, then it is a perpetual miracle, even if it is credible that the spaces through which these leaps occur are smaller than can be explicated by their ratio to magnitudes known by us. And these kinds of spaces are taken in geometry to be points or null spaces, so that motion, although metaphysically interrupted by rests, will be geometrically continuous—just as a regular polygon of infinitely many sides cannot be taken metaphysically for a circle, even though it is taken for a circle in geometry, on account of the error being smaller than can be expressed by us. (A VI 3, 568–569/LLC 409)

The idea is that even though the extended spaces through which the leaps occur are always finite, and take a finite time, such spaces and times are so small as to be "unassignable". It is in this way that an infinite polygon can be taken for a circle. Thus even if motion is "metaphysically" interrupted by unassignable leaps (like the unassignable differences of his calculus), it will still be "geometrically continuous". We will come back to this analogy with the infinite polygon below, as well as this justification of the vanishing difference between how things change discontinuously in reality and a true geometrical continuity (which for Leibniz is purely ideal).

But, secondly, from these premises Leibniz draws a startling conclusion in the *Pacidius* dialogue, namely that bodies *do not act*, and therefore *do not even exist*, between changes of state. For, he argues, given that "there is no moment of change common to each of two states, and thus no state of change either", then "if it is supposed that things do not exist unless they act, and do not act unless they change, the conclusion will follow that things exist only for a moment and do not exist at any intermediate time" (A VI 3, 557). "Hence," he concludes, "it follows that proper and momentaneous actions belong to those things which by acting do not change." (566) Thus bodies exist at every assignable moment, but they do not exist at the unassignable times between these moments. There is therefore *no temporal continuant to which the changes or actions can be ascribed*.

In the dialogue, Leibniz uses this consequence to prove the necessity of "a superior cause which by acting does not change, which we call God" (567), "whose special operation is necessary for change among things" (568–569). So here, after all his innovations, he has arrived at (more accurately, returned to[8]) a version of continuous creation not unlike those he will later criticize in the *Theodicy*. As such it seems vulnerable to the same criticisms. There are no temporal continuants, just isolated instances of God's creating bodies here, there and everywhere, but only for a moment at a time.

But Leibniz's use of the plural in referring to "those *things* which by acting do not change" hints at a different solution. This is in fact the view he wants to hold, as he believes that bodies contain their own individual principles of activity. He starts to develop this view in earnest when he arrives in Hanover in 1677. In what appears to be a draft of his intended book on physics written in 1678, Leibniz writes:

> certainly if we consider matter alone, ... no moment will be assignable at which a body will remain identical with itself, and there will never be a reason for saying that a body ... is the same for longer than a moment. (A VI 4, 1399/LLC 245)

In order for there to be something in body that retains its identity through time, Leibniz urges, it must contain a substantial form as its "principle of unity and of duration" (1399). Such a form is modelled on the human Ego, which retains its form while having a succession of different perceptions. In a passage discussing change from the *Divisio terminorum ac enumeratio attributorum* of 1683–1685, Leibniz writes:

> The only difference that occurs when everything else remains the same, and makes there be no contradiction of any kind when the same things are said to be both contiguous and separate, is the difference of *time*. But whether those things are really the same that we think to be so is a matter for a more profound discussion. It is enough that there are some things that remain the same while they change, such as the Ego. (A VI 4, 562/LLC 267)

Here we see the Ego clearly identified as Leibniz's model for "those things which by acting do not change" (A VI 3, 566/LLC 211). We shall have to say more about how this is supposed to constitute a solution to the problem of the continuum below.

But what about the other difficulty concerning the enduring states that bodies are supposed to have between changes? In a piece dating from April–October 1686, ("*Dans les corps il n'y a point de figure parfaite*"), after alluding to his analysis of change in the *Pacidius* according to which a body exists only at the assignable moment it changes state, Leibniz notes that this analysis still presupposes enduring states between the changes. He then makes the intriguing suggestion that all such enduring states must be understood to be "vague":

> Now I believe that what exists only at a moment has no existence, since it begins and ends at the same time. I have proved elsewhere that there is no middle moment, or moment of change, but only the last moment of the preceding state and the first moment of the following state. But that supposes an enduring state. Now all enduring states are vague, and there is nothing precise about them. For example, one can say that a body will not leave some such place greater than itself during a certain time, but there is no place where the body endures that is precise and equal to it. One can thus conclude that there is no moving body of a definite shape.... (A VI 4, 1613–1614/LLC 297)

---

[8] In (Arthur, 2009) I sketch an account of Leibniz's first theory of the continuum from the early 1670s, prior to his immersion in Hobbes's philosophy, where he gives a version of continuous creation of just this kind.

This is related to a marginal note Leibniz had made to himself on the first draft of the *Pacidius*: "Why not say rather that the conclusion that things exist only at a moment, and do not exist at any intermediate time, will follow if it is supposed that things do not exist unless they act and do not act unless they change?" (A VI 3, 558/LLC 191). Not only do moving bodies of a definite shape not exist: strictly speaking, between changes they do not exist at all! Leibniz is interpreting "state" in its root sense of stasis, and arguing that only what acts, and thus changes, exists.

But let us stay with the idea of enduring states. Given Leibniz's identification of a substance's states with perceptions, this means that *perceptions* are also taken by him to be enduring, rather than strictly instantaneous. This is confirmed in the continuation of the passage from the *Divisio terminorum* about change and the Ego quoted above:

> But if someone contended that not even I endure beyond a moment, he cannot know whether he himself exists.[9] For this he knows only by experiencing and perceiving himself. But every perception needs time, and so either he persists during the whole time of his perception, which suffices for us, or he himself does not perceive, otherwise he would persist only for a moment, namely, for that moment alone at which he exists. (A VI 4, 562/LLC 267)

Now the crucial feature of the soul and its perceptions that is not possessed by a body (as conceived by the Cartesians) is *memory*. It is this that links the perceptions together and forges the self-identity and persistence of substances through their changes. As Leibniz wrote in the *Definitiones notionum metaphysicarum atque logicarum* of mid-1685,

> Certainly those things which lack [substantial] forms are no more one entity than a pile of logs, indeed they are no more real entities than a rainbow or a mock-sun. Certainly they do not persevere the same for longer than one moment, whereas true substances persist through changes; for we experience this in ourselves, for otherwise we would not be able even to perceive ourselves, since each of our perceptions involves a memory. (A VI 4, 627–628/LLC 273)

So we have a clear contrast between substances which persist through the time of their perception, and bodies and other "quasi-substances" which do not persist for longer than a moment. The perceptions of these substances, moreover, are such that not all the changes occurring within them are perceived. All this coheres with Leibniz's doctrine of *petites perceptions* in the *Nouveaux Essais*:

> There is at every moment[10] an infinity of perceptions within us, unattended by awareness or reflection, that is to say, changes in the soul itself, of which we are unaware, because these impressions are either too small and too numerous, or too unvarying, so that they are not sufficiently distinctive on their own. ... but that does not prevent them from having their effect when they are combined with others, and from making themselves felt, at least confusedly, in the aggregate. (Preface; A VI 6, 53)

---

[9] This, of course, is a criticism of Descartes's *cogito*. You cannot argue that you exist because you are thinking without presupposing that you continue to exist long enough to form the thought, and then to remember it.

[10] Here a moment is to be understood as an arbitrarily short duration.

The idea is that a perception of however short a duration will still contain other perceptions within it. Even if a perception appears as one continuous state because of the limitations of sense—it is *vague*!—it is in fact infinitely divided into other smaller perceptions lasting for discrete durations. The instants that are actually assigned are the endpoints of these durations, change being the aggregate of its existence in one state at one instant and its existence in a contradictory state at the next. These perceptions, Leibniz says, *constitute* the individual's self-identity:

> These insensible perceptions also indicate and constitute the same individual, which is characterized by the traces or expressions they preserve of the previous states of this individual, thereby connecting these with its present state; and even when this individual itself has no sense of these traces of previous states. (*Nouveaux Essais*, A VI 6, 55)

In this connection Leibniz refers to his Law of Continuity, according to which "any change always passes from the small to the large, and vice versa, through the intermediate, in respect of degrees as well as of parts" (*Nouveaux Essais*, A VI 6, 56). "All of this supports the judgement that noticeable perceptions come by degrees from those that are too small to be noticed." (A VI 6, 56–57). Since every perception of the same individual substance preserves traces or expressions of its previous states, it follows that the whole series of states forms a continuum, with each state or perception arising by degrees from the previous ones.

Now such a continuum is not the ideal continuum of mathematics, since it is constituted by the states or perceptions, which are divided from one another by actual changes of state, as opposed to the instants of the ideal continuum, which mark positions of possible changes. The consecutive states themselves are touching, so that they form a contiguum, rather than a true continuum.

Leibniz makes precisely this point in a fragment of uncertain date recently discovered and transcribed by Osvaldo Ottaviani, *Locus et tempus sunt continua.*[11] I will be analysing this text in detail in what follows, for it throws great light on Leibniz's metaphysics of change. It begins:

> Place and time are continua, matter and change are contigua. A continuum does not have actual parts, except those that are assigned by an actual division. Nor does a line consist of points, or time of instants, even though there is no part of a line in which there is not an actual point, nor any part of time in which there is not an actual instant. A continuum, like a line and time, are ideal things like numbers; and in fact they are orders of possibles in which actuals are designated. Place or space is the order of possible co-existents; time is the order of possible changes or of incompatible states. Each order is continuous since nothing can be conceived as interposing that is not contained in it. (*Locus et tempus sunt continua*, LH 37, 5, Bl. 134^{r})

---

[11] LH 37, 5, Bl. 134. Ottaviani and I are intending to publish a transcription and English translation of this manuscript together with other related pieces on the metaphysics of the infinite. There are no obvious external criteria for dating for this manuscript. One could speculate a date of composition of around October 1705 on the basis of its content: the claim that the actual instants of time contain "just as many fulgurations of the divinity, ôr states of the universe" evokes what Leibniz wrote to De Volder in October 11, 1705: "Time is also resolved into unities of duration through actual changes, ôr into just as many creations infinite in number" (GP II 279/ LDV 327)—as well as what he wrote to Sophie ten days later, quoted here below.

It is interesting here how Leibniz can talk about "actual parts" in the continuum. For an ideal continuum (such as a line or time) does not have any actual parts at all, so that the phrase "except those that are assigned by an actual division" seems problematic. Although it is true that the points of an ideal continuum, as an order of possible divisions, designate where actual divisions may occur, still, if such divisions are made, then it is no longer continuous.[12] So an actually divided continuum seems to be different from an ideal or mathematical continuum. We find the same variant meaning of 'continuum' as something containing determinate divisions in the passage from "Primary Truths", that I quoted from near the beginning of this essay. There Leibniz claims that

> a continuum is not divided into points, nor is it divided in all possible ways—not into points, since points are not parts but boundaries, and not in all possible ways, since not all creatures are in a given thing, but there is only a certain progression of them *ad infinitum*. Thus if you bisect a straight line and then any part of it, you will set up different divisions than if you trisect it. ("Primary Truths", [1689], A VI 4, 1648)

The "continuum" here cannot be the ideal mathematical continuum, since it does not consist in all possible parts, but is rather actually divided into a certain progression of determinate parts, albeit an infinite progression of smaller and smaller ones. In a previous publication I had suggested that a continuum in this latter sense of containing a particular infinite progression governed by a law of progression "is what Leibniz refers to on occasion as the *physical continuum*" (Arthur, 2018, 279). In support I cited his explanation in his letter to Des Bosses of 24 January, 1713:

> on the hypothesis of mere monads, the infinitude of the physical continuum would depend ... on the principle of sufficient reason, since there is no reason for limiting or ending [the progression], or for its stopping anywhere; whereas the Mathematical Continuum consists in mere possibility, like numbers, and therefore necessarily contains infinitude in its very concept. (GP II 474/LDB 299)

Thus whereas a physical continuum consists in an infinite progression of determinate actual parts, united and determined by the law of progression, a mathematical continuum consists only in potential parts.[13]

Leibniz's discussion in his *Locus et tempus sunt continua* supports this analysis. First he stresses that the parts of the ideal continuum are *potential*, in contrast with the actual divisions into parts that occur in actuality. In this respect the parts of ideal place or time are analogous to the fractions into which we can divide unity in

[12] Cf. what Leibniz wrote to De Volder (19 January, 1706): "But continuous quantity is something ideal, and pertains to possibles and to actuals insofar as they are possible. The continuum, that is, involves indeterminate parts, whereas in actuals there is nothing indefinite—indeed, in them whatever division can be made, is made." (GP II 282/LDV333).

[13] Also in line with this definition of the physical continuum is the wonderful image Leibniz gives in the *Pacidius* when he compares its divisions with the folds in a sheet of paper or tunic: "Accordingly the division of the continuum must not be considered to be like the division of sand into grains, but like that of a sheet of paper or tunic into folds. And so although there occur some folds smaller than others infinite in number, a body is never thereby dissolved into points or minima." (A VI 3, 555/LLC 185)

infinitely many different ways, but from which it cannot be regarded as composed. But "the comparison of number with place or time is when they are considered in themselves and abstracted from the things existing in them." The parts of matter, by contrast, are actual parts that are individuated by their differing motions:

> These parts [of unity] are therefore potential not actual, and so also are the parts of a line, and a line can no more be composed of points than a number from numerical minima. And in fact to every section of a line into parts there corresponds proportionally a section of unity into parts. Meanwhile in a spatial or ideal line infinitely many actual points can be assigned, actually distinct from one another, and endowed with different motions, which do not compose the line. And these points will be extremities of the parts into which the line is actually divided by the variety of motions in matter. (*Locus et tempus sunt continua*, LH 37, 5, Bl. 134^r)

So here the "actual points" are described as extremities of the parts into which matter is actually divided. This suggests that, as in the *Pacidius*, the extremities of two contiguous parts of matter are touching but distinct, because they belong to parts of matter distinguished by their differing motions.

But in the immediate continuation of this passage Leibniz proceeds to discuss the division of a given line *AB* into two equal parts by "an actual point *C*", with *AC* and *BC* likewise to be divided into two equal parts by "actual points *D* and *E*, and so on, always taking actual points in the middle of the part by division". This clearly suggests that the "actual points", while still distinguished by their differing motions, are now to be regarded as being precisely *not* contiguous. For, assuming such a continued bipartition of the line as proceeding to infinity—"as it certainly can in fact proceed, since in nature no two points can be assigned in the universe which have precisely the same motion"—it will follow that between any two such actual points—distinguished by their different motions—there will always be another:

> Nor will two assignable points running into each other with different motions ever touch one another, since other points are always interposed. The same holds for time, the actual instants of which contain just as many fulgurations of the divinity, ôr states of the universe. (LH 37, 5, Bl. 134^r)

Leibniz then appeals to the same example of infinite bisection or trisection that he had given in "Primary Truths":

> For since actual divisions can be instituted in actually infinitely many different ways, e.g. by infinite tripartitions or mixed bipartitions and tripartitions, or by any other partitions whatever, just as if every line were cut in the extreme and middle ratio, it is clear that the simplest, which proceeds by perpetual bipartitions, is not determined as unique. Hence it could be understood that what is actual of place and of time is made up of [*conflatum*] points or instants ôr is an aggregate of them; but not place itself and time itself, which are continuous and potential things. Namely, just as place could be divided in a different way, as by tripartitions, so also could time; which are no more the aggregates of points or instants, or of the lines into which they can be resolved, than number is the aggregate of the fractions into which it can be broken down. (*Locus et tempus sunt continua*, LH 37, 5, Bl. 134^v)

Here it seems that "what is actual" of place are the assignable points individuated by their differing motions (more accurately, endeavours), separated by unassignable gaps. This, Leibniz notes, invites a difficulty, that of "how a whole body could be impelled if points are not touching each other immediately". It is here that he introduces a distinction between physical and mathematical continuity. But it is interestingly different from the way I had phrased it above, and this worth a sizable digression for the further light it throws on his views on the mereology of change:

> But that must be said to be equivalent to a continuum [in] which there are no two actually assignable points between which another cannot be actually located. This is physical continuity, not geometrical[14]; geometric points are endpoints of the continuum, not elements of it; but physical points are the elements of mass. Both are indivisibles. But mass is not something continuous, but a conjunction; for there is no separation of its parts, ôr no interval can be assigned between two of its parts in which there is not some part of it. But if mass were something substantial, not phenomenal, we would arrive at infinitely small real things, intervals of nearest points or instants. (*Locus et tempus sunt continua*, LH 37, 5, Bl. 134^v)

A physical continuum, according to the definition of the first sentence, is essentially a dense aggregate of physical points or "elements of matter"; and likewise, a physical continuum of duration will be an aggregate of what is actual in time, namely the changes in it. But this claim that there are no two points between which there are no others appears to contradict the account of the physical continuum given above, where the points bounding the intervals *are* in fact immediately next to one another, the last of one with the first of the next, with these intervals forming a contiguum, not a continuum.[15] Leibniz notes this objection himself at the end of the fragment: "It can be objected that in every order where there is prior and posterior, there are two things immediately next to one another; the issue comes down to an examination of this rule. If it is accepted, it will turn out that the latter determinations [concerning the physical points not touching] would not be brought into play" (LH 37, 5, Bl. 134^v).

Vincenzo De Risi (2019, 127, n. 30) has noted a similar apparent discrepancy in views Leibniz expressed on other occasions. For example, in a letter to Bernoulli in the Fall of 1698 Leibniz stated that "even if in motion all the points are gone through, it does not however follow that there are two points infinitely close to one another, and even less that there are two next to one another" (GM III 536). Yet in his letter to De Volder of 11 October 1705, as already noted above, he asserts that "we cannot conceive three continuous points in the same straight line; but two may be conceived: the endpoint of one straight line [segment] and the endpoint of a second, out of which the same whole is constituted" (GP II 278/LDV 327). But we need to take into account the immediately preceding sentence in that letter: "Endpoints of a line and unities of matter do not coincide." This is the same distinction that we see in the above passage from *Locus et tempus sunt continua*: the points we can conceive as next to one another are "endpoints of the continuum", i.e. those of contiguous line

[14] After this declaration, the rest of Leibniz's explication is written in a darker ink, as are many interpolations and clarifications he has interjected. His marginal comment, however, that "The whole of the latter is an intelligible explanation of the phenomena, not of substances…" is written in the original fainter ink, so the darker ink commentary and corrections would seem to have been added afterwards.

[15] For a pellucid treatment of Leibniz's changing views on contiguity and continuity, as well as an explanation of how it is impossible to represent contiguity in point-set topology, see (Levey, 1999).

segments, "geometric points". But they are not physical points, and it is these that are held to be "elements of mass" or "unities of matter".[16]

This answer, however, creates further difficulties.[17] What is the status of the distinct endpoints of contiguous segments? If these *geometric* points are points in the mathematical continuum, they should be identical, not separate and next to one another. And what, on the other hand, are these physical points that Leibniz describes as "elements of mass" and "indivisible"? If these are the physical points he refers to elsewhere, such as in his *Systeme Nouveau*, then they cannot be indivisible in the strict sense. For those physical points "are indivisible only in appearance, whereas mathematical points are exact, but are nothing but modalities" (GP IV 483/WFT 149). A physical point is not strictly indivisible; rather it is an arbitrarily small portion of matter, distinguished by its differing motion (i.e. endeavour). Since every such point is distinguished from any other, no matter how close, by a differing endeavour, no two of these physical points can be actually next to each other.

Significantly, I believe, Leibniz sidesteps the first difficulty in *Locus et tempus sunt continua* by abandoning the appeal to endpoints and instead concentrating the analysis on *midpoints* of successive intervals. He argues that if one bisects a line and each part of the line, "always taking actual points in the middle of the part by division", then such an "actual simultaneous division proceeds to infinity, since in nature no points can be assigned which have precisely the same motion" (LH 37, 5, Bl. 134^{r}).[18] The midpoints of each of these subdivided lines will be individuated by a different motion (strictly speaking, by a different *conatus* ôr endeavour), but between any two such physical points there will always be further points differentiated from them by their own motions; so "no two of [these] assignable points running into each other with different motions will ever touch one another, since other points are always interposed" (LH 37, 5, Bl. 134^{v}). Analogously, the "actual instants" of time marking actual changes would also always have further instants interposed. "The same holds for time," Leibniz writes, "the actual instants of which contain just as many fulgurations of the divinity, ôr states of the universe" (LH 37, 5, Bl. 134^{v}).[19]

In his October 1705 letter to De Volder, Leibniz also describes the "unities of matter" as not touching, but explains how their denseness is related to the "order of changes":

---

[16] De Risi wisely notes, however, that "The apparent oscillation in his views here may perhaps be accounted for by saying that the passage addressed to Bernoulli concerns ideal objects, while the passage to De Volder deals with the real world." (De Risi 2019, 127, n. 30).

[17] I owe thanks here to Vincenzo De Risi for suggesting I needed to describe these difficulties explicitly.

[18] Leibniz stresses that there is nothing unique about taking midpoints or "perpetual bipartitions into two equal parts". "For since actual divisions can be instituted in actually infinitely many different ways, e.g. by infinite tripartitions or mixed bipartitions and tripartitions, or by any other partitions whatever, just as if every line were cut in the extreme and middle ratio, it is clear that the simplest, which proceeds by perpetual bipartitions, is not determined as unique". This echoes his demonstration that, in finding quadratures, it is not necessary to decompose the area under a curve into equal subintervals. See Knobloch (2002) for details.

[19] —here we might have expected Leibniz to have said "The same holds for time, between any two actual instants of which are contained just as many fulgurations…"; I will come back to that below.

> One unity is not touched by another, but there is a perpetual transcreation in motion. This is, namely, that when a thing is in such a state that it by continuing its changes through an assignable time there would have to be a penetration at the next time afterwards, each and every point would be in another place, as the avoidance of penetration and the order of changes demands. (GP II 278/LDV 327)

This mention of "transcreation" refers us back to Leibniz's Paris notes, where the notion first occurs. Leibniz introduces it (and its synonym, "transproduction") in the manuscript *Infinite Numbers* of c. 10 April, 1676, and then makes it the centrepiece of the dialogue *Pacidius Philalethi* with which I began my analysis.[20] Strictly continuous motion, he declares in *Infinite Numbers*, is impossible, and must occur "by a leap", so that "motion is nothing but *transcreation*" (A VI 3, 500/LLC 93). The fact that in all three places—in the dialogue, in *Locus et tempus sunt continua*, and in his October 1705 letter to De Volder—the transcreation-cum-denseness account occurs *alongside* the other account of the continuum in which there are contiguous sections, strongly suggests that Leibniz regards the two accounts as compatible. My suggestion is going to be that the account in terms of contiguous sections or states pertains to how the continuum appears in the imagination: successive states are really distinct, because contradictory, but when we represent them in the imagination, we do so using geometry, the science of the imagination. We use the image of contiguous line segments to represent successive states when in fact the states are vague: in reality the states no more have precise boundaries than bodies have precise shapes.[21] This also relates closely to the idea that we represent differences in the calculus as if they are indivisible elements, when in fact they are unassignable.

We can gain further insight into Leibniz's reasoning about these matters by returning to the context in which he formulated his idea of transcreation. In "On the Secrets of the Sublime" of February 1676, he speculates that "a perfectly fluid matter is nothing but a multiplicity of infinitely many points, ôr bodies smaller than can be assigned", leaving a "metaphysical vacuum" (A VI 3, 473/LLC 47). "A metaphysical vacuum is an empty place however small, only true and real", constituting the difference between a true continuum and a perfect fluid. The latter is "not a true continuum, even if space is a true continuum"; rather it is "discrete, ôr a multiplicity of points". This supposes that both the points ("bodies smaller than can be assigned") and the "metaphysical vacuum" are actual infinitesimals. But in that case, "any part of matter would be commensurable with any other", as would the circle be to the square (LLC 49). Leibniz suggests he should look into this more carefully. This he does two months later in "Infinite Numbers", which begins with a thorough examination of this issue. But there he comes to the conclusion that unassignable things—like a line smaller than any assignable, or the "metaphysical vacuum"—are not actual infinitesimals, but *fictions*. There are no such things *in rerum natura*, even though they express "real truths": "these fictitious entities," he writes, "are excellent

[20] In Infinite Numbers of c. 10 April, 1676, Leibniz concludes that continuous motion is impossible, and must occur "by a leap", so that "motion is nothing but transcreation" (A VI 3, 500/LLC 93).

[21] For an informative discussion of Leibniz's treatment of vagueness, see Samuel Levey's (2002), and for application to the idea of precise boundaries in matter see the same author's (2010) and (2012).

abbreviations of expressions, and for this reason extremely useful" (A VI 3, 499/LLC 89–91).

In other words, even though it is "really true" that matter is discrete—it is a multiplicity of physical points, each distinguished from the others by a different endeavour—the difference between it and a true continuity is infinitesimal. The "metaphysical vacua" separating these points are so small that, however small one takes them to be, they are smaller still. Clearly these "metaphysical vacua" evoke the "metaphysical" interruption of motion by rests referred to in the *Pacidius*, referenced earlier, an interruption which still leaves the motion "geometrically continuous—just as a regular polygon of infinitely many sides cannot be taken metaphysically for a circle, even though it is taken for a circle in geometry, on account of the error being smaller than can be expressed by us" (A VI 3, 569/LLC 409). Leibniz had elaborated on this topic in "Infinite Numbers":

> The circle—as a polygon greater than any assignable, as if that were possible—and other things of that kind, are fictive entities. So when something is said about the circle, we understand it to be true of any polygon such that there is some polygon in which the error is less than any assigned amount *a*, and another polygon in which the assignable error is less than any definite assigned amount *b*. But there will be no polygon in which it is less than all assignable errors *a* and *b*, even if it can be said that polygons somehow approach such an entity in order. ... And even though this ultimate polygon does not exist in the nature of things, one can still give an expression for it, for the sake of abbreviation of expressions. ... For entities of this kind, i.e. polygons whose sides do not appear distinctly, are made apparent to us by the imagination, whence there arises in us afterwards the suspicion of an entity having no sides. (A VI 3, 498, 499/LLC 89, 91)

It is by extending this analysis to motion that Leibniz arrives at his idea of transcreation (or transproduction):

> For something to become another thing is for something to remain which pertains to it rather than to the other thing. But this is not always matter. It can be the mind, understanding a certain relation; for instance, in *transproduction*, even though everything is new, still, by the very fact that this transproduction happens by a certain law, continuous motion is imitated in a way, just as polygons imitate the circle. And hence one is said to come out of the other, by a similar abuse, as it were, of the imagination. (A VI 3, 503/LLC 99)

Thus it is the imagination that fills in the gaps between the points or instants, giving rise to the illusion of a perfect uniformity and true continuity: "in the mind there is thought of uniformity, yet no image of a perfect circle: instead we apply uniformity to the image afterwards, a uniformity we forget we have applied after sensing the irregularities" (A VI 3, 499/LLC 91). Almost thirty years later, Leibniz gives a striking illustration in a letter to Sophia, Electress of Hanover:

> Matter appears to us as a continuum, but it only appears so, as does actual motion. It is similar to how alabaster powder appears to make a continuous fluid when one boils it over the fire, or how a spoked wheel appears continuously translucent when it turns with great enough speed; without one being able to distinguish the locations of the spokes from the empty spaces between them, our perception unites the separate places and times. (to Electress Sophia, 21st October (NS), 1705; GP VII 564)

This evokes the doctrine of *petites perceptions*, where the irregularities in a given perception are unnoticed and it is sensed as smooth and uninterrupted. Either we do not sense the gaps (because of the limitations of sense perception), or we forget we have sensed them, and apply uniformity to the image afterwards.[22] Thus not only the image of a perfect circle, but also the geometric representation of states as smooth and uninterrupted between changes, are products of the imagination. This helps to clarify Leibniz's discussion of physical continuity in the *Locus et tempus sunt continua*. It is because there is always further matter between any two physical points that "no interval can be assigned between two of its parts in which there is not some part of it", so that between any two such assignable points there is always another. But the interval between any two physical points is unassignable, and cannot be something real.[23] If, on the other hand, "mass were something substantial, not phenomenal", he argues there, "we would arrive at infinitely small real things, intervals of nearest points or instants." Analogously with time, the actual instants at which change is taking place are separated by "vague" states in which further changes are occurring, but of which we are not conscious. As Leibniz wrote in "Infinite Numbers", "they are not on this account any less sensed by our consciousness [*a nobis consciis*]. Rather we forget them, just as we are oblivious of the things we dream." (A VI 3, 499/LLC 91).

There is therefore a very strong correlation in Leibniz's thinking between the fictional nature of infinitesimals and the insubstantiality of matter (and the phenomenality of states). In *Locus et tempus sunt continua* this prompts him to add:

> So here is a popular way of explaining the issue, suitable for those who do not grasp that material things are only phenomenal. For if the things are real, we will resolve space into a multiplicity of points, and time into a multiplicity of instants. Should we not therefore say that in fact between any two actual instants or points another must be interposed, and that we never arrive at two of them that are unassignably distant, between which nothing actual is interposed? this could be said on the subject of the continuum. (LH 37, 5, Bl. 134^{v})

This is consistent with the marginal note Leibniz had made on the first sheet, before entering into the discussion of physical continuity: "The whole of the latter is an intelligible explanation of the phenomena, not of substances. That is, everything

---

[22] For a thorough discussion of the relation between sensation and imagination in Leibniz's thought, I refer the reader to Lucia Oliveri's forthcoming book. She explains (inter alia) how minute perceptions make available to the mind the infinite actual modifications of discrete matter that are first unified into distinguished sensations, then synthesized by the imagination into perceptible wholes, with shapes and sizes determined by confused sensations.

[23] This seems to undermine De Risi's contention (2019, 128) that "it is possible that these unassignable spaces are to be regarded as actually infinitesimal vacua between parts of matter. They would thus tend to push Leibniz toward a non-Archimedean geometrical system." In support of it, he construes a passage from Leibniz's reading notes on Froidmont as "clearly represent[ing] an opening toward non-Archimedean considerations" (128, n. 33); but on my reading it is a view that Leibniz finds in Froidmont and quotes for the purpose of rejecting it, which he then does. But see the whole footnote; indeed, see the whole of De Risi's excellent article for an authoritative treatment of Leibniz's views on the continuity of space.

is offered to the mind as if this were so; and one monad makes up the deficit of the phenomena of the other" (LH 37, 5, Bl. 134^r).[24]

Coming back to the apparent discrepancy between Leibniz's statements about the denseness of points, then, we can say this. Leibniz presents two different models of physical continuity, without acknowledging any incompatibility between them. On one model, the continuum is actually divided into contiguous parts. Having a line stand for the continuum, it is divided into contiguous line segments each of which has a different endeavour at a given instant. Here the points in the continuum are "extremities of the parts into which the line is actually divided by the variety of motions in matter"; they are *geometric points*, indivisibles as the endpoints of the subintervals into which the continuum is divided, and no more than two of them will next to one another. On the second model, the points in the continuum are *physical points*, "bodies smaller than can be assigned", each distinguished from the others by a different endeavour; in this case, between any two such points there will always be others: they will be dense.[25] I suggest the following interpretation of how it is that Leibniz regards these two models as compatible. If the intervals (the contiguous line segments) of the first model are regarded as real, then because of the actually infinite division Leibniz takes as definitive, they will have to be actual infinitesimals. The assumption of actual infinitesimals, however, leads to contradiction. In fact the intervals are unassignable, so small that no error is derivable from neglecting their extension. This corresponds with the fact that we perceive bodies and changes as continuous, even if what is actual in them is discrete.[26]

This still leaves moot the status of the "geometric points" of the first model, as pointed out by De Risi. They subsist in a kind of untenable middle ground between the physical points and truly mathematical points. A point in the mathematical continuum marks the place where it an be divided. If a division is actually made, there will be two disjoint actual parts, each with its own endpoints, so the rightmost of one is contiguous with the leftmost of the other. These boundaries are limitations of the extended, they are modifications of each extended part, and specific to it. If on the other hand no such division is made, but only considered as possible, the two endpoints of those potential parts mark the same point of possible division of what is still continuous—the actual, contiguous endpoints occur at a single point of the mathematical continuum.

[24] We might also relate this claim to Leibniz's assertion in a piece from the late 1670s, "A Body is not a Substance", that "if we only say this, that bodies are coherent appearances, this puts an end to enquiry about the infinitely small, which cannot be perceived. But this is also a good place for that Herculean argument of mine, that all those things which are such that it is impossible for anyone to perceive whether they exist or not, are nothing." (A VI 4, 1637/LLC 261).

[25] Cf. De Risi (2019, 131): "Leibniz, in fact, was obliged to attempt to characterize an infinite sequence of elements (i.e. the boundaries of bodies) within the continuum of space, which is dense in this latter but does not exhaust it (since, in addition to the boundaries, there is also matter in space)."

[26] Ted McGuire, in his perceptive essay (1976), drew attention to the non-continuity of phenomenal extension and change (see esp. 306–307), but equated it with what he calls the "perceptual continuum", which "comes into existence in discrete chunks" (311; see also 306, 310–311, 314). This contradicts Leibniz's claims that matter and change are perceived as continuous, with the irregularities smoothed out by the imagination.

This accords with Leibniz's mature definitions in the *Specimen geometriae luciferae* and the *In Euclidis πρῶτα*, where two parts of a continuous line have a point in common.[27] The contiguous endpoints, I suggest, must therefore be regarded as artifacts of the representation. Insofar as a change is represented as occurring at an actual instant, this is a boundary between the two contradictory successive states; but since these states are vague, so too will be the boundary. In *Locus et tempus sunt continua* Leibniz finesses this by taking intermediate points within the vanishing intervals. This side-stepping of the contiguous points is consistent with his claims in the *Pacidius* and beyond that "there is no moment of change common to each of two states, and thus no state of change either, but only an aggregate of two states, old and new" (A VI 3, 566/LLC 211). Yet, "all enduring states are vague, and there is nothing precise about them" (A VI 4, 1613–14/LLC 297).

So the analogy between matter and change comes down to this. There is no assignable point in space at which there is not a unit [*unitas*][28] of matter that is not actually moved with a different motion (endeavour): this is a body of arbitrary smallness, a physical point. Similarly, there is no assignable state in which there is no actual change occurring. Perceptions are always of a finite duration (even though this involves some abstraction); but because meanwhile other things are changing, and these changes must be reflected in every monadic state, these states must in fact be of vanishingly small duration, so that they are momentaneous. The duration of any created thing must be an aggregate of such momentaneous states, produced by the changes of state at each actual instant. As Leibniz explained further to Electress Sophia in his letter of October 1705:

> And one can thus conclude that a cluster [*un amas*] of matter is not a true substance, that its unity is only ideal, and that (setting aside extension) it is only an aggregate, a cluster, a multitude of an infinity of true substances, a well-founded phenomenon, without ever violating the rules of pure mathematics, but always containing something besides. And one can conclude also that the duration of things, or the multitude of momentaneous states, is the cluster of an infinity of flashes [*d'eclats*] of Divinity, each of which at each instant is a creation or reproduction of all things, having no continual passage, strictly speaking, from one state to another. (GP VII 564)

This echoes the "just as many fulgurations of the divinity, ôr states of the universe" of *Locus et tempus* (LH 37, 5, Bl. 134ᵛ), and is in turn echoed by Leibniz in his essay *Monadologie*,

> all the created or derivative monads are productions of God, and arise, so to speak, by the continual fulgurations of the Divinity from moment to moment (§47, GP VI 614)

It should not be forgotten, however, that transproduction "happens by a certain law" (A VI 3, 503/LLC 99), this being a necessary condition for actually dense states to be perceived as a continuous motion, or for a polygon of arbitrarily many sides

[27] See De Risi (2019) for quotations and a pellucid discussion.

[28] Leibniz's use of the term *unitas* here is certainly problematic, since it confuses the unities that are in matter—the monads—with bodies of arbitrary smallness, which monads certainly are not. There is an analogous problem with his describing the momentaneous states as *unitates*.

to be depicted by the imagination as a true circle. Leibniz makes precisely this point in his *Specimen Geometriae luciferae* of around 1695, one of his most important and original essays in geometry. In a passage noted already by Ernst Cassirer in his *Leibniz' System* (1902, 183), and more recently praised by De Risi (2019, 131–133) for its prescience in distinguishing denseness from true continuity, Leibniz writes:

> From these considerations the nature of continuous change can also be understood: it does not truly suffice for it that between any states that you choose an intermediate one is found, for other progressions can be thought of in which such an interpolation may be made perpetually, so that [this state of affairs] cannot be conflated with something continuous; instead, it is necessary that a continuous cause can be understood that is operating at every moment ... And such changes can be understood in respect of place, species, magnitude, velocity, and indeed also of other qualities that do not belong to this consideration, like heat and light ... (*Specimen Geometriae luciferae*, GM VII 287)[29]

The mention here of the "continuous cause operating at every moment" takes us back to the idea of continuous creation. As I argued above, in the *Pacidius* Leibniz argued that the discontinuity in motion and change demonstrated the necessity of "a superior cause which by acting does not change, which we call God" (567), who produces the changes among things by his "special operation" (568–569). Subsequently, the simple substances that he posits in things are supposed to perform that role, with each of these substances producing its changes from its own store, and consequently the changes in composite things too. But the discreteness of actual changes creates a profound difficulty for this conception, one that is well stated by McGuire: "For if change in the temporal nature of substantial existence can only be characterized by means of a discontinuous series of momentary states, simple substances cannot truly endure" (McGuire, 1976, 316).

If I am right, the key to resolving this difficulty is the recognition that Leibniz is modelling the continuity of substance on his (successful) rendering of continuous transitions in his calculus. The law of the series of monadic states is an extremely complex encoding of the future behaviour of the substance, giving its state at any instant. But this law (essentially equivalent to the complete concept of the individual substance) remains the same, while the substance corresponding to it produces its states in accordance with the law. The individual substance is the subject of change and the source of its own momentaneous states, and is what remains the same through these changes. It is "the thing which by acting does not change". As Leibniz writes to De Volder:

> The succeeding substance will be considered the same as the preceding substance as long as the same law of the series or of simple continuous transition persists, which makes us believe in the same subject of change, or monad. (January 21, 1704/LDV 291)

This can usefully be compared with the discussion in the *Specimen Geometriae luciferae* in which the above distinction between density and continuity occurred.

---

[29] The dating of the *Specimen* is also difficult. De Risi suggests that internal clues suggest that it is closely connected with other manuscripts datable as from the mid-1690s, and certainly after 1693. See (De Risi 2019, 131) for discussion.

There Leibniz discusses how a solid figure can be constituted as an "aggregate of all states of a certain continuous transformation":

> We may understand something continuous not only in simultaneously existing things, and in fact not only in space and time, but also in a transformation [*mutatione*] and in the aggregate of all states of a certain continuous transformation. For instance, if we assume that a circle is continuously transformed, and goes through all the species of ellipses preserving its magnitude, then the aggregate of all these states, that is to say of all the ellipses, may be conceived as continuous even if the ellipses are not in contact and do not exist together but one of them is produced by the other. (GP VII 285; De Risi, 2019, 132)

We may even form a solid figure from all these ellipses, says Leibniz, "that is to say, a solid whose sections parallel to the base are all those ellipses taken in order".[30] Likewise, one might say, when a moving body is at a new physical point at each different assignable instant, its motion may be regarded as continuous—even though the points are not touching—because its existence at each point is produced by a perduring cause, the monads internal to the body. In the same way, too, so may a continuous duration be aggregated from momentary states. These "momentary states" or intervals between changes correspond to the differences between successive values of variables, the differentials of Leibniz's calculus. In forming an integral, one has a variable stand for a value (such as the location of a point) at each assignable instant, and the particular functional relationship allows one to calculate the quantity of the aggregate that results from summing all the differences. This, I suggest, is why Leibniz feels entitled to describe physical continuity as constituted by the dense physical points or actual instants, and equally as the aggregate of the intervals between these points or instants. The actual instants at which changes are occurring are densely ordered, just as are the physical points on an infinitely divided line. But when you calculate an integral as a sum of differences, you do so by means of a function that gives values at each assignable instant.[31] As Leibniz explains to the Electress Sophia,

> Thus although matter consists in a cluster [*Amas*] of simple substances without number, and although the duration of creatures, as well as actual motion, consists in a cluster of momentaneous states, nevertheless it must be said that space is not composed of points, nor time of instants, nor mathematical motion of moments, nor intension of extreme degrees. It is just that matter, the course of things, and finally any actual composite, is a discrete quantity, whereas space, time, mathematical motion, intension or the continuous increase that we conceive in speed and in other qualities, and finally everything that gives an estimate that extends to possibilities, is a quantity that is continuous and indeterminate in itself, or indifferent to the parts that we can take in it, and those that are actually taken in nature. (GP VII 562)

Nevertheless, despite the lack of uniformity in actual changes, the mathematics of the continuous is still applicable, as Leibniz explains to Bayle in 1702, replying to the latter's article "Rorarius" in his *Critical Dictionary*:

---

[30] As De Risi remarks, "From a mathematical point of view, Leibniz seems to be remarkably close to the modern idea of a continuous fibration of a manifold" (De Risi 2019, 132).

[31] The idea that Leibniz's anticipation of the notion of a mathematical function lies at the heart of his metaphysical innovations was one of the main contentions of Ernst Cassirer in his (1902).

> This inclusion of the possible with the existent makes a continuity which is uniform and indifferent to every division. It is true that in nature we never find perfectly uniform changes, such as are required by the idea of motion that mathematics gives us, any more than we find actual figures that are precisely like those we learn about in geometry, because the actual world does not remain in an indifference of possibilities, but arises from actual divisions or multiplicities, whose results are the phenomena that present themselves, and which are varied in their smallest parts. Yet the actual phenomena of nature are ordered, and must be so, in such a way that one never encounters anything in which the law of continuity, or any of the other most exact rules of mathematics, is violated. (GP IV 568)

As we have seen, the imagination "supplies the deficit" between physical and ideal continuity, filling in the gaps between the physical points. That is the continuum as perceived. If we agree to call this the "perceptual continuum", then this is perfectly continuous. But this is not the same as the series of perceptions that is generated in each substance. According to Leibniz this is a physically continuous series of perceptions, consisting in successive finite perceptions, each apparently uniform, although divided within by changes of which the perceiver is unaware, "because these impressions are either too small and too numerous, or too unvarying" (A VI 6, 53), with this division proceeding internally without limit. But the continuity of these perceptions, what makes them perceptions of the same individual, is provided by the law of the series, which provides the basis for each state containing traces of the previous states of that individual, its physical memory. For this, as we have seen, "it is necessary that a continuous cause can be understood that is operating at every moment" (GM VII 287).[32]

Now, one might object that if all the states are of finite duration, this still only gives us a discontinuous transition. But the idea is that the states are of an assignable duration only on a limited level of discrimination—they are vague! Increasing the resolution reveals further subdivisions, without limit, so that they can be treated as "unassignable". This has a parallel with how Leibniz conceives the transition from an infinite series of terms differing by finite amounts to a truly continuous sum or integral. In a letter to Bernoulli in 1702, he writes:

> For instance, $1/3 + 1/8 + 1/15 + 1/24 + 1/35$ etc. or $dx/(xx - 1)$, with x equal to 2, 3, 4, etc., is a series which, taken wholly to infinity, can be summed, and dx is here 1. For in the case of numbers the differences are assignable. (GM V 356)

He then contrasts it with an integral of the same form, where

> x and y are not discrete terms but continuous ones, i.e. not numbers that differ by an assignable interval but the abscissas of a straight line, increasing continuously or by elements, that is, by unassignable intervals, so that the series of terms constitutes the figure… (GM V 357)

---

[32] The idea that the continuity of the series is grounded in the continuous operation of a cause could be seen as circular, a "pulling a rabbit from the hat", as an anonymous referee has wryly noted. But as I see it, the point is that the changing modifications presuppose the continued existence of something permanent, which Leibniz likened to the law of a series; and on the other hand, the continuity of the series of states or modifications produced is justified mathematically by an argument that is essentially equivalent to an ε-δ ∈; justification of the continuity of a function in modern mathematics.

On his syncategorematic interpretation of infinitesimals,[33] such "unassignable intervals" correspond to assignable ones that may be made so small that the resultant error is smaller than any that is pre-assigned. Leibniz calls these "incomparable" to finite ones.

> We must at the same time consider that these incomparable common magnitudes themselves, being not at all fixed or determined, can be taken to be as small as we wish in our geometrical reasoning, and so have the effect of the rigorously infinitely small. For if any opponent tries to contradict this statement, it follows from our calculus that the error will be less than any error that could be assigned, since it is in our power to make this incomparably small magnitude small enough for that purpose, inasmuch as we can always take a magnitude as small as we wish. (to Pierre Varignon, 1702; GM IV 92)

Analogously, then, this analysis can be applied to any continuous transition: all changes in such a transition are in actual fact discrete, as are the intervals between these changes. But since there are always changes in every interval at a further level of discrimination, there are no intervals so small that further change is not occurring within them. Consequently, this can be modelled by the calculus. Thus, taking an "actual instant" to be one at which change occurs, then although we never arrive at two instants separated by an actually unassignably distance, the intervals between them "can be taken to be as small as we wish", and have the effect of being rigorously infinitesimally small. The resulting duration will be the integral of such momentaneous states. It does not matter that the partition of the duration is into a particular progression of instants, and not into all possible instants, because the difference of this physical continuity from a rigorous geometrical continuity can be rendered smaller than any pre-assigned difference, and will consequently be (mathematically, if not metaphysically) null.

This analysis, I believe, resolves the difficulties noted by Russell and other commentators about the apparent incompatibility of Leibniz's assertions of the discreteness of the actual and the universal applicability of his Law of Continuity. Let me summarize its main theses:

- Every duration is divided into a series of successive, contiguous states.
- These successive states are mutually contradictory.
- Change is the aggregate of two successive, contiguous states.
- (Geometric) instants are the endpoints of states, so that there are never more than two of them immediately next to each other in the same time.
- All enduring states are vague. This means that no change is discernible within them, at a certain level of discrimination; and that they have no precisely determined boundaries.
- But during any state, further changes are actually occurring, even if they are not discernible on a given level of discrimination.
- This entails that the changes are dense within any interval. But they are not continuous, since they are separated by unassignably small states.

---

[33] For accounts of Leibniz's syncategorematic approach to infinitesimals, see Ishiguro (1990), Arthur (2008), Levey (2008) and especially Arthur (2013), Rabouin (2015), and Arthur and Rabouin (2020).

- Any given state is actually infinitely divided by these changes within it into further states (where the infinity is understood syncategorematically).
- The states, understood as intervals between changes, are analogous to the infinitesimal differences (differentials) of Leibniz's calculus: they are "unassignable", momentaneous, that is, of a finite but arbitrarily short duration.
- These unassignables, however, cannot be understood as actual infinitesimals, since the notion of the actually infinitely small contains a contradiction. They are rather, fictions, abbreviated ways of speaking, which are nevertheless of the greatest utility.
- The duration of any thing is therefore divided into an actual infinity of such momentaneous states, separated by actual changes.
- Provided these states and their changes are generated by a law of progression, they constitute a physical continuum, consisting in an infinite progression of momentaneous states separated by changes.[34]

This account of change also gives us fascinating insight into Leibniz's deep metaphysics and some of his motivations for it. First, there is the claim in the *Pacidius* that "things exist only when they act, and do not act if they do not change". This analysis of change implies that bodies, understood geometrically à la Descartes, *do not act*, and therefore *do not even exist*, between changes of state. Understood in this way, bodies are not temporal continuants: they exist at every assignable instant, but they do not exist at the unassignable times between these instants. So if the world consisted only of material bodies (as it did for Hobbes, for instance), they (and indeed the whole universe) would stand in need of being continuously created by God at each assignable instant. In order for there to be action in bodies, they would have to contain certain other "things which by acting do not change", producing their changes. As we saw, Leibniz took the human ego or self as his model for such active temporal continuants when he rehabilitated the derided substantial forms or primary entelechies of the scholastics. Such an entelechy is an enduring continuous cause within body, the "law of its series", actively producing its series of states through appetition. The states are conceived by Leibniz as perceptions, that is, representations of the rest of the universe. The perceptions of any one substance are connected by (not necessarily conscious) memory, so that each perception contains traces of every preceding one, and emerges from them "in degrees of the infinitely small" in a continuous process of change governed by the law of succession. As far as they can be perceived, these states are vague, and involve abstraction from the changes that are actually produced during them, but of which we are oblivious. Even so, these changes of state are always occurring, discernible at a more profound level of discrimination. For even if there is no change apparent in a thing in a given state, nevertheless, since other things change all around it while it is in this state, it modifies its relations with them, and these changes are reflected in its minute subliminal perceptions. Since

[34] In Appendix 1, A 1.5 of my recently published book, *Leibniz on time, space, and relativity* (Arthur, 2021), I give a mathematical rendition of the continuity of time based on the preceding reflections. Assuming its success, this constitutes a corroboration of the consistency of Leibniz's theory as embodied in the theses described here.

each state reflects the changes happening in all the others with which it is compatible, each extended state is further divided without end. It follows that the states are of vanishingly small duration: they are momentaneous. There are so many states that the difference from a true mathematical continuum can be made smaller than any assignable error by increasing the resolution of the analysis. Monadic duration is thus constituted by an actually infinite aggregate or cluster of momentary states, each produced by successive "fulgurations of the divinity" in accordance with the law of the series of each monad.

References

Arthur, R. T. W. (2008). Leery bedfellows: Newton and Leibniz on the status of infinitesimals. Arthur, R. T. W. (2008). In U. Goldenbaum & D. Jesseph (Eds.), *Infinitesimal differences: Controversies between Leibniz and his contemporaries* (pp. 7–30). De Gruyter.

Arthur, R. T. W. (2009). Actual infinitesimals in Leibniz's early thought. In M. Kulstad, M. Laerke, & D. Snyder (Eds.), *The philosophy of the Young Leibniz, Studia Leibnitiana Sonderhefte* (Vol. 35, pp. 11–28). Franz Steiner.

Arthur, R. T. W. (2013). Leibniz's syncategorematic infinitesimals, smooth infinitesimal analysis, and second order differentials. *Archive for History of Exact Sciences, 67*, 553–593.

Arthur, R. T. W. (2018). *Monads, composition and force: Ariadnean threads through Leibniz's labyrinth*. Oxford University Press.

Arthur, R. T. W. (2021). *Leibniz on time, space, and relativity*. Oxford University Press.

Arthur, R. T. W., & Rabouin, D. (2020). Leibniz's syncategorematic infinitesimals II: Their existence, their use and their role in the justification of the differential calculus. *Archive for History of Exact Sciences, 2020*, 1–43. https://doi.org/10.1007/s00407-020-00249-w

Cassirer, E. (1902). *Leibniz' system in seinen wissenschaftlichen Grundlagen*. N. G. Elwert'sche Verlagsbuchhandlung.

De Risi, V. (2007). *Geometry and Monadology: Leibniz's* analysis situs *and Philosophy of space*. Birkhäuser.

De Risi, V. (2019). "Leibniz on the continuity of space." In V. De Risi (Ed.), *Leibniz and the structure of sciences: Modern perspectives on the history of logic, mathematics, epistemology* (pp. 111–169). Boston Studies in the Philosophy of Science. Springer.

Ishiguro, H. (1990). *Leibniz's philosophy of logic and language* (2nd ed.). Cambridge University Press.

Knobloch, E. (2002). Leibniz's rigorous foundation of infinitesimal geometry by means of Riemannian sums. *Synthese, 133*, 59–73.

Leibniz, G. W. (1710). *Theodicy,* trans. E. M. Huggard. Routledge and Kegan Paul, 1951; reprint edition, La Salle: Open Court, 1985.

Leibniz, G. W. (1923–). Sämtliche Schriften und Briefe. In Berlin-Brandenburgische (Ed.), *Akademie der Wissenschaften and the Akademie der Wissenschaften zu Göttingen*. Akademie-Verlag; cited by series, volume and page, e.g. A VI 2, 229.

Leibniz, G. W. (1849–1863). *Leibnizens Mathematische Schriften*. In C. I. Gerhardt (Ed.). Asher and Schmidt; reprint ed. Hildesheim: Olms, 1971. 7 vols; cited by volume and page, e.g. GM II 316.

Leibniz, G. W. (1875–1890). *Der Philosophische Schriften von Gottfried Wilhelm Leibniz*. In C. I. Gerhardt (Ed.). Weidmann; reprint ed. Hildesheim: Olms, 1960. 7 vols; cited by volume and page, e.g. GP II 268.

Leibniz, G. W. (1981). *New essays on human understanding*, ed. and trans. Peter Remnant and Jonathan Bennett. Cambridge University Press.

Leibniz, G. W. (1998). *Philosophical texts*, trans. and ed. R. S. Woolhouse & R. Francks. Oxford University Press; cited as WFT.

Leibniz, G. W. (2001). *The labyrinth of the continuum: Writings on the continuum problem, 1672–1686*, ed., sel. & trans. R. T. W. Arthur. Yale University Press; cited as LLC.

Leibniz, G. W. (2007). *The Leibniz-Des Bosses correspondence.* Selected, ed. and trans. with an introductory essay, by Brandon Look and Don Rutherford. Yale University Press; cited as LDB.

Leibniz, G. W. (2013). *The Leibniz-De Volder correspondence*, trans. and ed. with an Introduction, by Paul Lodge. Yale University Press; cited as LDV.

Leibniz, G. W. (unpublished). *Locus et tempus sunt continua.* LH 37, 5, Bl. 134^{r}-134^{v}.

Levey, S. S. (1999). Matter and two concepts of continuity in Leibniz. *Philosophical Studies, 94*(1–2), 81–118.

Levey, S. S. (2002). Leibniz and the sorites. *The Leibniz Review, 12*, 25–49.

Levey, S. S. (2005). Leibniz on precise shapes and the corporeal world. In D. Rutherford & J. A. Cover (Eds.), *Leibniz: Nature and freedom* (pp. 69–94). Oxford University Press.

Levey, S. S. (2008). Archimedes, infinitesimals and the law of continuity. In Goldenbaum & Jesseph (Eds.) (pp. 107–133).

Levey, S. S. (2010). *Dans les corps il n'y a point de figure parfaite*: Leibniz on time, change and corporeal substance. *Oxford Studies in Early Modern Philosophy, 5*, 146–170.

Levey, S. S. (2012). On time and the dichotomy in Leibniz. *Studia Leibnitiana, 44*(1), 33–59.

Machamer, P., & Turnbull, R. G. (Eds.). (1976). *Motion and time, space and matter.* Ohio State University Press.

McGuire, J. E. (1976). *'Labyrinthus Continui': Leibniz on substance, activity and matter.* In P. Machamer & R. G. Turnbull (Eds.) Motion and time, space and matter (pp. 290–326). Ohio State University Press.

Rabouin, D. (2015). Leibniz's rigorous foundations of the method of indivisibles, or how to reason with impossible notions. In V. Jullien (Ed.), *Seventeenth-century indivisibles revisited* (pp. 347–364). Science Networks. Historical Studies. Birkhäuser.

Russell, B. (1900). *A critical exposition of the philosophy of Leibniz.* Cambridge University Press; 2nd edition 1937, reprinted London, Routledge, 1992.

Is Leibniz's 'Lex Iustitiae' a Logical Law?

Enrico Pasini

1 The Analysts' Justice

For some years Leibniz kept a diary. It did not work for him, but the result is now an interesting reading, at least for scholars. On August 9, 1696, he recorded that Baron von Bodenhausen, in a letter from Florence, had asked him to kindly explain how he understood the "rules of justice in analytical calculations" that he had mentioned in a previous letter to the same Bodenhausen, while discussing a mathematical problem.[1]

The year before, Bodenhausen, who was a dilettante mathematician of some ability, had offered to help his correspondent in the study of differential equations. Leibniz, who was willing to outsource tiresome algebraic manipulations, proposed to him in mid-1695 a couple of problems (A III 6, 323f.) that Bodenhausen tackled with, in his own words, "good will and scarce ability," finally begging for Leibniz's compassion (A III 6, 550). In his annotations to Bodenhausen's homework, Leibniz (A III 6, 553f.) mentioned a "law of justice", *lex justitiae*, and even introduced the adjective *justitianae*, and then *justitiariae*, to describe equations that respect that 'law'—apparently he was still making up at least some of this language. Subsequently, in a letter to Bodenhausen of December 1695, Leibniz would suggest that, in order to solve one of the proposed problems, they had better choose the less troublesome (*bequemsten*), most promising equations, that is, the *justitiariae* ones, by means of which certain parameters could be directly investigated "denn darinn

[1] "Hr. Baron von Bodenhausen begehrt durch Schreiben von Florenz, ich möchte ihm expliciren, wie ich *justitiae regulas in calculo analytico* bei einen von mir in vorigen erwehnten Exempel verstehe" (Leibniz 1847, I, 4, 191). If not otherwise noted, translations are mine.

E. Pasini (✉)
ILIESI/CNR, Rome, Italy
e-mail: enrico.pasini@unito.it

University of Turin, Turin, Italy

F. Ademollo et al. (eds.), *Thinking and Calculating*, Logic, Epistemology, and the Unity of Science 54, https://doi.org/10.1007/978-3-030-97303-2_16

observirt man abermahls *justitiam*," i.e., since in relation to them 'justice' is consistently respected. One of the equations was "justitiaria per se", the other two only if taken together ("si simul sumantur"). Overall, he added, he found "considerationem justitiae vel homoeoptoseos"—attention to justice and 'homeoptosis'—to be very useful (A III 6, 582). No clarification was provided.

We know now that these ideas had been already put down in an ambitious writing of the same year, known to us as *Mathesis universalis*. There Leibniz had written:

> it must be noted that some formulas observe the law of justice, so that every letter in them appears in the same way: as is the case in rectangles [i.e. first-degree binomials] or simple combinations, [...] and in their powers.[2]

Such are, for instance, ab, a^2b^2. These would be, in the language of the correspondence, *iustitiariae per se*. Other formulas observe the laws of justice only when several similar are properly added together: while a^2b handles a differently than b, if we add together $a^2b + ab^2$, "the injustice is corrected, and in the composed formula both letters enjoy equal rights."

Leibniz adds inspiredly that this shows how useful are justice and piety in every domain,[3] and indeed the expression *lex iustitiae* has an obvious moral and juridical (maybe even jusnaturalistic) sound.[4] In its most classical treatment, i.e. the Fifth Book of Aristotle's *Nicomachean Ethics*, justice as a particular disposition or virtue can be of two kinds: 'distributive' justice, that allots goods in equal proportion to some quality of individuals, whereas simple equality of treatment is the distinguishing feature of the other kind: 'corrective' or 'retributive' justice. Both fit loosely with the features of Leibniz's algebraic law, which is not just a principle of harmony or equilibrium; it is a rule of equal or uniform attribution—hence the name. Of course

---

[2] "Denique notandum est, quasdam formas servare legem justitiae, ita ut quaelibet in iis litera se habeat eodem modo, ut fit in rectangulis seu combinationibus simplicibus, [...] et in harum potentiis" (GM VII, 64).

[3] "Ceterae formae leges justitiae non observant nisi plures similes addantur inter se, ex. gr. quadrato simplex a^2b aliter tractat a quam b: si tamen in unum addantur $a^2b + ab^2$, corrigitur injustitia, et in formula hac composita ambae literae aequali jure utuntur. [...] ut suo loco patebit, justitia (quemadmodum et pietas) ad omnia utilis est, ut etiam in calculo Algebraico ejus simulacrum prosit" (GM VII, 64).

[4] Aquinas, e.g., had called *lex iustitiae* the rule according to which God's will is both righteous and just (ST I^a, q. 21, a. 1 ad 2). Ulpian's third principle, *suum cuique tribue*, in the sense "enter into a society with [others] in which each one can keep what is his," will be labeled by Kant *lex iustitiae* in the *Metaphysische Anfangsgründe der Rechtslehre* (AK VI, 237; see Tomassini 2018). But neither use of the expression is connected to our present object in a relevant way, nor shall we consider the various independent uses of the expression in Leibniz's political philosophy and jurisprudence (on which see Bouveresse 1999, 138–139). In Ausin (2005, 109–110) we find this overingenious and at the same time imprecise suggestion: "Although Leibniz called this principle of physics and mathematics 'lex iustitiae,' he did not explicitly apply it to the normative domain" and yet he produced deontic descriptions in terms of gradual notions. "The core of this type of analysis is the principle of graduation, according to which, when two presumptions of fact are similar, their juridical treatment must also be similar. This idea is clearly due to the above-mentioned Leibnizian principle of transition or continuity (*lex iustitiae*)." The *lex justitiae* would thus be applicable to the domain it metaphorically derived from—or the reverse, as for what can be inferred from this analysis.

there is no virtue implied toward others, no duty or due, and this denomination is based instead on a rather vague analogy—that sort of analogy that takes place by way of metaphor (McInerny, 1968, 82–83). Thanks to this vagueness the word *homoeoptosis*, that implies equality or similarity between falls, can be introduced, as we have seen, as synonymous with this 'justice.'[5] And as we shall see, both the 'law of justice,' that Leibniz claimed as his own invention—keen as he was on both finding and formulating rules, principles, and laws—, and homoeoptosis, appear in other texts that have variously to do with different kinds of natural knowledge and formal procedures.

After receiving that rather cryptic letter, Bodenhausen was truly perplexed and, having solved some of the other problems, in July 1696 he got back to Leibniz, as we already know, asking for clarifications. He excused himself for not being able to figure out this *justitia analytici*, not to mention other "avantigeusen combinationen und finessen":

> I do not understand the rationale of this justice, be it 'as such' or 'taken together', nor the definition or form of this justice. I only find an analogy in the terms and dimensions of the 1st and 2nd equations. So I feel like a blind man, and do not yet grasp nor figure how and where it belongs; I beg for an explanation, that perhaps will teach me all the rest.[6]

Nothing issued out of this plea, and in November he again asked for *explicationem oder definitionem justitiae analyticae* (A III 7, 187). In Leibniz's letter of December 28, he finally received the required explication:

> What I have devised as 'justice' in analysis is not something strictly necessary, but maybe, as they say, useful for greater convenience. Anyway, to explain this kind of Justice we can understand it thus: just as the administration of justice concerning people admits no 'respect of person' [i.e. no favoritism], so here all letters are treated on the same foot; and indeed at times all of them without distinction, at times only some of them with their equal and the others again with their equal.[7]

---

[5] On the face of it, the abstraction (*-osis*) of *homoeoptotos*, Latinized from ὁμοιόπτωτος, "similiter cadens" (Göckel 1615, 154), a term used by grammarians for words with a similar declension and for the connected effects of alliteration and assonance; an homoeoptoton being different from an homoeoteleuton in that the latter concerns words and the former grammatical endings. Indeed πτῶσις means "falling, fall", and in Aristotelian logic it can designate moods of syllogisms and inflexions of properties. An abstruse term: yet Leibniz was using *homoeoptotos*, in a different sense, already in 1673 (A VII 1, 61; see De Risi 2007, 163).

[6] "Nun verstehe ich nicht *rationem istius justitiae vel per se, vel simul sumtae*, ja nicht *definitionem* oder *formam hujus justitiae* sondern nur eine *analogiam terminorum et dimensionum* in 1. et 2. aeqv. so ich wie ein blinder fühle, aber noch nicht faße noch rangire wie v. wo sich gehöret; bitte umb erklärung, so vielleicht alles übrige mir lehren wird" (A III 7, 37).

[7] Leibniz was still playing with the language of practical philosophy and law: "Was ich de *justitia Analytica* gedacht, ist zwar nicht eben *de necessitate*, aber vielleicht *ad melius esse*, wie man redet, dienlich. Diese arth von justiz inzwischen in etwas zu erclaren so verstehe solche: wenn gleichwie in der justiz gegen Menschen keine *acceptio personarum*, also hier die *literae* auff gleichen fuß tractirt werden und zwar zu zeiten alle ohne unterschied, zu zeiten etliche mit ihres gleichen und andere wieder mit ihres gleichen" (A III 7, 250).

Although Leibniz's language was still burdened with allegory, some points had been clarified. He had in mind a precept of algebraic artistry, a precept or a set of directions for the performance of analytical operations. The allegory wasn't becoming more precise, and maybe more precision wasn't needed for the moment: the language of 'justice' mainly seemed to be allusive of opportunity in practical matters, and of equal treatment of any kind. Moreover, although Leibniz could not distinctly account for them, there were different levels, or entities, involved: there were a maxim that recommended to respect a law, the law itself, and a property that either derived from the law or that the law was derived from. There also were, in these texts, some hints that the law could be much more general than the algebraic precept that indicated how to apply it to formulas. But it is also important to underline that Leibniz clearly shied away from presenting any of this as "strictly necessary."

In Leibniz we find different kinds of more or less general laws, both of necessary and contingent nature: laws of mathematics, of logic, of metaphysics, of nature easily turn out in his writings. It would be easy to suggest here that our 'law of justice' could be in itself, or be based on, a logical law that applies to mathematics insofar as it is dependent on logic itself. Now, for Leibniz, logic applies indeed to mathematics, but in a peculiar sense, that has to do neither with Aristotelian subalternation of disciplines, although he occasionally makes use of it, nor with contemporary projects of rewriting mathematics according to methodological rules or traditional logic, which he also knows of, and sometimes appreciates, but only to a certain extent. For instance, discussing the logical validity and the nature of mathematics in the Preface to his *Logistica*, the Jesuit mathematician and astronomer G.-F. de Gottignies wrote:

> in the whole of Mathematics, nothing is found, nothing is determined, if not by mathematical, i.e. demonstrative reasoning, according to the laws of Logic: namely, by legitimate syllogisms or enthymemata.[8]

Leibniz was not always sympathetic to such projects: while he appreciated the efforts towards a better and possibly more rigorous foundation of geometry, in the domain of analysis he feared instead the impact of restrictions on mathematical progress. In his main answer to Nieuwentijt, he wrote:

> I recognize that I myself attach great importance to those who endeavour to bring carefully all the demonstrations back to their first principles, and to have constantly devoted to this all my efforts. But for all that, I do not incite to hinder the art of invention on account of too many scruples, nor to reject under this pretext the best discoveries, by depriving ourselves of their advantages; and of this I have in the past tried to convince Father Gottignies and his disciples who were punctilious on the principles of Algebra.[9]

So when we are considering a purported logicism of Leibniz's as regards the foundation of mathematics, or its 'laws', we must be cautious. We can be confident, instead, that Leibniz has a hierarchy of disciplines in mind, where the more abstract and general are superordinate:

---

[8] "in tota Mathesi, nihil invenitur, nihil stabilitur, nisi per Mathematicos sive demonstrativos discursus, secundum Logicae leges institutos, hoc est, per legitimos syllogismos, aut enthymemata" (Gottignies 1675, f. π4r).

[9] GM V, 322. Translated in (Mancosu 1996, 90, *quid vide* for Gottignies at pp. 89–90, 102).

> Just like Logistics, or the general Science of magnitude (of which Algebra is a part) is subordinate to the universal Art of characters and finally to Logic itself, in its turn Logistics has under itself Arithmetic and Geometry, as well as those disciplines called mixed mathematics and Mechanics.[10]

If we look at this upside-down, it means that general precepts and rules of a subordinate art are written in the terms of a superordinate one: "In fact, the definite numbers of Arithmetic follow the laws of the indefinite numbers that are handled by Algebra, and they receive from Algebra the very canons [i.e. general rules and principles] of their operations."[11] The last point too could be projected upward, in the sense that algebra, in turn, can receive from the *speciosa generalis* or *characteristica universalis*—that is, indeed, universal logic—the canons, or meta-rules, of its operations. And among such canons the 'law of justice' could find a suitable place.

2 Logical Laws and Laws of Justice

Our starting point was that according to some contentions of Leibniz's there is a *justitia analytici*, that depends on, or is the ground for, a *law* of equalitarian justice toward symbols.[12] Analysis is wider than algebra, and it is, in the end, a part or a component of logic; and so it can be wondered whether this law might be a 'logical' law. This can mean different things: a logical law in the contemporary sense, whatever it may be; a logical law in the sense in which this expression could have been used in Leibniz's time; a logical law in Leibniz's own view. The latter may simply mean that this law is part of the vast domains that are assigned to logic in Leibniz's view. But maybe the first point to be checked is how for Leibniz and fellow logicians of his time it is acceptable to speak of 'laws' of logic.[13]

In the seventeenth century the idea of 'logical laws' is not conspicuous—differently from during the late eighteenth and nineteenth century—nor does anything particular seem to be founded on it; but it is indeed widespread and appears in

[10] "Quemadmodum autem Logistica vel Generalis de magnitudine Scientia (cuius pars Algebra est) Speciosae Generali et ipsi postremo Logicae subordinata est, ita vicissim sub se habet Arithmeticam et Geometriam et Mechanicen et Scientias quae mistae Matheseos appellantur" (GM 7, 51).

[11] "Nam numeri definiti Arithmeticae sequuntur leges numerorum indefinitorum quos Algebra tractat et ipsos suos ex ea operationum canones petunt" (GM 7, 51).

[12] Couturat (1901, 228), who maintains the principle of reason to be a principle of symmetry, identifies the law of justice with a mathematical principle of symmetry as well: "ce que Leibniz appelle par métaphore la *loi de justice*, et ce que les mathématiciens modernes nomment le principe de symétrie." He clearly does not do justice to the law, so to say.

[13] Leibniz does acknowledge logical laws, both existing and of his own invention: e.g. respectively the "laws of oppositions" (GP VII, 212) in syllogistic theory mentioned in the *Difficultates quaedam logicae*, and the 'law of expressions', according to which—the idea of the expressed thing being composed of the ideas of other things—the expression of the thing should be composed of the characters of those other things ("Lex expressionum haec est: ut ex quarum rerum ideis componitur rei exprimendae idea, ex illarum rerum characteribus componatur rei expressio"; A VI 4, 916).

writings by more or less eminent logicians whom Leibniz knew well and mostly admired.

For instance, in his *Metalogica*, bishop Caramuel y Lobkowitz (1654, 15) testified to the existence of logical laws—at least of those of his own production—in this passage: "Ad novissimam [Logicam] spectant omnes illi arguendi et concludendi modi quibus natura duce usi sunt illustrissimi viri [...] quos ego primus ad regulas et leges reduxi." Namely, the new logic contains 'rules and laws' that regulate the natural logical abilities of great men 'in every kind of argumentation and inference'—perhaps the amplest possible scope. In Caramuel's *Leptotatos*, it seems that such laws should be more closely connected with analysis and derivation rules: "King John II of Castile had to proceed according to what was required and proofed. We too must weigh every passage in demonstrations, and perfect our analyses with the true laws of Logic."[14] And in his *Moralis seu Politicae logicae liber secundus* there is an even stricter identification with derivation rules: "Videntur consequentiae [...] juxta bonas Logicae leges ex datis Praemissis profluere." But "Logicae leges" also prescribe how to set forth a definition (Caramuel y Lobkowitz, 1680, 282–283; 201).

The prescript "Syllogistice disserunto" ('discourse by syllogisms!') that we can find among the "Logicae leges" listed by Reneccius in his *Artificium disputandi* (Reineck, 1611, 207), alludes to the same connection, although these 'laws' are in fact precepts for disputation. Another writer well known to Leibniz, Johann Heinrich Alsted, writes in his *Logicae systema harmonicum* that a fallacy can be refuted by censoring the conclusion and pointing out at the same time which "logical law" has been violated.[15]

This rules or laws are, in these passages, clearly intended as part of conventional, more or less Aristotelian logic: in this sense, in a letter to More of 1649, even Descartes uses the expression "vulgares logicae leges"[16] in the sense of the rules of definition (AT V, 69), resorting, as it is typical of his approach to controversy, to an appeal to 'normal' academic philosophy. In Early Modern strands of juxtaposition and, sometimes, even identification of logic and method, there is a wider space for the invention and development of 'logical laws.' And in this respect, a curious fact is the existence in the Ramist logical corpus of a 'law of justice', that applies also to mathematical knowledge and is one of the few elements that Ramus was inclined to keep from Aristotle's *Organon*.

Ramus introduced three laws that generalized the characters that, in his logic, marked the propositions, and in particular the axioms, of true science and thus

[14] "Debuit ille [Johannes Rex] procedere secundum adlegata, et probata. Debemus etiam Nos librare Probationum momenta: et ad verae Logicae leges Resolutiones elimare" (1681, 7). At p. 6, Caramuel had quoted a celebrated anonymous *romance*: "El rey Don Juan el Segundo / Turbado toma la pluma, / Para firmar la sentencia, / de Don Álvaro de Luna". The king wrote *trepida manu*, as the young Caramuel (*etiam Nos*) does when he first devises the full project of his logical and theological works.

[15] "Hic solutio fit, reprehensione consenquentiae, et simul ostensione, quae lex Logica violata sit de consequentia syllogistica formali" (Alsted, 1628, 778).

[16] One can juxtapose with this passage Leibniz's *Dissertatio de conformitate fidei cum ratione*, where he refutes any demonstration "quæ ad vulgatissimas Logicae leges exacta non sit" (D I, 102).

concerned the treatment of scientific content inside, rather than across, disciplines: axioms shall be necessary and universal, according to the law of truth (*lex veritatis*); a disciplinary field being thus defined, all parts of the discipline shall be homogeneous, according to the law of justice; and its development shall be, according to the law of wisdom (*lex sapientiae*), continuous—thus assuring the completeness of the disciplinary field. They were indeed held by Ramus to be logical laws, and a presentation quite suitable for our intent is found in his *Praefatio in Scholas physicas*:

> The logical laws concerning the matter and form of the art must be attended to, so that all matter be κατὰ παντὸς, καθ'αὑτὸ, καθ'ὅλου πρῶτον[17]: that is, of all, by itself, foremost universally. The first is the law of truth, so that no instruction or proof appears in the art, unless it is true in a fully necessary way. [...] The second law has a wider provision: that any proposition in the art not only be valid for all and necessarily, but homogeneous as well, like a limb that belongs to one and the same body.[18]

Thus Ramus's law of justice had two sides: on the one hand it entailed a prohibition, on the other hand it was a principle of systematic organization. The fundamental Ramist doctrine that no elements of one art can belong to another art derived from this law, as it applied to principles, or axioms, and to the basic concepts of each discipline.[19] Applied to mathematics, the law had the obvious effect of walling up both geometry and arithmetic, in a way that could jeopardize even Euclid's treatment of numbers in the *Elements*, let alone the development of algebraic techniques; it prescribed

> that in Arithmetic nothing geometric be done, nor in Geometry anything arithmetic; otherwise there would be in arithmetic an un-arithmetic geometrical element, and an un-geometric arithmetical element in Geometry. So anything logical in physics would be so much unphysical as something physical in logic. [...] This is the law of justice, the most just to govern the purposes of the art, and to each his own.[20]

---

[17] Ramus's starting point was Aristotle's *Posterior Analytics*, I, 4, 73a: 'predicated of all' or 'in every instance', 'per se' or 'essential', and 'commensurately universal'. These characters, that Ramus already reproduced in his first *Dialectique* (Ramée, 1555, 84), were modified by him and transformed into three laws, that did not apply only to logical premises and principles. For a precise comparison between Aristotle and Ramus on the three laws, see (Risse, 1986). Ramus saw the same Aristotelian scheme behind a section of Proclus's *Commentary on Euclid* (I, 11), "ubi terminos mathematicae materiae definit". He concluded that these were Aristotle's own *logical laws*: "istas leges Procli de elementis mathematicis examinandis et probandis Aristotelis logicas leges esse, et ex Aristotelis libris depromtas, ut elementa sint necessaria, homogenea, propria" (Ramée, 1567, 340–342).

[18] "Logicae leges illae de materia formaque artis ante oculos habendae sunt, ut materia omnis sit κατὰ παντὸς, καθ'αὑτὸ, καθ'ὅλου πρῶτον· de omni, per se, universaliter primum. Prima lex est veritatis, ne ullum sit in arte documentum, nisi omnino necessarioque verum. [...] Secunda lege cavetur amplius, ut artis decretum sic non tantum omnino, necessarioque verum, sed homogeneum, et tanquam corporis ejusdem membrum" (Ramée, 1569, f. Cc5r).

[19] The difficulty to reconcile this autonomy of sciences with some obvious features of scientific knowledge inspired unorthodox 'systematic' interpretations of the second law (Angelini, 2008, 18; 90; 142).

[20] "nec in Arithmetica fit quicquam geometricum: nec in Geometria arithmeticum, secus geometricum, in Arithmetica fuerit ἀνάρριθμον, arithmeticum in Geometria ἀγεωμέτρητον. Sic logicum

Ramus's laws had great diffusion and many of his followers would reproduce them, often verbatim, in their works.[21] Although Leibniz was not precisely an avid reader of Ramus, nor of his faithful disciples, he could find the three laws in Alsted's writings, where they were treated under different viewpoints and approaches. In Alsted's *Encyclopaedia* there was a paragraph "De axiomate vero et falso", on true and false axioms, that partly summarized, partly extended, a section of Ramus's *Dialectique* of 1576[22]: an axiom is true when it declares the thing as it is, and this either necessarily or contingently. For the truth and falsity of axioms he formulated 'rules', or 'canons', among which Ramus's three laws were mentioned:

> Precepts of disciplines, since they are their matter, must be probed according to the three degrees of necessity,[23] that are wisely called the axiomatic laws, because they govern the axioms of disciplines. [...] Thereon the three propositions of syllogisms—major, minor, and conclusion—must be inspected according to these three laws, that Ramus called the law of truth, the law of justice, and the law of wisdom.[24]

Yet already in book 8 of his *Compendium logicae harmonicae*, Alsted had dedicated a chapter to these laws, that he labeled there "Leges methodicae": laws that govern the form of things to be taught, "quae dirigunt formam rerum docendarum." They were the usual three: "Lex generalitatis" or "coordinationis", on the succession of general and particular; "lex colligationis", or "coordinationis", on the necessity of proper transitions; and of course the law "homogeniae, quae iubetur ut partes tractationis respiciant subjectum idem eundemque finem"—the law of homogeneity,

---

in Physica tam fuerit ἀφύσικον, quam physicum in Logica sit ἄλογον. [...] Haec justitiae lex est ad regendos artium fines, and to give to each its own" (Ramée, 1569, f. Cc5r).

[21] Take for instance Pierre Gaultier Chabot, who had been a pupil of Ramus's and Omer Talon's. He wrote commentaries on Horace, and the separation of disciplines was essential to his poetics, since poetry, in his view, must be studied successively and separately through dialectic, grammar and rhetoric (Verhaart, 2014). He faithfully reproduced the second law in the Preface to his *Praelectiones*: "Habes, erudite Lector, quid de vera legitimaque distinctione primarum artium vere utiliterque sentiendum sit; sicut Logica lex καθ' αὑτὸ in primis praescribit, ut singulae artes habeant certum et proprium quoddam subjectum ad interpretandum, suisque praeceptis informandum, quo praecipue inter se differant, maximeque contineantur" (Chabot, 1587, f. †3r). For an ample picture of some freer lines of reception, see Hotson (2011).

[22] See the Appendix in Bruyère (1996, 85–86).

[23] Alsted had just distinguished between different inferential connections: those necessary but not essential, those necessary and essential but not reciprocate or convertible, and those that are all three, and so entail identity: "*Tres gradus necessitatis axiomaticae in se continent argumenta consentanea, sed diversimode.* Primus continet consentanea necessaria, licet non sint essentialia: secundus, necessaria, sed essentialia, licet non aequalia: tertius, necessaria, essentialia, et aequalia, seu reciproca, τὰ ἀντιστρέφοντα, seu ἀντιστραμμένα, convertibilia" (Alsted, 1649, 430).

[24] "In disciplinis enim praecepta, tanquam materia, examinari debent iuxta tres istos necessitatis gradus: qui scite vocantur leges axiomaticae, quod regant axiomata disciplinarum. [...] Deinde in syllogismo tres propositiones, maior, minor, et conclusio debent examinari iuxta tres istas leges: quae Ramo dicuntur lex veritatis, lex iustitiae, et lex sapientiae" (VIII, ch. 3, sect. 2; Alsted, 1649, 430).

dictating to respect in every part of a tractation the same and unique subject and purpose (Alsted, 1615, 122–123).[25]

3 Algebraic Laws, Homogeneity and Homeoptosis

It is on this vast background that François Viète famously subjects algebra to, again, homogeneity, by means, again, of a law. Algebra seems to be the answer in the quest for a 'universal mathematics': a symbolism that can treat all kinds of problems, everywhere relations between abstract quantities take the place of comparisons between magnitudes. Yet Viète oscillates between the opening entailed by this universality, and the need to keep the order that is necessary to a solid foundation of the art. The third chapter of his *Isagoge in artem analyticam* is titled *De lege homogeneorum, et gradibus ac generibus magnitudinum comparatarum.* This 'law of homogeneous [quantities]' is the "prime and perpetual law of equalitions and proportions", that dictates: "homogeneous terms must be compared with homogeneous terms."[26]

Consequently, addition and subtraction must respect homogeneity. Multiplication and division produce different kinds (*genera*) of quantities, that are not homogeneous, and even 'non geometrical': for instance, "Square times cube equals square-square [...] Cube times square-square equals square square-cube". But Viète's symbolism—*B plano* or *D solido* for b^2, d^3—would impose homogeneity anyway.[27]

In his *Mathesis universalis*, after the simple forms that we have examined above, Leibniz introduces 'compound forms' (e.g. $x^2 + y^2$). These forms too are subject to laws: "There are two laws which can be observed or violated in this composition: one is the Law of homogeneity, which was established by Viète, the other is the Law of justice, that I introduced."[28] It is plain, Leibniz continues, that Viète's law of homogeneity can be ignored when some quantity is taken as unit; yet, although Descartes inaugurated and often used this device, Leibniz prefers to side with Viète, that is, to enforce homogeneity in order to secure geometric interpretability.[29]

---

[25] Alsted was indeed repeating, or plucking from, his former teacher (Hotson, 2000, 11–12) Polanus's 'methodical laws', among which also this law *homogeniae*, of homogeneity, one of the "leges de forma artis." It prescribes "ut omnia praecepta sint inter se cognata, et ad eiusdem artis essentiam pertineant: et vetat heterogeniam et confusionem praeceptorum diversarum artium." For instance it is irrational (ἄλογον) to use a logical rule in grammar (Polanus, 1593, f. **1r).

[26] "Prima et perpetua lex aequalitatum seu proportionum, quae, quoniam de homogeneis concepta est, dicitur lex homogeneorum, haec est: Homogenea homogeneis comparari" (Viète, 1591, f. 4v; Viète, 1983, 15).

[27] "Latus in Cubum facit Quadrato-quadratum [...] Cubus in Quadrato-quadratum, facit Quadrato quadrato-cubum" (Viète, 1591, f. 5v). See (Bos, 2001, 125; 147–151).

[28] "Duae autem sunt leges quae in hac compositione observari vel violari possunt: una est Lex Homogeneorum, quam tulit Vieta, altera est Lex Justitiae, quam ego introduxi" (GM VII, 65).

[29] Eventually Leibniz will describe Viète's law as the "vulgar" algebraic one, in comparison to his own "transcendental law of homogeneity" (in a short writing in the 1710 *Miscellanea Berolinensia*; GM V, 382). For Leibniz's attitude to homogeneity in algebra and the calculus, and in relation to

Viète's is, in Leibniz's view, the most important of the two algebraic laws. Nonetheless, even if the law of justice is less necessary than the law of homogeneity, it is not less useful:

> not only it serves to examine calculations in a deeper and more exquisite manner, and warns of mistakes that would otherwise easily creep in; it shows as well how what we have found by calculations about one quantity, can be written, immediately and without calculation, about another one, on the basis of the principle of likeness, or of the same relationship.[30]

Conformity to the law also allows the use of abbreviated notations, and favors the invention of very general theorems, "beneficio justitiae inter literas observatae" (*ib.*)—thanks to the justice observed between letters, that can be observed for all letters in the formula or just for some of them.

As for equations, their treatment shows Leibniz's ability in extending the applications of his law:

> Equations [...] respect justice in two ways. First, if all the quantities being brought on one side, and a null on the other side, the resulting formula, that is equal to zero, respects justice; second [...] if, when the equation is written as a formula and both letters are not treated by that itself in the very same way, but what is done with one can be done with the other and vice versa, simply exchanging the signs.[31]

In fact, ways to develop the law of justice abound. In another writing Leibniz shows how the law of justice can be a guideline in the trasformation of systems of equations, and at the same time be observed in its usual acceptation, so that the law would, so to say, double up.[32]

In sum, the law of justice is both useful and versatile: signs of a quite powerful law. But, again and again, Leibniz refrains from considering his law of justice (as well as Viète's law) as 'necessary'; indeed his own is the less necessary (*minus necessaria*) of the two. It is, plainly, a law of forms in the sense—a sense that we have already met—of a canon, that dictates at a basic level how well-formed algebraic formulas can be fitted to maximize utility in the practice of the art. The level of application

symbolic calculations in general, see (Bos, 1974, 33–34); (Serfati, 2001). About Leibniz's oscillations between geometric and algebraic methods, and between synthesis and analysis, see several contributions in Panza and Roero (1995).

30 "Porro lex justitiae etsi minus necessaria sit quam lex homogeneorum, tamen non minus est utilis; non tantum enim inservit ad calculi examen ulterius et exquisitius, erroresque alias facile irrepente praecavet, sed etiam modum ostendit, id quod de una quantitate per calculum venati sumus, de alia statim scribendi sine calculo, ex principio similitudinis seu ejusdem relationis" (GM VII, 66).

31 "Aequationes [...] duobus modis justitiam servant, uno: si omnibus quantitatibus ab una parte positis et nihilo posito in altera, oritur formula observans justitiam, quae formula est nihilo aequalis; altero modo observatur justitia in aequatione, si quidem aequatione ad formulam redacta ambae literae non tractantur actu ipso eodem modo, quod tamen de una nunc factum est, fieri potest de altera, et vice versa, quod contingit simplice mutatione signorum" (GM VII, 67). The examples are respectively $x^2 + y^2 = 0$, $x^2 + x = y^2 + y$.

32 "In hac calculandi ratione observatur duplex justitiae Lex, una est ut notae unius aequationis originalis, eodem modo tractentur ac notae alterius aequationis originalis; [...] Altera lex justitiae est, ut coefficientes ipsius x, eodem modo tractentur ut respondentes coefficientes ipsius a". See (Knobloch, 1980, 271).

of this law, that is, the lieu of the precept or maxim that orders to respect the law, is the composition of any sort of algebraic formulas. It seems anyway that such a law must take its force and validity from one or more higher levels.

The very fact of being a law of forms could be sufficient to recognize to this law a logical character: rules pertaining to the form of expressions made of characters are evidently part of the art of characters, the *speciosa generalis*. But the first leap of generalization is due to its being based on a general property—a property of forms, yes, but ultimately of relations. The apparent identification of 'analytical justice' and 'homeoptosis' that we met in Leibniz's annotations concealed the fact that the latter is a concept much less metaphoric than his 'algebraic justice', and far more general.

The first important treatment of homoeoptosis is in Leibniz's *De ortu, progressu et natura algebrae*, a mathematical manifesto of 1685–1686. The text begins with *mathesis universalis*, that is the perfect theoretic locus for general properties concerning the structure of expressions. It is divided into algebra and *ars combinatoria*, where algebra is not the *ars inveniendi* any more: quite obviously it is an analytical art, an art of analysis. Nonetheless, the structure of this writing is based on a close parallel between algebra and logic. Logic has simple terms and their relations (*habitudines*, in form of propositions), then syllogisms to demonstrate propositions, and finally a method that brings operations together in view of an end. Algebra has numbers and their relations, i.e. equations and proportions, that are quasi-propositions; then derivation rules, and a method that, so to say, lays out precepts for the purpose of invention and discovery.

Numbers, we read, come in different kinds, and so do their relations. There are simple relations, like equality (in algebra, 'equation'), homogeneity, commensurability, ratio, proportion. Other relations are "compound," when several homogeneous quantities are introduced to express the relation between two quantities. As for compound relations, Leibniz adds,

> It is of great use to investigate which [quantities] are homoeoptotous, or similarly related: in fact *homeoptosis* is to relation as proportion, or similarity of ratios, is to ratio; f.i. sine and cosine in the circle are homeoptotous to the radius.[33]

The mathematical example would be quite elementary from a mathematical point of view, while it might be difficult for those philosophers who are not used to go beyond simple arithmetic examples, but is nonetheless interesting. It seems to imply, at least, that Leibniz was anticipating an intuitive understanding of the property on the part of his virtual readers, and that he intended the similarity of such relations, as it was customary for him with regard to concepts, intentionally and not extensionally: consequently, it might be the case that the use of a fancy, Greekish name was an ersatz for the lack of a proper intentional definition.

Leibniz writes again about relations, *habitudines*, and their similarity or homeoptosis, in the *Specimen geometriae luciferae* of 1695: another programmatic writing that, moreover, is roughly contemporary with *Mathesis universalis* and the exchange

[33] "magni usu est dispicere quaenam sint Homoeoptata seu similiter relata, ut enim analogia seu similitudo proportionum est ad proportionem, ita *homoeoptosis* ad relationem; ex. gr. sinus rectus et sinus complementi in circulo sunt homoeoptata ad radium" (GM VII, 208).

with Bodenhausen. It concerns geometry, theory of the continuum, transformations of geometrical entities. Relations are "the habitudes of things to each other," in general, while ratio or proportion are a very simple species of relations.[34] If, moving from this hint, we were to further generalize what we found about homeoptosis in algebra in the writings that we have already considered, extending it to the formal representation of any sort of relations, we would end, more or less, with what follows:

> Homeoptosis and heteroptosis can be considered also with regard to relations. Of course if there is a relation between the homogeneous things A, B, C, and each of these three things is in the relation in the same way, so that, swapping their place in the formula, nothing else may arise but the previous relation, then the relation will be an absolute Homoeoptosis; but it can also happen that only some of the homogeneous things that fall under the relation are in it homeoptotely, for example A and B, although C is in it differently than A or B. And this Homoeoptosis is of great importance in reasoning.[35]

This shows, among other things, that the vocabulary of homeoptosis was fully developed at a time when Leibniz still tried to build a similar jargon for 'analytical justice', attesting to an originary entitlement of the former concept to the role of a general property of which the 'law of justice' would be an application. And, differently from other explanations, and notwithstanding that the writing is unrestrictedly mathematical, here homeoptosis was presented as a general property of relations, that does not depend on the particular application or instantiation (as, in algebraic formulas, the distribution of powers); as a relation between relations, so to say, that was of peculiar importance when reasoning about things related.

Relations are notoriously, in Leibniz's thought,[36] a crossing point of metaphysics and logic. Although, as a metaphysical law, it is not easy to properly define its content and significance, the law of justice has indeed, alongside with algebra and logic, some role in metaphysics. A feeble but clear indication of this character is provided by Leibniz in the *Tentamen Anagogicum*, that is, his "Anagogical Essay in the Investigation of Causes". It concerns the role of teleology in nature, as it is exemplified by a principle of economy, one of those that Rescher (1981, 50–51) calls 'minimax' principles. It is, according to a well known Leibnizean distinction, an 'architectonic' principle:

> Geometric determinations introduce an absolute necessity, the contrary of which implies a contradiction, but architectonic determinations introduce only a necessity of choice whose contrary means imperfection—a little like the saying in jurisprudence: *Quae contra bonos mores sunt, ea nec facere nos posse credendum est*. So there is even in the algebraic calculus what I call the law of justice, which greatly aids us in finding good solutions. (Loemker, 484)

---

[34] "etiam aliquid dicendum est de Relatione sive habitudine rerum inter se, quae multum a ratione seu proportione differt, quippe quae tantum una aliqua ejus species est simplicior" (GM VII, 287).

[35] "Potest etiam in relationibus spectari homoeoptosis et heteroeoptosis. Nimirum si sit relatio quaedam inter res homogeneas A, B, C, et una quaeque harum trium rerum eodem modo se habeat, ita ut permutando eorum locum in formula, nihil aliud a priore relatione oriatur, tunc relatio erit absoluta quaedam Homoeoptosis; potest tamen et fieri, ut quaedam tantum rerum homogenearum in relationem cadentium se habeant homoeoptote, verbi gratia A et B, licet C aliter quam A vel B se habeat. Atque haec Homoeoptosis maximi est in ratiocinando momenti" (GM VII, 287).

[36] On this subject, given the purpose of the present volume, I'll refer the reader exclusively to the fundamental (Mugnai, 1992, 2012).

Thus 'analytical justice' built upon homeoptosis is not, all in all, just an empty metaphor; rather the metaphorical denomination, in this case, mirrors the nature of its *denominandum*, as it locates the corresponding 'law' inside the framework of moral necessity and architectonics, of God's choice, of the best of all possible worlds, that is, of the main pillars of Leibniz's philosophy. This apparently elevates homeoptosis to a general feature of the created universe, that is connected to the complex criteria (like the balance of simplicity and variety) which, according to Leibniz, measure its perfection.

If we ignore the oddity of the epithet, this situation is similar to that of the property, and principle, of identity. But the latter, like the law of contradiction, is of 'logical', or 'metaphysical' necessity—indeed it is the root of it, and depends on God's intellect, of which it is the fundamental law; whereas our 'law of justice' or principle of homeoptosis can be 'metaphysical' in its widest application, but is not of metaphysical necessity: its validity, as far as we can see from these texts, does not derive from the principle of identity but from the principle of the best, and from that moral necessity that guides God in choosing a certain world for the act of creation. Precisely for this reason, it can also be strictly connected to a regulating principle of nature, or of reasonings about nature.

In Leibniz's *Initia rerum mathematicarum metaphysica*, a writing that is mostly famous, one would say, for the treatment of homogeneity and a peculiar use of homogony, i.e. the property of being "homogonous,"[37] we find a passage concerning the law of justice and the law of continuity. The starting point is a typical Leibnizean stance: "the whole doctrine of Algebra is an application to quantities of the Art of Combinations, i.e. of the abstract doctrine of Forms, which is the Art of Characters in general, and pertains to Metaphysics."[38] A *hinc* follows, then another: twice in this passage an inference is drawn from this single premise. The first is the familiar introduction of the law of justice:

> hence in [algebraic] calculations not only the law of homogenous [quantities] is observed profitably, but also the law of justice, according to which between those things that, in what is given or assumed, have certain relations, will have the same relations in what is sought or results correspond to the relations and they are treated in the same way to the extent that this is advantageous in action. In general, it can be said that when what is given follows an order, what is sought proceeds orderly as well.[39]

---

[37] In current English, it is a term of botanics that means 'having similar reproductive organs.' For Leibniz it means 'having same genesis'—in line with what he could find, e.g., in Goclenius's lexicon of philosophical Greek ("ὁμόγονα dicuntur, quae simul fiunt et esse desinunt"; Göckel, 1615, 153). Alongside with feebler cognate terms like συγγενής (in the *Specimen geometriae luciferae*, GM VII, 287), the term already appears, according to a friendly tip I received from Vincenzo De Risi, in at least one mid-1680s manuscript. Yet in the *Initia* we find, it seems to me, the first and only deliberate and systematic treatment of this concept.

[38] "Notandum est etiam, totam doctrinam Algebraicam esse applicationem ad quantitates Artis Combinatoriae, seu doctrinae de Formis abstractae animo, quae est Characteristica in universum, et ad Metaphysicam pertinet. [...] Hinc in calculo", etc. (GM VII, 24).

[39] "Hinc in calculo non tantum lex homogeneorum, sed et justitiae utiliter observatur, ut quae eodem modo se habent in datis vel assumtis, etiam eodem modo se habeant in quaesitis vel provenientibus,

The second concerns the law of continuity: "From this the law of continuity also results". Leibniz reminds us that he was the first to enunciate it, and that on its basis "the law of equal things is a special case of the law of unequal things"; it "always holds whenever one category ends in an opposite quasi-species."[40] Right after a nearly uncharted law, comes the most bustling one. Both are issued from the same origin, the 'doctrine of forms': possibly, be it called homeoptosis or otherwise, there is a property of formal systems and relations, on which both the law of continuity and the law of justice depend—just as, under the same headings of moral necessity and perfection, the principle of indiscernibles depends on the principle of reason.

At this point, we seem to be losing track of exactly what the 'law of justice' is.[41] The fact that the 'law of justice' is a practical law—a law of the trade, if one may say so—and nonetheless a logical law seems puzzling. Of course, the question is not whether there is some overarching history connecting the practical and the logical side of 'justice' in the sense analysed here. The question is rather how this amphiboly impinges on our understanding of Leibniz's conception of laws, and of logic. The coexistence, in Leibniz's approach, of logical laws both in a stricter sense and in a more practical sense seems to imply a different vision and role of logic than, say, Couturat would have allowed for.

4 Conclusion

Do such laws exist in logic as logical laws in a qualified sense? And: do such laws exist because the cognitive nature of finite rational beings is so determined as to be capable of knowing eternal truth precisely because of the well-intended structure of the created world, so that atomist physics and euristic laws of analytic proceedings, although not necessary or possibly untrue, equally have citizenship in this best and richest of all possible worlds?

It is not that we are directly applying divine goodness to algebra. We know in fact that for Leibniz, although even atomist physics is possible and, up to a point, it may work well, true natural science requires some metaphysical foundation. The following passage shows impressively how much this tenet is tied with demonstration:

---

et qua commode licet inter operandum eodem modo tractentur; et generaliter judicandum est, datis ordinate procedentibus etiam quaesita procedere ordinate" (GM VII, 24–25).

[40] "Hinc etiam sequitur Lex Continuitatis a me primum prolata, qua fit ut lex quiescentium sit quasi species legis in motu existentium, lex aequalium quasi species legis inaequalium, ut lex Curvilineorum est quasi species legis rectilineorum, quod semper locum habet, quoties genus in quasi-speciem oppositam desinit" (GM VII, 25).

[41] I'd like to thank an anonymous reviewer for pointing this out.

> Although also universal logic, or the art of demonstrations, cannot be despised with impunity by the natural scientist, I would maintain for certain that without some metaphysical demonstrations the true laws of motion cannot be demonstrated.[42]

'Demonstrations everywhere' implies the art of demonstrations—logic—everywhere. This aptly corresponds to the description of Leibniz's *ars characteristica* project in the *Elementa rationis*: "to devise certain formulas or general laws, to which every kind of reasoning can be uniformized."[43] And since the beginning we have seen that, for Leibniz, 'analytical justice' and homeoptosis are applicable as well as useful, and the deriving principles or laws are valid, in every domain and at any level. This, again, could be easily recognized as a characteristic of logical principles in a wide sense. And it can be argued that in Leibniz's views, explicitely or implicitely, logic itself supports this wider sense and its validity.

In the *Initia rerum mathematicarum metaphysica*, in reality, any metaphysical surcharge is brought back—while heading towards mathematics—to logical structures. Homogony, for one, is not grounded upon a mathematical relation, if not for the fact that it is introduced concerning mathematical (geometrical) entities, and Leibniz recurs to it in mathematical contexts. Plane figures and segments are not homogeneous, the latter are not parts of the former, and no transformation can produce one out of the other. But when a plane figure is produced, also possible or actual segments are produced: the genesis of the former entails the genesis of the latter. This is in its essence a logical relation, and in consideration of the title of the writing, it can be deemed logico-metaphysical.

In being a relation of relations, also homeoptosis patently reveals its nature as a logical relation. Yet in some forms it is valid, or extant, only inside mathematics; in other forms it is located on metaphysical or physical, briefly: philosophical grounds. But in the end it can be argued that a law concerning homeoptosis, or grounded upon it, should be in itself, in its ultimate nature, a logical law only in the wider sense that we have just introduced.[44]

It is now evident that a positive answer to the title question of this paper does not entail a renovated pan-logicist interpretation of Leibniz's thought: it would rather aid in shedding light on the nuanced role that logical laws can have in Leibniz's peculiar view of the constitution of the present universe and of possible worlds, beyond the well-known and obvious role of the principles of identity and non-contradiction. Nonetheless the 'law of justice'—that, as we have seen, differently from them depends on moral rather than metaphysical necessity—has in common with them to be amphibian between logic and metaphysics, with formal applications,

---

[42] "Quanquam logica quoque ipsa generalis, seu ars demonstrandi impune a physico contemni non possit; imo compererim ego, sine quibusdam demonstrationibus metaphysicis non posse demonstrari veras motuum leges" (A III 3, 372).

[43] "Atque hoc ipsum est, quod ego nunc agito, excogitare formulas quasdam sive leges generales, quibus omne ratiocinationis genus astringi possit" (A VI 4, 719).

[44] And since it doubles as a universal law, and as an algebraic precept or maxim, in this last role it is more of a logical nature, than it is a law.

a logical nature, and a metaphysical import; and of being, ultimately, a sort of logical law in its own right.

References

Alsted, J. H. (1615). *Compendium logicae harmonicae, exhibens universum bene disserendi modum juxta principia peripateticorum & Rameorum celebriorum*. s.n. [Corvinus].
Alsted, J. H. (1628). *Logicae systema harmonicum* (2nd ed.). s.n. [Corvinus].
Alsted, J. H. (1649). *Scientiarum omnium Encyclopaedia*, 4 vols. Huguetan & Ravaud.
Angelini, A. (2008). *Metodo ed enciclopedia nel Cinquecento francese*, 1, *Il pensiero di Pietro Ramo all'origine dell'enciclopedismo moderno*. Olschki.
Ausin, T. (2005). Weighing and gradualism in Leibniz as instruments for the analysis of normative conflicts. *Studia Leibnitiana, 37*(1), 99–111.
Bos, H. J. M. (1974). Differentials, higher-order differentials and the derivative in the Leibnizian calculus. *Archive for History of Exact Sciences, 14*(1), 1–90.
Bos, H. J. M. (2001). *Redefining geometrical exactness: Descartes' transformation of the early modern concept of construction*. Springer.
Bouveresse, R. (1999). *Perspectives sur Leibniz*. Vrin.
Bruyère, N. (1996). *Dialectique 1555: Un manifeste de la Pléiade*. Vrin.
Caramuel y Lobkowitz, J. (1654). *Metalogica, disputationes de logicae essentia, proprietatibus et operationibus continens*. J. Schönwetter.
Caramuel y Lobkowitz, J. (1680). *Moralis seu Politicae logicae liber secundus. Qui praecipue dialecticus vocari debet*. C. Corradi.
Caramuel y Lobkowitz, J. (1681). *Leptotatos, latine subtilissimus*. C. Corradi.
Chabot, P. G. (1587). *Praelectiones [...] quibus Quinti Horatii Flacci Venusini vatis poemata lyrica triplici artificio; dialectico, grammatico, et rhetorico accurate explicantur*. L. Ostenius.
Couturat, L. (1901). *La logique de Leibniz*. Alcan.
de la Ramée, P. (1555). *Dialectique*. Wechel.
de la Ramée, P. (1567). *Prœmium mathematicum*. Wechel.
de la Ramée, P. (1569). *Scholae in liberales artes: Quarum elenchus est proxima pagina*. Episcopius.
De Risi, V. (2007). *Geometry and monadology: Leibniz's analysis situs and philosophy of space*. Birkhäuser.
Göckel, R. (1615). *Lexicon philosophicum Graecum, opus sane omnibus philosophiae alumnis valde necessarium*. Hutwelcker.
Gottignies, G.-F. (1675). *Logistica siue Scientia circa quamlibet quantitatem demonstratiue discurrens cui mathematicum nullum problema insolubile nullum theorema indemostrabile*. Varese.
Hotson, H. (2000). *Johann Heinrich Alsted 1588–1636: Between renaissance, reformation, and universal reform*. Oxford UP.
Hotson, H. (2011). The Ramist roots of Comenian Pansophia. In S. J. Reid & E. Wilson (Eds.), *Ramus, pedagogy and the liberal arts: Ramism in Britain and the wider world* (pp. 227–252). Ashgate.
Knobloch, E. (1980). *Der Beginn der Determinantentheorie: Leibnizens nachgelassene Studien zum Determinantenkalkül (Textband)*. Gerstenberg.
Leibniz, G. W. (1847). *Gesammelte Werke*. In G. H. Pertz (Ed.). Hahn.
Mancosu, P. (1996). *Philosophy of mathematics and mathematical practice in the seventeenth century*. Oxford UP.
McInerny, R. M. (1968). *Studies in analogy*. Nijhoff.
Mugnai, M. (1992). *Leibniz's theory of relations*. Steiner.

Mugnai, M. (2012). Leibniz's ontology of relations: A last word? In D. Garber & D. Rutherford (Eds.), *Oxford studies in early modern philosophy VI* (pp. 171–208). Oxford UP.
Panza, M., & Roero, C. S. (Eds.). (1995). *Geometria, flussioni e differenziali. Tradizione e innovazione nella matematica del Seicento.* La Città Del Sole.
Polanus, A. (1593). *Logicae libri duo, juxta naturalis methodi leges conformati* (2nd ed.). Corvinus.
Reineck, J. (1611). *Artificium disputandi praeceptis logicis et exemplis theologicis dispositum atque expositum* (2nd ed.). Berger.
Rescher, N. (1981). *Leibniz's metaphysics of nature: A group of essays.* Reidel.
Risse, W. (1986). Petrus Ramus und sein Verhältnis zur Schultradition. *Revue des sciences philosophiques et théologiques, 70*(1), 49–65.
Serfati, M. (2001). Mathématiques et pensée symbolique chez Leibniz. *Revue d'histoire des sciences, 54*(2), 165–221.
Tomassini, F. (2018). Kants Auffassung der Ulpian-Formeln in der „Einleitung in die Rechtslehre". In A. Falduto & H. F. Klemme (Eds.), *Kant und seine Kritiker* (pp. 405–416). Olms.
Verhaart, F. (2014). Horace and Ramist dialectics: Pierre Gaultier Chabot's (1516–1598?) commentaries. In K. A. E. Enenkel (Ed.), *Transformations of the classics via early modern commentaries* (pp. 15–46). Brill.
Viète, F. (1591). *In artem analyticem isagoge. Seorsim excussa ab opere restitutae Mathematicae analyseos, seu, Algebra nova.* J. Mettayer.
Viète, F. (1983). *The analytic art.* Tr. T. R. Witmer. Kent University Press.

Leibniz among the Nominalists

Calvin G. Normore

Gottfried Wilhelm Leibniz was, by his own telling, a Nominalist. Exactly what sort of Nominalist he was is a disputed question in part because of disagreement about what Nominalism was and what Leibniz thought it was and in part because of disagreement about his own views.

1 The Medieval Nominalist Tradition

The *nominales* were a twelfth century school perhaps founded by and certainly associated with Peter Abelard.[1] There were people calling themselves *nominales* at least until the early thirteenth century and while memory of the school apparently dimmed considerably after the non-logical work of Aristotle became available there were thirteenth century references to the *nominales* in Albertus Magnus, Aquinas, and Bonaventura. There are no reference to them again of which I am aware until very near the end of the fourteenth century when, perhaps because of the reference in Albert, the term appears again in the correspondence of Jean Gerson and in Denis the Carthusian. By early in the fifteenth century there were again thinkers identifying themselves and being identified by others as *nominales*.

By the late fifteenth century there was beginning to be something of a canonical history of a Nominalist tradition. We can see it in the 1474 letter by the *soi-disant* Nominalist masters at the University of Paris to the King of France in response

[1] For texts and literature about the *nominales* see Vivarium, 30 (1), May 1992. For an argument that they were Abelard's students see my "Peter Abelard and the School of the Nominales" in that issue, pp. 80–96.

C. G. Normore (✉)
Department of Philosophy, University of California Los Angeles, 321 Dodd Hall, Los Angeles, CA 90095, USA
e-mail: normore@humnet.ucla.edu

F. Ademollo et al. (eds.), *Thinking and Calculating*, Logic, Epistemology, and the Unity of Science 54, https://doi.org/10.1007/978-3-030-97303-2_17

to the banning of the teaching of Nominalist doctrines in the university. There the list of great Nominalists of the past begins with Ockham, Gregory of Rimini and Buridan and those on it are explicitly contrasted with "Aristotle and his commentator Averroes, Albertus Magnus, St. Thomas Aquinas, Aegidius Romanus, Alexander of Hales, Scotus, Bonaventura, and other realists (*reales*)."[2]

There are no twelfth century figures on this list but some time in the next century Abelard, the *princeps nominalium* as Walter Map called him, and his teacher Roscelin of Campiegnes are added and by the time of Aventinus' *Annales Bojorum* (1580) there is a canonical list of Nominalists stretching from Roscelin into the sixteenth century. To it a few names are added by Jean Salabert in his *Philosophia Nominalium vindicata* of 1651 and Salabert's list is adopted and added to by Obadiah Walker in his 1673 *Ars rationis maxima ex parte Nominalium.*

Leibniz knew this historiography unusually well. He studied the history of the Dukes of Bavaria closely and made significant discoveries about it and he read Walker's *Ars rationis.*[3] Thus when he speaks of nominalists we can safely assume it is this tradition he has in mind.

2 Nominalism

But what of Nominalism? What, by Leibniz and more generally in the second half of the seventeenth century, would be thought to bind these figures together? Why the twelfth century *nominales* were originally so called remains a bit of a mystery. Aquinas and Bonaventura both claim it is because they held the 'doctrine of the unity of names' i.e. that variation in case or tense did not produce variation in signification.[4] By Gerson's time Nominalism was strongly associated with the doctrine that knowledge (*scientia*) involved a grasp of *propositiones* rather than of *res*. Late in the fifteenth century the authors of the 1474 letter claimed that "Those doctors are called nominalists who do not multiply things that are principally signified by terms according to the multiplication of terms. Realists, on the other hand, are those who contend that things are multiplied with the multiplication of terms... Also, nominalists are called those who apply diligence and study to know all the properties of terms

[2] See Lynn Thorndike's introductory remarks to his translation of DuPlessis d'Argentre (1755, I, ii, 286–88) in Thorndike (1975, 355).

[3] For Leibniz's interest in the history of the dukes of Bavaria and his knowledge of Aventinus' work see Antognazza (2009, esp. 289).

For Leibniz's knowledge of Walker's *Ars rationis* see Mugnai (2012), who notes that in the Herzog August Library in Wolfenbüttel there is a copy "written by Obadiah Walker (1616–1699), on the first page of which Leibniz has made the curious remark 'I suspect that the author is Wilkins, because I see that he is quite well acquainted with natural sciences and mathematics; moreover he makes frequent use of examples from theology".

[4] For Bonaventura cf. *Sent.* I, d.41, art. 2 q.2. For Aquinas, *Summa Theologiae*, I q.15 ad. 2.

from which depend the truth and falsity of speech, and without which there can be no perfect judgment of the truth and falsity of propositions."[5]

In our own time Nominalism has been variously understood as the doctrine that there are no abstract objects and the doctrine that there are no universals *in re* and some have claimed that through the tradition runs a thread insisting that differences in the truth-conditions of sentences need not reflect differences in ontology—that there are more truths than truth-bearers.[6]

All of these possibilities were available to Leibniz and there is no reason to think he would have felt a need to characterize Nominalism by one rather than by the others. Indeed he seems to some extent at least to have embraced them all.[7] A key to why may be found in the little manuscript titled *De Abstracto et Concreto* which Massimo Mugnai has edited and analyzed.

3 Abstract and Concrete Terms

Although the distinction between abstract and concrete terms is in the first instance grammatical it has immediate semantic consequences. As Leibniz points out, in a sentence like "A/The wise (one) is wealthy" (*Sapiens est dives*), containing only concrete terms, nothing prevents 'wise' and 'wealthy' from picking out the very same thing(s) but the surface reading of the sentence "Wisdom is wealth" (*Sapientia est divitia*), where the terms are abstract, supposes two things (*entia*), wisdom and wealth, which are quite different.[8]

This difference is at the heart of fourteenth century and subsequent medieval nominalist theories of predication. It seems to have been universally agreed that a proper name like 'Socrates' picks out Socrates. As writers like Buridan and Ockham read it a sentence like "Socrates is wise" is true just in case 'wise' also stands (inter alia) for Socrates. Hence it does not on its face commit one to anything other than Socrates. "(A) wise (one) is wealthy" is true just in case 'wealthy' stands for one or more of the things 'wise' stands for—i.e. for particular substances—and does not commit one to anything other than substances. Contrasting with this 'two-name' or

[5] "Illi Doctores Nominales dicti sunt qui non multiplicant res principaliter signatas per terminos secundum multiplicationem terminorum, Reales autem qui e contra res multiplicatas esse contendunt secundum multiplicitatem terminorum... Item Nominales dicti sunt qui diligentiam et studium adhibuerunt cognoscendi omnes proprietates terminorum a quibus dependet Veritas & falsitas orationis, & si ne quibus non potest fieri perfectum judicium de veritate & falsitate propositionum" (Du Plessis d'Argentre, 1755, I, ii, 286; tr. Thorndike, 1975, 355).

[6] For discussion of this thread see Normore (1987) and Rauzy (2004)

[7] He does characterize Nominalism in his Preface to his 1670 edition of Nizolius' *Antibarbarus philosophicus* where he writes "They are Nominales who think everything other than singular substances to be bare names. They therefore wholly reject the reality of abstractions and universals" but there is no reason to think he takes this to be exhaustive.

[8] Leibniz points out that sentences involving distinct abstract terms need not always be false; he cites "Wisdom is (a) virtue" (*Sapientia est virtus*).

'identity' theory of predication is another ('Realist') one on which the underlying semantic structure of "Socrates is wise" would be more perspicuously rendered by something like "Socrates has Wisdom" (or "Wisdom inheres in Socrates") and the underlying semantic structure of *Sapiens est dives* by something like "A thing in which wisdom inheres is a thing in which wealth inheres"—a sentence which appears to talk about, besides individual substances, such items as Wisdom and Wealth.

Not all abstract terms cause trouble for Nominalists. Though they need reject singular abstract terms since these will name abstract objects they can and did accept some abstract terms when these could be treated as common names for items they thought needed in the ontology. Thus Ockham can speak of (a) paleness or of many palenesses. Still in general what Leibniz will call metaphysical or real abstract terms—whether singular like 'paleness'—or plural like palenesses suggest ontological commitment. If a Nominalist can do without them so much the better.

With concrete terms matters are more complicated. Nominalists have no trouble with sentences like "Socrates is (a) human" because they can treat both 'Socrates' and 'human' as absolute terms i.e. terms whose whole semantic function is to signify or refer to in the one case Socrates and in the other all of the humans. On this picture these terms are names. Just as 'Socrates' is a proper name so 'human' is a common name. Such names do not signify things because those things meet some condition beyond simply existing. They are candidates to be what Leibniz will call primitive terms.

Matters are otherwise with terms like 'wise' and 'wealthy'. If Socrates is wealthy it is because he meets some condition which is not met simply by his being. The sentence 'Socrates is wealthy' is true only if 'wealthy' stands for Socrates and 'wealthy' stands for Socrates just in case such a further condition is met. The post-thirteenth century Nominalist tradition will express this by saying that 'wealthy' is a connotative term. At this point things get a bit more complex. Ockham and Buridan might claim that a term like 'wise' will signify (and so in a normal context stand for) Socrates only if a particular quality instance or trope—a wisdom trope perhaps—is appropriately connected with him and will claim that 'wise' signifies substances and connotes wisdom tropes.[9] However few if any in the Nominalist tradition will admit wealth tropes. How then is the semantic structure of "Socrates is wealthy" to be represented?

[9] This entails a commitment to such tropes and Ockham, Buridan and most of their followers thought that orthodox Theology and perhaps even the best Physics required some such tropes because they thought that there were compelling reasons to posit distinct items in some of the Aristotelian Categories of Accident. Ockham, for example, thought that the orthodox understanding of the Eucharist required that there be real colour tropes, individual whitenesses for example, and hence that 'whiteness' (*albedo*) was an absolute term standing for them. Buridan agreed and thought that Physics required that there be also individual quantities and magnitudes for which 'quantity' and 'magnitude' stand. See Normore (1985) for further discussion.

4 Connotative Terms and the Language of Thought

There has long been dispute about whether the Nominalist tradition admits primitive connotative terms and it is in part a dispute about the aspirations implicit in the tradition.[10] On one understanding, pioneered by John Trentman, Ockham, at least aspirationally, regarded thought as a logically perfect language containing all and only what was required to completely account for what there is.[11] In particular he thought that in the Language of Thought the work of connotative terms such as 'wise' and 'wealthy' could be (and so was) done by complexes of absolute terms and syncategoremata, what Leibniz would call 'particula'. On this picture the Language of Thought has neither primitive categorematic relational terms nor terms in oblique cases.

As Massimo Mugnai points out Leibniz shared this aspiration. In Mugnai's words "From a reading of the *De lingua rationali* it appears that Leibniz intended the construction of an artificial language composed of root terms, auxiliary terms and copula" where "the *lingua rationalis...* requires the division of linguistic expressions into root expressions—substantives, proper nouns and adjectives—and auxiliary expressions".[12]

5 Ontology

An ideal language should not seem to commit one to anything there is not.

Medieval Nominalist theorists agreed in their rejection of the Realist thought that in a categorical sentence predication expressed some sort of ontological tie between things stood for by the subject term and things stood for by the predicate, insisting that in a true affirmative sentence the subject and predicate stand for (at least some of) the same things and that predication simply indicated this. They disagreed somewhat, however about what there was for such terms to stand for. They agreed that there were substances and their essential parts—parcels of matter and individual substantial forms—but not about what there was beyond that. In general Ockham admits as beings only individual substances and their parts and instances of two species of quality. Each of these instances is a *res* in its own right not dependent on a substance for its identity and only naturally dependent on a substance for its existence. As Ockham analyzes a piece of bread, it is a substance composed of an individual (chunk of) matter and an individual form and 'having' a number of individual quality instances (tropes), such as a colors, which 'inhere' in it. Ockham maintains that after the consecration in the Eucharist the quality instances which

[10] For detailed discussion of the issue in Ockham see Panaccio (2004). For discussion of the issue in Buridan see King (unpublished).

[11] Trentman (1970).

[12] Mugnai (1990, 67).

used to inhere in the bread exist but do not inhere in anything. Buridan's ontology is more liberal, admitting as well individual quantities.

Leibniz' ontology is more parsimonious than was traditional among medieval Nominalists because, while they were prepared to admit as *res* really distinct from every other not only individual substances and their essential parts but individual qualities and in some cases individual quantities, Leibniz will admit only substances. As he writes in *De Accidentibus*:

> There is no need to raise the issue whether there are various realities in a substance that are the fundaments of its various predicates (though, indeed, if it is raised, adjudication is difficult). It suffices to posit only substances as real things [*res*] and to assert truths about these.[13]

If we suppose that by 'It suffices' Leibniz means that only substances and truths about them are required for an expressively complete Ideal Language, this puts Leibniz in the company of the theory Buridan in his *Questiones in Librum Metaphysicorum* attributes to Aristotle on which to say that a quality, such as a paleness, exists (*Albedo est*) is just to say that something is pale (*Aliquid est album*).[14] Since both 'something' and 'pale' are correctly predicated of substances, on such a theory what is expressed by sentences apparently committing one to qualities could be expressed without such commitment. A similar strategy could be adopted for the other Aristotelian accidental categories. Buridan himself, however, rejected this theory and to my knowledge no one in the canonical Nominalist tradition accepted it. Instead, admitting individual qualities and (sometimes) quantities they committed themselves to one version or another of what came to be known as the doctrine of real accidents.

It is this doctrine of real accidents that is the primary target of *De Accidentibus.* As Mugnai shows Leibniz begins by considering various ways in which accidents might be realities. They might be constituents of the substance itself but then, given what would now be called 'the indiscernibility of identicals', the substance would not persist through any change of accidental predicates—which Leibniz regarded as absurd. Alternatively (and this was the Medieval Nominalist strategy) the accident might be a thing in its own right. If so Leibniz suggests then "Socrates is pale" would either be equivalent to "There is Socrates and there is a paleness" which does not seem to capture the sense that it is Socrates who is pale, or "There is Socrates with an accidental feature and a paleness with an accidental feature such that when all that is the case Socrates is pale." In that case however all accidental change would reduce to the generation or destruction of a real thing—which is nearly as implausible. Leibniz' solution is to reject the problem and insist that accidental changes are simply changes in the truth values of certain sentences—changes which are primitive.

Whether or not we admit particulars in some categories other than substance (and so admit absolute terms standing for them) it does seem that if we are confined to posit

---

[13] Transl. Mates (1986, 171), citing Mugnai (1976, 133ff.).

[14] Buridan, *Questiones in Metaphysicam Aristotelis*, IV q. 6. Buridan reads Aristotle as denying any ontological commitment to beings in accidental categories. For some discussion see Normore (1985, 194–95).

only individual substances and accidental individuals there are non-synonymous sentences whose terms pick out only the very same ones. A challenge any Medieval Nominalist faces is how to account for this.

6 Appellation and Predication

Here Buridan's approach is more sophisticated than Ockham's and seems to have been taken up more widely within the tradition. Buridan and Ockham agree that in the Mental Language there are absolute categorematic terms and syncategoremata (what Leibniz will call *particula*). Therefore if they can account for the behavior of syncategoremata such as 'not' they can explain why "Socrates is human" and "Socrates is not human" differ in truth value even though their categorematic terms are exactly the same. What though of sentences whose categorematic and syncategorematic terms are exactly the same? What for example of "Abraham is the father of Isaac" which is true and "Isaac is the father of Abraham" which is false? Ockham and Buridan both speak of such constructions as 'father of Isaac' as connotative terms and as connoting items they admit into their ontologies, but Buridan also speaks of such terms as 'appellative' and as expressing how the things connoted are related to the things for which the terms stand.

Buridan's doctrine of appellation emerges from a rather strained analysis of the claim that a predicate term 'appellates its form'. He writes:

> In this way, therefore, I say that an appellative term found in a sentence appellates its form, i.e. those things which it connotes or of which it is appellative, and it appellates them as adjacent to something else, either past or present or future or possible, i.e. the thing for which that term supposits or for which a substantive nominative term construed with it would supposit for if it were the subject or predicate of the sentence.[15]

For Buridan a connotative term connotes items in the ontology but it appellates them as related to the items for which the term stands; in the most general case as having a connection with (*adiacentia*) or being disconnected from the items for which the term stands. Thus for Buridan every connotative term involves a relation.

Now, as Paul Spade has pointed out, in his *Sophismata* Buridan explicitly claims that "in accordance with the different positive ways of connecting [*modos positivos adiacentiae*] connoted things to the things for which terms stand there arise different modes of predicating. For example, how, how much, when, where, how this is related to that, and so on. From these different modes of predicating, the different categories are taken."[16]

---

[15] Transl. King (1985, 160).

[16] "Secundum diversos modos positivos adiacentiae rerum appellatarum ad res pro quibus termini supponunt, proveniunt diversi modi praedicandi, ut in quale, in quantum, in quando, in ubi, in quomodo hoc se habet, hoc ad illud, etc. Ex quibus diversis modis praedicandi, sumuntur diversa praedicamenta..." (Buridanus 1977, Chapter 5, remark 3, p. 62); quoted in Spade (1990, 606).

We have here in Buridan what we do not find, as far as I know, in Ockham, an explicit general connection between connotation and predication. Like Ockham, Buridan will claim that connotative terms connote things, and like Ockham he would be happy to replace 'inesse' by 'praedicari', but for Buridan how terms connote can only be expressed by propositions. The stage is set for Leibniz's introduction of his *abstracta logica*—which he claims to be contracted propositions

7 *Abstracta Logica*

Massimo Mugnai has discovered and emphasized the importance of the *abstracta logica* in Leibniz thinking. Leibniz's distinguishes two kinds of abstract term. One he consistently calls 'Logical', the other sometimes 'Metaphysical' and sometimes 'Real'. Metaphysical or Real abstract terms are nouns like 'wisdom' or 'wealth' which on their face commit one to abstract objects such as wisdom or wealth. On the other hand Logical abstract terms, expressions in Latin like *esse sapientem* or *esse hominem* which we might render in English with a participle and an adjective as in 'being wise' or 'being human', contain nothing which would apparently commit one to such objects. It would seem then that if one can express what is required for connotation using only such terms one will have significantly advanced the Nominalist project.

As Mugnai points out Leibniz puts his *abstracta logica* to several uses and one of them is indeed to eliminate the metaphysical or real abstract terms from the ideal logical language. Mugnai points us to Leibniz's approving quote of Horace's remark *virtus est vitium fugere.*[17] The point can be generalized: *humanitas est esse hominem,* and *sapientia est esse sapientem.* On such a picture for there to be humanity is just for something to be human and for there to be wisdom is just for something to be wise. In the regimented language none but concrete terms such as 'Socrates' and 'wise' appear and since these are understood to stand for individual substances there is no ontological commitment to anything else.

As Mugnai notes Leibniz seems to have been proud of having introduced *abstracta logica* into Philosophy.[18] And perhaps he did introduce them into the Philosophy of his time but their use to replace what he called Metaphysical or Real abstract terms goes back at least to Abelard who claimed that the 'common cause' of the term 'homo' applying to several humans was not some universal thing but their each having a status, that of *esse hominem.*

There has been little consensus about whether Abelard's talk of status is ontologically committing. He, himself, denies that statuses are *res* but he also insists they can be causes and, although accusative plus infinitive constructions like *esse hominem* or *esse sapientem* are not nouns, they can be subjects and predicates in

[17] Mugnai (2017, 120).

[18] "Sed a me alia quaedam abstracta posteriora concretis in philosophiam introducuntur, verbi gratia τὸ esse sapientem ut Horatius ait virtus est vitium fugere" (Mugnai 1986, 129).

categorical sentences. What of Leibniz's *abstracta logica*? How does he understand them to function semantically and do they commit ontologically? Since, as Buridan's doctrine of appellation indicates, even the use of connotative terms which are not explicitly relational involves a relation, it is to the theory of relations that we might turn for an answer.

8 Relations

In "Leibniz's Ontology of Relations: A Last Word?" (2012) Mugnai shows clearly a crucial difference between the approaches to relations taken in the later Middle Ages and the Early Modern period by thinkers usually classified (then and now) as Nominalist and Realist.

As Mugnai explains, thinkers like Walter Burley who are usually classified as Realists take typical binary asymmetric relations to supervene on what they term *fundamenta* and *termini*; real monadic accidents inhering in the subject and the correlate of the relation respectively. Thus, the relation *father of* holding between Abraham and Isaac has Abraham as its subject, Abraham's act of generating Isaac as its foundation, Isaac as its correlative and the effect in Isaac (perhaps his existence) as its *terminus*. The particular instance of *father of* is real but not, in medieval terminology, absolute, it supervenes on the items just mentioned and ceases to exist if any of them does.[19]

Later Medieval Nominalists approach matters differently—though they do not always speak with one voice. In general theirs are 'flat ontologies' which admit only particulars which are really distinct from one another and could by the absolute power of God exist apart from each other. Such ontologies have no trouble with certain sorts of symmetrical relations. Ockham, for example, claims that for two things to be essentially similar is just for them both to exist. Thus two humans such as Ockham and Buridan are similar to each other if both exist. If either ceases to exist the remaining one ceases to be similar to the deceased one but there is no change in it—only a change in the sentences that are true of it. Again, if Socrates and Buridan are similar in being brown it is because each is brown and the two brownnesses are essentially similar i.e. the term 'similar', picks them out not singularly but plurally. That said, the Nominalist tradition has no place for relations in the ontology. In what is apparently the first work we have from his pen, *De Relativis*, Buridan argues, as both Leibniz and Bradley were later to do, that if one posits a relation to explain how two things are connected one raises at once the question of how the relation is connected to each of them and so embarks on a vicious infinite regress. If, as Ockham

[19] "Therefore, one must know that to have a relation we need a subject, a foundation, a *terminus a quo* and a *terminus ad quem*. Hence, everywhere there is a real relation, there are five things, i.e. the relation itself, the subject of the relation, the foundation of the relation and the two extremes between which the relation subsists" (Burley, *De Relativis*; transl. Mugnai 2012, 165).

seems sometimes to suggest, Theology requires positing real relations the result is a mystery beyond our ken.

It seems then that the basic nominalist strategy for dealing with the Aristotelian Categories other than substance and quality must be semantic ascent. Where Realists like Burley have an ontological item, a relation, supervening on various (other) items in the ontology, Nominalists like Ockham will have a term in a mental or spoken language which stands for and connotes items already in the ontology and, as Mugnai has noted, although Ockham does not refuse the language of one thing inhering in another he analyses it in terms of predication, writing in *Summa Logicae* I, cap. 10 that "per inesse intelligitur praedicari".

For Ockham and those in the Nominalist tradition after him connotative terms signify the items of which they can be correctly predicated and consignify or connote or appellate others. This is expressed linguistically (and on the 'Ideal Language' interpretation mentioned above is represented in thought) by a phrase in which there is an absolute term in the nominative signifying the items of which the connotative term can be predicated while the connotation is captured either by terms in other cases or by sentences involving them. Thus, for example Ockham will define 'pale' (*album*) as 'thing having a paleness' (*res habens albedinem*) where 'paleness' (*albedinem*) is in the accusative. Thus to eliminate real relations one has to account for oblique cases. This is central to Leibniz's project.

As Massimo Mugnai points out, "In his reflections on 'oblique terms' Leibniz constantly expresses the conviction that, in a certain sense, they imply complex sentences, and therefore cannot be fully 'explained' without reference to several interconnected propositions."[20] Mugnai sees in Leibniz' thought a certain development beginning with his early assertion that inflections and oblique cases of nouns and adjectives (that is cases other than the nominative) should be forbidden in a logical calculus—and so in an Ideal language—to later admissions that they are in fact ineliminable. Mugnai goes on, however, to suggest that Leibniz was ambivalent about this and that "From the [later] *Analysis particularum...* there emerge both the proposal to replace prepositions and cases with special *particulae* and the conviction that the *particulae*, and in a subordinate way the cases, may be replaced by complex propositions in which neither cases nor prepositions are present."

I propose we read Leibniz here slightly differently—as arguing that one can do without prepositions in the Ideal Language if one has cases and can do without cases if one has prepositions. His proposal is to seek a notation which could be interpreted either way.

Leibniz seems never to have found a notation which satisfied him but in Chapters 5 and 6 of *Articulating Medieval Logic* Terence Parsons proposes an approach which he argues persuasively to involve nothing that would not be available to medieval logicians like Buridan and Ockham and which would seem to fulfill Leibniz's Project. I recommend it to the reader.

---

[20] Mugnai (1990, 63).

9 Conclusion

I suggest that Leibniz can be seen as not merely a tentative or tepid Nominalist but as a serious standard-bearer of a tradition he inherited. Like his forebears in that tradition he sought an Ideal Language which was expressively at least as powerful as any natural language and which committed one only to substances and truths, and so rejected universals, abstract objects, real accidents and real relations. From this perspective when he characterizes himself as a Nominalist provisionally he is not characterizing the project he has in hand but his confidence that it can be fully carried out and when he distinguishes his position from that of Hobbes whom he said was "more than a Nominalist" (*plusquam nominalis...*) he is criticizing Hobbes not for his ambition (than which nothing could be more Nominalist) but for his failure to see that without accepting objective truths the Nominalist project could not be carried out.[21]

If this is right then on two crucial issues about which Massimo Mugnai has done more than anyone else to clarify Leibniz' views, namely the nature of relations and of syncategorematic expressions, Leibniz's views are continuous with the Medieval Nominalist tradition and perhaps owe more to it than even Massimo has suggested. For showing us that and for much more we are deeply indebted to him.

References

Antognazza, M. R. (2009). *Leibniz: An intellectual biography*. Cambridge University Press.
Aventinus. (1580). Johannes Turmair, *Annalium Boiorum lib. VIII*. Basel.
Buridanus, J. (1977). *Sophismata.* In T. K. Scott (Ed.), *Grammatica speculativa* (Vol. 1). Frommann Holzboog.
Du Plessis d'Argentre. (1755). *Collectio judiciorum de novis erroribus*. Cailleau.
King, P. (1985). *John Buridan's logic*. Reidel.
King, P. (unpublished). Between Logic and Psychology: Jean Buridan on Mental. http://individual.utoronto.ca/pking/presentations/Buridan_on_Mental.pdf
Loemker, L. E. (Ed.) (1969). *G. W. Leibniz. Philosophical papers and letters*. Kluwer.
Mates, B. (1986). *The philosophy of Lebniz: Metaphysics and philosophy of language*. OUP.
Mugnai, M. (1976). *Astrazione e realtà: Saggio su Leibniz*. Feltrinelli.
Mugnai, M. (1986). Leibniz: De Abstracto et Concreto. *Studia Leibnitiana, 18*, 127–131.
Mugnai, M. (1990). A systematical approach to Leibniz's theory of relations and relational sentences. *Topoi, 9*, 61–81.
Mugnai, M. (2012). Leibniz's ontology of relations: A last word? *Oxford Studies in Early Modern Philosophy, 6*, 171–207.

[21] "Hobbes seems to me to be more than a nominalist. For not content like the nominalists, to reduce universals to names, he says that the truth of things itself consists in names and what is more, that it depends on the human will, because it allegedly depends on the definitions of terms, and definitions depend on the human will. This is the opinion of a man recognized as among the most profound of our century, and as I said, nothing can be more nominalistic than it" (*Preface to an Edition of Nizolius*, transl. Loemker 1969, 128 modified).

Mugnai, M. (2017). Review of Marko Malink, Anubav Vasudevan, "The Logic of Leibniz's Generales inquisitiones de analysi notionum et veritatum." *The Leibniz Review, 27*, 117–137.
Normore, C. (1985). Buridan's ontology. In J. Bogen & J. E. McGuire (Eds.), *How things are: Studies in predication and the history of philosophy and science* (pp. 189–203). Reidel.
Normore, C. (1987). The tradition of mediaeval nominalism. In J. Wippel (Ed.), *Studies in medieval philosophy* (pp. 201–217). Catholic University of America Press.
Panaccio, C. (2004). *Ockham on concepts.* Ashgate.
Parsons, T. (2014). *Articulating medieval logic.* OUP.
Rauzy, J.-B. (2004). How to evaluate Leibniz's nominalism? *Metaphysica, 5*(1), 43–58.
Spade, P. V. (1990). Ockham, Adams and connotation: A critical notice of Marilyn Adams, William Ockham. *The Philosophical Review, 99*, 593–612.
Thorndike, L. (1975). *University records and life in the middle ages.* W.W. Norton.
Trentman, J. (1970). Ockham on mental. *Mind, 79*, 576–650.
Walker, O. (1673). *Ars rationis maxima ex parte ad mentem nominalium.* Oxford.

Modern Logic and Its Applications

Oskar Becker and the Modal Translation of Intuitionistic Logic

Stefania Centrone and Pierluigi Minari

1 Introduction

"Heyting's intuitionistic propositional calculus **IPC** can be soundly and faithfully translated into the classical modal system **S4**": this is—rephrased in the now current terminology—the well known main result, comprehensive of a conjecture later proved to be true in McKinsey and Tarski (1948), that is contained in the short, deservedly celebrated paper published in 1933 by Kurt Gödel with the title *An interpretation of the intuitionistic propositional calculus.*[1]

Yet, the idea of a modal translation of intuitionistic logic was not new: three years earlier, in the *Appendix to Part I* of his essay *On the Logic of Modalities*[2], Oskar Becker not only had seriously considered that very idea, but he had also actually tried—although unsuccessfully—to realize it at a formal level.

Now, the fact is that Gödel was aware of Becker's aim: indeed, in 1931 he had reviewed[3] *On the Logic of Modalities.* In the *Review*, Gödel is pretty accurate in describing Becker's main intent of extending Lewis's "Survey system" to a modal system with a linearly ordered, finite number of positive irreducible modalities (we will say more on this in Sect. 3) and in drawing attention to some weak points in Becker's formal "experiments" as well. On the other side, he is rather hasty and

[1] Godel (1933).

[2] Becker (1930) (here quoted according to the original pagination). The forthcoming volume Centrone and Minari (2022b) contains the first English translation of Becker's *Zur Logik der Modalitäten*, together with an extensive commentary.

[3] Godel (1931).

S. Centrone
Institut für Philosophie, FernUniversität in Hagen, Hagen, Germany
e-mail: stefania.centrone@fernuni-hagen.de

P. Minari (✉)
Dipartimento di Lettere e Filosofia, Università degli Studi di Firenze, Firenze, Italy
e-mail: pierluigi.minari@unifi.it

F. Ademollo et al. (eds.), *Thinking and Calculating*, Logic, Epistemology, and the Unity of Science 54, https://doi.org/10.1007/978-3-030-97303-2_18

dismissive in commenting the *Appendix to Part I* and Becker's explicit intent of translating intuitionistic logic into the "Survey system":

> In conclusion the author discusses, from a formal as well as a phenomenological standpoint, the connections that in his opinion obtain between modal logic and the intuitionistic logic of Brouwer and Heyting. It seems doubtful, however, that the steps here taken to deal with this problem on a formal plane will lead to success. (Gödel, 1931, 217)

It is all the more surprising that in his *An interpretation of the intuitionistic propositional calculus* Gödel does indeed mention Becker (once), but only concerning a certain modal axiom (see below) introduced in *On the Logic of Modalities*; and not—*quite unfairly*—for having (at least) anticipated the idea of a modal translation of intuitionistic logic.[4]

The aim of the present note is to reconsider this unjustly neglected contribution by Oskar Becker, and other interesting ones as well.

2 Gödel's Result

Of course, it is not our intention to diminish the importance of Gödel's own result. So, to start with, let us briefly summarize Gödel's key accomplishments in Godel (1933). The target modal system $\mathfrak{S}$ introduced by Gödel features a language containing a modal operator B—*beweisbar*, intended to mean 'provable by any correct means'– in addition to the usual Boolean connectives $\neg, \vee, \wedge, \rightarrow$, and is axiomatically presented as an extension of the classical propositional calculus **CPC** by means of four postulates, namely three axiom schemas[5] ($\mathfrak{S}$.1)–($\mathfrak{S}$.3) and a new rule of inference ($\mathfrak{S}$.4):

($\mathfrak{S}$.1) $\mathsf{B}\alpha \rightarrow \alpha$
($\mathfrak{S}$.2) $\mathsf{B}\alpha \rightarrow (\mathsf{B}(\alpha \rightarrow \beta) \rightarrow \mathsf{B}\beta)$
($\mathfrak{S}$.3) $\mathsf{B}\alpha \rightarrow \mathsf{BB}\alpha$
($\mathfrak{S}$.4) from α to infer $\mathsf{B}\alpha$

$\mathfrak{S}$ coincides, up to the use of 'B' in place of '$\Box$', with Lewis's system **S4** *recast in the now familiar axiomatic presentation.* It is worth to remark that this is the first time ever that the "user-friendly" format "**CPC** + specific modal rules and axioms" for the axiomatic presentation of a (classical) modal system, which has become standard since the 1950s[6], is adopted. Indeed, **S4** had been *officially* introduced one

[4] Likewise, no mention of Becker is found in A. S. Troelstra's "Introductory note to *An interpretation of the intuitionistic propositional calculus*", in Feferman et al. (1986, 296–299). Notice that a modal translation of intuitionistic logic was in a sense foreshadowed also by Ivan Orlov in (Orlov, 1928). The paper, written in Russian, remained however unknown outside the Soviet Union for a very long time: Orlov's contributions were indeed "rediscovered" and came to be known and discussed only in the 1990's, see Chagrov and Zakharyashchev (1992) and Došen (1992).

[5] Actually, in Gödel's paper the axioms are not given in schematic form, and the rule of substitution is assumed.

[6] Thanks to Feys (1950), Prior (1955) and, in particular, Lemmon (1957).

year earlier in the *Appendix II* of Lewis and Langford (1932)[7] in the much less perspicuous "Lewis-style" axiomatization not constructed as an extension of **CPC**, which has been in force during the early development of modal logic up until the 1940s.

Henceforth we use the same symbols ($\neg, \wedge, \vee, \rightarrow$) for the intuitionistic and the classical connectives, and replace throughout B with $\Box$ and $\mathfrak{S}$ with **S4**.

Gödel's translation $(\ldots)^G$ from the formulas of the language $\mathcal{L}_I$ of **IPC** into the formulas of the language $\mathcal{L}_\Box$ of **S4** is inductively defined as follows:[8]

- $p^G := p$, where p is a propositional atom
- $(\neg\beta)^G := \neg\Box\beta^G$
- $(\beta \wedge \gamma)^G := \beta^G \wedge \gamma^G$
- $(\beta \vee \gamma)^G := \Box\beta^G \vee \Box\gamma^G$
- $(\beta \rightarrow \gamma)^G := \Box\beta^G \rightarrow \Box\gamma^G$

Next, Gödel claims (without giving any proof[9]) that the translation is *sound*, that is for all $\alpha \in \mathcal{L}_I$:

$$\vdash_{\mathbf{IPC}} \alpha \quad \Rightarrow \vdash_{\mathbf{S4}} \alpha^G$$

and conjectures that the translation is also, as we usually say today, *faithful*, that is for all $\alpha \in \mathcal{L}_I$:

$$\vdash_{\mathbf{S4}} \alpha^G \quad \Rightarrow \vdash_{\mathbf{IPC}} \alpha$$

Fifteen years later the conjecture was indeed cleverly solved in the positive by J. C. C. McKinsey and A. Tarski[10] by introducing a suitable algebraic-semantics characterization for **S4** together with the alternative translation $(\ldots)^*$:

- $p^* := \Box p$, where p is a propositional atom
- $(\neg\beta)^* := \Box\neg\beta^*$
- $(\beta \wedge \gamma)^* := \beta^* \wedge \gamma^*$
- $(\beta \vee \gamma)^* := \beta^* \vee \gamma^*$
- $(\beta \rightarrow \gamma)^* := \Box(\beta^* \rightarrow \gamma^*)$

---

[7] Lewis and Langford (1932) is not mentioned by Gödel. He says that $\mathfrak{S}$ is equivalent to "Lewis's system of strict implication", that is the *Survey system* [Lewis (1918), emended in Lewis (1920) and eventually named '**S3**' in the mentioned *Appendix II*), supplemented by "Becker's axiom" $\Box(\Box\alpha \rightarrow \Box\Box\alpha)$.

[8] Gödel also indicates as variants $(\neg\beta)^G := \Box\neg\Box\beta^G$ and $(\beta \wedge \gamma)^G := \Box\beta^G \wedge \Box\gamma^G$.

[9] A syntactical proof is indeed straightforward (and tedious).

[10] McKinsey and Tarski (1948), Theorems 5.2 and 5.3 (the latter for Gödel's variant of the translation $(\ldots)^G$), which almost immediately follow from Theorem 5.1 that states the soundness and faithfulness of their own translation $(\ldots)^*$. The two translations $(\ldots)^G$ and $(\ldots)^*$ are indeed easily proved to be related as follows: for all $\mathcal{L}_I$-formulas α, $\vdash_{\mathbf{S4}} \Box\alpha^G \leftrightarrow \alpha^*$, hence $\vdash_{\mathbf{S4}} \alpha^G \Leftrightarrow \vdash_{\mathbf{S4}} \alpha^*$.

Gödel's accomplishments, beyond their intrinsic conceptual and technical interest, and their more or less immediate consequences as well[11], have played an important role in a number of subsequent developments and investigations. Let us just briefly mention some of the most significative ones.

First of all, Saul Kripke's invention of the relational semantics for intuitionistic logic was actually inspired by his own possible-worlds semantics for modal logics[12] together with "the known mappings of intuitionistic logic into the modal system **S4**"[13]—indeed, an intuitionistic model is exactly the pre-image of a **S4** model under the McKinsey-Tarski translation.

Next, Gödel's remark that the modal operator $\Box$ of **S4** *cannot* be read as 'provable in a given formal system' like e.g. **PA** (first-order Peano arithmetic), together with his 1931 incompleteness theorems, paved the way in the 1970s to the birth of the *provability logics*, a new family of modal logics, relevant also from a foundational perspective, which feature instead a box-like operator meant to capture in an abstract, modal setting the structural properties of the notion of *provability in* **PA** (or in another fixed, typically arithmetical, theory).[14]

S4 is the standard system of (propositional) *epistemic* logic, $\Box$ being interpreted as a *knowability* operator. In the 1980's, Gödel's translation was successfully extended from logic to mathematics. Starting with seminal results by G. Mints (1992) and N. Goodman (1984), intuitionistic versions of various formal mathematical theories like arithmetic (**HA**), type theory and set-theory were shown to be faithfully G-translatable into suitable corresponding *intensional*, or *epistemic*, mathematical theories formalized with (quantified) **S4** as underlying logic—see Shapiro (1985) for a survey.

Finally, Gödel's way of capturing the intended semantics of the intuitionistic logical operators through a modal, **S4**-like notion of 'provability by any correct means' motivated in the 1990's, starting with the work of S. Artemov, the elaboration of the *logic of proofs* **LP**, a sort of explicit version of **S4** in whose syntax *proof terms*, representing (classical) proofs, become first class citizens, and its subsequent generalization to the so called *justification logic*.[15]

[11] E.g. the *disjunction property* for **IPC**, which follows from the translation theorem together with Gödel's conjecture that $\vdash_{\mathbf{S4}} \Box\alpha \vee \Box\beta$ implies $\vdash_{\mathbf{S4}} \Box\alpha$ or $\vdash_{\mathbf{S4}} \Box\beta$, later proved in McKinsey and Tarski (1948). The first "official" proof of the *disjunction property* for **IPC** was given by Gerhard Gentzen in 1935, via cut-elimination (Gentzen, 1935).

[12] Kripke (1963, 1965b).

[13] Kripke (1965a), 92.

[14] See Artemov and Beklemishev (2004) for a survey.

[15] See Artemov and Fitting (2019).

3 Oskar Becker and the Search for a "System of Closed Modalities"

Oskar Becker's 1930 essay *On the Logic of Modalities* sets out explicitly as an attempt to deal with issues pertaining to modal logic by supplementing the method provided by the 'calculus of logic' (in the Russell and Lewis tradition) with the 'phenomenological' method[16]—an enterprise that appears from the outset not to be easy at all.

> The concurrent usage of these methods of research, which are so different as to their essence and to their methodological technique, could appear to be questionable and is, actually, not bare of difficulties. Nevertheless, it seems to be unavoidable, if one does not want to end up in two "polar" unilateralities, namely, the mathematical combination of mostly empty concept-constructs, on the one hand, and the quite shortsighted description of obvious, more or less arbitrarily assembled concrete cases, on the other hand, which latter has been called, jokingly, "empiricism of the apriori". (Becker 1930, 1)

Indeed, we might better say that in *On the Logic of Modalities* Becker pursued two loosely related goals. The first one, more technical in character, was to find axiomatic conditions that reduced to the finite the number of logically non-equivalent combinations arising from the iterated application of the operators "not" and "it is impossible that (...)" in Lewis's *Survey system.* The second one, more philosophically oriented and, in a sense, much more ambitious, was to treat the logic of modalities from a phenomenological perspective and to understand, from this perspective, the philosophical and logical-ontological problems underlying, and posed by, Intuitionism.

In the present paper we focus exclusively on Part I of the essay, entitled *On the rank order and reduction of logical modalities* and dealing with the first of the above mentioned goals, and (in Sect. 4) on the *Appendix to Part I* entitled *The logic of modalities and the Brouwer-Heyting "intuitionistic" logical calculus.*

[16] Oskar Becker (Leipzig 1889–Bonn 1964) is often remembered as one of the most prominent students of Edmund Husserl. He graduated in mathematics in 1914 (Becker, 1914), and in 1922 he wrote under Husserl's supervision his *Habilitationsschrift, Contributions Toward a Phenomenological Foundation of Geometry and Its Physical Applications* (Becker, 1923). In 1927 Becker published what is considered to be his masterpiece, *Mathematical Existence* (Becker, 1927), in the *Jahrbuch für Philosophie und phänomenologische Forschung* (he was, together with Martin Heidegger, Moritz Geiger, Alexander Pfänder, Adolf Reinach and Max Scheler a member of the editorial board of this journal). In 1952—when the study of modal logic was already well beyond its pioneering era—Becker would come back to this subject with the monograph *Investigations on the Modal Calculus* (Becker, 1952), perhaps too old-fashioned for the time, cp. (Martin, 1969). For a complete bibliography of Becker's works see Zimny (1969).

The "Survey System", Alias S3, in a Nutshell

The "System of Strict Implication", or "Survey system", the real object of the investigations of Part I, was introduced by Lewis in *A Survey of Symbolic Logic* (1918), and by him emended two years later[17] after Emil L. Post's discovery that the original system proved the (strict) equivalence between the negation of p and the impossibility of p, thus *collapsing* into classical logic. It eventually received the now familiar name '**S3**', which we will use henceforth, in the already mentioned *Appendix II* of Lewis and Langford (1932).

As we said, the formal language and the style of axiomatization employed by Lewis in the *Survey* and followed by Becker in his essay are different from the now current ones.

As primitives, they take the unary operators "$-$" and "$\sim$", respectively for *negation* and *impossibility*, and the binary operators "$\times$" and "$=$", respectively for *conjunction* and *strict equivalence*.[18]

In turn, Lewis's (and Becker's) axiomatization of the system is *not* given as an extension of an axiomatic calculus for classical logic by means of additional axioms and inference rules. Actually, it is not at all trivial to prove that all classical tautologies are theorems of this axiomatization of **S3**.[19]

The equivalent, now more familiar axiomatization[20] of **S3**—in a language having as primitives the boolean connectives and the modal operator $\Box$, with $\Diamond\alpha$ defined as $\neg\Box\neg\alpha$—looks as follows:

Axioms and axiom schemas:

- all classical tautologies
- $\Box\alpha \rightarrow \alpha$ (schema T)
- $\Box(\alpha \rightarrow \beta) \rightarrow \Box(\Box\alpha \rightarrow \Box\beta)$ (schema K^+)

Inference rules:

$$\frac{\alpha \rightarrow \beta \qquad \alpha}{\beta}\ \text{MP } (\textit{modus ponens})$$

$$\frac{\alpha}{\Box\alpha}\text{RN}^- \text{ provided } \alpha \text{ is an (instance of an) axiom (schema)}$$

Notice that the necessitation rule (RN^-) comes in a *restricted* form. Indeed, it turns out that **S3** is not closed under the unrestricted necessitation rule (*RN*) and so that *it is not a normal modal system*. For instance, denoting by "$\top$" any classical tautology, say $p \rightarrow p$, $\Box\top$ is a theorem of **S3**, while by contrast $\Box\Box\top$ is not a

[17] Lewis (1920).

[18] Thus "$-\alpha$", "$\sim\alpha$", "$\alpha \times \beta$", "$\alpha = \beta$" correspond, respectively, to "$\neg\alpha$", "$\neg\Diamond\alpha$", "$\alpha \wedge \beta$", "$\Box(\alpha \leftrightarrow \beta)$" in the now current notation. The other logical boolean and "strict" operators, in particular "$\subset$" (*material implication*), "$<$" (*strict implication*) and "$+$" (boolean disjunction) are instead *defined* in the expected way.

[19] See Lewis and Langford (1932, 136 ff).

[20] Introduced in Lemmon (1957).

theorem. Indeed, one can prove that **S3** extended with the axiom $\Box\Box\top$ is equivalent to the *normal* system **S4** mentioned in the previous Section.

In the following we will give an idea of Becker's investigations on **S3** by rephrasing them—for the reader's convenience—in terms of the $\Box$-language and the above axiomatization, while making reference to the Lewis-style setting only when it necessary to understand certain motivations underlying Becker's proposals.

Becker's "Completions" of S3

A *nested modality* (in short, *modality*) is a possibly empty string of $\Box$ and $\neg$, which prefixed to a formula yields another formula:

$$\_\ (\text{empty modality: } \_\alpha \equiv \alpha),\ \neg,\ \Box,\ \Box\Box,\ \neg\Box,\ \neg\Box\neg\Box\Box\neg,\ \ldots$$

or, in Lewis's and Becker's formalism, a possibly empty string of $\sim$ and $-$:

$$\_,\ -,\ \sim-,\ \sim-\sim-,\ -\sim-,\ -\sim\sim-\sim,\ \ldots$$

In the $\{\Box, \neg\}$ notation, a modality is *positive* (*negative*) iff it has an even (odd) number of occurrences of the negation; the other way around in the $\{\sim, -\}$ notation.

Given two modalities $\mathsf{m}_1, \mathsf{m}_2$, these are said to be **S3**-*equivalent* iff, for all formulas α,

$$\vdash_{\mathbf{S3}} \Box(\mathsf{m}_1\alpha \leftrightarrow \mathsf{m}_2\alpha)$$

while m_1 is said to be *stronger* than m_2—in symbols: $\mathsf{m}_1 \rightarrowtail \mathsf{m}_2$—iff

$$\vdash_{\mathbf{S3}} \Box(\mathsf{m}_1\alpha \rightarrow \mathsf{m}_2\alpha) \text{ for all } \alpha, \text{ and } \nvdash_{\mathbf{S3}} \Box(\mathsf{m}_2\alpha \rightarrow \mathsf{m}_1\alpha) \text{ for some } \alpha$$

Trivially, the number of modalities is infinite. But: how many *irreducible*, i.e. pairwise non equivalent, or "logically distinct", *modalities* are there in Lewis's **S3**?

Lewis did not consider this question, and Becker starts his investigation with the (implicit) conjecture that **S3** has indeed an infinite number of irreducible modalities—a conjecture which in 1939 William Parry will prove to be false: **S3** has exactly 42 irreducible modalities[21]. Becker's aim is to (axiomatically) extend **S3** in such a way that the set of irreducible modalities of the resulting system **S** be

(i) *finite*, and
(ii) possibly such that the positive modalities are *linearly ordered* by the above relation $\rightarrowtail$ (and, dually, the negative modalities as well),[22]

---

[21] Parry (1939).

[22] The 21 positive (resp. negative) irreducible modalities—after Parry's result—are indeed *not* linearly ordered w.r.t. to $\rightarrowtail$.

so that, in other words, any two positive (negative) modalities of **S** would be comparable with respect to logical strength.[23]

> Before going into the actual meaning of the reduction problem of infinitely many nested modalities, which arise through the iteration and composition of the symbols "∼" and "−", we present a purely formal investigation, by which Lewis's system becomes a closed system, thanks to the addition of a further axiom. This can be done in several ways. Here we will consider two of them.
>
> The assumptions introduced by Lewis are (apparently) not sufficient to obtain a closed system of irreducible modalities.
>
> Therefore we add to Lewis's axioms the new axiom 1.9:
>
> $$(1.9) \qquad -(\sim p) < \sim(\sim p)$$
>
> This represents, to some extent, an *ad hoc* choice of the additional axiom. Later on we will make different choices and draw the corresponding conclusions. (Becker 1930, 11)

The first extension of **S3** proposed by Becker is thus

$$\mathbf{S3}' \ := \ \mathbf{S3} + \text{ the schema: } \Box(\Diamond\alpha \rightarrow \Box\Diamond\alpha)$$

which he calls the "Six modalities system". Indeed, he gives a correct and detailed proof of the fact that $\mathbf{S3}'$ has exactly 6 irreducible modalities:

- positive modalities: $\Box$, $\Diamond$, _ ("factual" truth),
- negative modalities: $\neg\Box$, $\neg\Diamond$, $\neg$ ("factual" falsity),

and that they are linearly ordered, as to logical strength, as follows

$$\Box \rightarrowtail \_ \rightarrowtail \Diamond \text{ (positive)}, \quad \neg\Diamond \rightarrowtail \neg \rightarrowtail \neg\Box \text{ (negative)}$$

It is not difficult to prove that $\mathbf{S3}'$ is equivalent to the normal modal system **S5**, officially introduced for the first time, and thus named, in the *Appendix II* of Lewis and Langford (1932)[24].

While the characteristic axiom schema of $\mathbf{S3}'$, alias **S5**, represented according to Becker more or less the fruit of a formally convenient choice, intuitionistic logic is behind the finding of one of the two characteristic schemas of the second "completion" proposed by him, the "Ten modalities system" (here $\mathbf{S3}''$).

As we said, the exploration of the connection between intuitionistic and modal logic is one of Becker's aims. In this context he is thus naturally led, in particular, to interpret the intuitionistic negation ("¬")—which is *stronger* than classical

[23] It looks like Becker was supposing that (i), possibly together with (ii), would also imply the existence of a decision procedure for the extended calculus.

[24] Here Becker is acknowledged for having introduced the characteristic schema of the system. In fn. 1, p. 492, the Authors also mention a letter sent by M. Wajsberg to Lewis in 1927, containing "the outline of a system of Strict Implication with the addition of the postulate later suggested in Becker's paper".

negation—in modal terms, as *impossibility* ("$\sim$") or, as he uses to say, *absurdity* (*Absurdität*).[25] By replacing, in the intuitionistic law

$$\text{(WDN)} \quad \alpha \to \neg\neg\alpha$$

"$\neg$" with "$\sim$" and "$\to$" with "<" (strict implication), one gets

$$\alpha < \sim\sim \alpha$$

"Truth — as he puts it[26] — implies the absurdity of the absurdity (but not conversely!)". The schema, in our $\Box$-notation, reads

$$\Box(\alpha \to \Box\Diamond\alpha)$$

which is the "boxed-version" of the modal schema *B* (*B* from "Brouwer's *Axiom*", as Becker called it, a name still current in the literature).

> One can now add (this is the weakest additional postulation we propose) "Brouwer's *Axiom*" to this setting
>
> $$(1.91) \qquad p = - - p < \sim\sim p$$
>
> [...] As an [additional] axiom we choose [...]:
>
> $$(1.92) \qquad \sim -p < \sim - \sim -p$$
>
> [...] If one postulates (1.91) $\times$ (1.92) one can thus set up a ten modalities calculus. (Becker 1930, 17–18)

Becker rightly realized that the addition of the schema $B^{\Box}$ alone was not sufficient to reduce to the finite the number of irreducible modalities, hence the addition of a second axiom schema, (1.92), corresponding in our notation to $\Box(\Box\alpha \to \Box\Box\alpha)$, that is the "boxed-version" of the modal schema 4, as it is currently named.[27] Thus

$$\mathbf{S3}'' := \mathbf{S3} + (B^{\Box}) : \Box(\alpha \to \Box\Diamond\alpha) + (4^{\Box}) : \Box(\Box\alpha \to \Box\Box\alpha)$$

Becker's claim, supported by an elaborate and detailed (putative) proof, is that this system has 10 irreducible modalities:

- positive modalities: $\Box$, $\Diamond\Box$, $\Diamond$, $\Box\Diamond$, _ ("factual" truth),
- negative modalities: $\neg\Box$, $\neg\Diamond\Box$, $\neg\Diamond$, $\neg\Box\Diamond$, $\neg$ ("factual" falsity),

and that they are linearly ordered, as to logical strength, as follows

---

[25] Becker refers explicitly to Heyting (1930), which contains the first (complete) presentation of intuitionistic logic as a formalized calculus. The paper was published in the same year of *On the Logic of Modalities*, but was circulating since 1928.

[26] Becker (1930, 17).

[27] This is "Becker's additional axiom" mentioned in Godel (1933).

$$\Box \rightarrowtail \Diamond\Box \rightarrowtail \_ \rightarrowtail \Box\Diamond \rightarrowtail \Diamond \ (+) \quad \neg\Diamond \rightarrowtail \neg\Box\Diamond \rightarrowtail \neg \rightarrowtail \neg\Diamond\Box \rightarrowtail \neg\Box \ (-)$$

While Becker was right in the claim that in **S3**$''$ any modality is equivalent to one of the above ten modalities, he unfortunately failed to realize that, by the combined effect of the two schemas, the two modalities $\Diamond\Box$ and $\Box\Diamond$ turn out to be equivalent, respectively, with $\Box$ and $\Diamond$. In other words, the "Ten modalities system" **S3**$''$ is equivalent to the "Six modalities system" **S3**$'$, thus boiling down to an alternative axiomatization of **S5**.[28]

In conclusion, we would like just to touch upon Becker's very interesting and elaborate attempt[29] to develop a more abstract approach to the problem of "completing" Lewis's calculus in such a way that in the resulting system, independently from the number (finite or infinite) of irreducible modalities, any two positive (or negative) modalities be comparable with respect to logical strength.[30]

Becker conveniently uses here as basic, elementary positive modality the operator "$\Box$" ("*N*" in his symbolism) and fixes a finite set $\mathcal{R}$ of "rules" which are intended to impose conditions concerning the *preservation* of relations of logical strength between $\{\Box, \neg\}$-modalities under juxtaposition.[31] Now, the point is that Lewis's calculus **S3**, while being closed under these rules, contains however *incomparable* positive modalities. So, this is Becker's very interesting idea, one should try to devise the "weakest possible axiomatic conditions" to be added to the $\mathcal{R}$-rules, in order to obtain a calculus in which all the positive modalities are pairwise comparable w.r. to logical strength. At the end of a rather complex argument, he arrives at the claim that a stepwise generalization of Brouwer's schema (in the form $B^{\Box}$, see above) provides an infinite number of axiomatic conditions which, added to **S3** together with the $\mathcal{R}$-rules, produce a calculus—let us call it **SM**—with the requested properties although with a possibly infinite number of irreducible modalities.

Unfortunately again, **SM** turns out to be equivalent to the system **S3**$'$, that is **S5**, as proved by Churchman in 1938.[32] Notwithstanding, whether **SM** collapses to **S5** also

---

[28] In his *Review*, Gödel indeed remarked that "it is nowhere shown that the [...] systems set up really differ from one another and from Lewis's system (in other words, that the additional axioms are not in fact equivalent and do not follow from Lewis's); nor, furthermore, that the six, or ten, basic modalities obtained cannot be still further reduced." (Godel 1931, 201).

[29] In §5 of Part I, Becker (1930, 25–30), entitled "On the Calculus of Modalities with least Requirements, which still yields a Linear Rank Order".

[30] See Centrone and Minari (2022b) for a detailed analysis and discussion.

[31] These rules have not been correctly interpreted and formalized in Churchman (1938), the first (and unique, as far as we know) paper where this experiment by Becker is detailedly analyzed. Incidentally, notice that the inference rules

$$\frac{\Box(\alpha \to \beta)}{\Box(\Box\alpha \to \Box\beta)} \quad \text{and} \quad \frac{\Box(\alpha \to \beta)}{\Box(\Diamond\alpha \to \Diamond\beta)}$$

known also in the current literature as *Becker's rules*, were given this name in Churchman (1938) because (uncorrectly) regarded as specific instances of one of the rules in $\mathcal{R}$. For a detailed discussion, see Centrone and Minari (2022a).

[32] Churchman (1938, 78 ff). The claim is indeed correct, although Churchman's proof thereof is not, because he did not formalize the system **SM** as Becker really intended it.

when based on a system *weaker* than **S3** (actually, weaker than **S2**) is an interesting question, open as far as we know, and worth to be investigated.

4 Becker's Idea of a Modal Interpretation of Intuitionistic Logic

The idea is exposed and elaborated in the mentioned *Appendix to Part I*[33], after an introductory short presentation of Heyting's intuitionistic calculus **IPC** and its relationship with the classical ("Russellian") calculus[34]:

> How is now Heyting's calculus related to the uncompleted and the completed Lewis's calculus?
>
> Firstly, the question of an appropriate "translation" of the symbols emerges. (Becker, 1930, 31)

More precisely, Becker's idea is to define a vocabulary translation associating to each intuitionistic logical operator ($\rightarrow, \vee, \wedge, \neg$) a corresponding logical operator of Lewis's calculus **S3**, in such a way that every theorem of the intuitionistic calculus **IPC** becomes, once transformed according to the translation, a theorem of **S3**. He tentatively considers three candidate translations[35]:

(T1) H: $\rightarrow, \vee, \wedge, \neg \Rightarrow$ L: $\rightarrow, \vee, \wedge, \neg$
(T2) H: $\rightarrow, \vee, \wedge, \neg \Rightarrow$ L: $<, \vee_s, \wedge, \sim$ (where $\alpha \vee_s \beta =_{df} \Box(\alpha \vee \beta)$)
(T3) H: $\rightarrow, \vee, \wedge, \neg \Rightarrow$ L: $\rightarrow, \vee, \wedge, \sim$

As to (T1), he observes that the T1-translation of every intuitionistic theorem is *obviously* a **S3**-theorem, because intuitionistic logic is included in classical non-modal logic, and the latter in turn is included in the Lewis's system. On the other side, he stresses that

> [...] this is a worthless triviality. Indeed, the purpose of a comparison between intuitionistic and modal logic can only be to make the deficits of the former with respect to the latter comprehensible by interpreting the intuitionistic notions by the specific modal-logical notions, that is Lewis's "strict" notions (strict implication, strict logical sum, impossibility). (*loc. cit.*)

Concerning the second translation (T2), which implements exactly the above proposal, he correctly observes that the T2-translation $\Box(\Box(p \vee p) \leftrightarrow p)$ of $p \vee p \leftrightarrow p$, which is one of the axioms of Heyting's calculus, *is not a theorem* of **S3**—otherwise, as he cleverly shows, the latter would collapse.

---

[33] Becker (1930, 30–35).

[34] Becker is aware of Glivenko's *double negation translation* of **CPC** into **IPC**, (Glivenko, 1929), and mentions this paper.

[35] Here we use a convenient mix of Lewis's symbolism and the current symbolism.

Finally, as far as the third of the proposed translations is concerned, Becker claims, without however giving a detailed proof, that the T3-translation of the **IPC** axiom $(\alpha \to \beta) \wedge (\alpha \to \neg\beta) \to \neg\alpha$, that is the modal formula

$$(*) \qquad (\alpha \to \beta) \wedge (\alpha \to \neg\Diamond\beta) \to \neg\Diamond\alpha$$

is not a theorem of **S3**. Indeed, it is not difficult to prove that Becker was right in claiming that $(*)$ is underivable in **S3**. Actually, it is possible to prove even more: every normal modal system containing the schemas T and $(*)$ collapses. This fact therefore implies that also with respect to the extended system(s) **S3′**/**S3″** the translation (T3) would boil down to the *trivial* translation (T1).

Becker thereby concludes his "translation-experiments" as follows:

> At this point a further investigation must begin, with the aim to assess *whether and which additions must be made to the extended Lewis's System (Calculus of 10 Modalities, Calculus of 6 Modalities) so that Heyting's Axiom (11)* [i.e. $(*)$ above] *holds*. Further problems can nevertheless arise because of the difference of the undefined notions in the Heyting's and the Lewis's System. The solution of these tasks and the overcoming of these difficulties shall be left to future work. (Ibid., 33.)

5 Conclusions

As we know, only three years later someone else, namely Kurt Gödel, did the "future work" and "found the solution of these tasks". Or rather, he *almost* found that solution, to be precise: Becker indeed was asking for a translation of **IPC** into **S3**, and Gödel provided a translation into the stronger system **S4**.

So, the very end of the story comes with Ian Hacking's 1963 paper[36], containing (no mention of Becker ..., but) the proof that the McKinsey-Tarski translation provides a sound and faithful translation of **IPC** *already into Lewis's* **S3**.[37] By the way, the McKinsey-Tarski translation is very close to Becker's (unsuccessful) translation (T2): apart from the essential fact that atomic formulas become boxed under the McKinsey-Tarski translation, the translation of the intuitionistic connectives is the same, except for the disjunction.

One might be tempted to underrate Becker's *formal* contributions in *On the Logic of Modalities*. On the contrary, and notwithstanding the many shortcomings, Becker's *pioneering* work, containing sophisticated insights and interesting technical solutions, has played an extremely important role in the *early* development of modal logic during the decade 1930–1940, as witnessed by the scientific contributions of other

[36] Hacking (1963). Hacking's proof-theoretical demonstration, which makes use of a normalization theorem for a suitable natural deduction presentation of **S3**, is rather convoluted. A much simpler semantical proof, obtained by exploiting a conjecture in Oakes (1999), can be found in Centrone and Minari (2022b).

[37] Gödel's translation $(\ldots)^G$ is instead not sound with respect to **S3**.

scholars who, at that time, referred to Becker's investigations and to the problems raised by him, and took them as a basis for further developments and researches.[38]

References

Artemov, S., & Beklemishev, L. (2004). Provability logic. In D. Gabbay & F. Guenthner (Eds.), *Handbook of philosophical logic* (2nd ed., Vol. 13, pp. 229–403). Kluwer.

Artemov, S., & Fitting, M. (2019). *Justification logic: Reasoning with reasons*. Cambridge University Press.

Becker, O. (1914). *Über die Zerlegung eines Polygons in exclusive Dreiecke auf Grund der ebenen Axiome der Verknüpfung und Anordnung*. F. A. Brockhaus.

Becker, O. (1923). Beiträge zur phänomenologischen Begründung der Geometrie und ihrer physikalischen Anwendung. In E. Husserl (Ed.), *Jahrbuch für Philosophie und phänomenologische Forschung VI* (pp. 385–560). Max Niemeyer Verlag.

Becker, O. (1927). Mathematische Existenz. Untersuchungen zur Logik und Ontologie mathematischer Phänomene. In E. Husserl (Ed.), *Jahrbuch für Philosophie und phänomenologische Forschung VIII* (pp. 439–809). Max Niemeyer Verlag.

Becker, O. (1930). Zur Logik der Modalitäten. In E. Husserl (Ed.), *Jahrbuch für Philosophie und phänomenologische Forschung XI* (pp. 497–548). Max Niemeyer Verlag (English translation in Centrone and Minari, 2022b).

Becker, O. (1952). *Untersuchungen über den Modalkalkül*. Westkulturverlag Anton Hain.

Centrone, S., & Minari, P. (2022a). *Becker's rule* is not Becker's rule. In A. Klev & S. Rahman (Eds.), *Festschrift for Göran Sundholm*. Springer, Cham.

Centrone, S., & Minari, P. (2022b). *Oskar Becker, on the logic of modalities (1930): Translation, commentary and analysis*. Springer, Cham.

Chagrov, A., & Zakharyashchev, M. (1992). Modal companions of intermediate propositional logics. *Studia Logica, 51*, 49–82.

Churchman, C. W. (1938). On finite and infinite modal systems. *The Journal of Symbolic Logic, 3*, 77–82.

Došen, K. (1992). The first axiomatization of relevant logic. *Journal of Philosophical Logic, 21*, 339–356.

Feferman, S., Kleene, S., Moore, G., Solovay, R., & van Heijenoort, J. (Eds.). (1986). *Kurt Gödel collected works. Volume I: Publications 1929–1936*. Oxford University Press.

Feys, R. (1937). Les logiques nouvelles des modalités. *Revue néo-scolastique de philosophie, 56*, 517–553.

Feys, R. (1938). Les logiques nouvelles des modalités (suite et fin). *Revue néo-scolastique de philosophie, 58*, 217–252.

Feys, R. (1950). Les systèmes formalisés des modalités aristotéliciennes. *Revue philosophique de Louvain, 48*, 478–509.

Gentzen, G. (1935). Untersuchungen über das logische Schließen. *Mathematische Zeitschrift, 39*, 176–210, 405–431.

Glivenko, V. (1929). Sur quelques points de la logique de M. Brouwer. *Acad. Roy. Belg. Bull. Cl. Sci., Sér. 5, 15*, 183–188.

Gödel, K. (1931). Besprechung von *Becker 1930*: Zur Logik der Modalitäten. *Monatshefte für Mathematik und Physik (Literaturberichte), 38*, 5–6. (Reprinted and translated in [Feferman et al., 1986], 216–217).

[38] In this regard, Feys (1937) and Feys (1938) should at least be added to the already mentioned works by Gödel, Lewis and Langford, Churchman and Parry.

Gödel, K. (1933). Eine interpretation des intuitionistischen Aussagenkalküls. *Ergebnisse eines mathematischen Kolloquiums, 4*, 39–40. (Reprinted and translated in [Feferman et al., 1986], 300–301).
Goodman, N. D. (1984). Epistemic arithmetic is a conservative extension of intuitionistic arithmetic. *The Journal of Symbolic Logic, 49*, 192–203.
Hacking, I. (1963). What is strict implication? *The Journal of Symbolic Logic, 28*, 51–71.
Heyting, A. (1930). Die formalen Regeln der intuitionistischen Logik. *Sitzungsberichte der Preussischen Akademie der Wissenschaften. Physikalisch-mathematische Klasse* (pp. 42–56, 57–71, 158–169).
Kripke, S. (1963). Semantical analysis of modal logic, I. Normal modal propositional calculi. *Zeitschr. f. Math. Logik und Grundlagen d. Math., 9*, 67–96.
Kripke, S. (1965a). Semantical analysis of intuitionistic logic. In J. Crossley & M. Dummett (Eds.), *Formal systems and recursive functions* (pp. 92–130). North-Holland Publishing Company.
Kripke, S. (1965b). Semantical analysis of modal logic, II. Non-normal modal propositional calculi. In J. W. Addison, L. Henkin, & A. Tarski (Eds.), *The theory of models. Proceedings of the 1963 international symposium at Berkeley* (pp. 206–220). North-Holland Publishing Company.
Lemmon, E. J. (1957). New foundations for Lewis modal systems. *The Journal of Symbolic Logic, 22*, 176–186.
Lewis, C. I. (1918). *A survey of symbolic logic.* University of California Press.
Lewis, C. I. (1920). Strict implication—An emendation. *Journal of Philosophy, Psychology and Scientific Methods, 17*, 300–302.
Lewis, C. I., & Langford, C. H. (1932). *Symbolic logic.* The Century Co. (Dover Publications, New York ^{2}1959).
Martin, G. (1969). Oskar Beckers Untersuchungen über den Modalkalkül. *Kant-Studien, 60*, 312–318.
McKinsey, J. C. C., & Tarski, A. (1948). Some theorems about the sentential calculi of Lewis and Heyting. *The Journal of Symbolic Logic, 13*, 1–15.
Mints, G. (1992). On Novikov's hypothesis (1978, Russian). In G. Mints (Ed.), *Selected papers in proof theory* (pp. 147–151). Bibliopolis, Naples and North-Holland Publishing Company.
Oakes, C. (1999). Interpretations of intuitionistic logic in non-normal modal logics. *Journal of Philosophical Logic, 28*, 47–60.
Orlov, I. (1928). The calculus of compatibility of propositions (in Russian). *Mathematics of the USSR, Sbornik, 35*, 263–286.
Parry, W. T. (1939). Modalities in the survey system of strict implication. *The Journal of Symbolic Logic, 4*, 137–154.
Prior, A. (1955). *Formal logic.* Oxford University Press.
Shapiro, S. (Ed.). (1985). *Intensional mathematics.* North-Holland Publishing Company.
Zimny, L. (1969). Oskar Becker—Bibliographie. *Kant-Studien, 60*, 319–330.

Reflecting and Unfolding

Andrea Cantini

Notionis distinctae primitivae non alia datur cognitio, quam intuitiva, ut compositarum plerumque cogitatio non nisi symbolica est. (G.W. Leibniz, 1684[1])

1 On Reflective Closure

Let us consider Gödel (1990, p. 151, vol II): why is a formalism incomplete? Well, contemplating a *fixed* system gives rise to new axioms, which are as evident as those we start with. But the extension process goes on forever and *can be iterated into the transfinite*. Hence, if we reflect upon a system, we are apparently left with an *infinite task*, and involved or committed to a number of statements which are to be made explicit, at least partially and potentially.

Are we then forced to use the so-called *ordinals*? In the last century this choice has been implemented in several directions, e.g. along the lines of the so-called *ordinal logics* in the sense of Turing, *predicative mathematics*, or else *progressions of theories*; and some conceptual as well as technical difficulties arise, which we list below.

I. Ordinals are *external* notions, and we have to cope with a *metabasis eis allo genos*. The difficulty shows that there are abstract higher-type notions as required by incompleteness.
II. We have to justify the view that ordinals or well-orderings are implicit in the given body of, say, arithmetical knowledge. This is technically much involved, as it essentially hinges upon the use of complex and abstract notions (higher notation systems, collapsing functions, etc.).

[1] *Meditationes de cognitione, veritate et ideis*, 1684.

A. Cantini (✉)
Dipartimento di Lettere e Filosofia, Università degli Studi di Firenze, Firenze, Italy
e-mail: andrea.cantini@unifi.it

F. Ademollo et al. (eds.), *Thinking and Calculating*, Logic, Epistemology, and the Unity of Science 54, https://doi.org/10.1007/978-3-030-97303-2_19

III. We have to pass from *closed* systems to *open-ended* systems. But how to express in a definite fashion this open-endedness?

Of course, it is impossible to explicitly and completely address I–III within the boundaries of this paper. We can nonetheless make an attempt to fix some points at least informally and intuitively. To this aim, let us recall that ordinals are *generalizations of numbers*, which were created by Cantor in order to control suitable topological processes, and in fact they are *reifications*[2] of *iteration procedures*, that go definitely beyond the realm of *finite* numbers into the so-called *transfinite*. Their role has become essential for assigning *invariants* to computations and proof-theoretic investigations, which involve *consistency*, as well as the *structure of proofs*. In general, one has to cope with order-theoretic structures, which in the standard cases must be *well-founded* (= no infinite regress is possible) and yet can be represented by elementary means, e.g. notation systems, i.e. *symbolic structures* (Leibniz). Moreover ordinals have been crucial—since the Thirties with Gentzen, and later with Schütte's school, Takeuti, the Bern school, Feferman, the investigations of Girard, Arai, Rathjen's "art of ordinal analysis" Rathjen (2006), etc., till most recent contributions Freund and Rathjen (2021).

The Kripke–Feferman Theory KF

The so-called theory of reflective closure over Peano Arithmetic PA emerges from the attempt to (partially) overcome difficulty I above. More precisely, if we try to make these ideas formally precise, we are naturally driven to formalize the clauses of Kleene's strong three valued semantical schema for *self-referential truth*, and hence to the well known KF-theory (over PA, as given in Feferman, 1991). For the reader's sake, let us summarize the essential points about KF.

Firstly, as a starting intuitive point, Kreisel (1970) suggested to study *principles of proof and definition that are recognized as valid, once we have understood certain given concepts*: the typical example was predicativism, as *the conception unfolding what is implicit in accepting the notion of natural number*. Formally, this view leads to articulating some form of self-reflection that developed along the so-called *autonomous transfinite progressions* of theories in the Sixties (see Feferman, 1968); which is not so natural and is technically involved, as it has to deal with well-foundedness by means of elementary methods.

If one avoids transfinite iterations along well-orderings, two routes have been opened (see Sect. 2).

The first route is to develop a direct *finite ordinal-free alternative* via the notion of *reflective closure* of a system, with its full-schematic variant. Let us recall the preliminary axioms for the simple non-schematic reflective closure. For the sake of a smoother formalization, we adopt some abbreviations.

[2] If a traditional philosophical terminology is here allowed.

- In general, we follow the conventional notations of Halbach (2011, pp. 32–33); hence we use the underscore-dot notation for naming expressions defining function symbols corresponding to logical operations and predicates of our formal language (e.g. $\underset{.}{T}$, $\underset{.}{\neg}$, $\underset{.}{\forall}$, $\underset{.}{\wedge}$, etc.);
- $F(s)$ is a shortening for $T(\underset{.}{\neg}\, x)$;
- s° stands for the *arithmetical value* of the term s (and this can be defined in the arithmetical language of Peano Arithmetic, see Halbach (2011, p. 32).

First of all, the language of KF expands the language of PA with a monadic predicate T for truth, and $F(s)$ is a shortening for $T(\underset{.}{\neg}\, x)$. The axioms of KF comprise PA, full induction schema, and

i. $\forall s\ \forall t \Big(\big(T(s \underset{.}{=} t) \leftrightarrow s^\circ = t^\circ \big) \wedge \big(F(s \underset{.}{=} t) \leftrightarrow s^\circ \neq t^\circ \big) \Big)$, and similarly for other predicates other than $=$, except for the special predicates T and F;

ii. $\forall s \Big(\big(T(\underset{.}{T} s) \leftrightarrow T(s^\circ) \big) \wedge \big(F(\underset{.}{T} s) \leftrightarrow F(s^\circ) \big) \Big)$;

iii. $\forall s \Big(\big(T(\underset{.}{F} s) \leftrightarrow F(s^\circ) \big) \wedge \big(F(\underset{.}{F} s) \leftrightarrow T(s^\circ) \big) \Big)$;

iv. $\forall x \Big(\mathrm{Sent}_{\mathsf{KF}}(x) \rightarrow \big(T(\underset{.}{\neg}\, x) \leftrightarrow F(x) \big) \wedge \big(F(\underset{.}{\neg}\, x) \leftrightarrow T(x) \big) \Big)$;

v. $\forall x\ \forall y \Big(\mathrm{Sent}_{\mathsf{KF}}(x \underset{.}{\wedge} y) \rightarrow \big(T(x \underset{.}{\wedge} y) \leftrightarrow T(x) \wedge T(y) \big)$
$\wedge \big(F(x \underset{.}{\wedge} y) \leftrightarrow F(x) \vee F(y) \big) \Big)$;

vi. $\forall v\ \forall x \Big(\mathrm{Sent}_{\mathsf{KF}}(\underset{.}{\forall} v x) \rightarrow \big(T(\underset{.}{\forall} v x) \leftrightarrow \forall t\ T(x[t/v]) \big)$
$\wedge \big(F(\underset{.}{\forall} v x) \leftrightarrow \exists t\ F(x[t/v]) \big) \Big)$.

Incidentally, it is worth mentioning that KF is intimately related to a logical *development of non-extensional concepts* (classification, operation) and *semantical investigations*; see Cantini (1996) for the connections with Aczel's Frege structures, explicit mathematics. Furthermore, KF has an interesting model theory (rich lattice theoretical results); it is also related to the standard fixed point theory $\widehat{\mathsf{ID}_1}$ (see Feferman, 1982); in turn, this leads towards the foundations of intuitionistic type theory (see Hancock's conjecture, again Feferman, 1982). KF can also be seen as a generalization of the standard theory of compositional truth, and, more naturally, as a generalization of a theory of a positive inductive definition of truth and falsity (see Halbach, 2011, §8.7).

Basic Results on KF

For the reader's sake, we add a reminder on significant fragments and their proof theory.

i. $\mathsf{KF}\lceil$ is KF with induction for T-free formulas;

ii. KF_c is KF with induction restricted to total predicates, i.e. such that $\forall x T(a(x) \vee T(\dot{\neg} a(x)))$;
iii. KF_p is KF with induction restricted to internal predicates.

KF and its variants can be compared with subsystems of second order arithmetic, see e.g. Simpson's monograph Simpson (1999). Indeed, let $(\Pi^0_1 - CA)_{<\lambda}$ stand for (an axiomatic version of) ramified analysis up to any level $< \lambda$ (λ being a constructive ordinal). Let CONS be the formal statement corresponding to the consistency of the truth predicate, i.e., to

$$\forall s(\neg(T(s) \wedge T(\dot{\neg} s))$$

Then:

Theorem 1

i. $\mathsf{PA} \equiv \mathsf{KF}\lceil + \mathsf{CONS} \equiv \mathsf{KF}_c$;
ii. $\mathsf{KF}\lceil$ has non-elementary speed-up over PA (see corollary 5.13 of Fischer [2014])[3];
iii. $(\Pi^0_1 - CA)_{<\omega^\omega} \equiv \mathsf{KF}_p + \mathsf{CONS}$;
iv. $(\Pi^0_1 - CA)_{<\varepsilon_0} \equiv \mathsf{KF} + \mathsf{CONS}$.

For the proof, see Cantini (1989).

The theorem is paradigmatic, in order to grasp at least some flavour of this research. There the function played by the ordinals is made transparent: ordinals are intended to classify truth-theoretic principles and to compare their strength. In fact, they measure how far we can proceed to iterate the arithmetical comprehension principle, which grants the existence of sets, e.g. up to ω^ω or higher up to ϵ_0.

Remark 1 As to the notion of *speed-up* and its precise definition, see Fischer (2014, Definition 2.3). This technical notion can be informally clarified by the intuition that, as already discovered by Gödel (1936) in the Thirties, adopting abstract notions and principles thereof—like truth or set—usually provide *sensible reduction in length* for (an infinite number of) already available proofs. More explicitly, there are arithmetical theorems, which are provable in $\mathsf{KF}\lceil$ with derivations, which are *much shorter* than those that can be found to exist in PA.[4]

Schematic Reflective Closure

Objection III of Sect. 1 above can be overcome with the idea that a schema does not apply to a *fixed language*, but *to any language which one comes to recognize*

[3] It is open if $\mathsf{KF}\lceil$ has non-elementary speed-up over its positive compositional fragments.

[4] I.e. if m is the length of a proof of A in $\mathsf{KF}\lceil$ then the length of a proof p of A in PA might require many iterations of the exponential $exp_2(x) = 2^x$, i.e. a tower $exp_2(exp_2(\ldots exp_2(m)))\ldots)$, in order to find an upper bound in terms of m to the length of a purely arithmetical proof of A in PA.

as embodying meaningful basic notions (see Strahm, 2017, p. 189). Informally, this suggests a rule which allows to make inferences from schemata accepted in the *original arithmetical language* to schemata of the *full* language. In particular, we can extend KF *by a suitable substitution rule* of the form

$$\frac{\varphi(P)}{\varphi(\hat{x}\psi(x))}$$

where φ is a formula of $\mathcal{L}_{\mathsf{PA}}$, the language of Peano arithmetic with an additional predicate symbol P, ψ is arbitrary (hence T, P can occur in ψ). $\varphi(\hat{x}\psi(x))$ stands as an abbreviation for the formula resulting from $\varphi(P)$ when each occurrence of a subformula $P(t)$ in $\varphi(P)$ is replaced by $\psi(t)$.[5]

The resulting system yields the *schematic reflective closure* $Ref^*(\mathsf{PA}(P))$, another way to characterize predicativity in the sense of Feferman and Schütte:

Theorem 2 (Feferman, 1991)

$$Ref^*(\mathsf{PA}(P)) \equiv (\Pi^0_1 - CA)_{<\Gamma_0}$$

Here Γ_0 is the well-known Feferman-Schütte upper bound of the predicative ordinals, and hence of those inductive methods that can be accepted predicatively and rely upon reflection and reflective closure. Is it possible to fix more structural routes? The answer is positive, as seen from the next section.

2 Feferman-Strahm's Unfolding

Objections can be raised against reflective closures of theories (see Strahm, 2017, p. 189): Feferman (2016, p. 282) states that *the resulting theories have still an air of artificiality*. More specifically:

1. it is questionable whether the truth axioms have the same evidence as the ground axioms without truth;
2. the formalization of syntax is an essential ingredient together with the coding machinery, and this somewhat conceals the ontological nature. Indeed, it tends to conceal the conceptual depth, yielding an air of taxonomy and mere linguistic reality.

It follows that, as stated in the opening abstract, a second operational route to the analysis of implicit commitment deserves due attention: one has to cope with (1)–(2), and the point is to attain *a syntax-free formalization* of the main notions. In other words, in order to understand the truth axioms, you assume suitable operations—mainly on numbers and finitary structures such as syntactical expressions, but

[5] $\hat{x}\psi(x)$ is the classical Russell-Whitehead notation for the class defined by $\psi(x)$; it is assumed that substitutions are correct, i.e. no confusion of variables can occur.

also quantifiers—which ought to be made explicit as *direct objects of our reflecting procedure.*

Unfolding Informally Presented

The idea is that, given a system S, *unfolding* S means to define a theoretical framework, where

- we have operations and predicates *understood as partial operations*;
- *logical* operations on propositions and predicates are given;
- operations may be applied to operations, so that self-application is not forbidden, i.e. a theory of untyped operations is assumed;
- partial recursive operations can be regarded as a possible model of our operations and hence non-extensionality holds.

Indeed, it turns out that the system of operations is at least as expressive as untyped lambda calculus and combinatory logic; and this step is a natural way to achieve independence *on encoding*. More explicitly, one assumes to live in an applicative structure, indeed a *partial combinatory algebra*, whose language consists of:

- application terms ts for expressing the application of t to s, $t\downarrow$ for expressing that t converges; application is *strict*, i.e. if ts is defined then both t, s are defined;
- a logic of partial terms and the relation $t \simeq s$ for expressing that $t = s$, whenever t, s are both defined;
- the so-called basic combinators K,S with standard axioms $Kts \simeq t$ and $Stsr \simeq tr(sr)$, whence λ abstraction and recursion operators become available; the system of operations is also equipped with pairing and projections, definition by cases.

For the reader's sake, we like to stress the informal reading, i.e. the natural effective interpretation of $xy \simeq z$ as: *the algorithm represented by x halts (i.e. converges) on the input y providing z as output*. For details, see Eberhard and Strahm (2015, pp. 158–160).

Unfolding Axiomatized

Let me now describe a few details for the unfolding of NFA —an acronym for *non-finitist arithmetic*—, which is regarded as the paradigmatic *ur*-system. The axioms of NFA itself are simply the usual ones for 0, sc (successor) and pd (predecessor), together with the *induction axiom*, given as $P(0) \wedge \forall x[P(x) \rightarrow P(sc(x))] \rightarrow \forall x(P(x))$, where P is a free predicate variable. The language of the unfolding of NFA adds a number of constants, the predicate symbol $N(x)$ (= "x is a natural number"), the predicate symbol $\Pi(x)$ (= "x is a predicate"), and the intensional membership relation $y \in x$ for x such that $\Pi(x)$.

Then the proper axioms of *the unfolding* U(NFA) *for non-finitist arithmetic* consist of the following groups[6]:

I. the axioms of NFA relativized to N, the collection of natural numbers;
II. the partial combinatory axioms, with pairing, projections and definition by cases;
III. an axiom for the characteristic function of equality on N;
IV. axioms for various constants in the domain Π of predicates, namely for the natural numbers, equality, the free predicate variable P, and for the logical operations $\neg$, $\wedge$, $\forall$ and inverse image of f along a[7];
V. an axiom (*join*)) for the *disjoint union* j(f) of a sequence f of predicates over numbers: whenever $f : N \to \Pi$, j(f) is the collection of all ordered pairs (u, v) where $u \in f(v)$, v being a natural number.

In analogy with the case of schematic reflective closure, *the full unfolding* U(NFA) is then obtained by applying the *substitution rule* $A(P)/A(B)$, where B is an *arbitrary formula* of the unfolding language.

The *operational unfolding* U_0(NFA) is simply obtained by restricting to axiom groups (I)–(III) and with the formulas B in the substitution rule restricted accordingly. In U_0(NFA) one successively constructs terms $t(x)$ intended to represent each primitive recursive function, by means of the recursion operator and definition by cases. By applying the substitution rule, it is then shown by induction on the formula $t(x)\downarrow$ that each such term defines a total operation on the natural numbers. Thus the language of PA may be interpreted in that of U_0(NFA) and hence—by application of the substitution rule once more—we have that PA itself is included in that system.

In U(NFA) the domain of predicates is considerably expanded by use of the *join operation*. Once we have established that a primitive recursive ordering $<$ satisfies the schematic transfinite induction principle $TI(<, P)$ with the free predicate variable P, we can apply the substitution rule, in order to carry out proofs by induction on $<$ with respect to *arbitrary formulas*. In particular, one can establish the existence of a predicate corresponding to the the so-called hyperarithmetical hierarchy[8] along such an ordering, relative to any given predicate p in Π, as naturally axiomatized by iterating arithmetical comprehension. Then by means of the usual arguments, if one has established in U(NFA) the schematic principle of transfinite induction along a standard ordering for an ordinal α, one can lift it upwards to a well-ordering of type $\varphi\alpha 0$,[9] and hence the same for each ordinal less than Γ_0, the upper bound for predicative reasoning. Thus U(NFA) can deal with the ramified analytic systems up to Γ_0 and it essentially matches up with the theory of Theorem 2.

6 For formal definitions and details, see Strahm (2017).

7 The inverse image of an operation f along a predicate a is—extensionally—the collection of all x such that $fx \in a$.

8 In essence a version of the ramified hierarchy in sense of Russell which is iterated up to the first non-recursive ordinal.

9 $\varphi\alpha\beta$ is the so-called Veblen hierarchy; for details, see Feferman (1991) and Rathjen (2006).

The main results of Feferman and Strahm (2000) are that (i) $\mathsf{U}_0(\mathsf{NFA})$ is proof-theoretically equivalent to PA and is conservative over it; (ii) $\mathsf{U}(\mathsf{NFA})$ is proof-theoretically equivalent to the union of the ramified analytic systems up to Γ_0 and is conservative over it.

In other words, $\mathsf{U}(\mathsf{NFA})$ is proof-theoretically equivalent to *predicative analysis* as characterized in Theorem 2. In addition, the intermediate system $U_1(NFA)$ without the join axiom (V) is proof-theoretically equivalent to the union of the ramified systems of *finite* level. While the unfolding of non-finitist arithmetic attains the limits of predicativism, it is proper to the essence of predicativism that *ascent towards more complex sets is only allowed through restricted quantification, i.e. quantifications over totalities that can be regarded as already given*, and hence only *predicative ordinals* are accepted, i.e. ordinals which can be recognized by exclusive appeal to notions that have already been secured and are generated from below, bottom-up.

Indeed, these ideas have a more general import: constructive theories—Martin-Löf type theory, and Myhill and Aczel set theory—*feature predicativity as a distinctive form of constructivity*. This may be condensed into a *constructibility* requirement for sets, which ought to be finitely specifiable in terms of uncontroversial primitive objects and simple operations over them. Historically, predicativity emerged at the beginning of the twentieth century as a component of an influential analysis of the paradoxes by Poincaré and Russell. According to this analysis, the paradoxes are caused by vicious circles in definitions; and adherence to predicativity was therefore proposed as a systematic method for preventing such problematic circularity.

Now a *subtler role* of ordinals occurs in the case of *impredicative* definitions, which can be analyzed up to a certain point via ordinal-theoretic methods, but with a *distinctive* feature: the unfolding of certain definitions and processes dealing with finite objects, e.g. proofs, requires us to climb up to and to compute with certain ordinals, which *surpass the first uncountable ordinal*. But then we have to collapse down onto a *countable* ordinal, in order to recover information on usual proofs and definitions! Typically—as in the case of the so called ordinal $\psi(\Gamma_{\Omega+1})$ defined in (Buchholtz, 2013; Buchholtz et al., 2016), which parallels the ordinal Γ_0 of predicative analysis—one defines a *countable* ordinal, which necessarily requires—to be defined—the *uncountable* ordinal Ω and $\psi(\Gamma_{\Omega+1}) < \Omega$!

Incidentally, let me conclude by mentioning that these investigations contribute to the research program of *metapredicativit*y (Jäger and the Bern school; see Ranzi & Strahm, 2019). The idea is that, while predicative methods build up objects from below, impredicative methods assume the existence of large objects in order to construct smaller ones. In the case of metapredicativity, one generalizes methods from predicative proof theory to investigate theories that are impredicative in the sense of Feferman and Schütte.

Unfolding Classified

It is interesting to look for generalizations, and hence to define the unfolding $\mathsf{U}(\mathsf{S})$ for given S. Indeed, for special choices, *significant theories* arise. Just to give concrete instances of this sort of phenomena, let us choose for S two standard theories: an axiomatization of *Polynomial Time Arithmetic* PTCA, as given, say, by Ferreira (1990) or Eberhard and Strahm (2015), and PRA, a standard version of *Primitive Recursive Arithmetic.* Now it is also known that the two systems can be rephrased as *applicative theories, as based on the notion of self-applicable operation,* and yet their computable content (the algorithms recognized as well-defined in the two systems) coincide—respectively—with the polytime operations and the primitive recursive operations. In other words, we are led to consider:

- FEA, a version of feasible arithmetic as defined by Eberhard and Strahm (2015, pp. 156–157). Roughly, FEA is an axiomatization of partial combinatory logic together with an underlying structure: the theory of binary strings (finite sequences of bits), as endowed with the natural operations of concatenation and string product and with a relation $\tau \leq \sigma$, meaning that the length of the string τ is less than the string σ.
- FA, the basic system of *finitist arithmetic*, which includes a basic theory of operations as based on partial application with the basic constructions for producing constant operations and substitutions, and includes standard partial combinatory logic, succcessor, predecessor, pairing operations and projections. As is well known, partial combinatory logic allows *self-application*; in order to understand it, it is enough to follow the informal reading $xy \simeq z$ as: *the algorithm coded by x applied to the input y converges and produces z as output*. Besides application, the intended universe includes natural numbers; hence three basic statements are possible: $N(t)$, $t = s$, $ts \simeq r$. The essential point is that the logical operations are now *restricted* to $\wedge$, $\vee$, $\exists$ (existential quantifiers). Provable propositions $A(x)$ are interpreted as verifying $A(n)$ for each natural number n, but *we do not have universal quantification* over the natural numbers as a logical operation. Nor do we have negation (except of numerical equations), which, when applied to existential formulas, could be interpreted as having the effect of universal quantification.
- U(FEA) = PTCA (Eberhard & Strahm, 2015); this is related to *feasibility*, and only *bounded quantification on binary strings* is allowed, while one has a weak notion of truth closed under bounded quantification, $\wedge$, $\vee$, positive prime conditions;
- U(FA) = PRA; this corresponds to *finitism* in the sense of Tait (Feferman & Strahm, 2000);
- Let BR be the so-called *bar rule*, i.e. the inference which roughly states: if $<$ is a well-founded ordering relation—in symbols WF($<$)—, then each instance of the transfinite induction schema TI($<$, B) on $<$ is allowed (B arbitrary formula of the full language). Then U(FA + BR) = PA; this is inspired by Kreisel's idea that PA is the *least upper limit* of finitism, as characterized by means of Gentzen's consistency proof via transfinite induction on a suitable natural well-orderings;
- U(ID_1), the unfolding of the elementary theory of inductive definitions, corresponds to overcoming the limits of standard predicativity towards metapredicativity and beyond; this yields is a theory of strength $\psi(\Gamma_{\Omega+1})$, but other equivalent systems can be found in Buchholtz et al. (2016).

Feferman and Strahm (2000, 2010), and Buchholtz et al. (2016) develop a suitable proof-theoretic analysis of the systems.

Let us mention an alternative strategy to the proof-theoretic results given in Feferman and Strahm (2000, 2010): they can be easily obtained by interpreting the unfolding systems into theories of *abstract truth over combinatory structures*, as outlined in Cantini (2016, 1996), thus establishing another bridge with *applicative systems*.

More explicitly, let PT be a theory of propositions and truth, which, roughly, consists of (i) the axioms for combinatory logic enriched by numbers; (ii) natural compositional axioms for truth T and propositions P; (ii) the schema of number theoretic induction (for details, see Cantini, 2016).

Now in PT the collections of propositional functions can simulate a rather rich *structure of types*; in particular it is closed under elementary comprehension and join, and it turns out:

Theorem 3
U(NFA) without substitution rule is interpretable into the system PT.

Remark 2 As to the import of U(NFA), observe that PT without substitution is known to have the same upper bound as ramified analysis of any level below ϵ_0 (ϵ_0 being the well-known proof-theoretic ordinal discovered by Gentzen in the thirties and associated to the consistency of PA). Of course, this means that U(NFA) is a highly non-trivial system and is worth considering.

Remark 3 The theories of propositions and truth we just cited are in a sense intermediate between reflective closures and unfolding, as they fully integrate the operational side, while making sense of an abstract notion of truth.

3 On the Implicit Commitment Thesis ICT

The implicit commitment thesis ICT (Dean, 2014b, p. 32) states that *in accepting a formal systems* S *one is also committed to additional resources not available in the starting theory* S *but whose acceptance is implicit in the acceptance of* S. Of course, this is only an informal statement and one might want to look for a more definite form of it. For instance, if we accept the axioms of a theory S together with logical inferences and principles, we can accept as basic the notion of proof in S. Eventually, we are led to accept formal proofs in S as sound (whence the formal consistency of S), which formally corresponds to the acceptance of a *reflection principle* RFN(S) for S:

- if S proves a statement φ, then φ is true.

As to RFN(S), either we regard it as a schema in the given formula φ, or else we explicitly add a truth predicate (see the next subsection).

Recently, this step has been questioned by Dean (2014b), essentially on the ground of the following argument: certain systems—incompleteness notwithstanding—are plausibly *complete* with respect to a given body of knowledge, typically PRA for *finitism* in the sense of Tait (1981), or PA with respect to *genuine first-order number-theoretic properties* (this is known as Isaacson's thesis Isaacson [1987]). Here by *genuine 1st order* it is meant that there are *no hidden higher order notions* around: or, put differently, the content is essentially arithmetical, no reference to the *notion of a set* of natural numbers.

But, e.g., if we accept Isaacson's thesis IT, we *cannot be committed* to a reflection principle RFN(PA) as implicit, since it goes far beyond the boundaries of PA, and hence the principle makes PA *unstable*, in the sense that it is naturally amendable, and also incomplete, against the thesis. Hence this fact conflicts with ICT; and it seems that we have to give up ICT or to modify ICT in some sense.

Furthermore ICT apparently yields *ontological* consequences. For instance, *consistency*, which follows from ICT, implies the existence of structures and models by the so-called arithmetized completeness theorem (see Dean, 2014a), and this fact apparently conflicts with the idea that truth is accepted as a *thin* notion, according to deflationism (see Cieslinski, 2017). Hence assuming ICT conflicts with deflationism, i.e. the thesis that truth is thin and unsubstantial.

ICT *Sharpened: Semantical Way-Out*

Let us try a fresh start. Taking ICT seriously leads to extend the ground system S with a notion of *truth*. Just to be concrete, choose S to be PA. Accepting PA, we are then committed to a theory of arithmetical truth CT(PA), which embodies the *standard compositional clauses*. Essentially, CT(PA) embodies the axioms of KF *restricted to the codes of* PA *-formulas*. Therefore, if we also allow full induction—namely also applied to formulas where T occurs—, we can prove the reflection schema (and hence the consistency) of PA in the form:

$$\forall x(Sent_S(x) \wedge Axiom_S(x) \rightarrow T(x))$$

But, if we accept the criticism above, CT(PA) is *inadequate*. We simply ought to consider weaker systems with restricted induction, e.g. CT(PA)$\lceil$, i.e. the extension of PA in the truth-theoretic language, as given above, which contains

1. the *compositional* axioms for truth, as applied to (codes of) PA-sentences;
2. the axiom stating that $\forall x(For_{PA}(x) \rightarrow T\,Ind(x)$ where $Ind(x)$ formalizes that x is the code of an arithmetical instance of the induction schema (for details see Halbach [2011]).

At this stage, it turns out that we can apply non-trivial results in Kotlarski et al. (1981), Leigh (2015), and Enayat and Visser (2015) about theories endowed with a truth predicate, implying that

Theorem 4
$\mathsf{CT}(\mathsf{PA})\lceil$ has the same arithmetical content as PA.

Recently, the theorem has triggered a new proposal, put forward by Nicolai and Piazza (2019). In essence, it amounts to stating that the implicit commitment IC has a composite nature, a *variable* component and an *invariable* one, the so-called *semantic core*, i.e. a set of of semantical principles about truth we are naturally committed to accept, and yet these principles are conservative over the ground system. More explicitly, the semantic core should include, besides the compositional truth axioms for PA-sentences, the axioms stating the truth of all propositional tautologies, the fact that the inference rules (modus ponens is enough) preserve truth, the truth of all its non-logical axioms.

However, it is not clear yet whether the theorem above can be strengthened to the effect that the conservation result keeps holding for the extended system (compare Nicolai & Piazza, 2019, p. 929, footnote).

Lastly, let us mention that another route is viable via KF, i.e. we can propose a notion of *reflective* semantic core. This means that we can regard truth as self-referential truth, and hence subscribe to the axioms in KF, including the axiom *LogT* formalizing the statement: all logical PA-axioms in the language of PA are true. This can easily be stated in a standard formalization of the syntax of PA:

- $\forall x(Ax_{PA}(x) \rightarrow T(x))$

Of course, we may choose $\mathsf{KF_c}$ as a variant of semantic core, i.e. a sort of reflective semantic core, i.e. $\mathsf{KF_c}$. However, adopting $\mathsf{KF_c}$ is not really a *panacea*: there remain problems similar to those of $\mathsf{CT}(\mathsf{PA})\lceil$, which are left open as well, e.g. it remains to be seen if $\mathsf{KF_c}$ with *LogT* is still conservative over PA.

And a crucial point ought to be clarified though: once truth as a resource is introduced, the reflection process should be applied to induction instances applied for *total* conditions. Why should it be so?

Ontological Way Out

Assume we consider the unfolding approach: can we avoid the previous difficulties? Well, we can rephrase ICT as an attempt at extracting genuine mathematical notions. Truth is no more accepted as a starting point, against a metatheoretical approach, but *operations* form the basic constituents.

Of course, operations are understood intensionally, as rules, and not defined via set-theoretical methods. Operations should be regarded as acting on the intended universe of individuals of the underlying theory, as well as on a domain of predicates and operations themselves. Both domains are included in a comprehensive domain V and the closure conditions ensure, as already stated, that we have a ground applicative structure, typically a partial combinatory algebra.

It also turns out that, as Strahm puts it, the basic aim for implicit commitment becomes wider in scope: given a schematic system S, *which operations and predicates, and which principles* concerning them ought to be accepted if one has accepted S? Eventually, elaborating the unfolding program directly leads to a confluence within the framework of the so-called applicative frameworks, that have been developed at length in the context of the so-called program of explicit mathematics (see Section 2 in Feferman [2016]).

4 Conclusion

Let us recall ICT, as stated by Dean (2014b, p. 32), in the form of a thesis:

> *Anyone who accepts the axioms of a mathematical theory S is thereby also committed to accepting various additional statements Δ which are expressible in the language of S but which are formally independent of its axioms.*

We wonder whether there might be a sensible restatement of Dean's ICT. We observed that *unfolding* induces a switch away from reflection principles: the problem is not quite to guess suitable reflection principles—as schemata or axioms—but to look for *new operations* or *predicates*, and corresponding *closure properties*, possibly in search of new tangible, significant features of the objects involved.

Hence we depart from the issue of implicit commitment and from a problem involving *the limits of the axiomatic method*, and we are moving to a problem of *mathematical content*. What is the effect of reflection principles on the kind of entities a given theory is calling into existence? In order to handle increasing logical complexity, we are pushed towards possibly higher-type operations, more than stronger and stronger kind of inferences rules: think of the use of the notion of universe in type theory (Martin-Löf, Feferman) and the emerging of abstract objects which concern proofs but are not merely logico-linguistic, as already envisaged in several contexts (Hilbert himself, Schütte and the Munich school, Girard).

As to the traditional ideas underlying ICT, and in order to make clear the open character of this note, we like to conclude with a few open questions: is there any objectual counterpart of the conceptual frame supporting a reflection principle somewhat implicit in accepting the axioms of PA? Is there a structural content of the idea of semantic core? What is the use of the semantic core? Does it help in making proofs more perspicuous or elegant or shorter?

References

Buchholtz, U. (2013). *Unfolding of systems of inductive definitions* [Ph.d. thesis, Stanford University].

Buchholtz, U., Jäger, G., & Strahm, T. (2016). Theories of proof-theoretic strength $\psi(\gamma_{\epsilon_{\Omega+1}})$. In *Concepts of proof in mathematics, philosophy, and computer science* (pp. 115–140). De Gruyter.

Cantini, A. (1989). Notes on formal theories of truth. *Mathematical Logic Quarterly, 35*, 97–130.
Cantini, A. (1996). *Logical frameworks for truth and abstraction*. North Holland.
Cantini, A. (2016). About truth and types. In R. Kahle, T. Strahm, & T. Studer (Eds.), *Advances in proof theory* (pp. 31–64). Birkhäuser-Springer.
Cieslinski, C. (2017). *The epistemic lightness of truth. Deflationism and its logic*. Cambridge University Press.
Dean, W. (2014a). Incompleteness via paradox and completeness. *The Review of Symbolic Logic, 13*, 541–592.
Dean, W. (2014b). Reflection and the provability of soundness. *Philosophia Mathematica, 23*, 31–64.
Eberhard, S., & Strahm, T. (2015). Unfolding feasible arithmetic and weak truth. In T. Achourioti, H. Galinon, J. Martinez, & F. F. Fujimoto (Eds.), *Unifying the philosophy of truth* (pp. 153–167). Springer.
Enayat, A., & Visser, A. (2015). New constructions of satisfaction classes. In T. Achourioti, H. Galinon, J. M. Martinez, & K. Fujimoto (Eds.), *Unifying the philosophy of truth* (pp. 321–335). Springer.
Feferman, S. (1968). Autonomous transfinite progressions and the extent of predicative mathematics. In *Logic, Methodology and Philosophy of Science III* (pp. 121–135). North-Holland.
Feferman, S. (1982). Iterated inductive fixed-point theories: Application to Hancock's conjecture. In *Patras logic symposion* (pp. 171–196). North-Holland.
Feferman, S. (1991). Reflecting on incompleteness. *Journal of Symbolic Logic, 56*, 1–49.
Feferman, S. (2016). The operational perspective: Three routes. In R. Kahle, T. Strahm, & T. Studer (Eds.), *Advances in proof theory* (pp. 269–289). Birkhäuser-Springer.
Feferman, S., & Strahm, T. (2000). Unfolding finitist arithmetic. *Annals of Pure and Applied Logic, 104*, 75–96.
Feferman, S., & Strahm, T. (2010). The unfolding of finitist arithmetic. *The Review of Symbolic Logic, 3*, 665–689.
Ferreira, F. (1990). Polynomial time computable arithmetic. In *Logic and computation* (pp. 137–156). American Mathematical Society.
Fischer, M. (2014). Truth and speed-up. *The Review of Symbolic Logic, 7*, 319–340.
Freund, A., & Rathjen, M. (2021). Derivations of normal functions in reverse mathematics. *Annals of Pure and Applied Logic, 172*, 1–49.
Gödel, K. (1936). Über die Länge von Beweisen. In K. Menger (Ed.), *Ergebnisse eines Mathematischen Kolloquiums* (Vol. 7, pp. 23–24).
Gödel, K. (1990). Remarks before the Princeton bicentennial conference on problems in mathematics. In S. E. A. Feferman (Ed.), *Collected works, vol. II: Publications 1938-1974* (pp. 150–154). Oxford University Press.
Halbach, V. (2011). *Axiomatic theories of truth* (1st ed.). Cambridge University Press.
Isaacson, D. (1987). Arithmetical truth and hidden higher order concepts. In J. Barwise et al. (Ed.), *Logic colloquium '85* (pp. 147–169). North-Holland.
Kotlarski, H., Krajewski, S., & Lachlan, A. (1981). Construction of satisfaction classes for nonstandard models. *Canadian Mathematical Bulletin, 24*, 283–293.
Kreisel, G. (1958). Ordinal logics and the characterization of informal notions of proof. In *Proceedings of the International Congress of mathematicians, 14-21 August 1958* (pp. 289–299). Cambridge University Press.
Kreisel, G. (1970). Principles of proof and ordinals implicit in given concepts. In A. Kino, J. Myhill, & R. E. Vesley (Eds.), *Intuitionism and proof theory* (pp. 489–516). North Holland.
Leigh, G. E. (2015). Conservativity for theories of compositional truth via cut elimination. *The Journal of Symbolic Logic, 80*, 845–865.
Nicolai, C., & Piazza, M. (2019). The implicit commitment of arithmetical theories and its semantic core. *Erkenntnis, 84*, 913–937.
Ranzi, F., & Strahm, T. (2019). A flexible type system for the small veblen ordinal. *Archive for Mathematical Logic, 58*, 711–751.

Rathjen, M. (2006). The art of ordinal analysis. In *International Congress of mathematicians* (Volume II, pp. 45–69). European Mathematical Society.
Simpson, S. (1999). *Subsystem of second order arithmetic*. Allen and Unwin.
Strahm, T. (2017). Unfolding schematic systems. In G. Jäger & W. Sieg (Eds.), *Feferman on foundations* (pp. 187–208). Springer.
Tait, W. (1981). Finitism. *The Journal of Philosophy, 78*, 524–546.

Metaphysical Modality, without Possible Worlds

Giorgio Lando

1 Introduction

In contemporary philosophy, there are two divergent understandings of metaphysical modality. Their divergence is a source of scepticism about metaphysical modality. It is indeed often unclear which of the two concepts is at stake in a specific discussion, allowing the sceptic the opportunity to attack metaphysical modality from two fronts. In this paper, I analyse these two views of metaphysical modality (the *absolutist* and the *essentialist*) and assess their main motivations and problems. In the light of my analysis of the two views, I also aim to show that *possible worlds* are not helpful in investigating metaphysical modality. At the end of the analysis I assess whether an absolutist or an essentialist is more entitled to use the expression "metaphysical modality", given a general principle governing the choice of philosophical lexicon.

This paper is structured as follows. In §2, I analyse the absolutist view, according to which metaphysical modality is the extreme variety of objective modality. Metaphysical necessity is then equated with *absolute necessity* and a proposition is deemed metaphysically necessary if and only if it is necessary for every variety of objective necessity. I observe that the absolutist conception does not confer any unity to metaphysical modality and is blind to its sources, and this last claim is reinforced in §3 by the analysis of McFetridge's thesis about logical and absolute modality. In §4, I scrutinise the essentialist view of metaphysical modality. According to it, metaphysical necessities are those that are explained or grounded by essences. Thus, the characterising mark of metaphysical necessity would not be its absoluteness, but its source or ground. In §5, with reference to some main theories of possible worlds, I argue that resorting to possible worlds is not helpful in analysing metaphysical necessity. In §6 I draw some conclusions about the preferable usage of phrases such as "it is

G. Lando (✉)
Dipartimento di Scienze Umane, Università degli Studi dell'Aquila, Viale Nizza 14, 67100 L'Aquila, Italy
e-mail: giorgio.lando@univaq.it

F. Ademollo et al. (eds.), *Thinking and Calculating*, Logic, Epistemology, and the Unity of Science 54, https://doi.org/10.1007/978-3-030-97303-2_20

metaphysically necessary that" or "it is metaphysically possible that" in philosophy, and I suggest that essentialism has an edge over absolutism on this terrain.

Some methodological premisses (and a personal note) are important and the subject matter of this introduction. First, it is not my purpose to clarify the notion of metaphysical modality by pinpointing uncontroversial examples of metaphysical necessities or metaphysical possibilities. In the course of the analysis, I obviously cite some examples of *prima facie* good candidates for these roles, but it is wrong to expect a clarification of the concept of metaphysical modality to deliver uncontroversial examples. There is no reason to expect any example of metaphysical modal truth to come for free once the concept is clarified, and even less to be *analytic* with respect to the concept of metaphysical modality.

I do not think that metaphysical modality is unique under this viewpoint. Examples do not come for free and are not analytic (and are not usually expected to do or be so) with respect to many other philosophical concepts. The analysis of knowledge in epistemology does not deliver, by itself, any uncontroversial instance of knowledge. The logical or semantic analysis of truth does not deliver, by itself, any uncontroversial truth (with the obvious exception of those truths that are part of the analysis itself, if it is a good analysis). The metaphysical analysis of properties does not deliver, by itself, any uncontroversial example of property. Nonetheless, in any philosophical discussion of knowledge, truth and properties, some examples are set forth. In many cases, they are mere heuristic tools for presenting a certain theory of knowledge, truth or properties. If we discovered (in contrast with a common example in philosophical discussions about truth) that snow is not really white, no aspect of any theory of truth by Alfred Tarski or Saul Kripke would thereby be refuted. The examples are not part of the theories, and even less are they analytic offsprings of the theories.

The case of metaphysical necessity is not different and should not be treated differently. The examples *are not* part of the theories of metaphysical necessity that I investigate. Perplexities and objections about the examples do not immediately or easily translate into perplexities and objections about the theories. I emphasise this point because *any* example of metaphysical modality will likely raise perplexities and objections. To anticipate some examples I will resort to throughout this paper, many readers will likely disagree with the claim that every human being is such that it is metaphysically necessary that he or she is human; or with the claim that it is metaphysically necessary that if a first entity is part of a second entity and this second one is part of a third, then the first is part of the third. Some examples of metaphysical non-necessity are arguably less controversial (for example, nobody doubts that it is *not* metaphysically necessary to abide by the speed limit on highways).[1] In any case, the vast disagreements about the positive examples of metaphysical necessity do not by themselves force or entitle the readers to reject the characterisations of metaphysical modality that shall be illustrated by these examples.

[1] Analogously, in epistemology it is not controversial that I do *not* know that 3 is identical to 4 and in the metaphysics of properties it is not controversial that the laptop I am using is *not* a property: also on this terrain the case of metaphysical modality is not different.

The disagreements about the examples are *mostly* beside the point. *Mostly* (and not always), because an ontological concern remains relevant, namely the concern that *nothing* at all performs the roles attributed to metaphysical modality by different conceptions of it discussed in the paper. However, this is not the topic of this paper. I do not address or refute the corresponding kind of global scepticism about the extension of metaphysical modality, according to which it may be a legitimate, adequately characterisable concept, but *nothing falls* under it.

Another premiss is that some features of metaphysical modality are not controversial, and I focus only on the controversial features. The uncontroversial features of metaphysical modality are uncontroversial in the literal sense that—as far as I know—nobody in the literature contests them; there is widespread consensus that, if a modality lack these features, then there is no good reason to label it as "metaphysical", no matter which among the various ways of interpreting this qualification of metaphysicality is chosen. These uncontroversial features are not sufficient to identify metaphysical modality, but put some constraints on its identification. It is for example uncontroversial that metaphysical modality is not deontic. In Daniel Nolan's example (2011, pp. 315–316), I am permitted by law and morality to make my table in the form of a square circle, but this is no indication that this action is metaphysically possible. It is also uncontroversial that metaphysical modality is not epistemic; it is not a matter of what can or must be the case given what a certain subject knows or what is knowable in general.

Metaphysical modality is uncontroversially *alethic*. This means that both of the following principles hold true ($\Box_{met}$ and $\Diamond_{met}$ express metaphysical necessity and possibility):

- $\Box_{met}\, p \rightarrow p$
- $p \rightarrow \Diamond_{met}\, p$

It is also uncontroversial that metaphysical necessities, possibilities or impossibilities do not depend on markedly local and specific hypotheses. If I block my bedroom door with a heavy wardrobe, burglars *cannot* pass through that doorway; but this impossibility is not metaphysical because it depends on various, rather specific circumstances (such as my having moved the wardrobe there and the burglar's lack of superior physical strength). Metaphysics is expected to be a distinctively general discipline, and marked locality or specificity is thus incompatible with metaphysicality.

Soon after the attention is restricted to alethic, not exceedingly local or specific modalities, the controversies begin, inasmuch as there are other widely discussed varieties of modality, such as nomic and logical modality, which are alethic, and neither highly local nor specific. Many philosophers expect metaphysical modality to collapse with one of these other alethic modalities, while others think that it is distinct from these other alethic modalities.

On a personal note, it is not by chance that I have chosen this specific topic for an essay in honour of Massimo Mugnai. Indeed, Massimo, Sergio Bernini, I and several other friends have discussed about possible worlds, metaphysical modality and many other related topics during hardly countable, challenging, chaotic and often funny seminars (locally known as *seminari del martedì*) organised by Massimo at the Scuola Normale Superiore in Pisa from 2009 to 2016.[2] For me, these seminars have been invaluable sources of intellectual stimuli, and are still active in my mind as repertories of philosophical ideas and fond memories. During the seminars, Massimo and Sergio were always sceptical of *both* metaphysical necessity *and* possible worlds; in general, their role in the seminars was to be sceptical of any theory under discussion, and in particular, when these theories belonged to metaphysics. In contrast, my usual role was to act as the *defence attorney* for contemporary metaphysics. In this paper, as regards metaphysical modality, I partially play my role, by concluding (in §6) that there are some good reasons to discuss metaphysical modality in philosophy despite many difficulties and ambivalences affecting the concept. Regarding possible worlds, I somewhat begrudgingly concede to Massimo (in §5) that they are not really useful for the specific purpose of understanding what metaphysical modality is.

2 Metaphysical Modality as Absolute Modality

According to the absolutist conception, metaphysical modality is absolute modality. This approach has been recently articulated and defended by Timothy Williamson (see in particular Williamson, 2016). The underlying idea is that when modal problems emerge in everyday life or in specific disciplinary contexts, we tend to countenance only some possibilities, due to the specificity of our interests and of the context; correspondingly, we tend to countenance, besides *absolute* necessities, also more specific necessities, which again depend on our interests or on the context. Some of these interests and contexts are quite important; in some cases, we want to ask what can and must happen under the presupposition that reality is governed by the scientific laws that actually govern it. In such a context, we are uninterested in dwelling on the hypothesis that scientific laws are violated, or different. In these cases, we resort to nomic modality.

In philosophy we are often more radical and want to prescind from specificities and contexts. This occurs in particular in metaphysics; how often it happens in metaphysics depends on how metaphysics is construed. For example, according to the influential scientistic conception of metaphysics emphatically advocated by Ladyman et al. (2007) (see also the essays in Ross et al., 2015), metaphysics has the exclusive purpose of unifying the scientific image of the world and therefore should not prescind at all from scientific laws. In contrast, when metaphysics is perceived as the investigation of an unconstrained field of possibilities (see Lowe, 2001 for a contemporary manifesto of this traditional view of metaphysics), we tend to be maximally liberal in admitting far-fetched possibilities and maximally careful in selecting necessities that are really absolute.

---

[2] Among the participants to the *seminari del martedì*, besides Massimo and Sergio, I am, in particular, grateful to Andrea Borghini, Giulia Felappi, Gabriele Galluzzo, Lorenzo Azzano, Francesca Poggiolesi, Stefano Di Bella and Andrea Strollo.

According to Williamson, metaphysical modality *qua* absolute modality is simply a special case of a broad range of objective (*i.e.*, alethic, non-deontic, non-epistemic) modalities. It is the extreme, *catch-all* case, in the following sense: a proposition is metaphysically possible "if and only if it has at least one sort of objective possibility" Williamson (2016, p. 455), while a proposition is metaphysically necessary if and only if it has every sort of objective necessity.

This absolutist conception of metaphysical modality makes scepticism about metaphysical modality collapse into the overall scepticism about objective modality. There is no good reason to be sceptical of *only* metaphysical modality. Given the other notions of objective modality, metaphysical modality is simply their extreme case, definable through a disjunction or a particular quantification in the case of possibility and through a conjunction or a universal quantification in the case of necessity. Thus, we may argue for scepticism about modality on an epistemological basis by expressing concerns about whether and how we manage to know if something is possible if it is not actual, but these concerns and the ensuing sceptical challenge are in no way specific to the extreme, catch-all variety of objective modality (*i.e.*, metaphysical modality).

Moreover, suppose that we contend that nomic modality (the kind of objective modality that does not allow any possibility in contrast to scientific laws, and for which scientific laws are necessary themselves) is adequately characterised and that there are no more comprehensive set of possibilities and narrower set of necessities. Given the absolutist conception of metaphysical modality, our contention *is not* a form of scepticism towards metaphysical modality. Instead, our contention can be expressed by saying that metaphysical modality *is* nomic modality, and therefore as adequately characterised as nomic modality is. Any further resistance specifically towards metaphysical modality risks being a mere matter of labels, and in particular the upshot of a distaste for the "metaphysical" label; but then, we should not have accepted to use "metaphysical modality" as equivalent to "absolute objective modality", as the absolutist conception suggests. Once we have made this terminological choice and claim that nomic modality is absolute objective modality, we are unentitled to be sceptical specifically of metaphysical modality.

The same occurs if we think that *logical modality* is absolute/metaphysical modality. Let us suppose that logical modality is adequately characterised, e.g. by claiming that a proposition p is logically possible if and only if p does not entail any contradiction; and that p is logically necessary if and only if $\neg p$ entails a contradiction.[3] Given this supposition, we are not entitled to be sceptical about metaphysical/absolute modality, which ends up being as adequately characterised and as legitimate as logical modality is: the above characterisation, if it is suitable for logical

[3] This characterization ends up being circular if *entailment* and/or *contradiction* are in their turn characterised in terms of logical modality. This circularity is avoided if entailment and contradiction are differently characterized (for example if entailment is characterised with respect to a specific logical system and contradictions are characterised syntactically as sentences of the form $p \wedge \neg p$). The focus of this paper is on metaphysical modality and for this reason I lay this problem aside in what follows.

modality, is suitable for metaphysical/absolute modality too, if logical modality *is* absolute modality.

Thus, the absolutist conception has an edge in the dialectics with the sceptical about metaphysical modality, inasmuch as this scepticism collapses into scepticism about objective modality in general. On the other hand, absolutism does not confer any kind of unity on metaphysical modality. It does not pinpoint any unitary, substantial feature of metaphysical modality and is utterly silent regarding the sources or grounds of metaphysical modality. It passes the buck to all the varieties of objective modality, of which some (*e.g.*, nomic modality) are in turn plagued by their own philosophical problems. In the absence of a substantial common element, the doubt might arise that metaphysical modality is not really unitary and is therefore a disjunctive and gerrymandered concept, perhaps on a par with *being a camel or a rhododendron.*

We might hope to attain a more unitary and explicative characterisation of metaphysical modality *by reflecting on what absoluteness is.* Generally, absoluteness is the lack of relativity. Bob Hale (most explicitly in Hale, 2012) has proposed to construe metaphysical necessity *qua* absolute necessity as the lack of relativity with respect to counterfactual hypotheses, *i.e.* as *counterfactual inevitability.* What would happen in any case, no matter what counterfactual hypothesis we consider, is metaphysically necessary. We could even define metaphysical necessity through a quantification in sentential position on counterfactual hypotheses, as follows ($>$ is the connective for counterfactual conditionals; let us assume that the values of the quantified variable q are propositions):

$$\Box_{met}\, p \equiv_{def} \forall q(q > p) \qquad \text{(Metaphysical Necessity)}$$

As in all the other characterisations of metaphysical modality, once metaphysical necessity is defined, metaphysical possibility can be easily defined in terms of it, as follows:

$$\Diamond_{met}\, p \equiv_{def} \neg\Box_{met}\neg p \qquad \text{(Metaphysical Possibility)}$$

Metaphysical modality *is not* reduced to something else by these definitions. First, counterfactuality is by itself a modal notion—what variety of modality? Since the *definiendum* is metaphysical modality, there are good reasons to expect the modality in the *definiens* to be metaphysical as well. Second, a presupposition of the adequacy of this definition is that, for metaphysically impossible propositions that are values of q, the resulting counterfactual conditional is a counterpossible conditional and is, as a consequence, trivially true.[4] This means that the only values of q that really matter are the *possible* propositions, but again, the possibilities at stake are plausibly

[4] In recent years, the claim that every counterpossible conditional is trivially true has been contested for a variety of reasons. See Berto et al. (2018) for a version of antitrivialism about counterpossibles and Williamson (2018) for a compelling defense of trivialism. Since the debate is not directly relevant for the characterisation of metaphysical modality, I simply assume Williamson's trivialism about counterpossibles in this paper.

metaphysical/absolute possibilities, so that the limits of the counterfactual scenarios in terms of which counterfactual inevitability is characterised end up presupposing a characterisation of metaphysical/absolute possibility.

This kind of irreducibility of metaphysical modality is perhaps unsurprising, given that it is generally very difficult to reduce modal notions to non-modal ones. However, irreducibility here provides evidence that this definition in terms of counterfactual inevitability will unlikely deliver the kind of unity of metaphysical modality that we have been looking for. The domain of the values of q is allowed to be highly heterogeneous, and the resulting domain of metaphysical necessities and possibilities is allowed to be no less heterogeneous.

To recap, metaphysical modality as *absolute* objective modality is easily characterisable in terms of other varieties of objective modality. Given these other varieties of objective modality, there is no specific reason to be sceptical *only* of metaphysical/absolute modality. This simple characterisation of metaphysical modality does not establish whether metaphysical modality collapses into a specifically and unitedly characterised objective modality, such as nomic or logical modality. Metaphysical/absolute modality can also be analysed by focusing on its non-relativity, which can be construed as counterfactual inevitability. This kind of superficial unification is compatible with a high degree of internal variety among different instances of metaphysical/absolute modality.

3 McFetridge's Thesis and the Sources of Modality

It is interesting to investigate more in depth how and why the absolutist conception of metaphysical modality fails to confer any unity to it and is in particular blind to the various sources of absolute/metaphysical modality. The blindness—a pivotal difference with respect to the essentialist conception, as we are going to see in §4—is already rather evident in the Williamsonian characterisation of metaphysical necessity: p is metaphysically necessary if and only if p is necessary for every variety of objective necessity. The varieties of objective necessity are allowed to be highly heterogeneous (for example, nomic modality is quite different from logical modality) and, thus, no real, unitary *source* of metaphysical modality is identified.

This blindness with respect to the sources of modality is a potential source of confusion also as regards the other forms of objective modalities. Thus, let us suppose that you think that nomic modality is absolute/metaphysical modality and that, as a consequence, scientific laws are absolutely/metaphysically necessary. Given the above characterization of metaphysical necessity, it follows that scientific laws are necessary for any variety of objective modality. Thus, scientific laws are, for example, logically necessary. This is in a sense to be expected: you contend that scientific laws are absolutely/metaphysically necessary and so you coherently deny that any other kind of objective modality disclose further possibilities; if another kind of objective modality were to disclose further possibilities, nomic modality would not be absolute/metaphysical.

Nonetheless, any pressure to thereby, implausibly, conclude that the negation of every scientific law entails a contradiction—in coherence with the characterisation of logical necessity we assumed in §2—should be resisted. The negation of—say—Coulomb's Law (according to which the magnitude of the electrostatic force of attraction or repulsion between two point charges is directly proportional to the product of the magnitudes of charges and inversely proportional to the square of the distance between them) does not entail any contradiction. This simply means that the source of this instance of metaphysical modality is not in logic: it is in science. Coulomb's Law, if it is taken to be absolutely/metaphysically necessary, is necessary for any variant of objective necessity, and so is both nomically and logically necessary; but this claim does not pinpoint in any way the source of its necessity. This source is connected with its being nomically necessary and not with its being logically necessary, but the absolutist conception of metaphysical necessity does not identify in any way the source of its necessity.

Analogously, suppose that you endorse the claim that logical necessity is absolute necessity. It is logically necessary that if Cristiano Ronaldo (CR henceforth) is a football player, then CR is a football player; indeed, the negation of this logical necessity entails a contradiction, namely that CR is a football player and CR is not a football player. Thus, it is also absolutely/metaphysically necessary that if CR is a football player, then CR is a football player. Given the Williamsonian understanding of absolute/metaphysical modality, it follows that it is also *nomically* necessary that if CR is a football player, then CR is a football player. This is in a sense to be expected: you contend that logical necessity is absolute/metaphysical necessity, and so you coherently deny that any other kind of objective modality discloses further possibilities; if another kind of objective modality were to disclose further possibilities, logical modality would not be absolute/metaphysical. However, this should not lead us astray about the source of the instance of metaphysical necessity at stake: it is not a law in any natural science that if CR is a football player, then CR is a football player. The source of this instance of metaphysical modality lies in logic, and the absolutist conception of metaphysical modality is blind to this source.

It is also interesting to make explicit the significance of the so-called McFetridge's Thesis (presented in McFetridge, 1990 and later systematised in Hale, 1996) with respect to this limit of absolutism. According to McFetridge's Thesis, the kind of logical necessity that characterises valid inferences (*i.e.*, the sense in which the premisses of a valid inference necessitate its conclusion) is such that no other kind of necessity is stronger: if McFetridge's Thesis is accepted, logical necessity is absolute, metaphysical necessity. More precisely, McFetridge's thesis states that, if the conditional corresponding to a valid inference (*i.e.*, a conditional whose antecedent is the conjunction of the premisses of the inference and whose consequent is the conclusion of the inference) is logically necessary, then there is no other, wider sense where it is possible that its antecedent is true and the consequent is false.

McFetridge's Thesis can be proven from the following, seemingly weak and reasonable assumptions ($\Box_{log}$ is an operator of logical necessity and $\Diamond$ is an arbitrary operator of possibility):

- (A1) If $\Box_{log}(p \to q)$, then $\Box_{log}((p \wedge r) \to q)$; and if $\Box_{log}(p \to q)$, then $\Box_{log}((r \wedge p) \to q)$
- (A2) $\Box_{log}(p \to p)$
- (A3) If $\Box_{log}(p \to q)$ and $\Box_{log}(p \to r)$, then $\Box_{log}(p \to (q \wedge r))$
- (A4) If $\Diamond p$ and $\Box_{log}(p \to q)$, then $\Diamond q$
- (A5) $\neg\Diamond(p \wedge \neg p)$

By keeping in mind that, when $\Box_{log}$ is prefixed to a conditional, the conditional corresponds to a valid inference, let us briefly see why (A1–A5) are weak and reasonable. (A1) expresses the principle that a valid inference cannot be disrupted by strengthening its premiss; (A1) is scarcely controversial, except in the context of relevant logic.[5] (A2) expresses the reflexivity of valid inference. (A3) corresponds to the contention that if two conclusions can be validly inferred from certain premisses, their conjunction can also be validly inferred from those premisses. (A4) claims that any kind of objective possibility is transmitted through valid inferences (*i.e.*, if the premisses of a valid inference are possible, its conclusion is possible too). (A5) states that for no reasonable candidate to the role of objective possibility, is a contradiction possible.

Given these premisses, it is possible to run the following argument, which proceeds from the assumption that a conditional is logically necessary to the conclusion that it is not possible (for any objective sense of possibility) that the antecedent is true and the conclusion is false:

1. $\Box_{log}(p \to q)$ (Assumption)
2. $\Diamond(p \wedge \neg q)$ (Assumption for *Reductio*)
3. $\Box_{log}((p \wedge \neg q) \to q)$ (1., A1)
4. $\Box_{log}(\neg q \to \neg q)$ (A2)
5. $\Box_{log}((p \wedge \neg q) \to \neg q)$ (4., A1)
6. $\Box_{log}((p \wedge \neg q) \to (q \wedge \neg q))$ (3., 5., A3)
7. $\Diamond(q \wedge \neg q)$ (2., 6., A4)
8. $\neg\Diamond(q \wedge \neg q)$ (A5)
9. $\neg\Diamond(p \wedge \neg q)$ (2., 7., 8., *Reductio*)

McFetridge's Thesis is a rather solid result, and the attempts to block the above argument go through relatively narrow paths, that are in need of an independent motivation: the adoption of relevant logic, thereby rejecting (A1); the adoption of a non-classical logic in which the *reductio* is not valid; or perhaps the rejection of (A4). This is not the topic of this paper: we are neither defending nor attacking the view that logical modality is absolute modality and the arguments in its support, but only discussing the proper understanding of this view when it is conjoined with an absolutist, Williamsonian conception of metaphysical modality.

What does McFetridge's Thesis show, if it is accepted, as far as absolute, metaphysical modality is concerned? McFetridge's Thesis shows that no modality is

[5] (A1) has two parts, inasmuch as it does not matter whether the additional premiss is postpended (first part) or prepended (second part) in the conjunction with the original premisses (the conjunction is the antecedent of the conditional).

more absolute than logical modality in the following sense: no *new* possibility that is foreclosed by logical necessity is then disclosed by another kind of objective modality; there is no more relaxed sense of possibility in which logical impossibilities are possible. Every instance of logical necessity is thus absolutely necessary. Let us now bring to the table the absolutist conception of metaphysical modality: absolute/metaphysical necessity is the conjunction of every objective necessity, *including logical necessity*; thus, every instance of absolute/metaphysical necessity is an instance of logical necessity. This means that, given McFetridge's Thesis and the absolutist conception, logical necessity and absolute/metaphysical necessity end up extensionally coinciding. As we have seen in §2, this does not translate into scepticism with respect to absolute/metaphysical modality. In contrast, we end up identifying absolute/metaphysical modality with an arguably rather well-defined modality (logical modality).

However, McFetridge's Thesis does not show that the source of every metaphysical necessity is logic; it does not show that any negation of a metaphysical necessity entails a contradiction. Every metaphysical necessity is logically necessary (in the sense that no other modality can disclose further possibilities), but—as it happens in general, when an absolutist conception of metaphysical modality is at play—no information is given about the source of various instances of metaphysical modalities: for some of them the source is arguably logic (as is the case for the necessity that if CR is a football player, then CR is a football player), but in other cases, as far as McFetridge's Thesis is concerned, the source could lie elsewhere.

The transitivity of parthood may be an absolute/metaphysical necessity even if no contradiction follows from the hypothesis that three entities are such that the first is part of the second, the second is part of the third, but the first is not part of the third. It may be absolutely/metaphysically impossible that a certain sound lacks any volume, even if no contradiction follows from the hypothesis of a volumeless sound, and it may be absolutely/metaphysically impossible that a human being becomes a cabbage even if no contradiction follows from this hypothesis. The absolutist conception is blind to the sources of metaphysical modality; and this blindness persists even when the absolutist conception is conjoined with the claim that metaphysical modality coincides with a specific kind of objective modality (such as logical modality, in the case of McFetridge's Thesis).

4 Metaphysical Modality as Essential Modality

I cannot go across the Tevere River in Rome *by swimming*. This impossibility is highly contingent; it depends on the fact that I am a poor swimmer and that the Tevere is a relatively large river, not a narrow and slow stream. It is definitely wrong to deem this impossibility *metaphysical*, precisely because it depends on circumstances and features of reality that are in turn contingent. Moreover, I cannot go from Lisbon through the Atlantic Ocean and reach New York in this way. This latter impossibility is *prima facie* rather solid and quite independent of my physical peculiarities. It is difficult to establish whether or not the impossibility is absolute. Perhaps the fact that I lack Herculean capacities, which would enable me to swim for 5,419 km (the distance from Lisbon to New York) *is not* in any sense contingent because it is rooted in what I am—in the fact that I am a human being and that human beings' strength and dimensions *cannot* in any case reach the required high levels. Moreover, the ocean's wideness and dangerousness are perhaps not contingent; there surely can be narrow and quiet watery basins, but arguably an *ocean* cannot be such.

If you think that the impossibility that I would cross the ocean by swimming is absolute, you may want to root it in my essence and/or in that of the ocean. You are not forced to do so. I will not discuss any argument for the specific claims that human beings and oceans have essences and that these essences are precisely connected with the features at stake in the above example.

The understanding of essences and of their role in grounding metaphysical modalities could be more minimal; for example, you can claim that it is impossible for me to be two or more entities, instead of merely one, and root this impossibility in my essence (a similar impossibility would plausibly hold for any individual and be rooted in its essence). A minimal example such as this about numerical oneness is also enough to appreciate the ensuing characterisation of metaphysical modality as rooted in essences. What is metaphysically necessary is rooted in *some* essences.

The literature about this understanding of metaphysical necessity (see, e.g., Fine, 1994; Hale, 1996) often discusses the problem of whether the grounds should be a single essence, various essences or the totality of the essences of all the entities. This debate is important if you are after a characterisation of *essential modality*, inasmuch as essential modality is plausibly specific to a certain entity or entities. What is essential for me (such as being human or lacking Herculean capacities) is not essential for the ocean; for this reason, in the attempts to formalise essential modality (see in particular Fine, 1995), the modal operators usually have an index for an entity (and these indices can be chained, thereby obtaining indices for multiple entities), by whose essence a certain instance of essential necessity is grounded.

However, metaphysical necessity—even in this second understanding of it, according to which it is grounded by essences—is not analogously *perspectival*. It matters that metaphysical necessity has a certain *general source*, namely the essences. However, it does not matter what entity or how many entities are the specific sources of an instance of metaphysical necessity. Some metaphysical necessities (such as: the fact that I am not identical to Jason Momoa) can be collectively grounded by more than one essence (such as: my essence and Jason Momoa's). In the technical terms of contemporary theories of metaphysical grounding, both essences *partially ground* this metaphysical necessity; neither of them individually, *totally grounds* it, but they collectively do so (see, *e.g.*, Fine, 2012; Raven, 2015; Rosen, 2010).

How general can an account of metaphysical modality as grounded by essences be? It may be suspected that such an account can only cover *de re* modalities and is constitutively unfit for *de dicto* modalities. The account *prima facie* seems more apt to capture what every human being possibly and necessarily does or is than what possibly and necessarily every human being does or is. It can aspire to capture the fact that every human being is necessarily human, but not the fact that necessarily every vixen is female and not even that necessarily every human being is human.

The gist of the problem is clear in these last two instances of typically *de dicto* modal claims—the essences of vixens and humans do not seem to play any role in these modal claims. In the former case, the fact that every vixen is female is the outcome of the definition of the word "vixen" or of the corresponding concept, and it is by virtue of this definition that this fact necessarily holds. In the latter case, it is a logical truth that every human being is human. Logical truths are typically formal or topic neutral. Thus, no specificity of human beings or of the meaning of "human" plays a role in this necessity; logic here is the only source of necessity.

The cases of the vixen and of the *de dicto* claim that necessarily every human being is human might seem extraneous to metaphysics. They involve a kind of necessity that abides by the minimal, uncontroversial constraints discussed in §1 (necessity is here neither deontic nor epistemic, but alethic, and is not the outcome of markedly local or specific circumstances). Nonetheless, they do not belong to metaphysics, and this is especially clear in the light of an *essentialist* conception of metaphysics. The essentialist conception of metaphysical modality as grounded by essences corresponds to the traditional view of metaphysics as the study of essences. Metaphysics would be the general investigation of the essences, natures or identities (these three labels tend to be interchangeable in the context of contemporary essentialism) of entities. It might be said that conceptual/analytic truths (*e.g.*, about vixens) and logical truths are necessary, perhaps absolutely so, but are not studied by metaphysics. We are delving into an understanding of metaphysical modality (the essentialist conception) that (in contrast to the absolutist conception, as discussed in §2) does not equate metaphysical modality with absolute modality. Thus, it is coherent and to be expected that conceptual/analytic truths and logical truths are not classified as metaphysical necessities by the essentialist conception, because they do not belong to metaphysics, and thus *are not metaphysical.*

While some instances of *de dicto* modality (*e.g.*, those above) plausibly do not belong to metaphysics as a discipline, some other instances do. Among the cited examples, the transitivity of parthood is a good candidate for the role of a necessary metaphysical principle, and its necessity is clearly *de dicto*, as the syntactic form of the following formulation shows (P expresses parthood):

$$\Box(\forall x \forall y \forall z(x\ P\ y \wedge y\ P\ z \rightarrow x\ P\ z))$$

The case of the transitivity of parthood is not special at all. When we say that grounding is an asymmetrical relation and that mental properties supervene on physical properties (and in general when we set forth supervenience claims), we use *de dicto* modalities within metaphysics. If metaphysical modality needs to mirror the subject matter of metaphysics, then the claim that every *de dicto* modality is correctly classified as non-metaphysical because it does not belong to the subject matter of metaphysics is simply false, and the restriction of metaphysical modalities to *de re* modalities is an inadvisable step.

Thus, some *de dicto* modal truths patently belong to the subject matter of metaphysics, while other *de dicto* modal truths—such as those that are in some way conceptual/analytic or logical—*prima facie* might be deemed *absolute but not metaphysical*, because they do not belong to the subject matter of metaphysics. Overall, the essentialist conception of metaphysical modality is confronted with a *prima facie* mismatch between the domain of metaphysical modal truths and the domain of modal truths that are plausibly grounded by essences. Moreover, it seems that the domain of modalities that are metaphysical but not *prima facie* grounded by essences (*e.g.*, the necessary transitivity of parthood) is in continuity—within the domain of *de dicto* modalities—with that of other modal truths that do not belong to metaphysics at all (*e.g.*, the necessity that every vixen is female).

In front of this *prima facie* mismatch, the defender of the essentialist account of metaphysical necessity can proceed in two broad ways: either by *widening the domain of essences and essence bearers* or by *narrowing the domain of application of its essentialist account of metaphysical necessity*. In the concluding section (§6) I tentatively explore the latter approach. In contrast, the former approach prevails in the contemporary essentialist literature (see both Fine, 1994; Hale, 2012). Let us then find out what this former approach of widening the domain of essences and essence bearers contends and what main difficulties it encounters.

Not even the staunchest supporter of essentialism thinks that there are some *pieces of concrete reality* whose essences account for the necessity of "every vixen is a female fox" or "every human being is human". In particular, everybody agrees that the essences of vixens and humans play no role in these necessities. Fine and Hale (here, I disregard many matters of detail about which they disagree) think that there are *other*, non-concrete entities whose essences explain or ground these metaphysical necessities.

In particular, they are happy to concede that there is nothing wrong in saying that the necessity about a vixen is analytic or conceptual (inasmuch as it relates to the definition of the concept *vixen* or of the corresponding predicate "vixen"); and that the necessity about human beings is logical (inasmuch as it is logically true).

However, in both cases, Fine and Hale think that the essences of some entities are involved and that it is thus correct to classify analytic or conceptual necessities and logical necessities as metaphysical necessities, given the understanding of metaphysical modality as grounded by essences. These are the essences of *meanings* or of *concepts* in the case of analytic or conceptual necessities and the essences of *logical objects* in the case of logical necessities.

This approach is ontologically burdensome. Concepts and meanings are problematic entities, whose characterisation is in turn the subject of many controversies. It is also doubtful, for reasons stemming from Quine (1951)'s notorious criticism of the synthetic/analytic distinction (see Rey, 2018 for an introduction to the current debate), that there is a domain of necessary truths that are grounded by meanings or concepts. Logical objects are even more troublesome, and less often discussed in recent philosophy. For this reason it is useful to dwell on logical objects.

Fine's and Hale's ideas about logical objects and their essences are in direct contrast with Wittgenstein's notorious thesis in the *Tractatus Logico-Philosophicus* that there are no logical objects (sect. 4.442); and in tension with the widely accepted contention that only *non-logical* terms stand for components of reality, while logical terms only express the ways in which the conditions of satisfaction or truth of complex formulas or sentences depend on those of the simpler formulas or sentences within them. In contrast, the idea that essentialists aim to defend here is that the various logical terms (connectives and quantifiers in particular) in some way correspond to entities. For example, conjunction as a logical connective would in some way correspond to conjunction$_{obj}$, a logical object. The essence of conjunction$_{obj}$ would ground the logical necessities involving conjunction, either totally or partially (partially if other logical objects are involved). Thus, consider the kind of necessity that is usually attributed to a *valid* logical inference, such as the following:

$$\frac{p \wedge q}{p}$$

The metalogical claim that p follows from $p \wedge q$ is metaphysically necessary (logical necessity being a subspecies of metaphysical necessity), and would be fully grounded by the essence of conjunction$_{obj}$. In the case of non-metalogical, logical truths in the object language, the essences of various logical objects would collectively ground their metaphysical necessity. This would also be the case for "every human being is human", whose necessity would be presumably grounded by the essences of something like universalquantifier$_{obj}$ and if-then$_{obj}$ (the standard formalisation in first-order logic of "every human being is human" being: $\forall x(Hx \rightarrow Hx)$).

Many problems affect this approach, which has never been developed in any detail, despite being advocated by some essentialists as a way to obtain a unitary account of metaphysical modality. Here are two potential concerns, which have not been so far adequately addressed:

1. The standard semantic analysis of logical language does not countenance these logical objects as referents (or as any other kind of semantic meaning) of logical expressions. Logical expressions are usually not expected to refer to anything at all. Their meanings are usually characterised by a clause in the recursive conditions of satisfaction or truth for a certain language. Thus, the essentialist either ends up relying on logical objects that lack any role in semantics; or is committed to adopt a non-standard semantic analysis of logical language, in which logical objects play a role.[6]
2. It is unclear to which degree of granularity we should distinguish different metaphysical necessities, which are grounded by the essences of different logical objects. As is customary in the theory of grounding, the distinctions to be drawn

[6] In the categorial grammars for natural languages *à la* Cresswell and Montague, connectives belong to categories and can be expected to semantically correspond to certain entities (usually functions). Also the attempt of Quine (1960) to devise a logical language in which connectives are systematically replaced by predicates might be revitalized for this purpose. The works in categorial grammar are usually scarcely explicit about the ontological import of categories, and to put Quine's proposal at the service of essentialism might seem sacrilegious. Thus, at least as far as I know, the attribution of essences to the objects at stake has never been investigated in the literature.

have to be *hyperintensional*; otherwise all logical truths risk collapsing into a single necessity to be grounded. It is difficult to know where to stop when drawing hyperintensional distinctions. Should we take the standard formalisation of a certain truth as a reliable guide in identifying the logical objects whose essences would ground its metaphysical necessity?

These unsolved problems are sufficient to motivate scepticism about this ambitious version of essentialism, which seeks grounding essences also for the most recalcitrant instances of absolute modality. In §6, I shall again turn to the more modest variant of the essentialist approach to metaphysical modality, which narrows the domain of application of essentialism and is content with accounting only for *some* instances of absolute modality. It is now time to ask whether some progress in the analysis of metaphysical modality can be made by means of the most successful tool for the semantic analysis of *modal logic*, namely possible worlds.

5 Possible Worlds and Metaphysical Modality

As discussed in §2, the absolutist conception of metaphysical modality corresponds to the idea that in metaphysics we are not interested in any restricted or local truth. Inasmuch as we are metaphysicians, we are—so to speak—maximally liberal in considering remote, unlikely hypotheses. In the attempt to make the absolutist conception explicit and precise, this lack of restriction can be construed in various ways, and we have considered one: Hale's analysis of absoluteness in terms of counterfactual inevitability. Given the analysis of modal notions in terms of possible worlds, there is another obvious way of construing absoluteness. We have at our disposal the set of all possible worlds (the *pluriverse*). We can consider either a proper subset of it or the whole set; if we consider the former, then we exclude some scenarios and are not as liberal as metaphysicians should be.

Thus, if I say that I cannot cross the Tevere River by swimming, I disregard many possible worlds at which my physical strength is greater than at the actual world, or the river's width or impetus is less. The restrictions usually pertain to minimum levels of similarity to the actual world under a certain respect, where the respect and the metrics of similarity are allowed to vary from case to case.

What does it mean in this context to *consider* all possible worlds (if the modality is metaphysical) and to *consider* only some of them (if it is not metaphysical)? It means—respectively—not to restrict and to restrict some *quantifiers*. These quantifiers can belong either to the metalanguage or to the object language, according to the specific variant of the theory of possible worlds that we resort to. In the standard, Kripkean semantics for modal languages (see, *e.g.*, Kripke, 1963), the truth conditions for modal sentences in the metalanguage include quantifiers over possible worlds (regardless of whether these sentences are *de re* or *de dicto*). In David Lewis' counterpart theory (Lewis, 1968), the quantifiers at stake are in the object language and *replace* modal operators. If the modality is *de dicto*, then there are quantifiers over possible worlds, which can be either unrestricted or restricted. If the modality is *de re*, then there are quantifiers over possible individuals, which are parts of possible worlds (the so-called counterparts of the actual individual to which the *de re* modality is attributed). In this case as well, the quantification over counterparts can be either unrestricted or restricted. In all these cases, the modality at stake would be metaphysical if and only if the quantifiers are unrestricted.

To assess the possible worlds' contribution to the analysis of metaphysical modality, it is instructive to compare the analysis of absolute/metaphysical modality in terms of possible worlds, on the one hand, with its analysis in terms of counterfactual inevitability (as discussed at the end of §2), on the other hand. For the purposes of this comparison, let us adopt the standard Lewis-Stalnaker semantic analysis of counterfactual conditionals as variably strict (Lewis, 1973; Stalnaker, 1968) in a simplified version that is sufficient here: a counterfactual conditional $p > q$ is true if and only if the consequent q is true at all the worlds at which the antecedent p is true which are similar at least to a certain degree and under a certain respect of similarity to the actual world. The full determination of the truth conditions of a counterfactual conditional therefore requires the specification of a minimum degree and of a respect of similarity. The expectation is that the context in which a counterfactual conditional is uttered can help determine the minimum degree and the respect.

According to the analysis of metaphysical necessity as counterfactual inevitability (as discussed at the end of §2), p is metaphysically necessary if and only if, for *every* q, $q > p$ is true. For each of the counterfactual conditionals obtained by replacing the variable q with a certain sentence, the Lewis-Stalnaker analysis can be applied by specifying a minimum degree and a respect of similarity or a context that in turn determines a minimum degree and a respect of similarity. In the analysis of metaphysical necessity, we have a universal quantification in sentential position on the antecedents of the counterfactuals, and thus no limits on the variation of the antecedents. On the other hand, the context is provided once and for all for the single claim of necessity whose absoluteness/metaphysicality has to be assessed. This means that the minimum degree and the respect of similarity are also determined once and for all by the context of the claim of necessity (they do not vary for the different values of the variable q). The context is presumably *metaphysics itself* (or more locally, a paper, a book, a seminar or a conversation on metaphysics), such that we are liberal in admitting similarities and we strive to be neutral as regards the respect of similarity, by privileging the sharing of some properties over the sharing of others only if these properties are privileged from a metaphysical viewpoint (*e.g.*, inasmuch as they are natural or fundamental properties).

Given this interpretation of the counterfactuals at stake, at the end of the day, are counterfactual inevitability and truth at all possible worlds equivalent to each other? Clearly, truth at all possible worlds entails counterfactual inevitability: if p is true at all possible worlds, then, given any q, p is true at all the worlds at which q is true, no matter how similar and under which respect these worlds are to the actual world. In the opposite direction, there are good reasons to *deny* that counterfactual inevitability entails truth at all possible worlds; that is, to deny that if for every q p is true at all the possible worlds at which q is true and which are similar to the actual world to a liberal degree under a metaphysically neutral respect, then *p* is true at every possible world.

The temptation might arise to argue in favour of this latter entailment from counterfactual inevitability to truth at all possible worlds based on the remark that in the metaphysical context, the minimum degree of similarity can be extremely low, thereby including the sharing of any property, including trivial properties such as *being one* or *being an individual*. Thus, the requirement that, in order for the counterfactual conditional to be true, the consequent has to be true at all *adequately similar* possible worlds at which the antecedent is true would be reduced to the requirement that the consequent is true at all the possible worlds at which the antecedent is true, because any similarity would be enough and any couple of worlds share properties, such as *being one* or even *being a world*. In counterfactual inevitability (*for every q*, $q > p$ *is true*) the position of the antecedent is universally quantified. It would turn out that if *p* is counterfactually inevitable, then *p* is true at all the possible worlds at which at least a value of *q* is true. For any world at least a value of *q* (*i.e.*, at least a proposition) is presumably true at it. Thus—it might be tempting to conclude—if *p* is counterfactually inevitable, then *p* is true at all possible worlds.

However, this line of argument in favour of the entailment from counterfactual inevitability to truth at all possible worlds should be resisted. Even in metaphysical contexts, it is not the case that any similarity with the actual world is sufficient to be in the domain of worlds, among those at which the antecedent of the counterfactual is true, which matter for the truth conditions of the counterfactual conditional. A major methodological component of metaphysics (and of many other subfields of philosophy) is constituted by *thought experiments*. In two famous thought experiments, we imagine for example that two indistinguishable spheres are at a one-meter distance from each other (as in Black, 1952); or that the wood planks constituting a ship are replaced one by one (as in the literature about the ship of Theseus).

As Williamson (2007, ch. 6) has convincingly shown, assessing a thought experiment in metaphysics (and in philosophy in general) is tantamount to assessing a counterfactual conditional: the antecedent is a description of the scenario that we are asked to imagine, while the consequent is a claim of philosophical interest. In assessing the counterfactual, we evaluate whether the claim of philosophical interest would be true or false, if the counterfactual scenario were the case. There is nothing trivial in this evaluation. We *should not* be *too* liberal in selecting, among the possible worlds at which the antecedent is true, those at which the consequent has to be true, in order for the counterfactual conditional to be true. We should only focus on those possible worlds that are significantly and substantially similar to the actual world. It is true that in metaphysics we should be *somewhat* liberal about the degree of similarity and neutral about the respects of similarity, but this does not mean that *anything goes*.

Inasmuch as it is false that anything goes, counterfactual inevitability does not entail truth at all possible worlds and the conception of metaphysical necessity as counterfactual inevitability does not collapse into the conception of metaphysical necessity as truth at all possible worlds. The former conception has the limits that

I have already underlined in §2 and is particularly unable to confer any unity to metaphysical modality.

The different and stronger conception of metaphysical necessity as truth at all possible worlds does not fare better than counterfactual inevitability from this viewpoint; it is not clear at all what kind of unity it may confer to metaphysical modality. Another serious, connected defect is that the limits of the domain of possible worlds are not independently settled. Which kind of limit does the attribute "possible" in "possible world" express? It risks expressing the fact that at all possible worlds all the absolute/metaphysical necessities hold true, and that at no possible world does any absolute/metaphysical impossibility hold true. If the purpose is to draw the limits of the entire domain of possible worlds, surely the only kind of modality that can turn out useful is the absolute one. If this were the case, the conception of absolute/metaphysical necessity as truth at all possible worlds would not be explicative at all. First, this conception does not pinpoint any unitary source of metaphysical modality; second, at the end of the day, it *presupposes* metaphysical modality.

The only hope for avoiding this outcome is the adoption of a *reductive* theory of possible worlds, which characterises possible worlds in a non-modal way. In David Lewis' modal realism (Lewis, 1986), possible worlds are giant individuals, which are characterised by being closed under relations of spatio-temporal distance. This characterisation of possible worlds does not presuppose metaphysical modality.

Nonetheless, even if you shoulder modal realism's heavy ontological costs,[7] you obtain no unitary account of absolute/metaphysical modality. From this viewpoint it is unsurprising that Lewis—as far as I know—never employed the phrase "metaphysical modality" in presenting his own theories, neither in *On the Plurality of Worlds* nor in any of his other works. A radical reductionist about modality, Lewis thought that any modal claim should be paraphrased away, in favour of quantifications over worlds (in the case of *de dicto* modalities) or parts of worlds (in the case of *de re* modalities). For him, absolute necessity and possibility can be reduced to quantification over worlds and their parts, where worlds are characterised non-modally. Absolute modality has no special *source*, and is not even especially bound to metaphysics as a discipline.

Lewis was *au fond* sceptical of modalities;[8] according to him, modalities can be treated in an acceptable way only by *replacing* them with non-modal notions. As shown in §2 and convincingly argued in Williamson (2016, §2), the main reasons to be sceptical of metaphysical modality are also reasons to be sceptical of objective modalities in general. Thus, inasmuch as Lewis was sceptical of modalities in general, it is unsurprising that he had no real use for the concept of metaphysical modality. We can interpret metaphysical modality in terms of Lewis' modal realism. In this way, *if* we are willing to shoulder the ontological costs of modal realism, we obtain a real, non-circular explanation of metaphysical necessity in terms of what occurs at a totality of certain individuals, with certain well-defined features (namely,

[7] Perhaps because you think that its explanatory benefits outbalance the costs, in coherence with Lewis' typical cost-benefit approach to philosophy. See Nolan (2015) on this matter.

[8] See for example Beebee and MacBride (2015) on this point.

closure under spatio-temporal distance). However, once we have made this move, metaphysical necessity no longer has any role to play.

Moreover, it should be observed that Lewis' reductionism about modality is notoriously silent about all the necessities that do not concern *spatio-temporal* entities. Presumably, logical truths and mathematical truths hold true at all possible worlds, that is, at all spatio-temporal closed individuals. However, how does modal realism explain these logical and mathematical necessities? Numbers or sets are not parts of possible worlds, inasmuch as any part of a possible world is at some spatio-temporal distance from all other parts of that world, and numbers and sets do not plausibly participate in any relation of spatio-temporal distance. Thus, the absolute necessity of logical and mathematical truths is only assumed (and not explained) by Lewis' modal realism.

Lewis was also a radical Humean, and consequently—*pace* some exegetical controversies about this (see for example Buras, 2006; Nencha 2017; Paul, 2006)—an adversary of necessary connections and essences. This means that on the one hand, logical and mathematical necessities are simply assumed to hold true at all possible worlds, without any real explanation. On the other hand, it is unclear how many other absolute necessities modal realism admits and accounts for. For a Humean, does anything hold true at all possible worlds and (in contrast to logical and mathematical truths) *really relate* to how possible worlds are, according to modal realism? Not many interesting examples come to mind, again confirming Lewis' coherence in his own overall approach in refraining from speaking of metaphysical necessity.

Arguably, some instances of metaphysical necessity are consequences of the tenets of modal realism itself. For example, inasmuch as possible worlds are spatio-temporal closed individuals, there is no *empty* world at which nothing exists (see Lewis, 1986, pp. 73–74); consequently, it is true at all possible worlds that at least one entity exists. Thus, it is absolutely necessary that at least one entity exists. However, the resulting conception of absolute/metaphysical modality ends up being quite disappointing, given that a) it fails to explain the absolute necessity of logical and mathematical truths and b) it only works for some instances of metaphysical necessities that are the outcomes of modal realism itself.

This overall picture suggests that modal realism offers no real benefit for the purpose of explicating absolute/metaphysical modality, coherent with the fact that it is a tool to *get rid* of modality in general, by analysing it in non-modal terms. Thus, we are back to the other, non-reductive theories of possible worlds, which have no ambition to get rid of modality (as well as metaphysical modality), inasmuch as the same notion of possible world is for them intrinsically and unavoidably modal. However, this also means that possible worlds are those where absolute/metaphysical necessities hold true and where absolute/metaphysical impossibilities do not hold true.

It is therefore doubtful that any explanatory analysis is obtained, and even more doubtful that we obtain any explanation of what metaphysical modality is. Possible worlds can be perhaps expected to be in a sense explanatory with respect to various concepts of modalities (including metaphysical modality) and in particular to their logic, if accessibility relations are countenanced. There is a well-known correspondence between the formal features of the accessibility relation and the axioms of modal logic; thus, many theorists of possible worlds would argue that the formal features of the accessibility relation among worlds for a certain variety of modality explain the fact that a certain axiom holds for this variety of modality.

The attribution of this kind of explanatory duty to possible worlds is also controversial, but we do not need to enter these controversies. Non-reductive theories of possible worlds may be useful to represent or even explain the logic of metaphysical modality, but, inasmuch as metaphysical modality is truth at every possible world, they are unhelpful in drawing the limit between what is metaphysically possible and what is not. This is especially clear if *impossible worlds* are countenanced.[9] It is beyond this paper's scope to assess the merits and the difficulties of the doctrine of impossible worlds. However, it should be noted that if there are good reasons to think that impossible worlds also exist, then again the limit between possible and impossible worlds risks consisting in the fact that possible worlds respect absolute/metaphysical necessities, while impossible worlds are such that at each of them at least one absolute/metaphysical necessity is false.[10] Again, this means that this limit is assumed and not analysed or explained by non-reductive theories of possible worlds. This holds true even independently of any commitment to impossible worlds, which only highlight the problem of delimiting the domain of possible worlds, inasmuch as—if impossible worlds also exist—there are other worlds from which the possible ones have to be distinguished.

Thus, on one hand, *non-reductive* theories of possible worlds do not explain what absolute/metaphysical modality—and in particular the limit between what is metaphysically possible and what is not such—consists in. On the other hand, in the main *reductive* theory of possible worlds (namely, Lewis' modal realism), many absolute necessities (*e.g.*, logical and mathematical necessities) are not really accounted for, and there are few and rather uninteresting others. Consequently, there are good reasons to think that the theories of possible worlds (both non-reductive and reductive ones) are not really helpful in characterising metaphysical modality. They do not bring any advantage over the conception of absolute/metaphysical necessity as counterfactual inevitability, and this conception is in turn already lacking, inasmuch as it does not confer any kind of unity on absolute/metaphysical necessity.

---

[9] See, *e.g.*, Rescher and Brandom (1980), Zalta (1997), Nolan (2014), and Priest (2016). See Berto and Jago (2018) for an overview and Berto and Jago (2019) for an in-depth systematisation.

[10] In the literature about impossible worlds, there are several contoversial attempts to draw the distinction between possible and impossible worlds, usually focused on logical possibility and not on metaphysical possibility. See Berto and Jago (2019, §1.4).

6 Conclusion: Terminological Issues

As seen in §2, the conception of metaphysical modality as absolute modality has an advantage regarding the dialectics with the modal sceptic, inasmuch as there is no specific reason to be sceptical of *absolute* objective modality. Perhaps there are good reasons to be sceptical of objective modalities in general (but these reasons are beyond the purpose of this paper). Nonetheless, once we have admitted some objective modalities in general, absolute/metaphysical modality can be easily characterised on the basis of them. However, the absolutist conception of metaphysical modality does not confer any unity on it, as is especially evident once absoluteness is analysed in terms of counterfactual inevitability.

The general problem of the absolutist conception seems to be its blindness to the sources or grounds of metaphysical modality. As seen in §4, the conception of metaphysical modality as modality grounded by essences directly addresses this problem, precisely because essences are pinpointed as the grounds at stake. We can doubt the existence of essences (and the general defence of essentialism is also beyond the purpose of this paper), but if they exist, they are excellent candidates for the role of grounds for absolute modalities.

However, we have found that for conceptual/analytic and logical modalities the idea that they are grounded by essences leads us to problematic commitments. In general, essences seem unfit to ground *de dicto* modalities. In some cases (*e.g.*, for logical necessities) and in open contrast with the absolutist conception, the essentialist might concede that they are absolute, but not metaphysical. The essentialist might try to enforce the typical essentialist metametaphysical stance, according to which metaphysics is the study of essences, so that where essences are not involved, the attribute "metaphysical" is misapplied. Once this position is assumed, the essentialist account of metaphysical modality cannot fail, inasmuch as any instance of necessity where essences are not involved becomes *ipso facto* non-metaphysical, precisely because essences are not involved.

This move risks being crafty and unconvincing on the part of the essentialist. Indeed, metaphysics is a *really practised* discipline, with a long historical pedigree. Nobody is allowed to decide out of the blue which doctrines or notions deserve to be called "metaphysical". Deciding it out of the blue is an especially deviant methodology if the purpose is to show that a certain conception of metaphysical x (in our case, metaphysical *modality*) is preferable to other rival conceptions of metaphysical x. The essentialist is dialectically not allowed to rule out all the counterexamples to her own conception of metaphysical modality by simply affirming that these counterexamples do not belong to metaphysics in the light of her own essentialist metametaphysics (to which the adversaries will unlikely subscribe—and in any case, are not forced to do so).

There seems to be no way to reconcile how absolutists and essentialists construe metaphysical modality. Here, I can do no more than try to assess who, between the two parties, is more entitled to speak about metaphysical modality or more precisely, to use the corresponding lexicon, that is, expressions in the vicinity of "it is metaphysically necessary that" and "it is metaphysically possible that". This assessment applies a quite simple rule of thumb, which I think to be widely applicable: a philosophically disputed term, *i.e.* a term that is employed by two or more philosophical parties in distinct and irreconcilable ways, is more appropriately used by that party (or those parties) that has (have) no adequate and already established *alternative term* for the same concept. The other parties can be content with the alternative terms, and avoid using the disputed ones: this will prevent confusions and (in the specific case) the misleading illusion that a single concept of metaphysical modality is discussed in the debate.

As far as absolutists are concerned, their jargon has two equivalent denominations for modality: "absolute" and "metaphysical". The adjective "absolute" adequately serves the purpose of expressing the notion at stake and is free of specific and potentially distracting connotations that—in contrast—the adjective "metaphysical" unavoidably carries. The existence of multiple labels for the same concept is a *waste* of linguistic resources. This waste is not innocuous, precisely inasmuch as one of the labels ("metaphysical") has diverging and potentially misleading connotations.

More specifically, it is methodologically inadvisable to use "metaphysical" for conceptual/analytic, logical and mathematical necessities and possibilities, unless we accept a commitment to an underlying, substantial doctrine that motivates this lexical choice. If we want to underline their absoluteness, we can call them "absolute". If we want to emphasise their sources, we cannot prescind from a theory about their sources. If we accept a commitment to analyticity as a source of necessity, those necessities that are so originated can be called "analytic". If we think that mathematics is an autonomous source of necessities, the ensuing necessities can be called "mathematical".

The usage of "metaphysical" for necessities and possibilities on the part of the essentialist is arguably more justifiable from the viewpoint of the rule of thumb for philosophically disputed terms I sketched above. While the absolutist has "absolute" as an alternative at her disposal, "essential" is not a good replacement for what the essentialist wants to express with "metaphysical". As shown in §4, since essentiality is always perspectival with respect to the bearers of certain essences, the essentialist needs a general term for qualifying those necessities that are grounded by some essences. There is no alternative, established label at the essentialist's disposal. Moreover, the usage of the term is justified by the fact that the concept of essence is undeniably central in the practice and the history of metaphysics.

This does not force the essentialist to be radical and to claim that all the modalities discussed in metaphysics are *ipso facto* grounded by essences. The essentialist needs specific arguments for this claim, and (as shown in §4) this is rather implausible in many cases. In the case of those modalities that belong to metaphysics but are not grounded by essences (in some of our examples, the necessity of the transitivity of parthood and the necessity of supervenience claims), there is no point in insisting that they are metaphysical because they belong to metaphysics. In most cases, there is no need to clarify through a specific adjective or predicate that a certain claim (including modal claims) belongs to a particular subfield of philosophy. If this need emerges, it can be served precisely by the predicate "to belong to metaphysics" and its derivatives (*e.g.*, "belonging to metaphysics").

This allows us to finally reserve expressions such as "metaphysical modality", "metaphysical necessity" and "metaphysical possibility" for modalities grounded by essences. It is important to reiterate that, as I emphasised in §1, this proposal is not committed to any specific example of metaphysical necessity or possibility. Thus, the claim that "metaphysical", when referring to modalities, is better reserved for what is grounded by essences for general reasons concerning the usage of philosophically disputed terms does not commit us to the specific essentialist claims that—say—belonging to a certain species is an essential feature of every organism or that a human being is essentially rational. These essences, if they indeed exist, would be sources of necessities. Every dog would be such that it is metaphysically necessary that it is a dog; every human being would be such that it is metaphysically necessary that he or she is rational. However, the characterisation of metaphysical modality as modality grounded by essences is not committed to any specific identification of essences.

References

Beebee, H., & MacBride, F. (2015). De re modality, essentialism, and Lewis's Humeanism. In B. Loewer & J. Schaffer (Eds.), *A companion to David Lewis* (pp. 25–39). Wiley Blackwell.

Berto, F., French, R., Priest, G., & Ripley, D. (2018). Williamson on counterpossibles. *Journal of Philosophical Logic, 47*(4), 693–713.

Berto, F., & Jago, M. (2018). Impossible worlds. In E. Zalta (Ed.), *The Stanford encyclopedia of philosophy. Fall 2018 edition.* Stanford University: Metaphysics Research Lab. https://plato.stanford.edu/archives/fall2018/entries/impossible-worlds/.

Berto, F., & Jago, M. (2019). *Impossible worlds.* Oxford University Press.

Black, M. (1952). The identity of indiscernibles. *Mind, 61*(242), 153–164.

Buras, T. (2006). Counterpart theory, natural properties, and essentialism. *The Journal of Philosophy, 103*(1), 27–42.

Fine, K. (1994). Essence and modality. *Philosophical Perspectives, 8*, 1–16.

Fine, K. (1995). The logic of essence. *Journal of Philosophical Logic, 24*(3), 241–273.

Fine, K. (2012). Guide to ground. In F. Correia & B. Schnieder (Eds.), *Metaphysical grounding* (pp. 37–80). Cambridge University Press.

Hale, B. (1996). Absolute necessities. *Philosophical Perspectives, 10*, 93–117.

Hale, B. (2012). What is absolute necessity? *Philosophia Scientiae, 16*(2), 117–148.

Kripke, S. (1963). Semantical considerations on modal logic. *Acta Philosophica Fennica, 16*, 83–94.

Ladyman, J., Ross, D., & Spurrett, D. (2007). In defense of scientism. In J. Ladyman & D. Ross (Eds.), *Everything must go: Metaphysics naturalized* (pp. 1–65). Oxford University Press.

Lewis, D. (1968). Counterpart theory and quantified modal logic. *The Journal of Philosophy, 65*(5), 113–126.

Lewis, D. (1973). *Counterfactuals.* Blackwell.

Lewis, D. (1986). *On the plurality of worlds.* Blackwell.

Lowe, J. (2001). *The possibility of metaphysics.* Clarendon Press.

McFetridge, I. (1990). Logical necessity: Some issues. In J. Haldane & R. Scruton (Eds.), *Ian McFetridge: Logical necessity & other essays* (pp. 135–154). Aristotelian Society.

Nencha, C. (2017). Natural properties do not support essentialism in counterpart theory. *Argumenta, 2*(2), 281–292.
Nolan, D. (2011). The extent of metaphysical necessity. *Philosophical Perspectives, 25*, 313–339.
Nolan, D. (2014). Hyperintensional metaphysics. *Philosophical Studies, 171*(1), 149–160.
Nolan, D. (2015). Lewis's philosophical method. In B. Loewer & J. Schaffer (Eds.), *A companion to David Lewis* (pp. 25–39). Wiley Blackwell.
Paul, L. (2006). In defense of essentialism. *Philosophical Perspectives, 20*, 333–372.
Priest, G. (2016). Thinking the impossible. *Philosophical Studies, 173*(10), 2649–2662.
Raven, M. (2015). Ground. *Philosophy Compass, 10*(5), 322–333.
Rescher, N., & Brandom, R. (1980). *The logic of inconsistency*. Blackwell.
Rey, G. (2018). The analytic/synthetic distinction. In E. Zalta (Ed.), *The Stanford encyclopedia of philosophy. Fall 2018 edition.* Stanford University: Metaphysics Research Lab. https://plato.stanford.edu/archives/fall2018/entries/analytic-synthetic/.
Rosen, G. (2010). Metaphysical dependence: Grounding and reduction. In B. Hale & A. Hoffmann (Eds.), *Modality: Metaphysics, logic and epistemology* (pp. 109–136). Oxford University Press.
Ross, D., Ladyman, J., & Kincald, H. (Eds.). (2015). *Scientific metaphysics*. Oxford University Press.
Stalnaker, R. (1968). A theory of conditionals. In N. Rescher (Ed.), *Studies in logical theory* (pp. 98–112). Blackwell.
Quine, W.v.O (1951). Two dogmas of empiricism. *The Philosophical Review, 60*(1), 20–43.
Quine, W.v.O. (1960). Variables explained away. *Proceedings of the American Philosophical Society, 104*(3), 343–347.
Williamson, T. (2007). *The philosophy of philosophy*. Blackwell.
Williamson, T. (2016). Modal science. *Canadian Journal of Philosophy, 46*(4–5), 453–492.
Williamson, T. (2018). Counterpossibles. *Topoi, 37*(3), 357–368.
Zalta, E. (1997). A classically-based theory of impossible worlds. *Notre Dame Journal of Formal Logic, 38*(4), 640–660.

Counterpart Semantics at Work: Independence and Incompleteness Results in Quantified Modal Logic

Francesco Belardinelli

1 Introduction

The model theory of quantified modal logic (QML) has always been a subject of debates due to the variety of semantical choices associated with it:

> From the semantical point of view, we are confronted with a number of decisions concerning the quantifiers, and these in turn prompt new questions about the semantics of identity, terms, and predicates. Since most of these choices can be made independently, the number of interesting quantified modal logics seems bewildering large. (Garson, 2001, p. 267)

From a philosophical perspective these choices need to be grounded in a consistent ontology. The metaphysics of modality is indeed devoted to investigate the ontology of possible objects and the meaning of counterfactual situations (Loux, 1979; Lowe, 1998), which in turn have influenced the development of the theory of quantified modal logic. For instance, the counterpart semantics for QML originated from the work of D. Lewis (1979) was motivated by his rejection of trans-world identity for individuals. On the other hand, the extension of Kripke semantics to the first order (Kripke, 1959, 1963), which is one of the mainstream semantical accounts for QML, implicitly assumes that it is possible to identify the same individuals across possible worlds.

Since the debate on the ontology of individuals and counterfactuals makes an ever-increasing use of arguments from quantified modal logic, the theoretical analysis of QML is useful not only from a technical point of view, but also from a philosophical perspective. In this respect, Kripke semantics has proved to be a valuable instrument to investigate the theoretical properties of QML systems. However, in the last few years a number of alternative semantical accounts, such as presheaves, categories and admissible semantics (Brauner and Ghilardi, 2007; Goldblatt, 2011; Skvortsov &

F. Belardinelli (✉)
Department of Computing, Imperial College, London, UK
e-mail: francesco.belardinelli@imperial.ac.uk

F. Ademollo et al. (eds.), *Thinking and Calculating*, Logic, Epistemology, and the Unity of Science 54, https://doi.org/10.1007/978-3-030-97303-2_21

Shehtman, 1993), has been put forward to solve some of the issues given by extending Kripke semantics to the first order.[1]

Our main contribution in this paper consists in extending the results of mutual independence for the Barcan formula BF and the necessity of fictionality N¬E in Goldblatt (2011) to normal modalities as strong as $S5$. Further, we make use of these independence results to prove the Kripke-incompleteness of a number of QML calculi based on free logic (Bencivenga, 2002) and containing BF. Specifically, we first introduce four well-known first-order modal principles:

BF	$\forall x \Box \phi \rightarrow \Box \forall x \phi$	Barcan formula
CBF	$\Box \forall x \phi \rightarrow \forall x \Box \phi$	converse of the Barcan formula
NE	$E(x) \rightarrow \Box E(x)$	necessity of existence
N¬E	$\neg E(x) \rightarrow \Box \neg E(x)$	necessity of fictionality

In the Kripke semantics for QML the converse of the Barcan formula CBF and the necessity of existence NE intuitively mean that existing individuals exist necessarily (i.e., they exist in all accessible worlds); whereas the Barcan formula BF and the necessity of fictionality N¬E state that non-existing individuals cannot exist (i.e., they are necessarily non-existent). As it is discussed in Corsi (2003):

> BF and CBF are often considered as dual principles: BF corresponds in Kripke semantics [...] to the condition that inner domains never increase, CBF to the condition that inner domains never decrease. We shall show that this duality is not intrinsic to the semantics of CBF and BF but rather depends on general features of Kripke semantics.

We follow Corsi's remark to show that this duality is not intrinsic to the meaning of BF and N¬E. In particular, we briefly review the completeness results available in the literature for QML systems containing the formulas above. Then, we extend the independence results in Goldblatt (2011) to QML systems on all normal modalities, including B and $S5$, by making use of counterpart semantics. This allow us to prove the incompleteness of QML systems based on free logic and containing BF.

The rest of the paper is structured as follows. In Sects. 2 and 4 we introduce the Kripke and counterpart semantics for QML respectively. In Sect. 3 we present a class of QML systems, and we briefly review the completeness results available in the literature for these calculi. In particular, we consider a class of QML systems based on free logic and containing BF, whose Kripke-incompleteness is proved in detail in Sect. 5. We conclude in Sect. 6 with a discussion of related work and some open problems.

[1] For instance, many first-order extensions of complete modal propositional calculi are not complete with respect to the corresponding classes of first-order Kripke frames Ghilardi (1991, 1992).

2 Kripke Semantics

In this section we introduce the Kripke semantics for first-order modal languages. Our presentation is based on Corsi (2002) and Fitting and Mendelsohn (1999).

We consider a first-order modal language $\mathcal{L}$ containing a numerable infinite set *Var* of individual variables $x_1, x_2, \ldots$; a numerable infinite set of n-ary predicate symbols $P_1^n, P_2^n, \ldots$, for $n \in \mathbb{N}$; the unary existence predicate symbol E; the propositional connectives $\neg$ and $\wedge$; the universal quantifier $\forall$; and the modal operator $\Box$. The presentation is restricted to languages without function symbols or the identity relation, as these features are not essential for the results in this paper[2]. Therefore the only terms $t_1, t_2, \ldots$ in language $\mathcal{L}$ are individual variables.

Definition 1 The first-order modal formulas in $\mathcal{L}$ are defined in Backus-Naur form as follows:

$$\phi ::= P^n(t_1, \ldots, t_n) \mid E(t) \mid \neg\psi \mid \psi \wedge \psi' \mid \forall x\psi \mid \Box\psi$$

The symbols $\vee$, $\rightarrow$, $\leftrightarrow$, $\exists$, $\diamond$ are introduced as standard. Further, if $y_1, \ldots, y_n$ are distinct variables, and $t_1, \ldots, t_n$ are individual terms, the substituted term $s[y_1/t_1, \ldots, y_n/t_n]$, also denoted as $s[\mathbf{y}/\mathbf{t}]$, is equal to t_j if $s = y_j$ for some $j \leq n$; otherwise, it is equal to s. We remind the reader that in the present setting, also $t_1, \ldots, t_n$ are variables, as variables are the only terms. However, we consider substitution for terms in general to keep open the possibility of extending the language so as to include other terms besides variables.

The substituted formula $\phi[y_1/t_1, \ldots, y_n/t_n]$, or $\phi[\mathbf{y}/\mathbf{t}]$, is inductively defined as follows:

$$\begin{aligned}
P^m(t_1, \ldots, t_m)[\mathbf{y}/\mathbf{t}] &= P^m(t_1[\mathbf{y}/\mathbf{t}], \ldots, t_m[\mathbf{y}/\mathbf{t}]) \\
E(t)[\mathbf{y}/\mathbf{t}] &= E(t[\mathbf{y}/\mathbf{t}]) \\
(\neg\psi)[\mathbf{y}/\mathbf{t}] &= \neg(\psi[\mathbf{y}/\mathbf{t}]) \\
(\psi \wedge \psi')[\mathbf{y}/\mathbf{t}] &= \psi[\mathbf{y}/\mathbf{t}] \wedge \psi'[\mathbf{y}/\mathbf{t}] \\
(\Box\psi)[\mathbf{y}/\mathbf{t}] &= \Box(\psi[\mathbf{y}/\mathbf{t}]) \\
(\forall x\psi)[\mathbf{y}/\mathbf{t}] &= \forall x(\psi[\mathbf{y}/\mathbf{t}])
\end{aligned}$$

where we assume without loss of generality that $x \neq y_i$ and $x \neq t_i$ for every $i \leq n$.

The definition of substitution is standard, but we report it here for comparison with substitution for the indexed languages introduced in Sect. 4. Finally, if $y_i = x_i$ for $i \leq n$, we simply write $\phi[t_1, \ldots, t_n]$ or $\phi[\mathbf{t}]$.

We provide a semantics to the language $\mathcal{L}$ by means of Kripke frames and models.

Definition 2 (K-frame) A *Kripke frame* is a tuple $\mathcal{F} = \langle W, R, D, d \rangle$ such that

- W is a non-empty set;

[2] We observe that in languages with identity, $E(t)$ is definable as $\exists x(x = t)$.

- $R \subseteq W \times W$ is a binary relation on W;
- D is a function such that for every $w \in W$, $D(w)$ is a non-empty set and wRw' implies $D(w) \subseteq D(w')$;
- d is a function such that for every $w \in W$, $d(w)$ is a subset of $D(w)$.

In the model theory of modal logic the set W is intuitively referred to as the *set of possible worlds*, while R is the *accessibility relation* between worlds. Each $D(w)$ is the *outer domain* of w, i.e., the interpretation domain for variables and predicate symbols; while each $d(w)$ is the *inner domain* of w, i.e., the variation domain for quantifiers. The constraint on outer domains guarantees that the interpretation of individual variables is always defined, that is, if wRw' and $\sigma : Var \rightarrow D(w)$ is a w-assignment from variables to individuals in $D(w)$, then σ is also a w'-assignment.

In the rest of the paper we consider the following classes of Kripke frames.

Definition 3 A K-frame $\mathcal{F}$ has *constant* (resp. *increasing*, *decreasing*) inner domains iff wRw' implies $d(w) = d(w')$ (resp. $d(w) \subseteq d(w')$, $d(w) \supseteq d(w')$). Otherwise, $\mathcal{F}$ has *varying* inner domains.

By Def. 2 a K-frame $\mathcal{F}$ can have only increasing, or constant outer domains, the latter whenever $\mathcal{F}$ is decreasing (or constant) and $D(w) = d(w)$ for every $w \in W$.

Definition 4 (K-model) A *Kripke model* based on a K-frame $\mathcal{F}$ is a pair $\mathcal{M} = \langle \mathcal{F}, I \rangle$ where I is a first-order interpretation such that

- if P^n is an n-ary predicate constant and $w \in W$, then $I(P^n, w) \subseteq D(w)^n$ is an n-ary relation on $D(w)$;
- for all $w \in W$, $I(E, w) = d(w)$.

To define the truth conditions for first-order modal formulas in K-models, for every variable x and element $a \in D(w)$, we define the variant σ^x_a of a w-assignment σ to be the w-assignment such that (i) $\sigma^x_a(x) = a$; and (ii) $\sigma^x_a(y) = \sigma(y)$ for every variable y different from x.

Definition 5 (Satisfaction) The satisfaction relation $\models$ with respect to a world $w \in \mathcal{M}$, a formula $\phi \in \mathcal{L}$, and a w-assignment σ is inductively defined as follows:

$$
\begin{array}{ll}
(\mathcal{M}^\sigma, w) \models P^n(t_1, \ldots, t_n) & \text{iff } \langle \sigma(t_1), \ldots, \sigma(t_n) \rangle \in I(P^n, w) \\
(\mathcal{M}^\sigma, w) \models E(t) & \text{iff } \sigma(t) \in d(w) \\
(\mathcal{M}^\sigma, w) \models \neg\psi & \text{iff } (\mathcal{M}^\sigma, w) \not\models \psi \\
(\mathcal{M}^\sigma, w) \models \psi \wedge \psi' & \text{iff } (\mathcal{M}^\sigma, w) \models \psi \text{ and } (\mathcal{M}^\sigma, w) \models \psi' \\
(\mathcal{M}^\sigma, w) \models \Box\psi & \text{iff for all } w' \in W, wRw' \text{ implies } (\mathcal{M}^\sigma, w') \models \psi \\
(\mathcal{M}^\sigma, w) \models \forall x\psi & \text{iff for all } a \in d(w), (\mathcal{M}^{\sigma^x_a}, w) \models \psi
\end{array}
$$

The truth conditions for formulas containing the symbols $\vee$, $\rightarrow$, $\leftrightarrow$, $\exists$, $\diamond$ are defined as standard. Also, the increasing outer domain condition on K-frames guarantees that it is always possible to evaluate modal formulas: if wRw' and σ is a w-assignment, then it is also a w'-assignment.

We say that a formula ϕ is *true at a world* w, or $(\mathcal{M}, w) \models \phi$, iff it is satisfied by every w-assignment σ; ϕ is *valid on a K-model* $\mathcal{M}$, or $\mathcal{M} \models \phi$, iff it is true at every world in $\mathcal{M}$; ϕ is *valid on a K-frame* $\mathcal{F}$, or $\mathcal{F} \models \phi$, iff it is valid on every K-model based on $\mathcal{F}$; ϕ is *valid on a class* $\mathcal{C}$ *of K-frames*, or $\mathcal{C} \models \phi$, iff it is valid on every K-frame belonging to $\mathcal{C}$.

In the following remarks we summarise the equivalences between the validity of the first-order modal principles presented in the introduction and specific conditions on Kripke frames. These equivalences express the intuitive meaning of these formulas in Kripke semantics, and will be used in the incompleteness results in Sect. 5. We refer to Fitting and Mendelsohn (1999), Hughes and Cresswell (1996) for details on proofs, in what follows we denote $D(w) \setminus d(w)$ as $\overline{d(w)}$.

Remark 1 For every K-frame $\mathcal{F}$,

$$
\begin{array}{llll}
\mathcal{F} \models CBF & \text{iff} & \mathcal{F} \text{ has increasing inner domains} & \text{iff} \quad \mathcal{F} \models NE \\
\mathcal{F} \models BF & \text{iff} & \mathcal{F} \text{ has decreasing inner domains} & \\
\mathcal{F} \models N\neg E & \text{iff} & wRw' \text{ implies } \overline{d(w)} \subseteq \overline{d(w')} &
\end{array}
$$

The converse of the Barcan formula CBF and the necessity of existence NE are semantically equivalent in Kripke semantics. Also, notice that if $\mathcal{F}$ has decreasing inner domains, then by the increasing outer domain condition, wRw' implies $\overline{d(w)} \subseteq \overline{d(w')}$. Therefore, any K-frame satisfying BF satisfies N¬E as well:

Remark 2 For every K-frame $\mathcal{F}$, if $\mathcal{F} \models BF$ then $\mathcal{F} \models N\neg E$

In general the converse is not true. However, if a K-frame $\mathcal{F}$ has constant outer domains, then it is easy to check that wRw' implies $\overline{d(w)} \subseteq \overline{d(w')}$ iff $\mathcal{F}$ has decreasing inner domains. Hence, the following result holds:

Remark 3 For every K-frame $\mathcal{F}$ with constant outer domains,

$$\mathcal{F} \models BF \qquad \text{iff} \qquad \mathcal{F} \models N\neg E$$

To conclude, in Kripke semantics BF entails N¬E. Further, in K-frames with constant outer domains, the two formulas are actually equivalent. The incompleteness results for QML systems based on free logic and containing BF in Sect. 5 show that this equivalence is semantical, but not proof-theoretical.

3 QML systems

In this section we introduce a class of QML systems based on free logic (Bencivenga, 2002). Then, we briefly review the completeness results available in the literature to set the stage for the independence and incompleteness results in Sect. 5. Besides BF, CBF, NE, and N¬E, we consider the following axioms and rules:

Taut	tautologies of classical propositional calculus
Dist	$\Box(\phi \to \psi) \to (\Box\phi \to \Box\psi)$
T	$\Box\phi \to \phi$
4	$\Box\phi \to \Box\Box\phi$
B	$\phi \to \Box\Diamond\phi$
MP	$\phi \to \psi, \phi \Rightarrow \psi$
Nec	$\phi \Rightarrow \Box\phi$
E-Ex	$\forall x\phi \to (E(y) \to \phi[x/y])$
E-Gen	$\phi \to (E(x) \to \psi) \Rightarrow \phi \to \forall x\psi$, where x is not free in ϕ

The axioms from *Taut* to *Nec* define the normal systems of modal logic; while axioms *E-Ex* and *E-Gen* formalise a sound theory of quantification for free logic (Bencivenga, 2002). For technical reasons we consider also the following postulates on the interaction between substitution and logical constants in our languages:

Subst	$P^n(t_1, \ldots, t_n)[\mathbf{y}/\mathbf{t}] \leftrightarrow P^n(t_1[\mathbf{y}/\mathbf{t}], \ldots, t_n[\mathbf{y}/\mathbf{t}])$
	$(\neg\psi)[\mathbf{y}/\mathbf{t}] \leftrightarrow \neg(\psi[\mathbf{y}/\mathbf{t}])$
	$(\psi \wedge \psi')[\mathbf{y}/\mathbf{t}] \leftrightarrow (\psi[\mathbf{y}/\mathbf{t}] \wedge \psi[\mathbf{y}/\mathbf{t}])$
	$(\Box\psi)[\mathbf{y}/\mathbf{t}] \leftrightarrow \Box(\psi[\mathbf{y}/\mathbf{t}])$
	$(\forall x\psi)[\mathbf{y}/\mathbf{t}] \leftrightarrow \forall x(\psi[\mathbf{y}/\mathbf{t}])$

where we assume without loss of generality that $x \neq y_i$ and $x \neq t_i$ for every $i \leq n$.

We can now introduce the QML systems.

Definition 6 The QML system $Q^E \cdot K$ includes the schemes of axioms *Taut, Dist, Subst, E-Ex*, and the inference rules *MP, Nec, E-Gen.*

We consider the standard definitions of *proof* and *theorem*: $S \vdash \phi$ means that ϕ is a theorem in the system *S*. The basic system $Q^E \cdot K$ can be extended by adding either BF, CBF, N¬E, or NE. Below we list all our QML systems with equivalences.

QML systems		
$Q^E \cdot K$		
$Q^E \cdot K$+BF		
$Q^E \cdot K$+N¬E		
$Q^E \cdot K$+CBF	$\equiv$	$Q^E \cdot K$+NE
$Q^E \cdot K$+CBF+BF	$\equiv$	$Q^E \cdot K$+NE+BF
$Q^E \cdot K$+CBF+N¬E	$\equiv$	$Q^E \cdot K$+NE+N¬E

Notice that each QML system containing CBF is equivalent to the corresponding system with NE, as the two formulas are proof-theoretically equivalent in $Q^E \cdot K$. So, in what follows we consider only QML calculi with CBF. Also, we do not consider the systems $Q^E \cdot K$+BF+N¬E and $Q^E \cdot K$+CBF+BF+N¬E, as their completeness can be proved by adding an extended version of the Barcan formula that first appeared in Thomason (1980). However, this is beyond the scope of the paper.

The QML systems above are defined on the modal base K. For obtaining systems on the stronger modal bases T, $S4$, B and $S5$, we just add a suitable combination of axioms T-B. For instance, $Q^E \cdot B$+CBF extends $Q^E \cdot K$+CBF with axioms T and B.

Given a QML system S and a K-frame $\mathcal{F}$, we say that *$\mathcal{F}$ is a K-frame for S*, or S-frame, iff $\mathcal{F} \models S$, i.e., $\mathcal{F} \models \phi$ whenever ϕ is a theorem in S. Further, $S \models \phi$ iff for every S-frame $\mathcal{F}$, $\mathcal{F} \models \phi$. By Remark 1 it is straightforward to obtain the following result, so we omit the proof.

Remark 4 A K-frame $\mathcal{F}$ is a frame for any QML system S in the first column iff $\mathcal{F}$ satisfies the constraints on inner domains in the second column:

QML system	*inner domain*
$Q^E \cdot K$	varying
$Q^E \cdot K$+BF	decreasing
$Q^E \cdot K$+N¬E	$wRw' \Rightarrow \overline{d(w)} \subseteq \overline{d(w')}$
$Q^E \cdot K$+CBF	increasing
$Q^E \cdot K$+CBF+BF	constant
$Q^E \cdot K$+CBF+N¬E	increasing, $wRw' \Rightarrow \overline{d(w)} \subseteq \overline{d(w')}$

In particular, by Remark 2 we have that $Q^E \cdot K$+BF $\models$ N¬E, but we will show that $Q^E \cdot K$+BF does not prove N¬E. This observation is the basis for the independence and incompleteness results in Sect. 5.

Further, by induction on the length of proofs it is easy to check that for every QML system S, if a formula ϕ is a theorem of S, then ϕ holds in the class of S-frames, that is, $S \vdash \phi$ implies $S \models \phi$. As a consequence, we state without proof the following soundness results.

Remark 5 (Soundness) Every QML system S is sound with respect to its class of S-frames.

As regards completeness, the proofs for most systems are well-known (Corsi, 2002; Fitting & Mendelsohn, 1999; Hughes & Cresswell, 1996), while the Kripke-incompleteness of $Q^E \cdot K$+BF and $Q^E \cdot K$+CBF+BF has been only partially addressed in Goldblatt (2011). Formally, a QML system S is Kripke-incomplete if there is no class $\mathcal{C}_S$ of K-frames such that $\mathcal{C}_S \models \phi$ iff $S \vdash \phi$.

In the following remark we summarise these (in)completeness results; we omit the completeness proofs as these would go beyond the scope of the paper, and refer to the works cited above.

Remark 6 (Completeness) The QML systems in the first column are complete with respect to the classes of K-frames satisfying the constraints on inner domains in the second column:

QML system	*inner domain*
$Q^E \cdot K$	varying
$Q^E \cdot K$+BF	incomplete
$Q^E \cdot K$+N¬E	$wRw' \Rightarrow \overline{d(w)} \subseteq \overline{d(w')}$
$Q^E \cdot K$+CBF	increasing
$Q^E \cdot K$+CBF+BF	incomplete
$Q^E \cdot K$+CBF+N¬E	increasing, $wRw' \Rightarrow \overline{d(w)} \subseteq \overline{d(w')}$

Similar soundness and completeness results hold also for the QML systems on modal bases T and $S4$ with respect to the corresponding classes of reflexive and reflexive, transitive K-frames respectively.

As to QML systems on modal base B, we have the following equivalence and completeness results. Also in this case we omit proofs.

Remark 7 (Completeness) The QML systems in the first column are complete with respect to the classes of reflexive and symmetric K-frames with constant outer domains, that also satisfy the constraints on inner domains in the second column:

QML system	*inner domain*
$Q^E \cdot B$	varying
$Q^E \cdot B$+BF	incomplete
$Q^E \cdot B + N\neg E \equiv Q^E \cdot B + CBF$ $\equiv Q^E \cdot B + CBF + BF$ $\equiv Q^E \cdot B + CBF + N\neg E$	constant

Notice that on modal base B all systems are equivalent, but $Q^E \cdot B$ and $Q^E \cdot B$+BF. The Kripke-incompleteness of $Q^E \cdot B$+BF is proved in Sect. 5. Similar results hold also for the QML systems on $S5$ with respect to reflexive, symmetric and transitive K-frames.

4 Counterpart Semantics

The counterpart semantics for QML can be seen as the result of two converging lines of research. The former, with a more philosophical flavour, originates from Lewis's *theory of counterparts* (Lewis, 1979). One of the key features of this account is the *notion of counterpart*, which follows from Lewis's rejection of trans-world identity. The latter line of research consists in a series of contributions on alternative semantics for QML based on sophisticated mathematical structures: categories, toposes, and presheaves (Brauner & Ghilardi, 2007; Ghilardi & Meloni, 1988; Skvortsov & Shehtman, 1993). Counterpart semantics generalizes these structures, while providing a formalisation of Lewis's theory of counterparts.

In Brauner and Ghilardi (2007), Corsi (2001, 2003, 2009), Kracht and Kutz (2001, 2002) various semantics for QML based on counterparts have been introduced. We build on these contributions and present counterpart frames for QML.

Definition 7 (c-frame) A *counterpart frame* is a tuple $\mathcal{F} = \langle W, R, D, d, C \rangle$ such that

- W, R and d are defined as in Def. 2;
- D is a function such that for every $w \in W$, $D(w)$ is a non-empty set;
- C is a function such that for wRw', $C_{w,w'}$ is a subset of $D(w) \times D(w')$.

The intuitive meaning of the elements in a counterpart frame is similar to Kripke frames. Moreover, each $C_{w,w'}$ defines the counterpart relation between the individuals in $D(w)$ and $D(w')$. A K-frames is then a c-frames where the counterpart relation is *total*, i.e., for all $a \in D(w)$, wRw' implies there is $b \in D(w')$ such that $C_{w,w'}(a, b)$; and it is the identity relation:

Lemma 1 Let $\mathcal{F} = \langle W, R, D, d, C \rangle$ be a c-frame. If (i) the counterpart relation is total; and (ii) it is the identity relation, then the restriction $\mathcal{F}' = \langle W, R, D, d \rangle$ of $\mathcal{F}$ is a K-frame.

Proof (Sketch of proof) Notice that (i) and (ii) are sufficient to ensure that $\mathcal{F}'$ has increasing outer domains. □

We need some definitions to introduce the relevant classes of counterpart frames to be analysed in the rest of the paper. First, $id = \{(a, a) \mid a \in D(w)\}$ is the identity relation on $D(w)$; the composition $C_{w,w'} \circ C_{w',w''}$ is the set of pairs (a, c) such that $C_{w,w'}(a, b)$ and $C_{w',w''}(b, c)$ for some $b \in D(w')$; the converse relation $\breve{C}_{w,w'}$ is the set of pairs (b, a) such that $(a, b) \in C_{w,w'}$. In what follows we focus on the following classes of c-frames:

Classical	for every $w \in W, d(w) = D(w)$
Existentially faithful	if wRw', $a \in d(w)$, and $C_{w,w'}(a, b)$, then $b \in d(w')$
Fictionally faithful	if wRw', $a \in \overline{d(w)}$, and $C_{w,w'}(a, b)$, then $b \in \overline{d(w')}$
Total	if wRw' and $a \in D(w)$, then there is $b \in D(w')$ such that $C_{w,w'}(a, b)$
Surjective	if wRw' and $b \in d(w')$, then there is $a \in d(w)$ such that $C_{w,w'}(a, b)$
Functional	if wRw', $C_{w,w'}(a, b)$ and $C_{w,w'}(a, b')$, then $b = b'$
Reflexive	R is reflexive
	and for every $w \in W$, $C_{w,w} \supseteq id$
Transitive	R is transitive and
	for every $w, w', w'' \in W$, $C_{w,w'} \circ C_{w',w''} \subseteq C_{w,w''}$
Symmetric	R is symmetric and
	for every $w, w' \in W$, $\breve{C}_{w,w'} \subseteq C_{w',w}$

We briefly compare Def. 7 with some other notions of counterpart frame appearing in the literature. In Brauner and Ghilardi (2007), Corsi (2001) the presentation is restricted to classical c-frames, which are existentially and fictionally faithful by definition. Non-classical c-frames are discussed in Kracht and Kutz (2002, 2001), but the authors assume the *counterpart-existence property*, which is tantamount to totality. Non-classical c-frames, in which the counterpart relation is not total, are considered in Corsi (2003).

We now introduce the counterpart models for first-order modal logic.

Definition 8 (c-model) A *counterpart model* based on a c-frame $\mathcal{F}$ is a pair $\mathcal{M} = \langle \mathcal{F}, I \rangle$ where I is a first-order interpretation defined as in Def. 4.

If we define satisfaction of modal formulas in counterpart models as in Section 2, we run into troubles. Since counterpart frames do not have increasing outer domains, if wRw' and σ is a w-assignment it does not necessarily follow that σ is also a w'-assignment. A number of approaches have been put forward to solve this issue, we refer to Corsi (2001) for a thorough discussion. In what follows we consider languages with indexed modalities Corsi (2009).

If $z_1, \ldots, z_n$ are distinct variables, and $s_1, \ldots, s_n$ are individual terms[3], then $\left[\begin{smallmatrix} s_1 \\ z_1 \end{smallmatrix} \cdots \begin{smallmatrix} s_n \\ z_n \end{smallmatrix}\right]$ is an indexed modal operator. We now define indexed formulas.

Definition 9 The first-order modal formulas in the language $\mathcal{L}_{ind}$ are defined in Backus-Naur form as follows:

$$\phi ::= P^n(t_1, \ldots, t_n) \mid E(t) \mid \neg\psi \mid \psi \wedge \psi' \mid \forall x \psi \mid \left[\begin{smallmatrix} s_1 \\ z_1 \end{smallmatrix} \cdots \begin{smallmatrix} s_n \\ z_n \end{smallmatrix}\right]\psi$$

where $z_1, \ldots, z_n$ appear free in ψ.

The symbols $\vee, \rightarrow, \leftrightarrow, \exists$ are defined as standard. The formula $\left\langle\begin{smallmatrix} s_1 \\ z_1 \end{smallmatrix} \cdots \begin{smallmatrix} s_n \\ z_n \end{smallmatrix}\right\rangle\psi$ is a shorthand for $\neg\left[\begin{smallmatrix} s_1 \\ z_1 \end{smallmatrix} \cdots \begin{smallmatrix} s_n \\ z_n \end{smallmatrix}\right]\neg\psi$. Also, for $n = 0$ we write $[\,]\psi$ and $\langle\,\rangle\psi$. Finally, $[z_1 \ldots z_n]\psi$ and $\langle z_1 \ldots z_n\rangle\psi$ stand for $\left[\begin{smallmatrix} z_1 \\ z_1 \end{smallmatrix} \cdots \begin{smallmatrix} z_n \\ z_n \end{smallmatrix}\right]\psi$ and $\left\langle\begin{smallmatrix} z_1 \\ z_1 \end{smallmatrix} \cdots \begin{smallmatrix} z_n \\ z_n \end{smallmatrix}\right\rangle\psi$ respectively. Intuitively, the formula $[z_1 \ldots z_n]\psi$ reads as "it is necessary for $z_1 \ldots z_n$ that ψ".

If $\phi \in \mathcal{L}_{ind}$, $y_1 \ldots, y_n$ are distinct variables, and $t_1, \ldots, t_n$ are individual terms, the substituted formula $\phi[\mathbf{y}/\mathbf{t}]$ is inductively defined as for the language $\mathcal{L}$, but for the case of the modal operator:

$$\left(\left[\begin{smallmatrix} s_1 \\ z_1 \end{smallmatrix} \cdots \begin{smallmatrix} s_n \\ z_n \end{smallmatrix}\right]\psi\right)[\mathbf{y}/\mathbf{t}] = \left[\begin{smallmatrix} s_1[\mathbf{y}/\mathbf{t}] \\ z_1 \end{smallmatrix} \cdots \begin{smallmatrix} s_n[\mathbf{y}/\mathbf{t}] \\ z_n \end{smallmatrix}\right]\psi$$

In particular we have that $([y_1 \ldots y_n]\psi)[\mathbf{y}/\mathbf{t}] = \left[\begin{smallmatrix} t_1 \\ y_1 \end{smallmatrix} \cdots \begin{smallmatrix} t_n \\ y_n \end{smallmatrix}\right]\psi$. As for non-indexed languages, we assume that substitution binds stronger than any other logical symbol.

---

[3] As mentioned in Sect. 2, in the present setting, also $s_1, \ldots, s_n$ are variables, as variables are the only terms. However, we consider substitution for terms in general to keep open the possibility of extending the language so as to include other terms besides variables.

Definition 10 The satisfaction relation $\models$ for a world $w \in \mathcal{M}$, a formula $\phi \in \mathcal{L}_{ind}$, and a w-assignment σ is defined as in Def. 5 but for the case of indexed modalities:

$$(\mathcal{M}^{\sigma}, w) \models \begin{bmatrix} s_1 & \ldots & s_n \\ z_1 & \ldots & z_n \end{bmatrix} \psi \text{ iff for every } w' \in W,\ w'\text{-assigment } \sigma', \text{ if } wRw' \text{ and } C_{w,w'}(\sigma(s_i), \sigma'(z_i)) \text{ for } i \leq n, \text{ then } (\mathcal{M}^{\sigma'}, w') \models \psi$$

Finally, a formula $\phi \in \mathcal{L}_{ind}$ is said to be *true at a world* w iff it is satisfied by every w-assignment; ϕ is *valid on a c-model* $\mathcal{M}$ iff it is true in every world in $\mathcal{M}$; ϕ is *valid on a c-frame* $\mathcal{F}$ iff it is valid on every c-model based on $\mathcal{F}$; ϕ is *valid on a class* $\mathcal{C}$ *of c-frames* iff it is valid on every c-frame belonging to $\mathcal{C}$.

Hereafter we consider the indexed version of the first-order modal principles introduced in Sect. 1:

BF_{ind}	$\forall z\, [x_1 \ldots x_n, z]\, \phi \to [x_1 \ldots x_n]\, \forall z \phi$	Barcan formula
CBF_{ind}	$[x_1 \ldots x_n]\, \forall z \phi \to \forall z\, [x_1 \ldots x_n, z]\, \phi$	converse of the Barcan formula
NE_{ind}	$E(x) \to [x]\, E(x)$	necessity of existence
$\text{N}\neg\text{E}_{ind}$	$\neg E(x) \to [x]\, \neg E(x)$	necessity of fictionality

We now list the equivalences between the validity of our first-order modal principles on a c-frame $\mathcal{F}$ and the constraints on $\mathcal{F}$.

Remark 8 For every c-frame $\mathcal{F}$,

$\mathcal{F} \models CBF_{ind}$	iff	$\mathcal{F}$ is existentially faithful	iff	$\mathcal{F} \models NE_{ind}$
$\mathcal{F} \models N\neg E_{ind}$	iff	$\mathcal{F}$ is fictionally faithful		
$\mathcal{F} \models BF_{ind}$	iff	$\mathcal{F}$ is surjective		

In counterpart semantics the converse of the Barcan formula is semantically equivalent to the necessity of existence, as it is the case in Kripke semantics. However, differently from Kripke semantics, it is not the case that the Barcan formula semantically entails the necessity of fictionality. In fact, there are surjective c-frames that are not fictionally faithful (see Sect. 5). This remark is the starting point for the incompleteness proof in the following section.

Finally, we notice that in counterpart semantics it is not the case that substitution commutes with the modal operator:

$$\models \begin{bmatrix} s_1 & \ldots & s_n \\ z_1 & \ldots & z_n \end{bmatrix} \phi \to [v_1 \ldots v_k](\phi[z_1/s_1, \ldots, z_n/s_n]) \tag{1}$$

$$\not\models [v_1 \ldots v_k](\phi[z_1/s_1, \ldots, z_n/s_n]) \to \begin{bmatrix} s_1 & \ldots & s_n \\ z_1 & \ldots & z_n \end{bmatrix} \phi \tag{2}$$

where $v_1, \ldots, v_k$ are all the variables occurring in $s_1, \ldots, s_n$ without repetitions.

The first formula is a validity, while the second is not. However, for total and functional c-frames we have the following result (Brauner & Ghilardi, 2007):

Lemma 2 Let $\mathcal{F}$ be a total and functional c-frame, then

$$\mathcal{F} \models \left[\begin{smallmatrix} s_1 \\ z_1 \end{smallmatrix} \ldots \begin{smallmatrix} s_n \\ z_n \end{smallmatrix}\right]\phi \leftrightarrow [v_1 \ldots v_k](\phi[z_1/s_1, \ldots, z_n/s_n])$$

Proof We only have to prove that (2) holds, so assume that $(\mathcal{M}^\sigma, w) \models [v_1 \ldots v_k](\phi[\mathbf{z}/\mathbf{s}])$, that is, if wRw' and $C_{w,w'}(\sigma(v_i), \sigma'(v_i))$ for every $i \leq k$, then $(\mathcal{M}^{\sigma'}, w') \models \phi[\mathbf{z}/\mathbf{s}]$. Equivalently, if wRw' and $C_{w,w'}(\sigma(v_i), \sigma'(v_i))$ for every $i \leq k$, then $(\mathcal{M}^{\sigma'\left(\begin{smallmatrix} z_1 \ldots z_n \\ \sigma'(s_1) \ldots \sigma'(s_n) \end{smallmatrix}\right)}, w') \models \phi$. To prove that $(\mathcal{M}^\sigma, w) \models \left[\begin{smallmatrix} s_1 \\ z_1 \end{smallmatrix} \ldots \begin{smallmatrix} s_n \\ z_n \end{smallmatrix}\right]\phi$, suppose that $C_{w,w'}(\sigma(s_j), \sigma'(z_j))$ for every $j \leq n$. Since $C_{w,w'}(\sigma(v_i), \sigma'(v_i))$ for every $i \leq k$, and $v_1, \ldots, v_k$ are all the variables occurring in $s_1, \ldots, s_n$, we have that $C_{w,w'}(\sigma(s_j), \sigma'(s_j))$ for every $j \leq n$. Further, by functionality and totality we obtain that $\sigma'(s_j) = \sigma'(z_j)$ for every $j \leq n$. By the assumption it follows that $(\mathcal{M}^{\sigma'\left(\begin{smallmatrix} z_1 \ldots z_n \\ \sigma'(z_1) \ldots \sigma'(z_n) \end{smallmatrix}\right)}, w') \models \phi$. Hence, $(\mathcal{M}^{\sigma'}, w') \models \phi$. This means that $(\mathcal{M}^\sigma, w) \models \left[\begin{smallmatrix} s_1 \\ x_1 \end{smallmatrix} \ldots \begin{smallmatrix} s_n \\ x_n \end{smallmatrix}\right]\phi$. $\square$

The results above will be used in the following section.

5 Independence and Incompleteness in Kripke Semantics

In this section we prove the independence of the Barcan formula and the necessity of fictionality for all normal modalities. From this result we derive that a class of QML systems based on free logic and containing BF is Kripke-incomplete. Specifically, in Sects. 5 and 5 we prove that BF and N¬E are mutually independent over QML systems based on modalities K, T, or $S4$. This result has already appeared in Goldblatt (2011) in a different setting. Here we provide a counterpart-theoretic proof, which is then extended in Sect. 5 to QML systems based on B and $S5$, which are not considered in Goldblatt (2011). The incompleteness proofs are inspired to similar results that first appeared in Corsi (2002), and make use of techniques and lemmas in Corsi (2003).

QML Systems over K

We start by proving that BF and N¬E are mutually independent in QML systems based on modality K, as it allows us to focus on the relevant notions and features of the construction, while discarding the details related to the specific modality considered. From this result we derive that the QML systems $Q^E \cdot K$+BF and $Q^E \cdot K$+CBF+BF are Kripke-incomplete, that is, there is no class of Kripke frames that validates all and only the theorems of $Q^E \cdot K$+BF (resp. $Q^E \cdot K$+CBF+BF). We immediately state the main result of this section.

Theorem 1 The Barcan formula BF and the necessity of fictionality N¬E are mutually independent in $Q^E \cdot K$, that is, $Q^E \cdot K$+BF $\nvDash$ N¬E and $Q^E \cdot K$+N¬E $\nvDash$ BF.

To prove Theorem 1 we need some preliminary results. Following Corsi (2003), Ghilardi (1990) we introduce a translation function from unindexed to indexed first-order modal languages.

Definition 11 For every formula $\phi \in \mathcal{L}$, the formula $\tau(\phi) \in \mathcal{L}_{ind}$ is inductively defined as follows:

$$\begin{array}{lll} \tau(P^m(t_1,\ldots,t_m)) & := & P^m(t_1,\ldots,t_m) \\ \tau(\neg\psi) & := & \neg\tau(\psi) \\ \tau(\psi \wedge \psi') & := & \tau(\psi) \wedge \tau(\psi') \\ \tau(\forall x\psi) & := & \forall x\tau(\psi) \\ \tau(\Box\psi) & := & [y_1 \ldots y_n]\tau(\psi) \end{array}$$

where $y_1, \ldots, y_n$ are all the free variables in ψ without repetitions and in alphabetical order.

We now prove a result on the relation between substitution and the translation function τ.

Lemma 3 Let $\mathcal{F}$ be an everywhere-defined and functional c-frame, and $\phi \in \mathcal{L}$, then

$$\mathcal{F} \models \tau(\phi)[\mathbf{y}/\mathbf{t}] \leftrightarrow \tau(\phi[\mathbf{y}/\mathbf{t}])$$

Proof The proof is by induction on the length of ϕ. The base case for ϕ atomic as well as the cases for propositional connectives are straightforward.

If $\phi = \forall x\psi$, then for every c-model $\mathcal{M}$ based on $\mathcal{F}$ and world w,

$$\begin{array}{ll} (\mathcal{M}, w) \models (\tau(\forall x\psi))[\mathbf{y}/\mathbf{t}] & \text{iff by definition of } \tau \\ (\mathcal{M}, w) \models (\forall x\tau(\psi))[\mathbf{y}/\mathbf{t}] & \text{iff by definition of substitution} \\ (\mathcal{M}, w) \models \forall x(\tau(\psi)[\mathbf{y}/\mathbf{t}]) & \text{iff by the induction hypothesis} \\ (\mathcal{M}, w) \models \forall x(\tau(\psi[\mathbf{y}/\mathbf{t}])) & \text{iff by definition of } \tau \\ (\mathcal{M}, w) \models \tau(\forall x(\psi[\mathbf{y}/\mathbf{t}])) & \text{iff by definition of substitution} \\ (\mathcal{M}, w) \models \tau((\forall x\psi)[\mathbf{y}/\mathbf{t}]) & \end{array}$$

where we assumed without loss of generality that $x \neq y_i$ and $x \neq t_i$ for every $i \leq n$.

For $\phi = \Box\psi$ we have that, for every c-model $\mathcal{M}$ based on $\mathcal{F}$ and world w,

$$
\begin{array}{ll}
(\mathcal{M}, w) \models (\tau(\Box\psi))[\mathbf{y}/\mathbf{t}] & \text{iff by definition of } \tau \\
(\mathcal{M}, w) \models ([y_1 \dots y_n]\ \tau(\psi))[\mathbf{y}/\mathbf{t}] & \text{iff by definition of substitution} \\
(\mathcal{M}, w) \models \left[\begin{smallmatrix} t_1 & & t_n \\ y_1 & \dots & y_n \end{smallmatrix}\right] \tau(\psi) & \text{iff by Lemma 2} \\
(\mathcal{M}, w) \models [v_1 \dots v_n]\ \tau(\psi)[\mathbf{y}/\mathbf{t}] & \text{iff by induction hypothesis} \\
(\mathcal{M}, w) \models [v_1 \dots v_n]\ \tau(\psi[\mathbf{y}/\mathbf{t}]) & \text{iff by definition of } \tau \\
(\mathcal{M}, w) \models \tau(\Box(\psi[\mathbf{y}/\mathbf{t}])) &
\end{array}
$$

where we assumed without loss of generality that $y_1, \dots, y_n$ are all the free variables in ψ, and $v_1, \dots, v_k$ are all the variables occurring in $t_1, \dots, t_n$ without repetitions. □

We make use of Lemma 3 to show that the translations of QML theorems according to τ hold in specific classes of c-frames.

Lemma 4 Let $\phi \in \mathcal{L}$ and let $\mathcal{F}$ be a total, functional c-frame,

- if $\mathcal{F}$ is also surjective, then $Q^E \cdot K$+BF$\vdash \phi$ implies $\mathcal{F} \models \tau(\phi)$.
- if $\mathcal{F}$ is also fictionally faithful, then $Q^E \cdot K$+N¬E$\vdash \phi$ implies $\mathcal{F} \models \tau(\phi)$.

Proof By induction on the length of the proof of ϕ. Notice that by Remark 8 the translations $\tau(BF) = BF_{ind}$ of the Barcan formula and $\tau(N\neg E) = N\neg E_{ind}$ of the necessity of fictionality hold in surjective and fictionally faithful c-frames respectively. So, we consider only axioms *E-Ex*, *E-Gen* and *Subst*, and refer to Corsi (2003) for the other postulates.

As regards *E-Ex*, we have to prove that $\mathcal{F}$ validates $\tau(\forall x\psi \rightarrow (E(y) \rightarrow \psi[x/y]))$, that is,

$$\mathcal{F} \models \forall x(\tau(\psi)[x/y]) \rightarrow (E(y) \rightarrow \tau(\psi[x/y]))$$

By Lemma 3, this is the case iff

$$\mathcal{F} \models \forall x(\tau(\psi)[x/y]) \rightarrow (E(y) \rightarrow \tau(\psi)[x/y]) \tag{3}$$

Notice that the formula in (3) is the indexed version of *E-Ex*. We remark that it holds in $\mathcal{F}$ as the interpretation of quantifiers is the same for Kripke- and counterpart frames.

As to *E-Gen*, suppose that $\mathcal{F}$ validates $\tau(\psi \rightarrow (E(x) \rightarrow \phi))$, where x does not appear in ψ, this means that

$$\mathcal{F} \models \tau(\psi) \rightarrow (E(x) \rightarrow \tau(\phi))$$

Since x does not appear in ψ, it does not appear in $\tau(\psi)$ either. Further, the typed version of *E-Gen* holds for c-frames (Corsi, 2003; Ghilardi, 1990). Hence,

$$\mathcal{F} \models \tau(\psi) \rightarrow \forall x(\tau(\phi))$$

As a result, $\mathcal{F}$ validates $\tau(\psi \rightarrow \forall x\phi)$.

The only non-trivial case for *Subst* is the modal operator. We prove that $\mathcal{F}$ validates $\tau((\Box\psi)[\mathbf{y}/\mathbf{t}] \leftrightarrow \Box(\psi[\mathbf{y}/\mathbf{t}]))$, that is,

$$\mathcal{F} \models \tau((\Box\psi)[\mathbf{y}/\mathbf{t}]) \leftrightarrow [v_1 \dots v_n]\tau(\psi[\mathbf{y}/\mathbf{t}])$$

where $v_1 \dots v_n$ are all the variables occurring in $t_1, \dots, t_n$ without repetitions.

By Lemma 3 this is the case iff

$$\mathcal{F} \models ([y_1 \dots y_n]\tau(\psi))[\mathbf{y}/\mathbf{t}] \leftrightarrow [v_1 \dots v_n]\tau(\psi[\mathbf{y}/\mathbf{t}])$$

by the definition of substitution,

$$\mathcal{F} \models \left[\begin{smallmatrix} t_1 \\ y_1 \end{smallmatrix} \dots \begin{smallmatrix} t_n \\ y_n \end{smallmatrix}\right]\tau(\psi) \leftrightarrow [v_1 \dots v_n]\tau(\psi[\mathbf{y}/\mathbf{t}])$$

and by Remark 2 this is equivalent to

$$\mathcal{F} \models \left[\begin{smallmatrix} t_1 \\ y_1 \end{smallmatrix} \dots \begin{smallmatrix} t_n \\ y_n \end{smallmatrix}\right]\tau(\psi) \leftrightarrow \left[\begin{smallmatrix} t_1 \\ y_1 \end{smallmatrix} \dots \begin{smallmatrix} t_n \\ y_n \end{smallmatrix}\right]\tau(\psi)$$

This completes the proof. □

If the system $Q^E \cdot K$+BF proved N¬E, then any total, surjective and functional c-frame would model $\tau(N\neg E) = N\neg E_{ind}$. But the latter fact is negated by the following lemma.

Lemma 5 There exists a total, functional and surjective c-frame $\mathcal{F}$ such that $\mathcal{F} \not\models N\neg E_{ind}$.

Proof Consider the c-frame $\mathcal{F}$ in Fig. 1 defined as follows:

- $W = \{w, w'\}$;
- $R = \{(w, w')\}$;
- $D(w) = \{a, a'\}$, $D(w') = \{b\}$;
- $d(w) = \{a\}$, $d(w') = \{b\}$;
- $C_{w,w'} = \{(a, b), (a', b)\}$.

By definition $\mathcal{F}$ is total, as both individuals a and a' in $D(w)$ have a counterpart b in $D(w')$; it is surjective, as b in $d(w')$ is a counterpart to a in $d(w)$; and functional. But $N\neg E_{ind}$ fails in $\mathcal{F}$ as it is not fictionally faithful: consider a c-model $\mathcal{M}$ based on $\mathcal{F}$, a

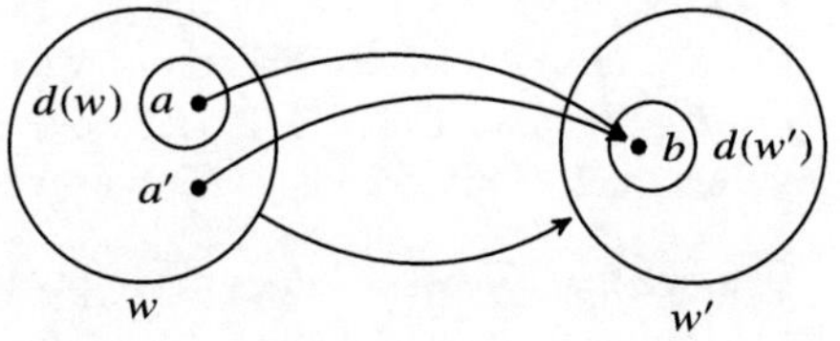

Fig. 1 A total, functional and surjective c-frame $\mathcal{F}$ such that $\mathcal{F} \not\models N\neg E_{ind}$

Fig. 2 A total, functional, and fictionally faithful c-frame $\mathcal{F}$ such that $\mathcal{F} \not\models BF_{ind}$

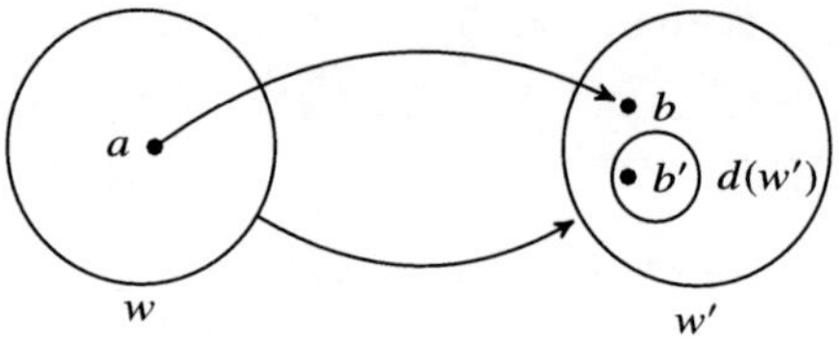

w-assignment σ such that $\sigma(x) = a'$ and $(\mathcal{M}^\sigma, w) \models \neg E(x)$, and a w'-assignment σ' such that $\sigma'(x) = b$. We have that $C_{w,w'}(\sigma(x), \sigma'(x))$ and $\sigma'(x) \in d(w')$, so $(\mathcal{M}^\sigma, w) \models \langle x \rangle E(x)$. Therefore, $(\mathcal{M}^\sigma, w) \models \neg E(x) \wedge \langle x \rangle E(x)$ and $\mathcal{F} \not\models N\neg E_{ind}$. □

We can prove a similar result for $Q^E \cdot K{+}N\neg E$ and *BF* as follows:

Lemma 6 There exists a total, functional, and fictionally faithful c-frame $\mathcal{F}$ such that $\mathcal{F} \not\models BF_{ind}$.

Proof Consider the c-frame $\mathcal{F}$ in Fig. 2 defined as follows:

- $W = \{w, w'\}$;
- $R = \{(w, w')\}$;
- $D(w) = \{a\}$, $D(w') = \{b, b'\}$;
- $d(w) = \emptyset$, $d(w') = \{b'\}$;
- $C_{w,w'} = \{(a, b)\}$.

By definition $\mathcal{F}$ is total, as $a \in D(w)$ has a counterpart b in $D(w')$; it is fictionally faithful, as $a \notin d(w)$ and $b \notin d(w')$; and functional. But BF_{ind} fails in $\mathcal{F}$ as it is not surjective: consider a c-model $\mathcal{M}$ based on $\mathcal{F}$ such that $I(P^1, w) = I(P^1, w') = \emptyset$, and a w-assignment σ such that $\sigma(x) = a$. We have that $(\mathcal{M}^\sigma, w) \models \forall x[x]P(x)$ trivially. Also, $C_{w,w'}(\sigma(x), \sigma'(x))$ for $\sigma'(x) = b$, and $b' \notin I(P^1, w')$. So, $(\mathcal{M}^{\sigma^{x}_{b'}}, w') \not\models P(x)$. Therefore, $(\mathcal{M}^{\sigma'}, w') \not\models \forall x P(x)$ and $(\mathcal{M}^\sigma, w) \not\models [\,]\forall x P(x)$. Thus, $\mathcal{F} \not\models BF_{ind}$. □

By Lemma 5 there is a total, functional, and surjective c-frame $\mathcal{F}$ that does not validate $\tau(N\neg E) = N\neg E_{ind}$. Hence, by Lemma 4 the system $Q^E \cdot K$+BF does not prove N¬E. Similarly, by Lemma 6 the system $Q^E \cdot K{+}N\neg E$ does not prove *BF*. As a result, *BF* and N¬E are mutually independent in $Q^E \cdot K$.

Finally, from Theorem 1 we derive the incompleteness of $Q^E \cdot K$+*BF*.

Theorem 2 The system $Q^E \cdot K$+*BF* is Kripke-incomplete.

Proof We have to show that every K-frame for $Q^E \cdot K$+*BF* validates $N\neg E$, but $Q^E \cdot K{+}BF \nvdash N\neg E$. By Theorem 1 we only need to prove that every K-frame for $Q^E \cdot K$+*BF* validates N¬E. Trivially every $Q^E \cdot K$+*BF*-frame $\mathcal{F}$ validates *BF*, and by Remark 2, $\mathcal{F}$ also validates N¬E as well. □

Notice that Theorem 1 does not imply that $Q^E \cdot K{+}N\neg E$ is Kripke-incomplete, as K-frames for $Q^E \cdot K{+}N\neg E$ do not necessarily validate *BF*.

Further incompleteness results

We now extend the independence and incompleteness results in the previous section. In what follows let S be any of the following QML systems: $Q^E \cdot K$+CBF, $Q^E \cdot T$, $Q^E \cdot S4$, $Q^E \cdot T$+CBF, and $Q^E \cdot S4$+CBF.

Theorem 3 The Barcan formula BF and the necessity of fictionality $N\neg E$ are mutually independent in every system S as above.

Proof Notice that Lemma 4 can be extended to every QML system S+BF (resp. S+$N\neg E$) with respect to the corresponding class of c-frames. For instance, if $Q^E \cdot S4{+}CBF{+}BF \vdash \phi$ then $\mathcal{F} \models \tau(\phi)$ whenever $\mathcal{F}$ is total, functional, surjective, existentially faithful, reflexive, and transitive. Similarly, if $Q^E \cdot S4{+}CBF{+}N\neg E \vdash \phi$ then $\mathcal{F} \models \tau(\phi)$ whenever $\mathcal{F}$ is total, functional, existentially and fictionally faithful, reflexive, and transitive. The proof is then immediate by Remark 8.

Further, we can extend Lemma 5 and 6 to the relevant classes of c-frames. As an example for Lemma 5, we can define a total, functional, existentially faithful, surjective, reflexive and transitive c-frame $\mathcal{F}$ such that $\mathcal{F} \not\models N\neg E_{ind}$ as in Fig. 3:

- $W = \{w, w'\}$;
- $R = \{(w, w'), (w, w), (w', w')\}$;
- $D(w) = \{a, a'\}$, $D(w') = \{b\}$;
- $d(w) = \{a\}$, $d(w') = \{b\}$;
- $C_{w,w'} = \{(a, b), (a', b)\}$, $C_{w,w} = \{(a, a), (a', a')\}$, $C_{w',w'} = \{(b, b)\}$.

Consider now a c-model $\mathcal{M}$ based on $\mathcal{F}$ and a w-assignment σ such that $\sigma(x) = a'$. We have $(\mathcal{M}^\sigma, w) \models \neg E(x) \wedge \langle x\rangle E(x)$, and therefore $\mathcal{F} \not\models N\neg E_{ind}$.

Similarly, for Lemma 6 we can define a total, functional, existentially and fictionally faithful, reflexive and transitive c-frame $\mathcal{F}$ such that $\mathcal{F} \not\models BF_{ind}$ as in Fig. 4:

- $W = \{w, w'\}$;
- $R = \{(w, w'), (w, w), (w', w')\}$;
- $D(w) = \{a\}$, $D(w') = \{b, b'\}$;
- $d(w) = \emptyset$, $d(w') = \{b'\}$;
- $C_{w,w'} = \{(a, b)\}$, $C_{w,w} = \{(a, a)\}$, $C_{w',w'} = \{(b, b), (b', b')\}$;

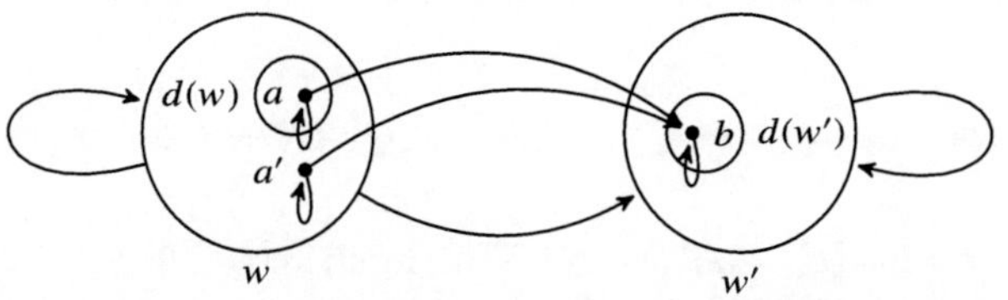

Fig. 3 A total, functional, existentially faithful, surjective, reflexive and transitive c-frame $\mathcal{F}$ such that $\mathcal{F} \not\models N\neg E_{ind}$

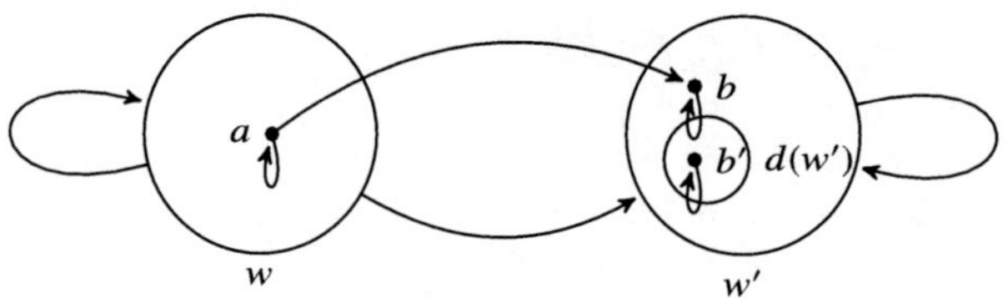

Fig. 4 A total, functional, existentially and fictionally faithful, reflexive and transitive c-frame $\mathcal{F}$ such that $\mathcal{F} \not\models BF_{ind}$

Also, consider a c-model $\mathcal{M}$ based on $\mathcal{F}$ such that $I(P^1, w) = I(P^1, w') = \emptyset$, and a w-assignment σ. We have that $(\mathcal{M}^\sigma, w) \models \forall x[x]P(x)$ but $(\mathcal{M}^\sigma, w) \not\models [\,]\forall x P(x)$. Thus, $\mathcal{F} \not\models BF_{ind}$.

Finally, by contraposition we obtain that $Q^E \cdot$ *S4+CBF+BF* $\nvdash N\neg E$ and $Q^E \cdot$ *S4+CBF+N¬E* $\nvdash BF$. Thus, *BF* and N¬E are mutually independent in $Q^E \cdot$ *S4+CBF*. □

From Theorem 3 we immediately derive the Kripke-incompleteness of every system *S+BF*.

Theorem 4 Every QML system *S+BF* is Kripke-incomplete.

Proof As for Theorem 2, we observe that every K-frame for *S+BF* validates N¬E, but *S+BF* $\nvdash N\neg E$. Because of Theorem 3, to derive incompleteness, we only need to prove that every K-frame for *S+BF* validates N¬E, which follows by Remark 2. □

Modal bases B and S5

In this section we prove the main original result of this paper. In Sect 3 we stated that the QML systems $Q^E \cdot$ *B+CBF+BF* and $Q^E \cdot$ *S5+CBF+BF* are Kripke-complete. Specifically, they both prove N¬E by using NE and axiom *B*. On the other hand, the following independence results hold for $Q^E \cdot B$ and $Q^E \cdot S5$:

Theorem 5 The necessity of fictionality $N\neg E$ is independent from the Barcan formula *BF* in $Q^E \cdot B$ and $Q^E \cdot S5$.

We prove this result by showing that neither $Q^E \cdot$ *B+BF* nor $Q^E \cdot$ *S5+BF* proves N¬E.

Lemma 7 The QML systems $Q^E \cdot$ *B+BF* and $Q^E \cdot$ *S5+BF* do not prove N¬E.

Proof We show that $Q^E \cdot$*S5+BF* $\nvdash N\neg E$; the result for $Q^E \cdot$*B+BF* trivially follows. Consider the c-frame $\mathcal{F}$ in Fig. 5 defined as follows:

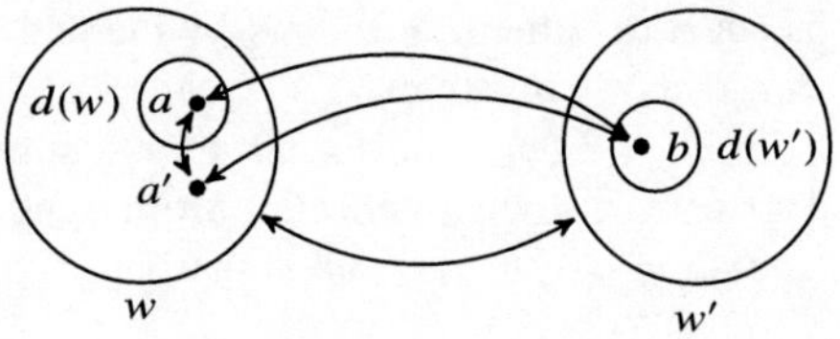

Fig. 5 A total, surjective, reflexive, symmetric, and transitive, frame $\mathcal{F}$. Reflexive arrows are omitted for clarity

- $W = \{w, w'\}$;
- $R = W^2$;
- $D(w) = \{a, a'\}$, $D(w') = \{b\}$;
- $d(w) = \{a\}$, $d(w') = \{b\}$;
- $C_{w,w'} = D(w) \times D(w')$, $C_{w',w} = D(w') \times D(w)$, $C_{w,w} = D(w)^2$, $C_{w',w'} = D(w')^2$.

By definition $\mathcal{F}$ is total, surjective, reflexive, symmetric, and transitive, but it is not functional as $C_{w',w}(b, a)$ and $C_{w',w}(b, a')$, so (2) is not valid on $\mathcal{F}$. We address this issue by defining a c-model $\mathcal{M} = \langle \mathcal{F}, I \rangle$ such that:

- for every predicate P^n, $I(P^n, w) = \{\mathbf{a} \mid \mathbf{a}$ is a n-tuple on the set $\{a, a'\}\}$; and $I(P^n, w') = \{\mathbf{b} \mid b \in D(w')\}$.

We now prove that the c-model $\mathcal{M}$ thus defined validates (2). □

Lemma 8 For every formula $\phi \in \mathcal{L}_{ind}$,

$$\mathcal{M} \models \left[\begin{smallmatrix} s_1 \\ z_1 \end{smallmatrix} \cdots \begin{smallmatrix} s_n \\ z_n \end{smallmatrix}\right]\phi \leftrightarrow [v_1 \dots v_k](\phi[z_1/s_1, \dots, z_n/s_n])$$

Proof We first prove that $(\mathcal{M}, w') \models \left[\begin{smallmatrix} s_1 \\ z_1 \end{smallmatrix} \cdots \begin{smallmatrix} s_n \\ z_n \end{smallmatrix}\right]\phi \leftrightarrow [v_1 \dots v_k](\phi[\mathbf{z}/\mathbf{s}])$. By definition this is the case iff for every w'-assignment σ',

$$(\mathcal{M}^{\sigma'}, w') \models \left[\begin{smallmatrix} s_1 \\ z_1 \end{smallmatrix} \cdots \begin{smallmatrix} s_n \\ z_n \end{smallmatrix}\right]\phi \leftrightarrow [v_1 \dots v_k](\phi[\mathbf{z}/\mathbf{s}])$$

that is,

for all $w^* \in W$ and w^*-assignments σ^*, if $w'Rw^*$ and $C_{w',w^*}(\sigma'(s_j), \sigma^*(z_j))$ for every $j \leq n$, then $(\mathcal{M}^{\sigma^*}, w^*) \models \phi$ iff for all $w^* \in W$ and w^*-assignments σ^*, if $w'Rw^*$ and $C_{w',w^*}(\sigma'(v_i), \sigma^*(v_i))$ for every $i \leq k$, then $(\mathcal{M}^{\sigma^*\left(\begin{smallmatrix} z_1 \dots z_n \\ \sigma^*(s_1) \dots \sigma^*(s_n) \end{smallmatrix}\right)}, w^*) \models \phi$

The case for $w^* = w'$ is straightforward, as the counterpart relation $C_{w',w'} = \{(b, b)\}$ is both total and functional.

If $w^* = w$, we prove the claim above by induction on the length of ϕ. Let ϕ be an atomic formula. The implication from left to right follows by (1). As to the converse, let ϕ be different from $E(x)$ and assume that $w'Rw$ and $C_{w',w}(\sigma'(s_j), \sigma(z_j))$ for every $j \leq k$, we have to show that $(\mathcal{M}^{\sigma}, w) \models \phi$. Notice that for every $j \leq k$ and $i \leq n$, $\sigma'(s_j) = b = \sigma'(v_i)$, therefore $C_{w',w}(\sigma'(v_i), \sigma(v_i))$ for $i \leq n$, as $\sigma(v_i)$ is either a or a'. Hence, $(\mathcal{M}^{\sigma\left(\begin{smallmatrix} z_1 \dots z_n \\ \sigma(s_1) \dots \sigma(s_n) \end{smallmatrix}\right)}, w) \models \phi$ by hypothesis. Further, each $\sigma(s_i)$

is equal to either a or a', and by the definition of the interpretation I for predicate constants we obtain that $(\mathcal{M}^{\sigma}, w) \models \phi$.

If $\phi = E(x)$, then for all w'-assignments σ', $(\mathcal{M}^{\sigma'}, w') \not\models [v](E(x)[\mathbf{z}/\mathbf{s}]) = [s]E(s)$. Thus, the implication from right to left is vacuously satisfied.

If $\phi = \neg\psi$, then for all w'-assignments σ',

$$(\mathcal{M}^{\sigma'}, w') \models [v_1 \ldots v_k](\neg\psi[\mathbf{z}/\mathbf{s}]) \Leftrightarrow (\mathcal{M}^{\sigma'}, w') \models [v_1 \ldots v_k]\neg(\psi[\mathbf{z}/\mathbf{s}])$$

$$\Leftrightarrow (\mathcal{M}^{\sigma'}, w') \not\models \langle v_1 \ldots v_k\rangle(\psi[\mathbf{z}/\mathbf{s}]) \quad (4)$$

$$\Leftrightarrow (\mathcal{M}^{\sigma'}, w') \not\models \left\langle {}^{s_1}_{z_1} \ldots {}^{s_n}_{z_n} \right\rangle \psi$$

$$\Leftrightarrow (\mathcal{M}^{\sigma'}, w') \models \left[{}^{s_1}_{z_1} \ldots {}^{s_n}_{z_n} \right] \neg\psi \quad (5)$$

where the step from (4) to (5) is obtained by the induction hypothesis.

The case for $\phi = \psi \wedge \psi'$ follows also by the induction hypothesis.

If $\phi = \left[{}^{t_1}_{y_1} \ldots {}^{t_m}_{y_m} \right]\psi$, then for all w'-assignments σ',

$$(\mathcal{M}^{\sigma'}, w') \models [v_1 \ldots v_k]((\left[{}^{t_1}_{y_1} \ldots {}^{t_m}_{y_m} \right]\psi)[\mathbf{z}/\mathbf{s}]) \leftrightarrow \quad [v_1 \ldots v_k]\left[{}^{t_1[\mathbf{z}/\mathbf{s}]}_{y_1} \ldots {}^{t_m[\mathbf{z}/\mathbf{s}]}_{y_m} \right]\psi \quad (6)$$

$$\leftrightarrow \left[{}^{v_1}_{y_1} \ldots {}^{v_k}_{y_k} \right](\psi[\mathbf{v}/\mathbf{t}[\mathbf{z}/\mathbf{s}]]) \quad (7)$$

$$\leftrightarrow [v_1 \ldots v_k](\psi[\mathbf{v}/\mathbf{t}[\mathbf{z}/\mathbf{s}]][\mathbf{y}/\mathbf{v}]) \quad (8)$$

$$\leftrightarrow [v_1 \ldots v_k](\psi[\mathbf{z}/\mathbf{t}][\mathbf{y}/\mathbf{s}]) \quad (9)$$

$$\leftrightarrow \left[{}^{s_1}_{y_1} \ldots {}^{s_n}_{y_n} \right](\psi[\mathbf{z}/\mathbf{t}]) \quad (10)$$

$$\leftrightarrow \left[{}^{s_1}_{z_1} \ldots {}^{s_n}_{z_n} \right]\left[{}^{t_1}_{y_1} \ldots {}^{t_m}_{y_m} \right]\psi \quad (11)$$

where step (6) follows by the definition of substitution; steps (7) and (11) follow by transitivity and reflexivity of the accessibility and counterpart relations; steps (8) and (10) are obtained by the induction hypothesis; while step (9) follows by reordering of variables.

Finally, let ϕ be equal to $\forall x\psi$, we have to prove that

for all w-assignments σ, for all $u \in d(w)$,
if $C_{w',w^*}(\sigma'(s_j), \sigma(z_j))$ for every $j \leq n$, then $(\mathcal{M}^{\sigma\binom{x}{u}}, w) \models \psi$ iff
for all w-assignments σ, for all $v \in d(w)$,
if $C_{w',w}(\sigma'(v_i), \sigma(v_i))$ for every $i \leq k$, then $(\mathcal{M}^{\sigma\binom{z_1 \ldots z_n, x}{\sigma(s_1) \ldots \sigma(s_n), v}}, w) \models \phi$

The implication from left to right follows from (1). As to the converse, assume that $u \in d(w)$ and $C_{w',w^*}(\sigma'(s_j), \sigma(z_j))$ for every $j \leq n$. Every $\sigma'(s_j)$ is equal to b, so we can find a w assignment σ'' such that $\sigma''(s_j) = \sigma(z_j)$ and $C_{w',w}(\sigma'(s_j), \sigma''(s_j))$ for $j \leq n$. Further, $C_{w',w}(b, u)$ for $u \in d(w)$. By hypothesis we have that $(\mathcal{M}^{\sigma''\binom{z_1 \dots z_n, x}{\sigma''(s_1) \dots \sigma''(s_n), v}}, w) \models \phi$, that is, $(\mathcal{M}^{\sigma^x_u}, w) \models \psi$.

The proof that $(\mathcal{M}, w) \models \left[{}^{s_1}_{z_1} \dots {}^{s_n}_{z_n} \right] \phi \leftrightarrow [v_1 \dots v_k](\phi[\mathbf{z}/\mathbf{s}])$ is similar, as a and a' are indistinguishable in w, but for the fact that $a \in d(w)$ and $a' \notin d(w)$.

The case for $w^* = w'$ is straightforward, as the counterpart relation $C_{w,w'} = \{(a, b), (a', b)\}$ is both total and functional.

If $w^* = w$, we prove the claim above by induction on the length of ϕ. Let ϕ be an atomic formula. The implication from left to right follows by (1). As to the converse, let ϕ be different from $E(x)$ and assume that wRw and $C_{w,w}(\sigma(s_j), \sigma(z_j))$ for every $j \leq n$, we have to show that $(\mathcal{M}^\sigma, w) \models \phi$. Notice that each $\sigma(s_j)$ is equal to either a or a', therefore $C_{w',w}(\sigma(s_j), \sigma(s_j))$ for every $j \leq n$. Hence, $(\mathcal{M}^{\sigma\binom{z_1 \dots z_n}{\sigma(s_1) \dots \sigma(s_n)}}, w) \models \phi$ by hypothesis. Further, each $\sigma(s_j)$ is equal to either a or a', and by the definition of the interpretation I for predicate constants we obtain that $(\mathcal{M}^\sigma, w) \models \phi$.

If $\phi = E(x)$, then for all w-assignments σ, $(\mathcal{M}^\sigma, w) \not\models [s]E(s)$. Thus, the implication from right to left is vacuously satisfied.

The inductive cases for the propositional connectives, the modal operator, and the universal quantifier are similar to above. This completes the proof. □

By using Lemma 8 we can prove the following version of Lemma 4.

Lemma 9 Let $\phi \in \mathcal{L}$, then $Q^E.S5 + BF \vdash \phi$ implies $\mathcal{M} \models \tau(\phi)$ □

Proof The proof is similar to Lemma 4. Axioms T, 4 and B hold because the c-frame $\mathcal{F}$ is reflexive, transitive and symmetric. Further, Lemma 8 is used to show that the translation of *Subst* according to τ holds in $\mathcal{M}$. □

Clearly, the c-model $\mathcal{M}$ above does not validate $\tau(N\neg E) = N\neg E_{ind}$, as $a' \notin d(w)$ but $C_{w,w'}(a', b)$ and $b \in d(w')$. Hence, $(\mathcal{M}^\sigma, w) \models \neg E(x) \wedge \langle x \rangle E(x)$ for $\sigma(x) = a$. Thus, by Lemma 9 the system $Q^E \cdot S5{+}BF$ does not prove $N\neg E$. As a result, Lemma 7 holds. □

Finally, by Lemma 7 we obtain Theorem 5. The following result immediately follows.

Theorem 6 The QML systems $Q^E \cdot B{+}BF$ and $Q^E \cdot S5{+}BF$ are Kripke-incomplete.

Proof The result follows from Theorem 5 by observing that N¬E holds in any K-frames for these QML systems by Remark 2. □

We compare the results for $Q^E \cdot B{+}BF$ and $Q^E \cdot S5{+}BF$ to those for $Q^E \cdot B{+}CBF{+}BF$ and $Q^E \cdot S5{+}CBF{+}BF$. In particular, we remarked that both $Q^E \cdot B{+}CBF{+}BF$ and $Q^E \cdot S5{+}CBF{+}BF$ are Kripke-complete, as they both prove N¬E.

6 Concluding Remarks

In Goldblatt (2011) the mutual independence of *BF* and N¬E is proved for the system *QES*4+*CBF*, i.e., the first-order extension of the modal logic *S*4 based on free logic and containing *CBF*. The proof can be extended to QML systems based on a modality as strong as *S*4.3+*M*. However, the independence proof here provided is different, as it makes use of counterpart semantics. This allows us to prove the independence of N¬E from *BF* for QML systems on modal bases *B* and *S*5. Further, we made use of these independence results to prove some incompleteness results that have not been explicitly considered in Goldblatt (2011), nor anywhere else in the literature to the best of our knowledge. In Garson (2005) the independence of *BF* from a system similar to $Q^E \cdot S5{+}N\neg E$ is tackled. However, the independence proof there provided is not correct.

The proofs given above make extensive use of counterpart semantics (Brauner & Ghilardi, 2007; Corsi, 2001), which has been further developed in a new setting in Corsi (2009). While we believe that our results still hold in the latter, more recent framework, we based our presentation on the former, older setting, possibly leaving the extension as future work.

Finally, the independence proof is inspired to Corsi (2003). However, there are differences, notably, we explicitly consider substitution as an axiom of QML systems. This has consequences on the commutativity of substitution and modal operators, as detailed in Lemmas 3 and 8.

In this paper we analysed Kripke and counterpart semantics for quantified modal logic. For each semantical account we checked the meaning of four principles well-known to the metaphysics of modality: the Barcan formula BF and its converse CBF, the necessity of existence NE, and the necessity of fictionality N¬E. We remarked that in Kripke semantics the Barcan formula semantically entails the necessity of fictionality, while in counterpart semantics the two principles are independent. We used this remark to give a counterpart-theoretic proof of the mutual independence of BF and N¬E for QML systems on a modality as strong as *S*4 (Goldblatt, 2011). Most importantly, we extended these independence results to QML systems on modal bases *B* and *S*5. Further, we proved an original incompleteness result, that is, there is a class of QML systems based on free logic and containing BF which is incomplete with respect to Kripke semantics.

In future work we aim at extending the techniques which have been successfully applied to the present independence and incompleteness results to other QML systems of interest. Specifically, the Kripke-(in)completeness of systems $Q^o \cdot K$+BF and $Q^o \cdot B$ (Corsi, 2002), which are based on Kripke's theory of quantification, is still an open (Belardinelli, 2006) problem.

Acknowledgements The work presented in this paper is based on the author's PhD thesis. Dr. Belardinelli would like to thank his PhD supervisor prof. M. Mugnai for his advice and guidance.

References

Belardinelli, F. (2006). Quantified modal logic and the ontology of physical objects. Ph.D. thesis, Scuola Normale Superiore.

Bencivenga, E. (2002). Free logics, pp. 147–196. Springer.

Brauner, T., & Ghilardi, S. (2007). First-order modal logic. In P. Blackburn, J. van Benthem, & F. Wolter (Eds.), *Handbook of modal logic* (pp. 549–620). Elsevier.

Corsi, G. (2001). Counterparts and possible worlds. A study on quantified modal logic. CLUEB. https://citeseer.ist.psu.eduhttps://corsi02counterpart.html

Corsi, G. (2002). A unified completeness theorem for quantified modal logics. *Journal of Symbolic Logic, 67*, 1483–1510.

Corsi, G. (2003). BF, CBF and Lewis semantics. *Logique et Analyse, 181*, 103–122.

Corsi, G. (2009). Necessary for. In C. Glymour, W. Wei, D. Westerstahl (Eds.), Proceedings of the 13th International Congress of Logic, Methodology and Philosophy of Science, pp. 162–184. College Publications.

Fitting, M., & Mendelsohn, R. L. (1999). *First-order modal logic*. Dordrecht: Kluwer.

Garson, J. (2001). Quantification in modal logic. In D. Gabbay & F. Guenthner (Eds.), *Handbook of philosophical logic* (Vol. 3, pp. 267–323). Dordrecht: Reidel.

Garson, J. (2005). Unifying quantified modal logic. *Journal of Philosophical Logic, 34*, 621–649.

Ghilardi, S. (1990). Modalità e categorie. Ph.D. thesis, Università di Milano.

Ghilardi, S. (1991). Incompleteness results in Kripke semantics. *Journal of Symbolic Logic, 56*, 516–538.

Ghilardi, S. (1992). Quantified extensions of canonical propositional intermediate logics. *Studia Logica, 51*, 195–214.

Ghilardi, S., Meloni, G. (1988). Modal and tense predicate logic. In F. Borceux (Ed.), Categorical algebras and its applications, pp. 130–142. Springer.

Goldblatt, R. (2011). *Quantifiers, propositions, and identity: Admissible semantics for quantified modal and substructural logics*. Cambridge: Cambridge University Press.

Hughes, G. E., & Cresswell, M. J. (1996). *A new introduction to modal logic*. London: Routledge.

Kracht, M., Kutz, O. (2001). The semantics of modal predicate logic II. In R. Kahle (Ed.), Intensionality, Lecture Notes in Logic, vol. 22. Association for Symbolic Logic, Los Angeles. www.math.fuberlin.de/user/kracht/

Kracht, M., Kutz, O. (2002). The semantics of modal predicate logic I. In F. Wolter, H. Wansing, M. de Rijke, M. Zakharyaschev (Eds.), Advances in Modal Logic, vol. 3. World Scientific Publishing.

Kripke, S. (1959). A completeness theorem in modal logic. *Journal of Symbolic Logic, 24*, 1–14.

Kripke, S. (1963). Semantical considerations on modal logic. *Acta Philosophica Fennica, 16*, 83–94.

Lewis, D. (1979). Counterpart theory and quantified modal logic. In M. Loux (Ed.), *The possible and the actual* (pp. 110–128). Ithaca, NY: Cornell University Press.

Loux, M. J. (Ed.). (1979). *The possible and the actual*. Ithaca, NY: Cornell University Press.

Lowe, E. J. (1998). *The possibility of metaphysics*. Oxford: Clarendon Press.

Skvortsov, D., & Shehtman, V. (1993). Maximal Kripke-type semantics for modal and superintuitionistic predicate logics. *Annals of pure and applied logic, 63*, 69–101.

Thomason, R. H. (1980). Some completeness results for modal predicate calculi. In K. Lambert (Ed.), *Philosophical problems in logic: Some recent developments* (pp. 56–76). Dordrecht: Reidel Publishing Company.

The Form of Practical Reasoning

Carla Bagnoli

Practical knowledge is aimed at the production of action, and it is an open question whether and how knowledge can generate action. An area of inquiry of special interest concerns practical thought and reasoning implicated in the production of moral action. The presumption of many objectivists engaging in this inquiry is that moral knowledge, understood as knowledge of the standard of right action, is a form of practical knowledge, capable of guiding and governing rational agents in what they do. The implication of this claim is that moral judgments are moral cognitions, that is, they have cognitive contents and constitutively aim at truth, though it is a matter of disagreement whether they are beliefs or special kinds of beliefs. What makes them special and raises the doubt that they cannot be beliefs is that moral judgments are understood to be motivating and action-guiding. If they are cognitions akin to beliefs, they are efficacious, that is, "practical" cognitions. Moral skeptics deny that there can be any objective standards of moral knowledge, precisely because they deny that moral judgements can be cognitions of this sort. Practical skeptics deny that there can be any practical truths. Instrumentalists are skeptical of the efficacy of reason, and thus oppose the rationalist construal of moral cognitions: they deny that practical reasoning and practical thought can directly guide action without and independently of the external intervention of desires or other conative states; and, hence, they deny that practical reasons can be efficacious. Of course, even objectivists admit that the practical truth and practical thought do not necessarily guide us. Indeed, as Henry Sidgwick remarks, this truism is the beginning of their enterprise: "We cannot help believing what we see to be true, but we can help doing what we see to be right or wise, and in fact often do what we know to be wrong or unwise: we are forced to notice the existence in us of irrational springs of action, conflicting with our knowledge and preventing its practical realization: and the very imperfectness of the connexion between our practical judgment and our will impels us to seek for

C. Bagnoli (✉)
University of Modena and Reggio Emilia, Reggio Emilia, Italy
e-mail: carla.bagnoli@unimore.it

F. Ademollo et al. (eds.), *Thinking and Calculating*, Logic, Epistemology, and the Unity of Science 54, https://doi.org/10.1007/978-3-030-97303-2_22

more precise knowledge as to the nature of that connexion".[1] The objectivist ethical theorist is thus in search of moral standards that could provide a firm basis of moral knowledge, understood as knowledge that generates action.

While the possibility of efficacious rational cognition is well grounded in the ancient traditions of practical thought, and has been endorsed by neo-Aristotelians and virtue-ethicists, in contemporary debates many have found this idea baffling, if not altogether inconsistent. There is a large consensus that any philosophical theory of knowledge which holds that moral judgments have cognitive content is vulnerable to J.L. Mackie's argument from queerness. Mackie's argument is that if we take moral judgments to be objective practical cognitions, we have to admit that they represent properties that are peculiar in that they are part of the fabric of the world or related to other natural properties, and yet intrinsically motivating. Such properties are queer from an ontological point of view, and from an epistemological point of view they are equally puzzling, since it is unclear how we come to know them.[2] This argument is directed against any objectivist ethical theory, and takes objectivity to be conceptually tied to ontology. On this argument, then, any objectivist theory is committed to a standard form of realism. Consequently, the argument construes the claim that moral rational cognitions are efficacious as implying that moral truths are the contents of moral beliefs which drive agents to action. Moral beliefs represent moral properties, which are part and parcel of the fabric of the universe, and yet intrinsically motivating, i.e., capable by themselves of driving people to act.

Thus understood, the inquiry about the practical import of moral cognition appears to be fundamentally misguided. But it should be noted that this critical approach is largely based on the tacit assumption that *reasoning* does not motivate. This seems to follow directly from the claim that the faculty of reason, by which agents determine what they should do, is inert. Critics of practical knowledge – such as Mackie – agree with David Hume that "reason is wholly inactive, and can never be the source of so active a principle as conscience, or a sense of morals".[3] At the same time, they conceive of cognitive judgments as beliefs aiming at truth, representing properties or facts of the matter, and hence, guided by the world.[4] Taken together, these two assumptions result in the view that in order to defend moral knowledge as practical knowledge, the objectivist must invoke a queer ontology, and indirectly assume that rational agents must be endowed with a special form of faculty, in virtue of which they detect moral properties and are moved to do the right thing. The argument from queerness – in one form or another – has been pervasive in meta-ethical debates. It represents the main argument against moral realism, and more generally,

[1] Sidgwick (1907, p. 5).

[2] Mackie (1977, pp. 38–44).

[3] Hume, *Treatise*, 3.1.1.11.

[4] It is questionable that the notion of cognition and belief deployed in this debate are akin to Hume's conception of them. Thus, it is a telling fact that this debate borrows from Hume only the narrow conception of reason and the simple view of the motivating function of moral judgments.

it captures the epistemological and ontological difficulties of defending an objectivist account of ethics.[5]

Recently, however, philosophers have rebutted this argument against ethical objectivism, by questioning the very premise that reason is inert.[6] Obviously, nobody claims that reason determines agents to do the right thing all the time. Agents capable of rationality do not always act rationally; besides, they often encounter obstacles and hindrances that overpower them. This is because they are not perfectly rational, and because they do not live in a rationally organized world. Furthermore, reason often does not function as it should, that is, it misfires or fails to determine action effortlessly, without any internal struggle, loss and suffering. But the point is that reason *can* drive action, and it often does: it is a capacity with a practical function.

This chapter illustrates how practical thought can be both cognitive and efficacious, according to an objectivist theory named Kantian constructivism. As argued in previous works, I believe that there are strong reasons to side with John Rawls in attributing to Kant a "constructivist" conception of practical reason, despite the problematic usage of the term "construction".[7] Exegetical questions aside, I shall focus on the Kantian approach to the logical form of practical thought, as it is understood to work in Kantian constructivism. I hope to illustrate that this is a live and fruitful option in contemporary debates in meta-ethics and meta-normativity. The claim that the logical form of practical thought is universal plays a crucial role in the objectivist Kantian account of reasons for action, and it is key to understanding why embodied agents endowed with reason, albeit imperfectly and limitedly, should act on reason rather than not. This claim explains and relates various embodied forms of action and agency, and identifies moral agency as the most complete form of rational agency, even though not the only one.[8]

In the Kantian tradition, the claim that moral knowledge is practical, and hence efficacious, is strictly tied to the claim that it is universal. Furthermore, and differently from other traditions of practical thought, the universality of practical cognitions is not explained in relation to the convergence of rational minds onto a shared object of thought. Rather, it is explained in relation to the minds of practical subjects. To say that moral cognitions have universal form means that they bind all rational

[5] Smith summarizes the predicament, here: "The problem is that ordinary moral practice suggests that moral judgments have two features that pull in quite opposite directions from each other. The objectivity of moral judgment suggests that there are moral facts (...) but it leaves totally mysterious how or why having a moral view is supposed to have special links with what we are motivated to do. And the practicality of moral judgments suggests just the opposite that our moral judgments express our desires. While this enables us to make good sense of the link between having a moral view and being motivated, it leaves mysterious (...) the sense in which morality is supposed to be objective", Smith 1994, p. 5.

[6] Foot (1978), Compare Darwall (1990, 257–68), Brink (1992), Korsgaard (1996).

[7] See Rawls (1980), and Rawls (2000). Kant denies that "construction" as understood in mathematics can be a viable method of investigation in moral philosophy. On the different notions of construction available in mathematics and ethics, see Bagnoli (2017, 2019, 2021).

[8] In a Kantian perspective, "some actions are more thoroughly actions than others", and a morally good agent's "actions are more truly active, more authentically her own, than those of agents who fall short of moral goodness", Korsgaard (2008, p. 2).

beings as such, insofar as they are rational.[9] My aim in this chapter is to underscore the effects of taking universality as the core claim of the Kantian conception of moral knowledge as practical knowledge, that is, as key to understand the practical significance of moral knowledge. The Kantian claim, as I understand it, is that the universal form of practical thought plays a crucial role in explaining the normative force of reasoning. This is an aspect of the fertility of Kant's defense of universality which has not been adequately appreciated in meta-ethics, and which emerges in the constructivist account of practical knowledge better than in other accounts.

1 The Debate about Universality

Kant argues that practical thought and practical reasoning center on universality. It is an outstanding question how these two claims are related. Universality is the mark of practical thought insofar as it is the mark of cognition, but it is also and importantly the feature that explains the categorical nature of moral obligations, that is, their unconditional and underivative authority.[10] Relatedly, and not trivially so, universality is also the most basic form of practical reasoning. The formula of universal law of the categorical imperative prescribes that we "act only in accordance with that maxim through which you can at the same time will that it become a universal law".[11] It is misleading to take this formula in isolation from the formula of humanity and the formula of the kingdom of ends, which are better understood as complementary, mutually specifying and completing.[12] Nonetheless, this formula is generally thought to identify a sort of decision procedure by which "practical subjects", that is, animals endowed with reason, can determine what to do. According to some scholars the formula of universal law is associated to the formula of universal law of nature exactly to make it clear that universality is a practical requirement of reason by which the deontic features of actions can be identified.[13] On this interpretation, Kant thought of universality as the form of moral reasoning, and it is precisely universality that allows reasoning to deliver its practical results. The form of practical reasoning is the principle of the categorical imperative, which is formal in the sense that it is not

[9] This claim does not preclude that there *will* be an object on which such rational minds converge, but the idea is that such an object is yet to be built, and the ground for such convergence is the agreement of agents endowed with rationality, see e.g. Kant DV 6: 357.

[10] On the relation between universality and authority, see e.g. O'Neill (2015), Bagnoli 2016, 2022. O'Neill argues that the categorical imperative is the supreme principle of reason in general. Korsgaard rephrases it as the supreme principle of normativity in general, Korsgaard (1996, p. 104).

[11] Kant G 4: 421.

[12] Kant's formula of humanity directs agents to "act in such a way that you treat humanity, whether in your own person or in the person of another, always at the same time as an end and never simply as a means", Kant G 4: 429. The formula of humanity is crucial completion, not so much because it helps derive the contents of moral obligations as because it identifies the basis of rational justification. See Korsgaard (1996, p. 123).

[13] See Timmons (2018), chapters 3 and 4.

in itself a substantive principle specifying right-making or good-making properties and yet it is deemed capable of determining the moral and deontic status of actions.

Kant's theses about the nature and practical import of the categorical imperative have been under attack since they appeared on the philosophical scene, but they continue to inspire new work concerning the theory of practical reasoning.[14]

Contemporary debates about the logical form of practical thought and its capacity to generate practical reasoning exhibit a sharp division in two opposing camps. Universalists side with Kant and argue that (i) specific moral truths have their source in universal practical principles, that (ii) universal principles justify the deontic status of particular actions, and that (iii) decisions are rationally justified insofar as they are based on the acceptance of universal principles. For the most part, universalists think of such an appeal to universality in analogy with logical deduction.[15] Particularists disagree, and suggest that practical thought does not have a universal logical form. Correspondingly, they take practical reasoning to rest on perception or singular judgment, and hold that practical reasoning should reflect the particularity and peculiarity of moral knowledge, hardly captured by abstract principles.[16]

To map the debate, it is useful to distinguish strong and weak forms of generalism and particularism.[17] *Strong moral generalism* holds either that all specific moral truths have their source in general moral principles, or that all reasonable or justified moral decisions and beliefs are grounded on or derived from the acceptance of general principles. *Weak moral generalism* holds either that at least some moral truths are grounded on general moral principles, or that at least some reasonable or justified moral decisions and beliefs are grounded on the acceptance of general moral principles. *Strong moral particularism* denies weak moral generalism while *weak moral particularism* denies strong moral generalism. Meta-ethical debates have almost exclusively focused on strong generalism and weak particularism, and their respective impact on practical reasoning. They have also largely assumed that practical reasoning generated by strong moral generalism takes a deductivist form.

As it appears, these debates are organized around a disagreement that concerns the role of universal principles in practical reasoning. However, both particularists and universalists take for granted that *if* universal principles play any role in practical reasoning, they play the role of the major premise in practical syllogism. Analogously, when considering Kant's appeal to universality, scholars have largely assumed that the categorical imperative functions as a major moral premise which is sufficient to derive moral conclusions about the deontic status of any action, given relevant data. This is a misleading way of conceiving the role of universality in Kant's account of practical thought and reasoning. In contrast to this assumption, I shall argue that

---

[14] See Timmons (2015), Wood (2017) and Bagnoli (2021).

[15] Hare (1952) and Donagan (1977).

[16] On practical induction, see Millgram (1997), and on moral perception Dancy (1984). The remaining part of the chapter will be devoted to transductive inference, as illustrated by Harman (2005).

[17] In this paragraph, I switch to the generalism/particularism dichotomy in order to keep a broader scope and include also theories that apply to general though not universal principles, as discussed in Sects. 3 and 4. The objections that hold against generalism hold *a fortiori* against universalism.

universality bears a larger and significantly different role in Kant's conception of practical knowledge, which best emerges in the constructivist account of practical reasoning understood as an alternative to deductivism. My proposal is that we refocus the discussion around Kant's robust notion of rational agency and practical reason, which lie at the core of his project.

From this constructivist perspective, a striking feature of the debates about universality sketched above is that they largely ignore rational agency. For the most part, the disagreement about the role of universal principles has had little to do with the features of action itself. Analytic philosophers dealing with practical inferences have set aside the problem of whether and how practical reason drives agents to action. They have assumed a broadly Humean picture, where the practical aspect of moral reasoning and moral judgment is taken care of by referring to desires. These philosophers devoted themselves to explaining how to apply universal principles in practice so as to sort things out in ethics. In other words, they have built practical reasoning on universality, leaving aside the Kantian account of practical knowledge.[18] For example, Nelson Potter writes: "A successful interpretation of the categorical imperative consists of an argument having only one premise (the categorical imperative), and whatever factual and causal empirical premises are needed, from which a conclusion concerning the moral rightness or wrongness of some particular kind of action follows" (Potter, 1975: 397). That is to say that the discussion in Kantian scholarship has mainly investigated the level of middle principles and their role in deductive principles. A significant aspect of this way of conceiving the application of the categorical imperative is that it takes the premises as mutually independent: the major premise states the universal principle and the minor premise states some data, presumably neutral, until they are combined with the major premise.

The reasons for this partial retrieval of Kant's argument for universality should be found primarily in the analytics' distaste for metaphysics, and their consequent aversion to transcendental idealism. Indeed, the requirement that a philosophical account of practical reasoning should be constrained by naturalism is a matter of descriptive plausibility. This is one reason why Kantian constructivism is useful: it promises to capture and vindicate the basic claims of Kant's argument for universality as generative of practical reasoning without importing any piece of objectionable metaphysics.

2 A Constructivist Approach to Practical Knowledge

As an interpretation of Kant's ethics, constructivism holds that practical truths can be explained as the objective result of practical reasoning without appealing to any cumbersome ontology. The categorical imperative is the form of practical thought in

[18] Richard M. Hare's treatment of practical inferences is paradigmatic: see Hare (1952, pp. 32–77). Interestingly, however, Hare attempts to attack the Cartesian view of inference as deduction from self-evident first principles by endorsing a Kantian route, see Hare (1952, p. 37).

two complementary senses: it is the model of practical reasoning and its constitutive norm, and it is also the constitutive norm of rational agency. The constructivist account of practical reason identifies a specific manner of agential efficacy, that is, a sort of efficacy earned through self-consciousness and self-reflection, which can be imputable as the agent's own doing. Correspondingly, practical knowledge cannot be plausibly construed as abstract knowledge encapsulated in universal principles to be "applied in practice", e.g., given the particular circumstances of the context of choice. Rather, practical knowledge indicates a specific kind of knowledge by which rational thinkers are transformed into rational agents: they become causally efficacious through intentions. To this extent, practical knowledge is always also knowledge of oneself as a practical subject, that is, a subject capable of affecting the world causally, by acting upon it on purpose and for a reason.

In order to vindicate this conception of practical knowledge, one must admit that the function of practical reason as opposed to speculative reason is not merely "recognitive", that is, meant to track things down, as suggested by the traditional metaphors of "vision" and "insight", but generative and productive.[19] This is where the notion of "construction" proves useful. The term construction is deployed to indicate that rational ends of action are the rational agent's constructions, and more precisely, constructions of reason. They acquire reality in virtue of the efficacy of reason. So, to say that reason is "practical" means that it is capable of generating or producing action by directly guiding the mind of rational agents. Differently than the metaphors of "invention" and "projection" used in the non-cognitivist tradition to illustrate the working of practical reason, the notion of construction implies that the activity of reason is *regulated by a plan*, in strict analogy with material construction.[20] Differently than what the metaphors of vision, invention, or projection suggest, rational construction is not whimsical and arbitrary. In fact, it importantly bears the mark of universality. And, finally, the fact that rational construction bears the mark of universality is the guarantee that it is a genuine cognition.

In order to account for the authority of moral obligations, and defend moral knowledge as a practical kind of knowledge, we need to show that knowledge of right action should *bring about* action: we need an account of how universality as the logical form of thought also governs the rational will. The Kantian claim is that the logical form of practical thought reflects the structure of rational action and rational agency. Thus, a successful theory of practical reasoning should explain that

[19] This would be a *theoretical* reconstruction of what practical knowledge is. Similarly, on this interpretation, this is the basis of Kant's attack to Leibniz on the one hand, and Hume on the other hand: they both think of reason as lacking productive powers: "Reason has no power to produce anything outside its representations, but serves merely to achieve a true representation of things that are there anyway. It simply tracks reality. Sharing this assumption about reason, dogmatic rationalism and skeptical empiricism confront each other as opposing positions", Engstrom (2013, p. 138). By contrast, Guyer takes a realist approach to Kant, when he talks about the "immediate recognition" of "the fundamental normative fact... that freedom has an 'inner value, i.e. dignity'" or "the normative recognition that this free will has an incomparable value," which sounds realist rather than constructivist in Korsgaard's sense, see Guyer (1998).

[20] On this analogy, cfr. O'Neill (1989).

reasoning bears some distinctive relations to agency in virtue of which it produces actions as opposed to events and other sorts of performances. It is in virtue of such relations to agency that reason can be directly action-guiding, and authoritative in the first person. The norms of reason represent the constitutive standards practical deliberation, which explain what it is for a subject to be a rational agent. Insofar as such constitutive standards are internal to rational agency they are in some sense necessary or, at least, inescapable.

Therefore, practical truths result from patterns of practical reasoning which are governed by universality. Universality is the *constitutive* norm of rationality; it is the governing plan for construction. To treat universality as a constitutive norm of rationality allows Kantians to say that the appeal to universality is neither constructed nor arbitrary because it fits the basic self-understanding and self-representation of rational agents. The capacity for thinking and acting on principles, on the form of universality, is a *constitutive feature of rational agency*, as rational agents *themselves* understand it.[21] The reflexive feature of the constitutive feature of rational agency is crucial to explain the conceptual relation between practical knowledge and knowledge of oneself as a rational agent. It is this connection that makes sense of acting on a reason, as something that expresses and manifests the identity of rational agents.

On this picture, rational agents are interested in reasoning and acting rationally because they have a constitutive interest in preserving their rational identity so as to avoid defeating themselves. The purported advantage of the Kantian view is that it faces upfront the predicaments of contingency, which are distinctive and peculiar to the human condition.[22] This conception of universality responds to problems that arise in coordinating actions among a plurality of inter-dependent agents, each acting on their own representation of the good.

Kantian constructivism represents a radical alternative to both traditional forms of particularism and universalism in the theory of practical reasoning. How radical this alternative picture is will become clear by comparison with competing theories of practical reasoning. Its distinctive feature, which sets it apart from competing theories, concerns the conception of the norm of universality, its foundation, and its role in practical reasoning.

[21] On the varieties of constitutivism, see Haase and Mayr (2019). A crisp formulation of the Kantian constitutivist claim is the following: "The principles of practical reason serve to unify and constitute us as agents, and that is why they are normative": for "the necessity of conforming to the principles of practical reason comes down to the necessity of being a unified agent... [which] comes down to the necessity of being an agent... [which in turn] comes down to the necessity of acting... [which] is our plight. The principles of practical reason are normative for us, then, simply because we must act," Korsgaard (2008, p. 19).

[22] Humean constructivists such as Sharon Street and Julia Driver argue that Kantian constructivism fails to make room for contingency. This charge underestimates the theoretical resources of Kantian constructivism in account for the predicaments of contingency, and that Humeans themselves ultimately do not meet this challenge, see Bagnoli 2022, chapter 3, (2019), cfr. Haase and Mayr (2019).

3 The Contested Role of Principles in Empirical Practical Reasoning

As mentioned above, when considering the practical role of universal principles in practical reasoning, philosophers have looked primarily at deductive inferences (e.g., Potter, 1975). However, this is a contestable approach, and recently, philosophers engaged in empirical psychology have questioned that the appeal to principles may be the starting point of a descriptively plausible and feasible account of practical reasoning. They have remarked that as a matter of fact, people do make sensible moral judgments and often seem reasonable in doing so, though they are incapable of explicitly stating the justificatory principles – even as *prima facie* or default principles – that might explain their judgments. This remark is offered as the empirical evidence showing that in grappling with practical problems real human agents do not use (general or universal) principles. So, if practical reasoning occurs at all, it is not akin to deductive practical inference. Since empirical practical thought is not governed by universal principles and yet real agents happen to succeed in their practical endeavor without any help from principled considerations, then the appeal to universal principles is proven dispensable.[23] Let us call this the *empirical argument* against principles.

This is an abridged version of the argument, but even in its extended form, the empirical argument for disposing of principled practical reasoning is too quick and problematic. The argument does not establish the conclusion because of its questionable methodology: it takes for granted that a sample of subjects' reports about how they solve a practical problem can establish the truth about the valid form of practical reasoning. Richard M. Hare challenges this methodology at an earlier stage of the debate about the form and possibility of practical inferences. In *The Language of Morals*, Hare remarks that the fact that people often do not openly appeal to practical principles does not immediately prove the truth of moral particularism, since in many instances people are sensitive to the relevant aspects of a situation and respond in ways that may indicate at least an implicit acceptance of general exceptionless principles.[24]

This kind of reply is paradigmatic of a way in which analytic philosophers have treated the features of universality and practicality of moral judgments, i.e., as logical and semantic features of judgments understood as linguistic constructs, rather than as the fundamental elements of cognition. This treatment, avowedly inspired by Kant, has little to do with Kant's own argument for universality as a mark of ethical judgment. First of all, for Kant, the notion of judgment is not a merely semantic notion; it is, rather, the most fundamental element of cognition. Secondly, the practicality of moral judgments has very little to do with its grammatical form, since such judgments are categorical imperatives and yet are expressed in assertive form. Finally, the universality and prescriptivity of ethical judgments are not understood as

[23] See, e.g., Haidt (2001).

[24] Hare (1952).

contrastive semantic features of moral language which require a differentiated explanation and justification. On the contrary, for Kant the universality of moral cognitions is supposed to explain and justify also their practical significance and normative authority. It is the universal form of reason that provides the normative force capable of driving rational agents. But this suggests that the form of reasoning should be different than that of a deductive inference. It also suggests that Kant's conception of practical reasoning does not need a piece of the machinery that Hare builds in order to vindicate practical reasoning. In order to establish that practical inferences lead to action, Hare needs to add a premise external to the practical inference itself, that is, that the reasoner must *sincerely assent* to the premises. The judgment about action follows from the universal and the particular premise, provided that the reasoner endorses the steps that leads to the conclusion. This is to say that sincere assent is the glue that keeps together the steps in practical reasoning, and explains its entire efficacy. In other words, the practical significance and normative authority of practical inference ultimately depends on a psychological mechanism.[25]

Secondly, according to Hare, in these cases, the understanding of the principles may be *practical* rather than theoretical, in the sense that they convey a form of know-how rather than propositional knowledge. People do not have to be able to formulate in words the principles on which they act. In fact, one can be driven to act on a principle without being aware that there is such a principle or that one is acting on it. Either way, the awareness of the underlying principle would play no role. This remark addresses and responds to the worry expressed by empirically minded philosophers, and yet it complicates matters as to what a practical inference is supposed to be. For Hare's denial that the awareness of principles plays any role in practical inference implies that the inference is reconstructed *ex-post*, e.g., as a way of making the first premise explicit. If so, then drawing a practical inference is merely a speculative exercise, which may still fulfill the normative purpose of justifying the action *ex post facto*, but plays no role *ex ante* in the reasoning that guides the agent's deliberation about what to do. It follows that this is not an account of a piece of rational deliberation in its practical function. Hare's treatment of practical inference is paradigmatic of an ambiguous way of using the term "practical", which can be used to cover both the deliberative (vs. speculative) and the normative (vs. descriptive). The issue is that in accounting for moral knowledge as practical knowledge we are mostly concerned with how moral obligations bind rational agents. If moral obligations are the result of practical inferences as Hare conceives of it, then moral obligations oblige only insofar as the agent sincerely assents. This is to say that moral obligations do not really oblige, since their obligating force wholly depends on the psychological mechanism of assent.

[25] Hare writes that universalization is a two-step process. The first step is to find a universal principle and the second "to actually hold it", Hare (1963, p. 219). Hare calls Kant into play: "It is necessary not merely to *quote* a maxim, but (in Kantian language) to will it to be a universal law", *ibidem*. But his formulation of the second step indicates a psychological mechanism of sincere assent akin to personal endorsement, a term that is now current in action theory.

4 Gilbert Harman and the Epistemic Argument for Transductive Inferences

By and large, current debates about practical reasoning reproduce this ambiguity about what makes reasoning "practical" and thus offer confusing views about how agents are guided in acting rationally. Gilbert Harman is a notable exception because he clarifies that "judgment" is not merely a linguistic expression but a cognition. Second, he defines general practical principles in terms of partial functions in the mathematical sense of "function", that is, as tools that map epistemic descriptions of cases into decisions or judgments about cases. This definition has the advantage of being clear, and clean of any metaphysical commitment. Moral generalism is equivalent to the claim that moral truths supervene on non-moral truths, a claim that can be defended without committing to any cumbersome and worrisome ontology.

Harman holds that, on a first approximation, a decision or judgment is based on a general principle if it arises out of a settled disposition to make decisions or judgments specified by the function/principle. He argues against generalism on the basis of two considerations that amount to an *epistemological argument*. First, one may have a settled disposition to reconsider what reasons there are to act one way rather than another: "Given an epistemic situation *E*, it would be wrong to do *d*, where you think *E* is highly unlikely to occur". Your settled dispositions might also be such that, if *E* were to occur, you would reconsider your principles and change your view about this sort of case. Second, when facing a practical problem, one might not have already a principle that serves as a reliable basis for making decisions.

Harman is right that these are very important epistemic considerations, and I shall come back to it in Sect. 5 discussing the importance of placing them in a broader context about the social and dynamic nature of practical reasoning. However, acknowledging a rational requirement to reconsider does not lead where Harman leads us. According to Harman, these epistemic considerations undermine any trust in generalism and supports some form of particularism. He explores an *analogy* with statistical learning theory, which takes seriously the claim that the methods of learning are basically methods for using data in order to train a classifier to do well on new cases. Harman's conjecture is that the debate between moral generalism and moral particularism should be framed as an instance of the question whether there can be a direct inductive inference —also called *transductive inference.*[26] A direct or transductive inference goes from data about previous cases to a classification of a new case. This classification does not involve first making an inductive generalization from data about previous cases and background assumptions, using inference to the best explanation, and then deducing a conclusion about the new case from that inductive generalization.[27] Likewise, moral transduction, *if* it occurs at all, leads to the acceptance of a specific moral conclusion which is not based on the prior

[26] Vapnik (1998); (2000, p. 293).

[27] Vapnik (2000, p. 293).

acceptance of a moral principle that covers that case. Acceptance of the specific moral conclusion is not based on acceptance of a new moral principle.

5 Problem: Does Transduction Convey Practical Knowledge?

Granted Vapnik and Harman's conjecture, there is a question that stands out: does transduction convey practical knowledge? It seems to me that the answer has to be negative. Moral transduction may solve the problem from the third-person perspective: how do we know what to do, morally speaking? In some such cases in which transduction works, it helps generating a new classification and action-description without relying on any principles. This possibility seems to indicate that moral knowledge does not necessarily depend on principles. If Harman is right, and there is an analogue to transduction in the moral domain, he has established that moral knowledge may be non-principled knowledge.

But is this a proof that *practical* knowledge can be non-principled? In other words, does Harman produce an argument which could be invoked against the Kantian claim that in order to be genuinely authoritative practical knowledge must be principled? It does not seem so because the argument provides no account of the relation between practical reasoning and the agent's self-conception.[28] Moral transduction delivers a moral conclusion which lacks first-person authority, and thus it does not count as a piece of practical knowledge in the sense relevant to the discussion. The conclusion is a piece of theoretical knowledge of what one ought to do in such and such cases, which can be retained without having knowledge of being the agent who has to do the action in question, and independently of the agent's cognition of her own efficacy. In other words, it is not the sort of knowledge that guides the agent's mind in action. Thus, it is not a *practical* account of moral knowledge. It is, rather, a theoretical account of what one ought to do, given the circumstances.

This characterization is consequential because it undercuts the normative force of the transductive inference. The objection can be pressed on Kantian grounds, although it could be argued from any other philosophical perspectives that distinguish between the practical and the theoretical functions of reason.[29] The transductive inference reaches a conclusion which qualifies as a justified moral cognition but it lacks normative authority, unless it is subsequently *endorsed* by the practical subject. The practicality of moral knowledge delivered by transduction, and hence its relevance for action, is mediated by the psychological mechanism of endorsement. The function of this presumed psychological mechanism is analogous to Hare's device of sincere assent. It is added to solve a problem that arises because the steps of the

---

[28] By contrast, see Korsgaard (1996, pp. 16–18).

[29] On the Kantian formulation, see Korsgaard (1996, p. 17). For instance, the distinction could be argued on Aristotelian grounds, compare Anscombe (1957), Engstrom (2009), Engstrom (2012). On the (limited) analogy with Anscombe, see Bagnoli (2013).

practical inference are not tied by any knowledge of oneself, nor do they transmit normative authority. In short, the Kantian complaint is that, on this view, the inference does not convey practical knowledge understood as knowledge of oneself as the agent of the action. The latter is the sort of knowledge required to be in place for reasoning to be practical, that is, authoritative in the first person.

One may ask: how does this count as an objection? But an account of practical inference that fails to explain how its conclusion is authoritative and guides agents in action is defective. Furthermore, even if one is content with an explanation of practical reasoning such that it explains action provided that a supporting psychological mechanism is in place, one must recognize that in such a case the capacity of practical inference of generating action is not only limited by the actual and effective presence of the mechanism of psychological endorsement; it is also conditional on it. As a consequence, the conclusion of the transductive inference is binding for rational agents only insofar as and to the extent that they endorse it. The model just described leads to deny the normativity of reasoning. Reasoning has no intrinsic normative authority or efficacy: it is binding and action-guiding only to the extent that the proper sort of endorsement is in place. On this view, then, the ultimate source of normative authority lies in the psychological make up of any individual subject.

This objection resonates with Kant' argument against heteronomous doctrines: they fail to vindicate the authority of moral obligations because they fail to capture the practical function of reason, that is, its self-governing efficacy.[30] Ultimately, their mistake is that they divorce practical reasoning and, in general, the inferential capacities, from rational agency. If reasoning is related only extrinsically to rational agency, its power to authorize and determine action is null.

6 The Requirement to Reconsider and the Form of Reasoning

While Harman's conjecture about moral transduction does not support a robust conception of practical knowledge, his remarks about the requirement to reconsider are worthy of further examination. They point to an aspect of reasoning that the debate centered on deductive inference ignores, that is, the fact that some paradigmatic instances of reasoning lead to revise or abandon the practical principle from which reasoning started off. The aim of reasoning is not to derive conclusions from universal premises, but to come to a better and deeper understanding of a practical problem, by re-describing the case at hand differently and more perspicuously than in its first characterization.

There are also important ethical, practical and political reasons for making room for defeasibility and reconsideration. Such reasons are importantly related to personal freedom, the capacity for self-determination by engaging in practical reasoning, and importantly, the capacity to revise one's life plans by revisiting past intentions

[30] GMS 4:441–43; KpV 5:35–41, 153, 157.

and decisions. Revising one's beliefs and reasons for action is a central task of rationality, which implicates one's inferential capacities, but it is also, and importantly, a kind of practical, ethical and political freedom.[31] This latter claim can be vindicated by a constructivist conception of practical reasoning.

7 The Constitutive Norm of Rational Agency

On the constructivist interpretation, practical reasoning does not take the form of deductive inference. Rather, the aim of practical reasoning is to come to a reasoned alteration of one's intentions, to borrow Gilbert Harman's definition.[32] The appeal to universal principles is meant to guarantee not only the coherent structure of our thought, but also and more generally a coherent account of deliberation over time and across personal relations.[33] The universal principles that guide rational thought and action are principles that real time and interdependent agents, endowed with similar emotional and rational capacities, could share. The constraint that such universal principles put on human practices makes reference to others, not as experienced advisors, spectators, or honorable representatives of an idealized community, but as peers and co-legislators. Universal principles ensure that human rational agents respond authoritatively to the demands of justification pressed by their peers. Revisions and changes demanded by peers are required to meet adequate standards of rational justification. Both the need of objectivity and the means to achieve it are grounded on shared practices of validation and mutual recognition.

On the Kantian account of practical knowledge, universality makes its appearance in two fashions: as the form of reasoning, and as the mark of objective cognition, which pertains to specific practical laws and principles of action (e.g., "Do not lie", "Do not kill", "Keep your promises", "Respect others", etc.). To say that universality is the form of (moral) reasoning commits to the view that it is the rational grounding for specific moral duties. It is notoriously a difficult exegetical, theoretical, and practical issue to show that the mere form of universality can be used to derive specific moral duties. Critics insist on this difficulty in the attempt to prove that Kant promises a decision procedure for determining the deontic status of moral actions, which he ultimately fails to provide: the categorical imperative cannot serve this purpose,

---

[31] "Free persons conceive of themselves as beings who can revise and alter their final ends and who give first priority to preserving their liberty in these matters", Rawls 1974, p. 641, and Scanlon 1975, p. 178. In this respect, Kantian constructivism entails a richer conception of rational agency than standard formal decision theories, namely, a conception of oneself and others as free and equal.

[32] For this definition of practical and moral reasoning, see Harman (1976), and Harman (2010). This definition differentiates between practical reasoning and a decision procedure. One has to define intentions appropriately in order to guarantee the practical significance of decisions. In fact, one may object that decisions are the appropriate conclusions of practical reasoning and differ from intentions.

[33] For an alternative account of the role of coherence in ethics, see Thagard (1999), and compare Millgram (2002).

and so its practical significance becomes problematic. However, it is doubtful that the categorical imperative is proposed as a decision procedure, and thus it is incorrect to hold Kant accountable for not providing a complete decision procedure.[34] The practical significance of the categorical imperative should be sought at a more general level. The constructivist interpretation proposes that the categorical imperative counts as basic methodological norm, which expresses the relevant conception of the rational constructors as free and equal rational agents. While formal, such a norm is practically significant because it structures rational deliberation and it is efficacious because it is rooted in one's self-understanding as a rational agent. Failing such a norm would be akin to failing to understand one's own actions as authentic and genuine productions of one's own agency. It would be a failure to attain unity and moral integrity via the flawed exercise of rational autonomy. Agents who consistently failed to act and think autonomously, according to the categorical imperative would be at a complete loss in understanding themselves as rational agents, or as agents at all.

To fully explain this thesis, we need to reconsider the role of universal principles in regard to rational agency. Understood as a methodological norm, the categorical imperative is not a criterion to identify the practical inferences that practical subjects could correctly draw, given relevant information. Likewise, the categorical imperative is not a classification method that allows the practical subject to classify together all those practical inferences that subjects could correctly draw given their deliberative set. Importantly, the practical subject is not (or, not only) a classifier. Rather, a practical subject is invested in action and interested in undertaking action on reasons. Universal principles work as structures of self-organization. To explain their function, the notion of construction as an analogue of the building trade is particularly useful.[35]

8 Rational Construction, and the Analogy with the Building Trade

Because of the appeal to universal principles, critics often assume that the Kantian account of practical reasoning is modeled on deduction. In the constructivist interpretation, practical reasoning sharply differs from deduction, and also from any like

---

[34] Ironically, this misreading is often imputed to John Rawls, see Wood 2017, Timmons 2015, chapters 4–5. Rawls offers a procedural representation of the requirements of practical reason as useful heuristic device, which has the advantage of avoiding the complications of Kant's transcendental idealism, while highlighting the key Kantian claim that the objective status of obligations depends on the nature of rational agents. However, Rawls also insists that it is misleading to think of it as a mere decision procedure, see Rawls 1989. For a critical assessment of this debate, see Bagnoli (2021).

[35] Kant uses the notion of construction (as in the building trade) when trying to explicate the activity of reason and its authority. Interestingly, the appeal to material construction parallel and complement the notion of trial and other forensic figures, also in the attempt to explicate the authority and autonomy of reason. Differently than other constructivists, I think "construction" or "trial" should be understood in strict analogy with the building trade in the former case, and forensic decision making in the latter, see Bagnoli (2017), compare O'Neill (2015).

form of reasoning privileged by foundationalist theories and often associated with deontological theories inspired by Kanty. In fact, the constructivist interpretation decisively opposes the deontological interpretation of Kantian ethics, bringing to the fore the constructive function of practical reason in the justification of moral obligation.

In constrast to the foundationalist approach, however, the constructivist interpretation does not say that the role of value is foundational. Indeed, the analogy with material construction serves the purpose of illustrating how the many different component of practical reasoning come into play, and provide objective standards without relying on any substantive foundation.

In this sense, Kantian constructivism is far more radical than many of its critics and supporters assume. It purports to be a genuine alternative to the extant forms of practical reasoning precisely because it does not start taking for granted a universal principle as a major premise, but it makes explicit a distinctive account of rational agency, which is embedded in ordinary self-conception. Against this background, Kantian constructivism explains the practical significance of practical knowledge by appealing to the self-awareness of rational agents. Principles do not merely rationalize *ex post*, but constitute basic forms of self-organization.

Constructivists hold that the reasoning can be correct because it tracks down some independent moral reality, but because it adequately expresses a shared conception of rational agency. To be engaged in any sort of rational decision, agents must conceive of themselves as free rather than determined, impeded, or coerced; and they should take themselves to have the same standing as any other rational agent. This conception of rational agency marked by freedom and equality is embedded in ordinary rational thinking, and works as the "basis of construction". That is, it specifies the practical standpoint from which rational agents relate to their action. But to identify the practical standpoint and specify the basis of the construction is not the same as laying down the foundations for moral obligations.

The issue of rational justification arises for practical subjects, finite, embodied and inter-dependent agents, capable of rational assessment and sensitive to reason. This qualification should serve as a warning against conflating constructivism and proceduralism, which mechanically transforms factual inputs and normative outputs. On the constructivist view, the correctness of practical reasoning resides in its conformity to the form of the categorical imperative, which is the only kind of reasoning capable of vindicating, respecting, expressing and manifesting rational agency. Consequently, a correct pattern of practical reasoning does not track an ontology that is independent of the abstract profile of rational agents, though it is independent of the stance of any one particular rational agent.

9 Conclusion

Kantian Constructivism has a decisive advantage over its competitors in contemporary meta-ethics and Kantian scholarship: it shows that Kant's theory of practical reasoning identifies and faces upfront the predicaments of finitude and embodiment, which are distinctive and peculiar to the human condition. It refocuses the debate on susceptibility to reasoning and social sensibility, which affords the cooperative resources to deal with interdependence and mutual vulnerability. In contrast to the approaches that center on instrumental, strategic or bounded rationality, Kantian constructivism establishes that the authority of practical reasoning does not depend on endorsement or assent, but it is anchored on features that are constitutive of human agency. The constitutivist strategy provides an explanation of the reason why human agents are bound by moral obligations as rational requirements.

The emphasis of this account on the bindingness of rational requirements uncovers a dimension of reasoning that is routinely ignored in the theory of practical reasoning, that is, its normative authority. Deductive, inductive, and transductive models of practical reasoning focus on validity, and leave untreated the issue of the authority of rational requirements. Constructivism approaches the problem by distinguishing between ideal and non-ideal conditions of rationality. Validity is a property of reasoning that warrants rational action in perfect conditions of rationality. Human agents do not operate under ideal conditions, and hence the question arises *for them* whether valid norms of rationality are subjectively authoritative. Constructivism explains the significance of practical reasoning, by exploiting the interdependence of mutually vulnerable and mutually accountable rational agents.

References

Bagnoli, C. (2013). Constructivism about practical knowledge. In C. Bagnoli (Ed.), *Constructivism in Ethics* (pp. 153–182). Cambridge University Press.

Bagnoli, C. (2018). Defeaters and practical knowledge. *Synthese, 195*(7), 2855–2875.

Bagnoli, C. (2021). The practical significance of the categorical imperative. *Oxford Studies in Normative Ethics, 11*, 178–199.

Bagnoli, C. (2022). *Ethical Constructivism.* Cambridge University Press.

Brink, D. (1992). A puzzle about the rational authority of morality. *Philosophical Perspectives, 6*, 1–26.

Darwall, S. (1990). Autonomist Internalism and the Justification of Morals. *Nous, 24*, 257–268.

Engstrom, S. (2009). *The form of practical knowledge.* Harvard University Press.

Engstrom, S. (2012). Bringing practical knowledge into view: Response to Bagnoli, Hill, and Reath. *Analytic Philosophy, 53*(1), 89–97.

Engstrom, S. (2013). Constructivism and practical knowledge. In C. Bagnoli (Ed.), *Constructivism in ethics* (pp. 133–152). Cambridge University Press.

Enoch, D. (2006). Agency, shmagency: Why normativity won't come from what is constitutive of agency. *Philosophical Review, 115*(2), 169–198.

Foot, P. (1978). Morality as a system of hypothetical imperatives. In *Virtues and vices and other essays in moral philosophy*, (157–173). University of California Press.

Haase, M., & Mayr, E. (2019). Varieties of constitutivism. *Philosophical Explorations, 22*(2), 95–97.

Haidt, J. (2001). The emotional dog and its rational tail: A social intuitionist approach to moral judgment. *Psychological Review, 108*, 814–834.

Harman, G. (1976). Practical reasoning. *Review of Metaphysics, 29*, 431–463.

Harman, G. (1986). *Change in View.* MIT Press.

Harman, G. (2010). Moral reasoning. In J. Doris et al. (Eds.), *The Moral Psychology Handbook*, Oxford University Press.

Harman, G. (2015). Moral Particularism and Transduction. *Philosophical Perspectives, 15*(1), 44–55.

Kolodny, N. (2005). Why be rational? *Mind, 114*(455), 509–563.

Korsgaard, C. M. (1996). *The sources of normativity*. Cambridge University Press.

Korsgaard, C. M. (2008). *Self-constitution*. Oxford University Press.

Korsgaard, C. M. (2008). *The constitution of agency: Essays on practical reason and moral psychology*. Oxford University Press.

O'Neill, O. (1975). *Acting on principle*. Columbia University Press.

O'Neill, O. (1989). *Constructions of reason*. Cambridge University Press.

O'Neill, O. (2015). *Constructing authorities: Reason, politics and interpretation in Kant's philosophy*. Cambridge University Press.

Potter, N. (1975). How to apply the categorical imperative. *Philosophia, 5*, 395–416.

Rawls, J. (1980). Kantian constructivism in moral theory: The Dewey lectures 1980. *Journal of Philosophy, 77*(9), 515–572.

Rawls, J. (2000). Barbara Herman and Christine M. Korsgaard. (Eds.),*Lectures on the history of moral philosophy*, Harvard University Press.

Thagard, P. (1999). Ethical coherence. *Philosophical Psychology, 11*, 405–422.

Von Wright, G. H. (1963). Practical inference. *The Philosophical Review, 72*, 159–179.

Massimo Mugnai: Selected Publications 1973–2021

The list does not include shorter reviews, newspaper articles and other publications aimed at a wider public.

1973

- Der Begriff der Harmonie als metaphysische Grundlage der Logik und Kombinatorik bei Johann Heinrich Bisterfeld und Leibniz. *Studia Leibnitiana* V/1: 43–73.

1976

- *Astrazione e realtà: Saggio su Leibniz*. Milano: Feltrinelli.

1978

- Bemerkungen zu Leibniz' Theorie der Relationen. *Studia Leibnitiana* X/1: 2–21.

1982

- La "Expositio Reduplicativarum" chez Walter Burleigh et Paulus Venetus. In *English Logic in Italy in the fourteenth and fifteenth centuries. Acts of the 5th European Symposium on Medieval Logic and Semantics. Rome, 10–14 November 1980*, ed. Alfonso Maierù. Napoli: Bibliopolis, 305–320.
- Review of Klaus Jacobi, *Die Modalbegriffe in den logischen Schriften des Wilhelm von Shyreswood und in anderen Kompendien des 12. Und 13. Jahrhunderts*. Leiden-Köln: Brill, 1980. *Annali dell'Istituto e Museo della scienza di Firenze* 7/1: 119–121.

1983

- Alle origini dell'algebra della logica. In *Atti del Convegno internazionale di storia della logica, San Gimignano, 4–8 dicembre 1982*, ed. V. Michele Abrusci, Ettore Casari, Massimo Mugnai. Bologna: CLUEB, 117–131.
- Higher Level Predicates e secundae intentiones nella logica della 'Neuzeit'. *Rivista critica di storia della filosofia* 38/1: 89–101.

F. Ademollo et al. (eds.), *Thinking and Calculating*, Logic, Epistemology, and the Unity of Science 54, https://doi.org/10.1007/978-3-030-97303-2

1984

- *Contraddizione e valore in Marx*. Bologna: Il Mulino.
- Review of Hans Burkhardt, *Logik und Semiotik in der Philosophie von Leibniz*. München: Philosophia Verlag, 1980. *Rivista di storia della filosofia* 39/2: 381–387.

1989

- Review of Giulio d'Onofrio, *Fons Scientiae. La dialettica nell'occidente tardo-antico,* Napoli: Liguori, 1986. *Journal of the History of Philosophy* 27/2: 302–303.

1990

- Leibniz's Nominalism and the Reality of Ideas in the Mind of God. In *Mathesis Rationis. Festschrift für Heinrich Schepers*, ed. Albert Heinekamp, Wolfgang Lenzen, and Martin Schneider. Münster: Nodus Publikationen, 155–167.
- "Voces constituent materiam, particulae formam orationis": osservazioni sulla filosofia del linguaggio leibniziana. *Lingua e stile* XXV/3: 337–350.

1992

- *Leibniz's Theory of Relations*. Stuttgart: Steiner.
- Segner Redivivo. *Rivista di storia della filosofia* 83/1: 155–161.
- Introduction. In Eric J. Aiton, *Leibniz*. Milano: Il Saggiatore, i-xix.

1993

- George Boole, *L'analisi matematica della logica*, ed. Massimo Mugnai. Torino: Bollati Boringhieri.

1994

- Ockham. In *Storia della filosofia. Vol. 2: Il Medioevo*, ed. Carlo Augusto Viano and Pietro Rossi. Roma: Laterza, 419–450.

1995

- Marxismo. In *La filosofia*, ed. Paolo Rossi. Torino: UTET, vol. IV, 231–280.

1996

- Leibniz. In *Storia della filosofia. Vol. 4: Il Settecento*, ed. Carlo Augusto Viano and Pietro Rossi. Roma: Laterza, 61–88.
- Review of Robert M. Adams, *Leibniz: Determinist, Theist, Idealist*, Oxford: Oxford University Press, 1994. *The Leibniz Review* 6: 78–98.

1997

- An Unpublished Latin Text on Terms and Relations. *The Leibniz Review* 7: 27–125.

1998

- Aspetti della tradizione logica in Germania dopo Leibniz. In *L'età dei Lumi. Saggi sulla cultura settecentesca*, ed. Antonio Santucci. Bologna: Il Mulino, 25–39.
- George Boole e lo psicologismo: la caratterizzazione delle leggi logiche in "An Investigation of the Laws of Thought". In *G. Boole, Filosofia, Logica, Matematica*, ed. Evandro Agazzi and Nicla Vassallo. Milano: Franco Angeli, 111–130.

1999

- Phänomene und Aussenwelt in der Leibnizschen Erkenntnistheorie. In *Wissenschaft und Weltgestaltung. Internationales Symposion zum 350. Geburtstag von Gottfried Wilhelm Leibniz vom 9. bis 11. April 1996 in Leipzig*, ed. Kurt Nowak and Hans Poser. Hildesheim: Olms, 145–154.

2000

- "Alia est rerum alia terminorum divisio": About an Unpublished Manuscript of Leibniz. In *Unità e molteplicità nel pensiero filosofico di Leibniz*, Simposio internazionale, Roma, 3–5 ottobre 1996, ed. Antonio Lamarra and Roberto Palaia. Firenze: Olschki, 257–270.
- G. W. Leibniz, *Nuovi saggi sull'intelletto umano*, in Leibniz, *Scritti filosofici*, ed. Massimo Mugnai and Enrico Pasini, vol. 2. Torino: UTET.
- G. W. Leibniz, *Saggi di Teodicea*, in Leibniz, *Scritti filosofici*, ed. Massimo Mugnai and Enrico Pasini, vol. 2. Torino: UTET.

2001

- *Introduzione alla filosofia di Leibniz*. Torino: Einaudi.
- Leibniz on Individuation: From the Early Years to the "Discourse" and Beyond. *Studia Leibnitiana* 33/1: 36–54.
- Leibniz in edizione critica: la recente pubblicazione delle *Philosophische Schriften* (1677-Juni 1690). *Rivista di storia della filosofia* 56/3: 459–468.
- Paolo Rossi, gli Zenonisti e Leibniz. *Rivista di storia della filosofia* 4: 1–4.

2002

- Gottfried Wilhelm Leibniz. In *Storia della Scienza. La rivoluzione scientifica*, ed. Sandro Petruccioli. Roma: Istituto della enciclopedia italiana, vol. 5, 323–328.
- Review of Jean-Baptiste Rauzy, *La doctrine Leibnizienne de la vérité. Aspects logiques et ontologiques*, Paris: Vrin, 2001. *The Leibniz Review* 12: 53–64.

2003

- Algebra della logica. In *Storia della Scienza*, ed. Sandro Petruccioli. Roma: Istituto della enciclopedia italiana, vol. 7, 174–179.
- Review of G. W. Leibniz, *The Labyrinth of the Continuum: Writings on the Continuum Problem 1672–1686*, Transl. Ed. and with an Introduction by Richard T.W. Arthur, New Haven: Yale University Press, 2001. In *The Leibniz Review* 13: 155–165.

2004

- Mondi possibili e spazio dei campioni. In *Filosofia e logica*, ed. Massimiliano Carrara and Pierdaniele Giaretta. Catanzaro: Rubbettino, 67–88.
- Negazione e intensioni leibniziane. *Rivista di estetica* 26/2: 75–89.
- Termini sincategorematici e 'cifra' in un passo della *Summa logicae* di Ockham. *Rivista di storia della filosofia* 49/2: 515–517.
- Michael Friedman, *La filosofia al bivio: Carnap, Cassirer, Heidegger*, trans. Massimo Mugnai. Milano: Cortina Editore.
- Review of Jan A. Cover and John O'Leary-Hawthorne, *Substance and Individuation in Leibniz*, Cambridge: Cambridge University Press, 1999. *The Leibniz Review* 14: 57–64.

2005

- Logic and Mathematics in the 18th Century: before and after Christian Wolff. In *Macht und Bescheidenheit der Vernunft*, ed. Luigi Cataldi Madonna. Hildesheim: Olms, 97–109.
- Leibniz on Substance and Changing Properties. *Dialectica* 59/4: 503–516.
- Review of Wolfgang Lenzen, *Calculus Universalis: Studien zur Logik von G. W. Leibniz*, Paderborn: Mentis Verlag, 2004. *The Leibniz Review* 15: 169–178.

2006

- On Leibniz's Logical Calculi. In *Logic and Philosophy in Italy, Some Trends and Perspectives*, ed. Edorardo Ballo and Miriam Franchella. Milano: Polimetrica, 215–227.

2008

- G. W. Leibniz, *Ricerche generali sull'analisi delle nozioni e delle verità e altri scritti di logica*, ed. Massimo Mugnai. Pisa: Edizioni della Normale.
- Essay Review of Glenn Harz, *Leibniz's Final System. Monads, Matter, and Animals*, London: Routledge 2006. *Studia Leibnitiana* vol. 38–39/1: 109–118.

2009

- Rota's Philosophical Insights. In *From Combinatorics to Philosophy. The Legacy of G. C. Rota*, ed. Ernesto Damiani, Ottavio D'Ancona, Vittorio Marra, and Fabrizio Palombi. Dordrecht: Springer, 241–249.
- "On extrinsic denominations" (LH IV, iii, 5a-e, Bl. 15): Transcription and English Translation. *The Leibniz Review* 19: 64–66.

2010

- Logic and Mathematics in the Seventeenth Century. *History and Philosophy of Logic* 31/4: 297–314.
- Leibniz and 'Bradley's Regress'. *The Leibniz Review* 20: 1–12.
- Leibniz's "Schedae de novis formis syllogisticis" (1715): Text and Translation. *The Leibniz Review* 20: 117–135.

- Review of Daniel Garber, *Leibniz: Body, Substance, Monad*, Oxford: Oxford University Press, 2009. *Journal of Philosophy* 107/12: 664–668.

2011

- Bolzano e Leibniz. *Discipline filosofiche* 21/2: 93–108.
- Leibniz o la morte di un difensore del "cristianesimo universale". *Rivista di storia della filosofia* 67/1: 141–152.

2012

- Leibniz's Ontology of Relations: A Last Word? In *Oxford Studies in Early Modern Philosophy*, vol. VI, ed. Daniel Garber and Donald Rutherford. Oxford: Oxford University Press, 171–208.
- Giovanni Vailati. In *Il contributo Italiano alla storia del pensiero*, ed. Michele Ciliberto. Roma: Edizioni dell'Enciclopedia Italiana, 538–543.
- Scienza e filosofia: Geymonat e Preti. In *Il contributo Italiano alla storia del pensiero*, ed. Michele Ciliberto. Roma: Edizioni dell'Enciclopedia Italiana, 759–769.
- Gerolamo Saccheri, *Logica dimostrativa*, ed. Massimo Mugnai and Massimo Girondino. Pisa: Edizioni della Normale.

2013

- *Possibile e necessario.* Bologna: Il Mulino.
- Leibniz, Kant e le modalità 'possibile' e 'necessario'. In *A Plea for Balance in Philosophy: Essays in Honour of Paolo Parrini*, ed. Roberta Lanfredini and Alberto Peruzzi. Pisa: ETS, 321–343.
- Leibniz e i futuri contingenti. *Rivista di storia della filosofia* 68/1: 191–210.

2015

- Essences, Ideas, and Truths in God's Mind and in the Human Mind. In *The Oxford Handbook of Leibniz*, ed. Maria Rosa Antognazza. Oxford: Oxford University Press, 11–26.
- Theory of Relations and Universal Harmony. In *The Oxford Handbook of Leibniz*, ed. Maria Rosa Antognazza. Oxford: Oxford University Press, 27–44.
- Ars characteristica, Logical Calculus, and Natural Languages. In *The Oxford Handbook of Leibniz*, ed. Maria Rosa Antognazza. Oxford: Oxford University Press, 177–209.

2016

- Ontology and Logic: the Case of Scholastic and Late-Scholastic Theory of Relations. *British Journal for the History of Philosophy* 24/3: 532–553.

2017

- Postfazione. In Ettore Casari, *La logica stoica*, 2nd ed. Pisa: ETS, 2017, 97–106.

- Review of Marko Malink and Anubav Vasudevan, "The Logic of Leibniz's Generales inquisitiones de analysi notionum et veritatum", *The Review of Symbolic Logic* 9/4 (2016): 686–751. *The Leibniz Review* 27: 117–137.

2018

- (with Fabrizio Amerini) Franciscus de Prato on Reduplication. In *Modern Views of Medieval Logic*, ed. Christoph Kann, Benedikt Loewe, Christian Rode, and Sara L. Uckelman. Leuven: Peeters, 1–14.

2019

- Leibniz's Mereology in the Essays on Logical Calculus of 1686–1690. In *Leibniz and the Structure of Sciences: Modern Perpsectives on the History of Logic, Mathematics, Epistemology*, ed. Vincenzo De Risi. Cham: Springer, 46–69.

2020

- *Il mondo capovolto: Il metodo scientifico nel Capitale di Marx*. Pisa: Edizioni della Normale.
- *Mereology in Medieval Logic and Metaphysics. Proceedings of the 21st European Symposium of Medieval Logic and Metaphysics*, ed. Fabrizio Amerini, Irene Binini, and Massimo Mugnai. Pisa: Edizioni della Normale.
- Introduction. In *Leibniz: Dissertation on Combinatorial Art*, trans. with introd. and comm. by Massimo Mugnai, Han van Ruler and Martin Wilson. Oxford: Oxford University Press.

2021

- Leibniz, *General Inquiries on the Analysis of Notions and Truths*, edited with an English translation by Massimo Mugnai. Oxford: Oxford University Press.